INDEX to Illustrations of ANIMALS and PLANTS

BETH CLEWIS

NEAL-SCHUMAN PUBLISHERS, INC.
New York   London

Published by Neal-Schuman Publishers, Inc.
23 Leonard Street
New York, NY 10013

Printed and bound in the United States of America

Library of Congress Cataloging-in-Publication Data

Clewis, Beth
      Index to illustrations of animals and plants / by Beth Clewis.
         p.   cm.
      Includes bibliographical references and index.
      ISBN 1-55570-072-1
      1. Natural history illustration--Indexes.   I. Title
   QH46.5.C54   1991
574' 022'2—dc20
                                                        90-23816
                                                          CIP

For Martin

Contents

Preface

The usefulness of any reference tool that indexes the contents of other books depends on the timeliness and availability of those books. For this reason, the present volume is offered as a supplement to two earlier illustration indexes: *Index to Illustrations of the Natural World* by John W. Thompson and Nedra Slauson (Gaylord, 1977), and *Index to Illustrations of Living Things Outside of North America* by Lucille Thompson Munz and Nedra Slauson (Scarecrow, 1981).

The original indexes direct readers to pictures of animals and plants in selected books. This edition follows a format similar to the original indexes, but primarily covers books published in the 1980s and includes animals and plants from around the world. Approximately 6,200 entries appear in *Index to Illustrations of Animals and Plants*, guiding readers to pictures in a total of 142 books.

Criteria for selection of books to index included currency, abundance of illustrations, with an emphasis on color or black-and-white photographs, and availability in public or medium-sized academic libraries. This availability was determined by checking holdings in the Virginia union catalog. While some specialized regional guides are included, especially for mammals of the United States, most of the books were selected because they are owned by numerous libraries. A few useful titles from the late 1970s are included, but no effort was made to go back and index books not included in the original indexes.

Most entries provide citations to at least two sources. A citation to a single source is given only when the organism also appears in one of the first two editions, or may otherwise hold some particular interest.

A major challenge in compiling this index was selecting the animals and plants to include. The availability of books limited coverage of some animals and plants, notably plants outside the United States. Coverage of other genera is purposefully representative rather than complete, such as beetles, other insects, and invertebrates. Because of space limitations, two types of organisms included in the original indexes are not included here: seashells, except when depicted as living mollusks, and cultivated or garden flowering plants. Readers will find the most complete coverage for mammals, reptiles and amphibians, birds, plants of the United States, and North American organisms in general.

In compiling this index I was struck by the amount of descriptive information contained in the indexed books. I hope that reference librarians will therefore consider using this book to track down narrative descriptions; as well as illustrations of organisms under study.

Finally, I wish to thank my husband Martin for his assistance with this project. Without his computer expertise I might not have been able to organize and index so much highly detailed information. I am also grateful to the librarians of the Henrico County Public Library for providing many of the books needed for this work.

Beth Clewis
Richmond, Virginia
December 1990

How to Use This Book

In the first index of the book the common name of a plant or animal is followed by its scientific name in parentheses. "See" and "See also" references refer the user to other common names used in the index, based on usage in the books indexed, and to related entries. In some cases the name used in the earlier editions is provided with a "See" reference to the name used here. Where one common name applies to two different kinds of organisms, an additional identification is given. For example: Locust, black (tree) (*Robinia pseudo-acacia*) and Locust, desert (insect) (*Schistocerca gregaria*).

The second index in this volume lists animals and plants by their scientific names to aid those who know only the Latin binomial name for their animal or plant, or who wish to track related organisms through the Index.

Citations in the Common Name Index include a three-letter code assigned to each source book, abbreviations indicating whether the illustration is in color or black-and-white, and page number or plate number. The Key to Abbreviations of Sources lists the codes alphabetically and gives the title. The Bibliography at the end of this book gives a full citation for each title.

Abbreviations used in the main entries include:

cp   color plate
p   black-and-white plate
bw   black-and-white illustration
absence of abbreviation indicates a color illustration
[page number]   page number in an unpaginated book
fig.   figure
sp.   species

Following is a typical entry:

Frog, African clawed (*Xenopus laevis*)
AAL 279(bw),281; ARA cp222; FTW cp78,p73; MIA 473; WRA fig.14

Illustrations for this organism appear in book AAL (*Audubon Society Encyclopedia of Animal Life*) on page 279 in black-and-white and page 281 in color;

ARA (*Audubon Society Field Guide to North American Reptiles and Amphibians*) on color plate number 222;

FTW (*Frogs and Toads of the World*) on color plate 78 and black-and-white plate number 73;

MIA (*Macmillan Illustrated Animal Encyclopedia*), a color illustration on page 473;

WRA (*A Field Guide to Western Reptiles and Amphibians*), an illustration in figure number 14.

Entries are arranged in word-by-word order with a hyphenated phrase considered as one word. For example, the following entries beginning with the word "water" are arranged in the order: Water lily, Waterbuck, Water-plantain, and Watersnake. Names beginning with an abbreviation such as St.-John's-wort are placed as if the abbreviated word were spelled out.

Key to Abbreviations of Sources

AAB   *Encyclopedia of Animal Biology*. R. McNeill Alexander, ed. NY: Facts on File, 1987.

AAL   *Audubon Society Encyclopedia of Animal Life*. Ralph Buchsbaum, et al. NY: Clarkson N. Potter, 1982.

ACF   *A Field Guide to Atlantic Coast Fishes of North America.*(Peterson Field Guide.) C. Richard Robins and G. Carleton Ray. Boston: Houghton Mifflin, 1986.

AGC   *Atlantic & Gulf Coasts*. (Audubon Field Guide.) William H. Amos and Stephen H. Amos. NY: Knopf, 1985.

ANB   *America's Neighborhood Bats*. Merlin D. Tuttle. Austin: University of Texas Press, 1988.

AOG   *Auks: An Ornithologist's Guide*. Ron Freethy. Poole, Dorset: Blandford Press, 1987; NY: Facts on File, 1987.

ARA   *Audubon Society Field Guide to North American Reptiles and Amphibians*. John L. Behler and F. Wayne King. NY: Knopf, 1979.

ARC   *Amphibians and Reptiles of the Carolinas and Virginia*. Bernard S. Martof, et al. Chapel Hill: University of North Carolina Press, 1980.

ARN   *Amphibians and Reptiles of New England: Habitats and Natural History*. Richard M. DeGraaf and Deborah D. Rudis. Amherst: University of Massachusetts, 1983.

ART   *Amphibians and Reptiles of Texas*. James R. Dixon. College Station: Texas A&M Press, 1987.

ASM   *Audubon Society Field Guide to North American Mammals*. John O. Whitaker. NY: Knopf, 1980.

AWC   *Audubon Society Book of Wild Cats*. Les Line and Edward R. Ricciuti. NY: Abrams, 1985.

AWR   *Wildlife of the Rivers.* William H. Amos. NY: Abrams, 1981.

BAA   *Arctic Animals: A Celebration of Survival.* Fred Bruemmer. Ashland, WI: NorthWord, Inc., 1986.

BAS   *Birds for all Seasons.* Jeffery Boswall. London: BBC, 1986.

BCF   *A Countryman's Flowers.* Hal Borland, with photographs by Les Line. NY: Knopf, 1981.

BCS   *Beneath Cold Seas: Exploring Cold-Temperate Waters of North America.* Photographs by Jeffrey L. Rotman, text by Barry W. Allen. NY: Van Nostrand Reinhold, 1983.

BMW   *The Audubon Society Book of Marine Wildlife.* Les Line, with text by George Reiger. NY: Abrams, 1980.

BNA   *A Field Guide to the Beetles of North America.* (Peterson Field Guide.) Richard E. White. Boston: Houghton Mifflin, 1983.

BOI   *Audubon Society Book of Insects.* Les Line, Lorus Milne, and Margery Milne. NY: Abrams, 1983.

BOP   *Birds of Prey.* (Birds of the World.) John P.S. MacKenzie. Ashland, WI: Paper Birch Press, Inc. 1986.

BWB   *The Audubon Society Book of Water Birds.* Les Line, Kimball L. Garrett, and Kenn Kaufman. NY: Abrams, 1987.

CGW   *A Country-Lover's Guide to Wildlife: Mammals, Amphibians, and Reptiles of the Northeastern United States.* Kenneth A. Chambers. Baltimore: Johns Hopkins, 1979.

CMW   *Mammals in Wyoming.* Tim W. Clark and Mark R. Stromberg. Lawrence, KS: University of Kansas Press, 1987.

COW   *Cranes of the World.* Paul A. Johnsgard. Bloomington, IN: Indiana University Press, 1983.

CRM   *Collins Guide to the Rare Mammals of the World.* John A. Burton and Bruce Pearson. Lexington, MA: Stephen Greene Press, 1988.

CWN   *Common Wildflowers of the Northeastern United States.* (New York Botanical Garden Field Guide.) Carol H. Woodward and Harold William Rickett. Woodbury, NY: Barron's, 1979.

DBN   *Diving Birds of North America.* Paul A. Johnsgard. Lincoln: University of Nebraska, 1987.

DBW   *Bears of the World.* Terry Dominco. NY: Facts on File, 1988.

DIE   *Insects Etc. An Anthology of Arthropods Featuring a Bounty of Beetles.* Painting by Bernard Durin, with a Literary Anthology introduced and selected by Paul Armand Gette. Entomological commentaries by Gerhard Scherer. NY: Hudson Hills Press, 1981; Munich: Schirmer/Mosel Verlag, 1980.

DNA   *Ducks of North America and the Northern Hemisphere.* John Gooders and Trevor Boyer. NY: Facts on File, 1986.

EAB   *Audubon Society Encyclopedia of North American Birds.* John. K. Terres. NY: Knopf, 1980.

EAL   *Encyclopedia of Aquatic Life.* Keith Banister and Andrew Campbell, eds. NY: Facts on File, 1985.

EOB   *Encyclopedia of Birds.* Edited by Christopher M. Perrins and Alex L.A. Middleton. NY: Facts on File, 1985.

EOI   *Encyclopedia of Insects.* Edited by Christopher O'Toole. NY: Facts on File, 1986.

ERA   *Encyclopedia of Reptiles and Amphibians.* Edited by Tim Halliday and Kraig Adler. NY: Facts on File, 1986.

FEB   *Eastern Birds.* (Audubon Handbook.) John Farrand. NY: McGraw-Hill, 1988.

FGH   *A Field Guide to Hawks: North America.* (Peterson Field Guide.) William S. Clark. Boston: Houghton Mifflin, 1987.

FGM   *A Field Guide to Mushrooms (North America).* (Peterson Field Guide.) Kent H. McKnight and Vera B. McKnight. Boston: Houghton Mifflin, 1987.

FTW   *Frogs and Toads of the World.* Chris Mattison. NY: Facts on File, 1987.

FWB   *Western Birds.* (Audubon Handbook.) John Farrand. NY: McGraw-Hill, 1988.

FWF   *Fall Wildflowers of the Blue Ridge and Great Smoky Mountains.* Oscar W. Gupton and Fred C. Swope. Charlottesville, Virginia: University Press of Virginia, 1987.

GEM   *Grzimek's Encyclopedia of Mammals.* (5 volumes.) Bernhard Grzimek, ed. NY: McGraw-Hill, 1990.

GMP   *Guide to the Mammals of Pennsylvania.* Joseph F. Merritt. Pittsburg, PA: University of Pittsburg for the Carnegie Museum of Natural History, 1987.

GSM   *A Field Guide to Southern Mushrooms.* Nancy Smith Weber and Alexander H. Smith. Ann Arbor: University of Michigan, 1985.

GWO   *Wild Orchids of the Mid-Atlantic States.* Oscar W. Gupton and Fred C. Swope. Knoxville, TN: University of Tennessee, 1986.

HHH   *The Herons Handbook.* James Hancock and James Kushlan. NY: Harper and Row, 1984.

HLB   *Hummingbirds - Their Life and Behavior: A Photographic Study of the North American Species.* Text by Esther Quesada Tyrrell, photographs by Robert A. Tyrrell. NY: Crown, 1985.

HSG   *Seawatch: The Seafarer's Guide to Marine Life.* Paul V. Horsman. NY: Facts on File, 1985.

HSW   *Seabirds of the World.* Photographs by Eric Hosking, text by Ronald M. Lockley. NY: Facts on File, 1983.

HWD   *Whales, Dolphins and Porpoises.* Richard Harrison and Michael Bryden, consulting eds. NY: Facts on File, 1988.

JAD   *Deserts.* (Audubon Society Nature Guide.) James A. MacMahon. NY: Knopf, 1985.

KCR   *A Field Guide to Coral Reefs of the Caribbean and Florida.* (Peterson Field Guide.) Eugene H. Kaplan. Boston: Houghton Mifflin, 1982.

KSW   *Seals of the World.* (Second edition.) Judith E. King. London: British Museum (Natural History); NY: Cornell University Press, 1983.

LBG   *Grasslands.* (Audubon Society Nature Guide.) Lauren Brown. NY: Knopf, 1985.

LOB   *The Lives of Bats.* Wilfred Schober. NY: Arco, 1984.

LOW   *Lizards of the World.* Chris Mattison. NY: Facts on File, 1989.

LSW   *Living Snakes of the World in Color.* John M. Mehrtens. NY: Sterling Publishing Co., 1987

MAE   *Encyclopedia of Animal Ecology.* Peter D. Moore, ed. NY: Facts on File, 1987.

MEM   *Encyclopedia of Mammals.* David MacDonald. NY: Facts on File, 1984.

MIA   *Macmillan Illustrated Animal Encyclopedia.* Philip Whitfield, ed. NY: Macmillan, 1984.

MIS   *The Audubon Society Field Guide to North American Insects And Spiders.* Lorus Milne and Margery Milne. NY: Knopf, 1980.

MNC   *Handbook of Mammals of the North-Central States.* J. Knox Jones and Elmer C. Birney. Minneapolis: U. of Minnesota Press, 1988.

MNP   *Mammals of the National Parks.* Richard G. van Gelder. Baltimore: Johns Hopkins Press, 1982.

MPC   *Pacific Coast.* (Audubon Society Nature Guide.) Bayard and Evelyn McConnaughey. NY: Knopf, 1985.

MPS   *Guide to Mammals of the Plains States.* J. Knox Jones, David M. Armstrong, and Jerry R. Choate. Lincoln: U. of Nebraska Press, 1985.

MSC   *The Audubon Society Field Guide to North American Seashore Creatures.* Norman A. Meinkoth. NY: Knopf, 1981.

MSW   *Snakes of the World.* Chris Mattison. NY: Facts on File, 1986.

NAB   *The Audubon Society Field Guide to North American Birds (Eastern Region).* John Bull and John Farrand. NY: Knopf, 1977.

NAF   *The Audubon Society Field Guide to North American Fishes, Whales and Dolphins.* Herbert T. Bosching, et al. NY: Knopf, 1983.

NAM   *The Audubon Society Field Guide to North American Mushrooms.* Gary H. Lincoff. NY: Knopf, 1981.

NAW   *The Encyclopedia of North American Wildlife.* Stanley Klein. NY: Facts on File, 1983.

NHA   *The Natural History of Antelopes.* Clive A. Spinage. London: Croom Helm, 1986.

NHP   *Natural History of the Primates.* J.R. Napier and P.H. Napier. Cambridge, MA: MIT Press, 1985.

NHU   *The Natural History of the USSR.* Algirdas Knystautas. NY: McGraw-Hill, 1987.

NMM   *Natural History of New Mexican Mammals.* James S. Findley. University of New Mexico Press, 1987.

NMT   *Mushrooms and Toadstools: A Color Field Guide.* U. Nonis. NY: Hippocrene Books, 1982.

NNW   *Nightwatch: The Natural World from Dawn to Dusk.* John Cloudsley-Thomas et al. NY: Facts on File, 1984.

NOA   *Nature of Australia: A Portrait of the Island Continent.* John Vandenbeld. NY: Facts on File, 1988.

NTE   *Audubon Society Field Guide to North American Trees (Eastern Region).* Elbert L. Little. NY: Knopf, 1980.

NTW   *The Audubon Society Field Guide to North American Trees (Western Region).* Elbert L. Little. NY: Knopf, 1980.

NWE   *Audubon Society Field Guide to North American Wildflowers (Eastern Region).* William A. Niering and Nancy C. Olmstead. NY: Knopf, 1979.

OBL   *Ocean Birds.* Lars Lofgren. NY: Knopf, 1984.

OET   *Oxford Encyclopedia of Trees of the World.* Bayard Hora, consulting ed. NY: Oxford, 1981.

ONA   *Field Guide to Orchids of North America.* John G. Williams and Andrew E. Williams. NY: Universe Books, 1983.

ONU   *Owls: Their Natural and Unnatural History.* John Sparks and Tony Soper. NY: Facts on File, 1989.

OTT   *Turtles, Tortoises and Terrapins.* Fritz Jurgen Obst. NY: St. Martin's Press, 1986.

PAB   *The Audubon Society Field Guide to North American Butterflies.* Robert Michael Pyle. NY: Knopf, 1981.

PBC   *Birds of the Carolinas.* Eloise F. Potter, James F. Parnell and Robert P. Teulings. Chapel Hill: University of North Carolina Press, 1980.

PCF   *A Field Guide to Pacific Coast Fishes of North America.* (Peterson Field Guide.) William N. Eschmeyer and Earl S. Herald. Boston: Houghton Mifflin, 1983.

PFH   *Plants and Flowers of Hawai'i.* S.H. Sohmer and R. Gustafson. Honolulu: University of Hawaii, 1987.

PGT   *The Pond.* Gerald Thompson, Jennifer Coldrey, and George Bernard. Cambridge, MA: MIT Press, 1984.

PMT   *Plants that Merit Attention: Volume I - Trees.* Janet Meakin Poore, ed. Portland, OR: Timber Press, 1984.

PPC   *Poisonous Plants: A Colour Field Guide.* Lucia Woodward. Newton Abbot, Devon: David & Charles, 1985.

PPM   *A Field Guide to Poisonous Plants and Mushrooms of North America.* Charles Kingsley Levy and Richard B. Primack. Brattleboro, VT: Stephen Greene Press, 1984.

RBB   *British Birds: A Field Guide.* Alan J. Richards. London: David & Charles, 1979.

RCK   *Coral Kingdoms.* Carl Roessler. NY: Abrams, 1986.

RES   *Rhinos: Endangered Species.* Malcolm Penny. NY: Facts on File, 1988.

RFA   *Furbearing Animals of North America.* Leonard Lee Rue III. NY: Crown, 1981.

RMM   *Rocky Mountain Mammals: A Handbook of Mammals of Rocky Mountain National Park and Vicinity.* Revised ed. David M. Armstrong. [Boulder, CO]: Colorado Associated University Press in cooperation with Rocky Mountain Nature Association, 1987.

RNA   *Reptiles of North America.* (Golden Field Guide.) Hobart M. Smith and Edmund D. Brodie. NY: Golden Press, 1982.

ROW   *Riches of the Wild: Land Mammals of South-East Asia.* Gathorne Gathorne-Hardy Cranbrook, Earl of Cranbrook. Singapore, New York: Oxford U. Press, 1987.

RUP   *The Underseas Predators.* Carl Roessler. NY: Facts on File, 1984.

SAB   *Encyclopedia of Animal Behavior.* Peter J.B. Slater, ed. NY: Facts on File, 1987.

SAS   *Seashore Animals of the Southeast: A Guide to Common Shallow-Water Invertebrates of the Southeastern Atlantic Coast.* Edward Ruppert and Richard Fox. Columbia: University of South Carolina, 1988.

SCS   *A Field Guide to Southeastern and Caribbean Seashores.* (Peterson Field Guide.) Eugene H. Kaplan. Boston: Houghton Mifflin, 1988.

SEF   *Eastern Forests.* (Audubon Society Nature Guide.) Ann Sutton and Myron Sutton. NY: Knopf, 1985.

SGB   *Simon and Schuster's Guide to Butterflies and Moths.* Mauro Daccordi, Paolo Triberti, and Adriano Zanetti. NY: Simon and Schuster, 1988.

SGI   *Simon and Schuster's Guide to Insects.* Ross H. Arnett andRichard L. Jacques. NY: Simon and Schuster, 1981.

SGM   *Simon and Schuster's Guide to Mushrooms.* Gary H. Lincoff. NY: Simon and Schuster, 1981.

SJS   *Sharks.* John D. Stevens, consulting ed. NY: Facts on File, 1987.

SOW   *Spiders of the World.* Rod Preston-Mafham and Ken Preston-Mafham. NY: Facts on File, 1984.

SPN   *Songbirds, How to Attract Them and Identify Their Songs.* Noble Proctor. Emmaus, PA: Rodale Press, 1988.

SSW   *Sharks of the World.* Rodney Steel. NY: Facts on File, 1985.

SWF   *Wildlife of the Forests.* Ann Sutton and Myron Sutton. NY: Abrams, 1979.

SWM   *Wild Mammals of Northwest America.* Arthur Savage and Candace Savage. Baltimore: Johns Hopkins University Press, 1981.

SWW   *Audubon Society Field Guide to North American Wildflowers (Western Region).* Richard Spellenberg. NY: Knopf, 1979.

TOW   *Turtles of the World.* Carl H. Ernst and Roger W. Barbour. Washington, DC: Smithsonian, 1989.

TSP   *A Field Guide to Tropical and Subtropical Plants.* Frances Perry and Roy Hay. NY: Van Nostrand Reinhold, 1982.

TSV   *Trees and Shrubs of Virginia.* Oscar W. Gupton and Fred C. Swope. Charlottesville: University Press of Virginia, 1981.

TTW   *Turtles and Tortoises of the World.* David Alderton. NY: Facts on File, 1988.

UAB   *The Audubon Society Field Guide to North American Birds (Western Region)*. Miklos D.F. Udvardy. NY: Knopf, 1977.

WAA   *Wildflowers Across America*. Lady Bird Johnson and Carlton B. Lees. NY: Abbeville Press, 1988.

WBW   *Wading Birds of the World*. Eric and Richard Soothill. Poole, Dorset: Blandford Press, 1982.

WDP   *Whales, Dolphins and Porpoises of the World*. Mary L. Baker. Garden City, NY: Doubleday, 1987.

WFW   *Western Forests*. (Audubon Society Nature Guide.) Stephen Whitney. NY: Knopf, 1985.

WGB   *Longman World Guide to Birds*. Philip Whitfield, ed. London: Longman, 1986.

WIW   *Insects of the World*. Anthony Wootton. NY: Facts on File, 1984.

WMC   *Mammals of the Carolinas, Virginia and Maryland*. Wm. David Webster, James F. Parnell, and Walter C. Biggs. Chapel Hill: North Carolina University Press, 1985.

WNW   *Wetlands*. (Audubon Society Nature Guide.) William A. Niering. NY: Knopf, 1985.

WOB   *The Wonder of Birds*. Robert M. Pyle, ed. Washington, DC: National Geographic, 1983.

WOI   *Wildlife of the Islands*. William H. Amos. NY: Abrams, 1980.

WOM   *Wildlife of the Mountains*. Edward R. Ricciuti NY: Abrams, 1979.

WON   *Wings of the North: A Gallery of Favorite Birds*. Candace Savage. Minneapolis: University of Minnesota, 1985.

WOW   *Sea Guide to Whales of the World*. Lyall Watson. NY: Dutton, 1981.

WPP   *Wildlife of the Prairies and Plains*. Kai Curry-Lindahl. NY: Abrams, 1981.

WPR   *Wildlife of the Polar Regions*. G. Carleton Ray and M.G. McCormick-Ray. NY: Abrams, 1981.

WRA   *A Field Guide to Western Reptiles and Amphibians*. (Second edition.) (Peterson Field Guide.) Robert C. Stebbins. Boston: Houghton Mifflin, 1985.

WTV   *Wildflowers of Tidewater Virginia*. Oscar W. Gupton and Fred C. Swope. Charlottesville, Virginia: University Press of VA, 1982.

WWD   *Wildlife of the Deserts*. Frederic H. Wagner. NY: Abrams, 1980.

WWF   *Watching Fishes: Life and Behavior on Coral Reefs*. Roberta Wilson and James Q. Wilson. NY: Harper and Row, 1985.

WWW   *Weird & Wonderful Wildlife*. Michael Marten, John May, and Rosemary Taylor. San Francisco: Chronicle Books, 1983.

Common Name Index

Aardvark (*Orycteropus afer*)
CRM 91; GEM IV:449, 452-457; MEM 467; MIA 29; SAB 22; WWW 170 (head only)

Aardwolf (*Proteles cristatus*)
CRM 149; GEM III:570, 571; MEM 159; MIA 99

Abalone, red (*Haliotis rufescens*)
MSC cp392, cp393

Acacia (*Acacia* sp.) See also Catclaw, Huisache
LBG cp159; OET 203, 204

Accentor (*Prunella collaris* or *montanella*)
EOB 362; NHU 104; WOM 88

Acouchi (*Myoprocta acouchy*)
GEM III:341

Addax (*Addax nasomaculatus*)
CRM 191; MEM 562; MIA 149

Adder See also Viper

Adder, death (*Acanthophis antarcticus*)
ERA 124; LSW 275

Adder, European or common viper (*Vipera berus*)
ERA 67; LSW 331-334; MIA 457; MSW p1

Adder, horned (*Bitis caudalis* or *cornuta*)
LSW 309, 310, 311

Adder, night (*Causus* sp.)
LSW 319, 320

Adder, puff (*Bitis arientans*)
LSW 306, 307; MIA 457

Adder's tongue, fetid (*Scoliopus bigelovii*)
SWW cp377

Agave See Century plant

Agouti (*Dasyprocta* sp.)
CRM 125; MIA 185

Agrimony (*Agrimonia* sp.)
CWN 83; NWE cp323

Ailanthus See Tree of Heaven

Albatross, black-browed (*Diomedea melanophrys*)
BMW 200-201; BWB 193-199; EAB 39, 41; EOB 47; HSW 75, 76; OBL 27, 222; WPR 147-149

Albatross, black-footed (*Diomedea nigripes*)
AAL 94(bw); EAB 40, 41; FWB 57; MPC cp548; OBL 55; UAB cp72

Albatross, Buller's (*Diomedea bulleri*)
AAL 94(bw); HSW 77, 78

Albatross, grey-headed (*Diomedea chrysostoma*)
BAS 41 (chick); BWB 188; WWW 28-29, 37 (head only)

Albatross, Laysan (*Diomedea immutabilis*)
BAS 21; EAB 43; EOB 45; FWB 56; OBL 210; UAB cp71

Albatross, light mantled sooty (*Diomedea palpebrata*)
EOB 46; MIA 201; WGB 11

Albatross, royal (*Diomedea epomophora*)
HSW 66

Albatross, short-tailed (*Diomedea albatross*)
EAB 43

Albatross, shy or white-headed (*Diomedea cauta*)
EAB 42; HSW 74; OBL 110

Albatross, wandering (*Diomedea exulans*)
BAS 20; BWB 196; EAB 42; EOB 44; HSW 69(bw); MIA 201; OBL 10, 30-31, 54; WGB 11

Albatross, waved (*Diomedea irrorata*)
BWB 196-197; HSW 68, 70-71; OBL 26, 182; WWW 71

Albatross, yellow-nosed (*Diomedea chlorhynchus*)
EAB 42; OBL 143

Alder, Arizona (*Alnus oblongifolia*)
NTW cp190; WFW cp86

Alder, black (*Frangula alnus*)
PPC 78

Alder, black or European (*Alnus glutinosa*)
NTE cp233; PMT cp[15]

Alder, common or hazel (*Alnus serrulata*)
FWF 151; NTE cp227; TSV 136, 137

Alder, mountain (*Alnus tenuifolia*)
NTW cp188, cp440; WFW cp88, cp144; WNW cp467

Alder, red (*Alnus rubra*)
NTW cp187; WFW cp89; WNW cp466

Alder, seaside (*Alnus maritima*)
NTE cp228, cp484

Alder, sitka (*Alnus sinuata*)
NTW cp192, cp441; WFW cp85

Alder, speckled (*Alnus rugosa*)
NTE cp229, cp372, cp616; NTW cp189, cp320; WNW
   cp468

Alder, white (*Alnus rhombifolia*)
NTW cp191, cp439; WFW cp87

Alewife (*Alosa pseudoharengus*)
ACF p12; MIA 503; NAF cp140, cp575

Alexanders, golden (*Zizia aptera*)
NWE cp349; SWW cp322

Alfilaria See Filaree

Algae, blue-green (*Anabaena oscillarioides*)
PGT 68

Algae, brown See Kelp

Algae, coralline (*Corallina* sp.)
AGC cp478, cp479; MPC cp505, cp506

Algae, green See under Chlamydomonas, Spirogyra,
   Volvox

Algae, red (*Batrachospermum* sp.) See also Irish Moss
PGT 75

Allamanda (*Mandevilla splendens*)
TSP 89

Allegheny-vine (*Adlumia fungosa*)
CWN 113

Alligator, American (*Alligator mississippiensis*)
AAL 212-213, 215(bw); AGC cp439; ARA cp256, cp259;
   ARC 143; ERA 143; MIA 415; NAW 251; RNA 209;
   WNW cp172

Alligator pear See Avocado

Alligator weed (*Alternanthera philoxeroides*)
WTV 25

Aloe (*Aloe* sp.)
TSP 45

Alpaca (*Lama guanicoe f.glama*)
GEM V:108-109; WWW 41(head)

Alpenrose (*Rhododendron ferrugineum*)
PPC 97

Alumroot (*Heuchera* sp.)
NWE cp5; SWW cp112; WFW cp415

Amanita See under Mushroom; See also Death cap,
   Destroying angel

Amberjack See also Yellowtail

Amberjack, greater (*Seriola dumerili*)
NAF cp554

Ambush bug (*Phymata* sp.)
MIS cp127; SGI cp54

Ameiva, giant (*Ameiva ameiva*)
RNA 93

Amoeba (various genera) See also Heliozoan
PGT 89; WWW 50

Amphioxus See Lancelet

Amphiuma, two-toed (*Amphiuma means*)
ARA cp15, cp18; ARC 52; ERA 26, 27; MIA 467

Anaconda (*Eunectes murinus* or *notaeus*)
AAL 238(bw), 241; AWR 174; LSW 24, 26; MIA 447

Anacua (*Ehretia anacua*)
NTE cp74, cp562

Anchovy (*Anchoa* sp.)
ACF p13; AGC cp403; NAF cp572; PCF p7

Anchovy, European (*Engraulis encrastiolus*)
MIA 503

Anchovy, Northern or Pacific (*Engraulis mordax*)
MPC cp295; NAF cp571; PCF p7

Anemone, blue (*Anemone oregana*)
SWW cp593; WFW cp523

Anemone, Carolina (*Anemone caroliniana*)
LBG cp139; NWE cp83

Anemone, desert (*Anemone tuberosa*)
JAD cp91; SWW cp71

Anemone fish See Clownfish

Anemone, sea See Sea anemone

Anemone, Western See Pasqueflower, mountain or
   western

Anemone, wood (*Anemone quinquefolia* or *nemorosa*)
CWN 71; NWE cp54; PPC 58; SEF cp441; WTV 14

Angel trumpets (*Acleisanthes longiflora*)
JAD cp104; SWW cp83

Angelfish, emperor or imperial (*Pomacanthus imperator*)
MIA 561; RCK 54

Angelfish, French (*Pomacanthus paru*)
AAB 72; ACF cp37; AGC cp362; KCR cp25; NAF cp332, cp335

Angelfish, grey (*Pomacanthus arcuatus*)
ACF cp37; AGC cp361; KCR cp25; NAF cp334; WWF cp33; WWW 36

Angelfish, king (*Pygoplites diacanthus*)
RCK 81; RUP 76

Angelfish, queen (*Holacanthus ciliaris*)
AAL 360; ACF cp37; AGC cp363; KCR cp25; MIA 561; NAF cp331, cp336; RCK 32; RUP 73; WWF cp26

Angelfish (other *Pomacanthus* sp.)
BMW 143; RCK 54, 158

Angel's trumpet (*Brugmansia* sp.)
TSP 45, 47

Angler-fish (various genera)
EAL 3, 94, 95, 97; MAE 111; MIA 531; WWW 76, 178, 180

Angwantibo See Potto, golden

Anhinga or water turkey (*Anhinga anhinga*)
AAL 100; AGC cp574; BAS 97; BWB 90, 91; EAB 44; FEB 71; MAE 89; MIA 207; NAB cp100; NAW 218; PBC 57; SEF cp264; WGB 17; WNW cp518; WOB 34-35

Anhinga, African (*Anhinga melanogaster*)
HSW 103

Ani, groove-billed (*Crotophaga sulcirostris*)
EAB 130; NAB cp578

Ani, smooth-billed (*Crotophaga ani*)
EAB 130; FEB 368; MIA 267; NAB cp577; WGB 77

Anise-tree (*Illicium* sp.)
NTE cp59, cp60, cp365

Anoa or dwarf water-buffalo (*Anoa* or *Bubalus* sp.)
CRM 185; GEM V:359, 369, 370; MIA 143

Anole, Carolina or green (*Anolis carolinensis*)
AGC cp440; ARA cp383, cp385; ARC 175, 176; ERA 94; LOW 123; MIA 417; RNA 103; SEF cp519

Anole (other *Anolis* sp.)
AAL 3, 226(bw); ARA cp384, cp389, cp390; ERA 90; LOW 78, 124(bw); NNW 29; RNA 103, 105

Ant, army (*Eciton* sp.)
SWF 178, 179; WIW 15(bw)

Ant, black (little) (*Monomorium minimum*)
JAD cp386; MIS cp321

Ant, carpenter (*Camponotus* sp.)
AAL 473; JAD cp388; MIS cp318; SGI cp281; WFW cp573

Ant, fire (*Solenopsis* sp.)
MIS cp313; SGI cp279

Ant, harvester (*Pogonomyrmex* sp.)
JAD cp387; LBG cp376; MIS cp317; SGI cp277

Ant, honey (*Myrmecocystus* sp.)
JAD cp384; SGI cp280

Ant, leaf-cutter (*Atta* sp.)
AAL 474(bw); MIS cp312; SWF 30, 31

Ant, red or mound (*Formica* sp.)
MIS cp316; SGI cp282

Ant, tree or weaving (*Oecophylla smaragdina*)
AAB 10; WIW 127

Ant, velvet See Wasp, velvet ant

Ant, wood (*Formica rufa*)
DIE cp26

Antbird, bicolored (*Gymnopithys leucaspis*)
MIA 299; WGB 109

Antbird, ocellated (*Phaenostictus mcleannani*)
MIA 299; WGB 109

Antbird, white-plumed (*Pithys albifrons*)
EOB 313

Anteater, giant (*Myrmecophaga tridactyla*)
CRM 89; GEM II:584-591; MEM 773; MIA 27; SWF 180-181(head only); WPP 174-175; WWW 98

Anteater, lesser (*Myrmecophaga tetradactyla*)
SWF 181

Anteater, scaly See Pangolin

Anteater, silky or dwarf (*Cyclopes didactylus*)
GEM II:595; MEM 774; MIA 27

Anteater, spiny See Echidna

Antechinus See Marsupial mouse

Antelope, American See Pronghorn

Antelope, beira (*Dorcatragus megalotis*)
CRM 195; MIA 151

Antelope, desert See Addax

Antelope, four-horned (*Tetracerus quadricornis*)
GEM V:359; MEM 544; MIA 141

Antelope, lyra See Topi

Antelope, roan (*Hippotragus equinus*)
GEM V:437; MEM 563; MIA 147; NHA cp8; WPP 66

Antelope, royal (*Neotragus pygmaeus*)
MEM 576; MIA 151

Antelope, sable (*Hippotragus niger*)
CRM 189; GEM V:439; MEM 562, 568; NHA cp4, 121(bw)

Antelope, saiga See Saiga

Antelope, Tibetan or Chiru (*Pantholops hodgsoni*)
GEM V:485

Antelope-brush (*Purshia tridentata*)
WFW cp112, cp200

Antlion (Myrmeleontidae)
MIS cp403(adult); SGI cp79

Antpitta (*Grallaria* and other genera)
AAL 155; EOB 317; MIA 299; WGB 109

Antshrike, barred (*Thamnophilus doliatus*)
AAL 155; EOB 317; MIA 299

Antshrike, great (*Taraba major*)
MIA 299; WGB 109

Antthrush, rufous-capped (*Foricarius colma*)
EOB 313

Antwren, white-flanked (*Myrmotherula axillaris*)
EOB 313

Aoudad See Sheep, barbary

Apache plume (*Fallugia paradoxa*)
JAD cp96; SWW cp41; WFW cp183

Ape, Barbary (*Macaca sylvanus*)
GEM II:205, 225-229; MIA 67

Ape, black (Sulawesi or Celebes) (*Macaca nigra*)
CRM 75; NHP 128 (bw)

Ape, night See Monkey, night

Aperea See Cavy, wild

Aphid (various genera)
AAB 42; SGI cp75; WWW 77

Apostlebird (*Struthidea cinerea*)
EOB 433; MIA 397; WGB 207

Apple, crab (*Malus sylvestris*)
LBG cp422, cp452; NTE cp133, cp456, cp570; NTW
   cp157, cp368, cp493

Apple, crab (various *Malus* sp.)
LBG cp415, cp424, cp451; NTE cp121, cp130, cp135,
   cp458, cp572; NTW cp149, cp361; PMT cp[80]-[84]

Arapaima or Pirarucu (*Arapaima gigas*)
AAL 324-325, 326(bw); AWR 186-187; EAL 43; MIA 501

Arawana (*Osteoglossum* sp.)
AWR 206(bw); EAL 43

Arborvitae See also Cedar, red (Western) and Cedar,
   white (Northern)

Arborvitae, Oriental (*Thuja orientalis*)
NTE cp31; NTW cp59

Arbutus, trailing (*Epigaea repens*)
BCF 167; CWN 175; NWE cp568; SEF cp486; WAA 158;
   WTV 123

Archerfish (*Toxotes* sp.)
AWR 211(bw); EAL 110; MAE 88; MIA 561

Argali (*Ovis ammon*)
CRM 203; GEM V:550; MEM 585; NHU 160

Argonaut See Nautilus, paper

Arkal See Urial

Armadillo, giant (*Priodontes maximus*)
CRM 89; MIA 29

Armadillo, larger hairy or Patagonian (*Chaetophractus
   villosus*)
GEM II:625; MEM 781; WWD 134-135

Armadillo, nine-banded (*Dasypus novemcinctus*)
AAL 41; ASM cp217, 218; MIA 29; MNP 172(bw); MPS
   112(bw); SEF cp598; WMC 101

Armadillo, pink or lesser fairy (*Chlamyphorus truncatus*)
CRM 89; GEM II:626; MIA 29

Armadillo, six-banded (*Euphractus sexcinctus*)
WPP 178

Armadillo, three-banded (*Tolypeutes tricinctus*)
CRM 89

Arnica (*Arnica* sp.)
PPC 60; SWW cp248; WAA 181; WFW cp443

Arrowhead or Sagittaria (*Sagittaria* sp.)
CWN 25; NWE cp149; PGT 53; SWW cp127, cp128; WTV
   47

Arrowwood (*Viburnum dentatum*)
NTE cp223, cp541, cp542; NWE cp213; SEF cp96, cp196;
   TSV 82, 83

Artist's fungus (*Ganoderma applanatum*)
FGM cp14; NAM cp518

Arum, arrow (*Peltandra virginica*)
CWN 47; NWE cp11; WNW cp310

Arum, water or wild calla (*Calla palustris*)
CWN 45; NWE cp109; PGT 43; WNW cp225

Ash, black (*Fraxinus nigra*)
NTE cp334; WNW cp435

Ash, blue (*Fraxinus quadrangulata*)
NTE cp326; PMT cp[53]

Ash, Carolina (*Fraxinus caroliniana*)
NTE cp325; WNW cp437

Ash, green (*Fraxinus pennsylvanica*)
NTE cp328; WNW cp436

Ash, mountain See Mountain ash

Ash, Oregon (*Fraxinus latifolia*)
NTW cp276; WFW cp106

Ash, singleleaf (*Fraxinus anomala*)
NTW cp121; WFW cp68

Ash, Texas (*Fraxinus texensis*)
NTE cp351; PMT cp[54]

Ash, velvet (*Fraxinus velutina*)
NTW cp274; WFW cp107

Ash, white (*Fraxinus americana*)
NTE cp329, cp492, cp602; SEF cp134, cp178; TSV 188, 189

Ash (other *Fraxinus* sp.)
NTE cp346; NTW cp275, cp277-279, cp298, cp530

Asp See Viper, asp

Aspen, bigtooth (*Populus grandidentata*)
LBG cp433; NTE cp191, cp599; SEF cp66

Aspen, quaking (*Populus tremuloides*)
LBG cp430; NTE cp184, cp620; NTW cp211, cp339; SEF cp64; WFW cp95

Ass See also Kiang

Ass, African (*Equus africanus*)
CRM 169; GEM IV:558, 582, V:628; MEM 484; MIA 125

Ass, Asiatic (*Equus hemionus*) See also Onager
AAL 71(bw); CRM 169; WWD 98

Assassin bug (various genera)
EOI 54; SGI cp53

Aster, calico (*Aster latiflorus*)
LBG cp135; NWE cp102

Aster, Engelmann (*Aster engelmannii*)
SWW cp70; WFW cp379

Aster, golden (*Chrysopsis mariana* or *villosa*)
CWN 257; SWW cp261; WFW cp440

Aster, golden (hairy) (*Chrysopsis camporum*)
LBG cp170; NWE cp292

Aster, Mohave (*Machaeranthera tortifolia*)
JAD cp61; SWW cp476; WAA 210

Aster, New England (*Aster novae-angliae*)
BCF 57; LBG cp246; NWE cp479, cp603; WAA 50

Aster, New York (*Aster novi-belgii*)
CWN 271; NWE cp609

Aster, panicled (*Aster simplex*)
LBG cp134; NWE cp99

Aster, salt-marsh (*Aster tenuifolius*)
AGC cp448; NWE cp96; SCS cp19

Aster, smooth (*Aster laevis*)
LBG cp293; NWE cp604

Aster, sticky (*Machaeranthera bigelovii*)
SWW cp481, cp596

Aster, wavy-leaved (*Aster undulatus*)
NWE cp480

Aster, white-topped (*Sericocarpus asteroides*)
CWN 277; WTV 35

Aster (other *Aster* species)
CWN 269-275; NWE cp97-cp101, cp481, cp608, cp610; SWW cp475; WTV 198

Auk, little See Dovekie

Auk, razorbill (*Alca torda*)
AGC cp522; AOG cp2; BWB 105; DBN cp13; EAB 49; EOB 210-211; FEB 64; HSW 148; MIA 251; NAB cp91; OBL 34; RBB 96; WGB 61

Auklet, cassin (*Ptychoraphus aleutica*)
DBN cp21; EAB 45; FWB 63; MPC cp542; UAB cp87

Auklet, crested (*Aethia cristatella*)
AOG cp7; BWB 106; DBN cp23, cp24, cp27; EAB 45; MIA 251; UAB cp86; WGB 61; WPR 141

Auklet, least (*Aethia pusilla*)
AOG cp7; BWB 106; DBN cp24, cp25, cp27; EAB 45; EOB 211; UAB cp89

Auklet, Parakeet (*Cyclorrhynchus psittacula*)
AOG cp9; BWB 106; DBN cp22; EAB 45; UAB cp88; WPR 141

Auklet, rhinoceros (*Cerorhinca monocerata*)
AOG cp8; BWB 107; DBN cp18, cp28; EAB 45; EOB 210; FWB 65; MPC cp541; UAB cp85

Auklet, whiskered (*Aethia pygmaea*)
BWB 107; DBN cp26, cp27; EAB 45

Avens (*Geum* sp.)
CWN 81

Avens, mountain (*Dryas* sp.)
SWW cp46; WAA 253

Avocado (*Persea americana*)
TSP 35

Avocet, America (*Recurvirostra americana*)
AGC cp577; BWB 182; EAB 51; FEB 155; FWB 140; NAB
cp244; NAW 101; PBC 137; UAB cp220, cp241; WBW
305; WNW cp547

Avocet, European (*Recurvirostra avosetta*)
MIA 245; RBB 62; WBW 307; WGB 55

Avocet, rednecked or Australian (*Recurvirostra
novaehollandiae*)
NOA 189

Axolotl (*Ambystoma mexicanum*)
MIA 465

Aye-aye (*Daubentonia madagascariensis*)
CRM 63; GEM II:71, 75; MIA 57; NHP 93(bw)

Ayu (*Plecoglossus altivelis*)
EAL 59

Azalea, alpine (*Loiseleuria procumbens*)
NWE cp454

Azalea, early or mountain (*Rhododendron roseum*)
TSV 100, 101

Azalea, flame (*Rhododendron calendulaceum*)
NWE cp380; SEF cp461; TSV 98, 99; WAA 13

Azalea, Western (*Rhododendron occidentale*)
SWW cp572; WFW cp208

B

Babirusa (*Babyrousa babyrussa*)
CRM 171; GEM V:44, 45; MIA 129; WWW 125

Baboon, anubis (*Papio anubis*)
GEM II:238, 242-243; WPP 110-111

Baboon, chacma (*Papio ursinus*)
MIA 69; NHP 133 (bw); WPP 108

Baboon, gelada (*Theropithecus gelada*)
AAL 3(head); CRM 77; GEM II:257; MEM 314, 372; MIA
69; NHP 138 (bw); WOM 101

Baboon, guinea (*Papio papio*)
MEM 373

Baboon, hamadryas (*Papio hamadryas*)
CRM 77; GEM II:237, 239(head only), 246); MIA 69; NHP
133; MEM 373, 395

Baboon, olive or common (*Papio cynocephalus*)
MEM 6, 371, 373, 376, 392; MIA 69; SAB 125

Baby blue eyes (*Nemophila menziesii*)
LBG cp284; SWW cp579; WAA 190; WFW cp516

Baby stars, false (*Linanthus androsaceus*)
LBG cp239; SWW cp448

Backswimmer (*Notonecta* sp.)
AAL 417; EOI 55; MIS cp98, cp99; SGI cp49; WIW 131
(bw); WNW cp345

Badger (*Taxidea taxus*)
ASM cp204; CMW 232(bw); GEM III:423-426; JAD cp514;
LBG cp51; MAE 77; MEM 131; MIA 89; MNC 265(bw);
MNP 263 (bw); MPS 277(bw); NAW 11 (head); NMM
cp[27]; RFA 228-233 (bw); SWM 112, 113; WPP 29;
WWD 120-121

Badger, Eurasian (*Meles meles*)
AAL 63(bw); CRM 141; MAE 52; MEM 131; MIA 89; NNW
38, 176; SAB 97

Badger, honey or ratel (*Mellivora capensis*)
CRM 141; GEM III:421; MEM 133; MIA 89

Badger (other genera)
GEM III:428(head only); MEM 130, 132; MIA 89

Bagworm (Psychidae)
MIS cp10

Baldpate See Wigeon, American

Balloonfish See Porcupinefish

Balloonflower (*Penstemon palmeri*)
JAD cp44; SWW cp523; WFW cp495

Balsam-root, arrowleaf (*Balsamorhiza sagittaria*)
JAD cp130; LBG cp174; SWW cp250; WFW cp442

Banana or plantain (*Musa* x *paradisiaca*)
TSP 127

Bandicoot (*Isoodon* and other genera) See also Bilby
CRM 21; GEM I:300, 302; MEM 847-849; MIA 21; NOA
37, 225; WWD 175

Bandy-bandy (*Vermicella annulata*)
LSW 281; MIA 455

Baneberry, common See Herb-Christopher

Baneberry, red (*Actaea rubra*)
BCF 175; CWN 67; NWE cp161, cp442; PPC 54; SEF
cp429, cp504; SWW cp108, cp427; WFW cp417

Baneberry, white See Doll's eyes

Banksia (*Banksia* sp.)
OET 223; TSP 45

Banteng (*Bos javanicus*)
CRM 187; GEM V:390-392; MIA 143

Banyan (*Ficus* sp.)
TSP 31

Baobab (*Adansonia digitata*)
OET 14

Barasingha, Indian (*Cervus duvauceli*)
CRM 177; GEM V:171, 616-617; WPP 56-57

Barberry (*Berberis vulgaris*)
LBG cp209; NWE cp362

Barber's pole See Candystick

Barbet, golden-throated (*Megalaima franklinii*)
EOB 292

Barbet, red-and-yellow (*Trachyphonus erythrocephalus*)
EOB 293

Barbfish See Scorpionfish

Barnacle (*Balanus* sp.)
AWR 157; BCS 95; BMW 27; MSC cp279

Barnacle, acorn (*Semibalanus balanoides*)
AAL 400; EAL 233

Barnacle, acorn (giant) (*Balanus nubilis*)
MPC cp462, cp466; MSC cp280, cp283

Barnacle, goose (*Lepas* sp.)
AAL 400; AGC cp241; BMW 30-31; EAL 228; MPC cp460; MSC cp288; SAS cpC11

Barnacle (other kinds)
AGC cp238-240, cp242; MPC cp461, cp463-468; MSC cp274-288

Barracuda, European (*Sphyraena sphyraena*)
HSG cp15b; RUP 90

Barracuda, great (*Sphyraena barracuda*)
ACF cp41; HSG cp15a; MIA 565; NAF cp583, cp584; WWF cp14

Barracuda (other *Sphyraena* sp.)
AAL 363; ACF p41; PCF cp35; WWW 108

Basil, wild (*Satureia vulgaris*)
CWN 223; NWE cp558

Basilisk (*Basiliscus* sp.)
AAL 222(bw), 223(bw); LOW 32(bw); MIA 419; WWW 103

Basketstar, Caribbean (*Astrophyton muricatum*)
MSC cp573; WWF cp61, cp62

Basketstar, Northern (*Gorgonocephalus arcticus*)
AGC cp201; MSC cp572

Bass, channel See Drum, red

Bass, harlequin (*Serranus trigrinus*)
KCR cp23

Bass, kelp (*Paralabrax clathratus*)
MPC cp238; NAF cp512; PCF cp29

Bass, largemouth (*Micropterus salmoides*)
AAL 354(bw); MIA 551; NAF cp92; WNW cp34

Bass, rock (*Ambloplites rupestris*)
MIA 551; NAF cp82; WNW cp45

Bass, sea See Sea bass

Bass, smallmouth (*Micropterus dolomieui*)
NAF cp90; WNW cp44

Bass, spotted (*Micropterus punctulatus*)
NAF cp91; WNW cp49

Bass, striped (*Morone saxatilis*)
ACF cp24; AGC cp373; MIA 549; MPC cp278; NAF cp94, cp510; PCF cp29; WNW cp33

Bass, white (*Morone chrysops*)
NAF cp95; WNW cp40

Bass, yellow (*Morone mississippiensis*)
NAF cp93

Basslet (*Lipogramma* and other genera)
ACF cp64; KCR cp24; RCK 6, 7

Basswood, American (*Tilia americana*)
NTE cp158, cp406, cp612; SEF cp72

Basswood, white (*Tilia heterophylla*)
NTE cp155; SEF cp71; TSV 60, 61

Bat, barbastelle (*Barbastella barbastellus*)
LOB 74(bw); MIA 51

Bat, big or common brown (*Eptesicus fuscus*)
ANB 56; ASM cp166; CGW 137; CMW 57(bw); GMP 104(bw), cp1; MIA 51; MNC 123(bw); MPS 106(bw); NAW 12(bw); WMC 81

Bat, big-eared (Rafinesque's) (*Plecotus rafinesquii*)
ASM cp144; MNC 131(bw); MPS 109(bw); WMC 97

Bat, big-eared (Townsend's) (*Plecotus townsendii*)
ANB 6; ASM cp145; CMW 66(bw); CRM 57; JAD cp463; MPS 110(bw); WMC 95

Bat, California myotis (*Myotis californicus*)
ASM cp160; CMW 38(bw); JAD cp458

Bat, cave myotis (*Myotis velifer*)
ASM cp171; JAD cp457; MPS 67(bw)

Bat, disc-footed (*Eudiscopus denticulus*)
CRM 55

Bat, evening (*Nycticeius humeralis*)
ANB 7, 58; ASM cp168; GMP 117(bw); MNC 129(bw);
   MPS 108(bw); WMC 93

Bat, fisherman (*Noctilio leporinus*)
LOB 48

Bat, fishing (*Pizonyx vivesi*)
AAL 34(bw); MIA 51

Bat, flying fox See Flying fox

Bat, free-tailed (big) (*Talarida macrotis*)
ASM cp151; JAD cp467; MNC 133(bw); MPS 111(bw)

Bat, free-tailed (Brazilian or Mexican) (*Talarida
   brasiliensis*)
ANB 26, 63; ASM cp152; CMW 71 (bw); CRM 57; JAD
   cp465; MNC 133(bw); MPS 111(bw); NMM cp[2]; WMC 98

Bat, free-tailed (others)
ASM cp149; GEM I:615; JAD cp466; LOB 80; MIA 47;
   WOI 187

Bat, fringed myotis (*Myotis thysanodes*)
ASM cp170; CMW 49(bw); MPS 66(bw)

Bat, fruit (Egyptian) (*Rousettus aegyptiacus*)
LOB 31(bw), 41; WWW 80

Bat, fruit (short-nosed) (*Cynopterus* sp.)
LOB 129; SWF 92

Bat, fruit (tube-nosed) (*Nyctimene* sp.)
LOB 43; MIA 39; WOI 140

Bat, fruit (other kinds)
AAL 33; CRM 47; LOB 42(bw)-46(bw)

Bat, ghost (*Macroderma gigas*)
CRM 49; MEM 803

Bat, ghost-faced (*Mormoops megalophylla*)
ANB 7; ASM cp154; JAD cp451

Bat, gray myotis (*Myotis grisescens*)
ASM cp165; CRM 53; MNC 109(bw); MPS 64(bw); WMC
   70

Bat, hammer-head (*Hypsignathus monstrosus*)
MEM 816, 817; MIA 39

Bat, hoary (*Lasiurus cinereus*)
ANB 6, 72; ASM cp178; CGW 137; CMW 62(bw); GMP
   114(bw), cp3; MNC 127(bw); MPS 107(bw); NMM cp[4];
   SEF cp596; SWM 9; WFW cp338; WMC 88

Bat, hognosed or bumblebee (*Craseonycteris
   thonglongyai*)
CRM 49; MEM 790; MIA 43

Bat, horseshoe (*Rhinolophus luctus*)
ROW 17(bw)(head only)

Bat, horseshoe (Greater) (*Rhinolophus ferrumequinum*)
CRM 49; GEM I:146-147; LOB 30(bw), 55(bw)

Bat, Indiana or social myotis (*Myotis sodalis*)
ASM cp176; CRM 53; GMP 91(bw); MNC 117(bw); MPS
   66(bw); WMC 72

Bat, Keen's myotis (*Myotis keenii*)
ASM cp163; CGW 137; CMW 51(bw); GMP 88(bw); MPS
   64(bw); WMC 71

Bat, leaf-nosed (California) (*Macrotus californicus*)
ASM cp147; JAD cp452

Bat, leaf-nosed (*Hipposideros* sp.)
LOB 49; MEM 791

Bat, little brown myotis (*Myotis lucifugus*)
ANB 52; ASM cp169; CGW 137; CMW 45(bw); GMP
   84(bw); LOB 131(bw); MIA 51; MNC 113(bw); MPS
   65(bw); NAW 60, 61; SWN 10; WMC 66

Bat, long-eared (*Plecotus* sp.)
AAL 34; LOB 31(bw), 73, 79(bw)

Bat, long-eared (*Myotis evotis*)
ASM cp161; CMW 53(bw); MPS 63(bw)

Bat, long-legged myotis (*Myotis volans*)
ASM cp162 ; CMW 47(bw); MPS 67(bw)

Bat, long-nosed (*Leptonycteris* or *Rhynchonycteris* sp.)
ANB front., 6; ASM cp164, cp174; JAD cp454, cp455; LOB
   132; MIA 41

Bat, long-tongued (*Choeronycteris* or *Glossophaga* sp.)
ASM cp180; JAD cp453; LOB 51(bw); WOI 187

Bat, mastiff (*Eumops* sp.)
ASM cp148, cp150, cp153; JAD cp468

Bat, mouse-eared (*Myotis myotis*)
AAB 109; CRM 53; GEM I:532, 544; LOB 136

Bat, mouse-tailed (*Rhinopoma* sp.)
LOB 47(bw)

Bat, natterer's (*Myotis natereri*)
LOB 78 (bw); MEM 790

Bat, noctule (*Nyctalus* sp.)
CRM 55; MIA 51; SAB 23

Bat, northern myotis (*Myotis septentrionalis*)
MNC 115(bw)

Bat, pallid (*Antrozous pallidus*)
AAL 34(bw)(head); ANB 14, 38, 60; ASM cp146; CMW
   68(bw); JAD cp464; MPS 110(bw); NMM cp[3]

Bat, pipistrelle (*Pipistrellus* sp.)
LOB 74 (bw); MIA 51

Bat, pipistrelle (Eastern) (*Pipistrellus subflavus*)
ASM cp159; CGW 137; GMP 101(bw); MPS 106(bw);
WMC 79

Bat, pipistrelle (Western) (*Pipistrellus hesperus*)
AAL 34(bw); ASM cp175; JAD cp461; MPS 105(bw)

Bat, red (*Lasiurus borealis*)
ANB 69; ASM cp157; CGW 137; GMP cp2; MIA 51; MNC
125(bw); MPS 107(bw); NAW 14; WMC 83, 84

Bat, Seminole (*Lasiurus seminolus*)
ASM cp158; GMP 111(bw); MPS 108(bw); WMC 86

Bat, silver-haired (*Lasionycteris noctivagans*)
ANB 79; ASM cp179; CGW 137; CMW 55(bw); GMP
98(bw), cp1; JAD cp459; MNC 119(bw); MPS 105(bw);
WMC 76

Bat, small-footed myotis (*Myotis leibii*)
ASM cp167; CMW 40(bw); GMP 95(bw); MNC 111(bw);
MPS 65(bw); WMC 74

Bat, Southeastern myotis (*Myotis austroriparicus*)
ASM cp177; MNC 107(bw); MPS 63(bw); WMC 68

Bat, Southwestern myotis (*Myotis auriculus*)
ASM cp173; JAD cp460

Bat, spotted or pinto (*Euderma maculatum*)
ANB 7, 83; ASM cp143; CRM 57; JAD cp462; NMM cp[1]

Bat, tent-making (*Uroderma bilobatum*)
ANB 7; MEM 806

Bat, vampire See Vampire bat

Bat, water (*Myotis daubentoni*)
LOB 75 (bw)

Bat, yellow (Northern) *Lasiurus intermedias*)
ASM cp155; WMC 90

Bat, yellow (Southern) (*Lasiurus ega*)
ANB 6, 75; ASM cp156

Bat, Yuma myotis (*Myotis yumanensis*)
ASM cp172; CMW 43(bw); JAD cp456; MPS 68(bw)

Bateleur See under Eagle

Batfish (*Ogcocephalus* sp. and other genera)
AAL 361; ACF p14; KCR cp25; MIA 561; NAF cp293,
cp294; PCF p13; RUP 42

Bay, loblolly (*Gordonia lasianthus*)
NTE cp119, cp438; PMT cp[56]; WNW cp460

Bayberry, odorless (*Myrica inodora*)
NTE cp61

Bayberry, Pacific (*Myrica californica*)
NTW cp139; WFW cp81

Bayberry, Southern (*Myrica cerifera*)
AGC cp453; NTE cp209; SEF cp100; TSV 174, 175; WNW
cp461

Bear, Asian or Tibetan black or moon (*Selenarctos
thibetanus*)
DBW 112, 117-121; GEM III:497; MEM 97; MIA 83; WOM
133

Bear, black (*Ursus americanus*)
ASM cp292, 293; CGW 145; CMW 204(bw); CRM 137;
DBW 18-35, 23 (Kermode); GEM III:500; GMP cp17; MAE
48; MEM 94, 95; MIA 83; MNC 247(bw); MNP 242(bw);
MPS 252(bw); NAW 15; NMM cp[24]; RFA 120-130(bw);
RMM cp[12]; SEF cp609; SWF 141, 143; SWM 89; WFW
cp359; WMC 185; WNW cp615

Bear, brown or grizzly (*Ursus arctos*)
ASM cp294, 295; AWR front., 139-142 (Kodiak); BAA
84(bw) (Kodiak); CMW 207 (bw); CRM 137; DBW 2, 12,
14, 53 (European), 36-62; GEM III:366-367, 384-385,
482, 483, 491-491, 495; MEM 88-91; MIA 83; MNC
301(bw); MNP cp16; MPS 271(bw); NAW 16, 17; RFA
102-117(bw); SWF 153-155; SWM 91, 93; WFW cp360;
WOM 68, 168, 169; WPR 97

Bear, moon See Bear, Asian black

Bear, polar (*Ursus maritimus*)
ASM cp296; BAA 1, 15, 71, 79, 80; BMW 220-223; DBW
4, 64-88; GEM III:484-485, 488-489; MEM 87, 92, 93;
MIA 83; NAW 18; NHU 73; RFA 133, 139 (bw); SWM 95;
WOI 44; WPR 62, 63, 188

Bear, sloth (*Melursus ursinus*)
AAL 59(bw); CRM 137; DBW 100-105; GEM III:502; MEM
97; MIA 59(bw)

Bear, spectacled (*Tremarctos ornatus*)
CRM 137; DBW 106-111; GEM III:505; MEM 97; MIA 83

Bear, sun (*Ursus* or *Helarctos malayanus*)
CRM 137; DBW 90-99; GEM III:504; MEM 97; MIA 83

Bearberry or kinnikinnick (*Arctostaphylos nova-ursi*)
CWN 175; NWE cp230, cp444; SEF cp507; SWW cp428,
cp558; WFW cp149, cp215; WOM 50; WPR 94(*A.rubra*)

Beardtongue See Penstemon

Beargrass (*Xerophyllum tenax*)
SWW cp116; WFW cp418

Beaugregory (*Pomacentrus leucostictus*)
ACF cp38; MIA 563; NAF cp323

Beautyberry (*Callicarpa americana*)
WTV 199

Beaver (*Castor canadensis*)
AAL 44; ASM cp224; AWR 144; CGW 177; CMW 140(bw);
GEM III:104-112; GMP 165(bw); MIA 165; MNC 183(bw);
MPS 180(bw); NAW 19; PGT 233; RFA 14-25(bw); RMM
cp[8]; SEF cp597; SWM 41, 44; WFW cp347; WMC 134;
WNW cp611

Beaver, Eurasian (*Castor fiber*)
CRM 103; MIA 165

Beaver, mountain (*Aplodontia rufa*)
ASM cp229; GEM III: 29, 31; MEM 611; MIA 165; MNP
    183(bw); WFW cp346; WNW cp607

Becard, rose-throated (*Pachyramphus* or *Platypsaris*
    sp.)
FEB 296; FWB 408; NAB cp430; UAB cp493

Bedbug (*Cimex lectularius*)
MIS cp67; WIW 73 (bw)

Bedstraw (*Galium* sp.) See also Cleavers; Madder, wild
CWN 209, 211; NWE cp182; SWW cp142; WFW cp416

Bee, alkali (*Nomia melanderi*)
SGI cp300

Bee assassin (*Apiomerus* sp.)
JAD cp376; MIS cp118, cp119

Bee, bumble See Bumblebee

Bee, carpenter (African) (*Xylocopa* sp.)
EOI 85; SGI cp305

Bee, digger (*Ptilothrix bombiformis*)
LBG cp373; SGI cp303

Bee, honey (*Apis mellifera*)
AAB 120; AAL 481; BOI 222-223; LBG cp375; MIS cp511;
    SEF cp404; SGI cp307, cp308; SWF 62; WIW 122

Bee killer (*Promachus fitchii*)
MIS cp401

Bee, leafcutter (*Megachile* sp.)
MIS cp490; SGI cp302; WIW 119

Bee, mason (*Osmia* sp.)
SEF cp408; SGI cp301

Bee, mining (*Andrena vicina*)
SGI cp298

Bee plant, Rocky Mountain (*Cleome serrulata*)
LBG cp260; SWW cp530

Bee plant, yellow (*Cleome lutea*)
JAD cp111; LBG cp204; SWW cp321; WAA 211

Bee, plasterer or mining (*Colletes* sp.)
EOI 119; SGI cp297

Bee, sweat (*Agapostemon texanus*)
SGI cp299

Bee-balm (*Monarda didyma*)
CWN 219; NWE cp434, cp537(*M.pectinata*)

Beech, American (*Fagus grandifolia*)
NTE cp152, cp528, cp601, cp614; SEF cp88, cp204; TSV
    134, 135

Beech, European (*Fagus sylvatica*)
NTE cp174; NTW cp196, cp337; PMT cp[49]; PPC 77

Beechdrops (*Epifagus virginiana*)
CWN 245; NWE cp393

Bee-eater, blue-cheeked (*Merops superciliosus*)
NHU 184

Bee-eater, European (*Merops apiaster*)
AAL 147; EOB 275; MIA 285; WGB 95; WOI 172

Bee-eater, rainbow (*Merops ornatus*)
EOB 273; WWD 42-43

Beefsteak fungus (*Fistulina hepatica*)
FGM cp14; NAM cp513; NMT cp6; SGM cp307

Beetle, bark (*Scolytus* sp.)
BNA 328(bw); SEF cp384; WFW cp567

Beetle, bear (*Paracotalpa* sp.)
SGI cp109, cp110

Beetle, bee parasite (Stylopidae)
BNA 269(bw); SGI cp154

Beetle, blister (desert) (*Cysteodemus wislizeni*)
BNA cp6; SGI cp150

Beetle, blister (fire) (*Pyrota* sp.)
BNA cp6; SGI cp152

Beetle, blister or oil (*Lytta* sp.)
BNA cp6; JAD cp382; WIW 79; WWD 155

Beetle, blister (other genera)
BNA 271-272(bw); MIS cp161, cp167, cp202; SGI cp153

Beetle, bombardier (*Brachinus* sp.)
EOI 68; MIS cp166; WNW cp353

Beetle, carpet (*Anthrenus* sp.)
BNA 195(bw); MAE 135; SGI cp133(larva)

Beetle, carrion (Silphidae)
BNA cp2, 121(bw); MIS cp174, cp223; SEF cp391; SGI
    cp92

Beetle, checkered (Cleridae)
BNA cp3, 209(bw); MIS cp141, cp142, cp144; SGI cp135,
    cp136

Beetle, click (Elateridae)
AAL 431(bw); BNA cp4, 174-177(bw); DIE cp31; EOI 65;
    MIS 186; SGI cp122-cp126

Beetle, cockchafer (*Melolontha melolontha*)
EOI 19 (head-on view)

Beetle, Colorado potato (*Leptinotarsa decemlineata*)
MAE 134; SGI cp182

Beetle, cucumber (*Diabrotica* sp.)
BNA 297(bw); SGI cp185, cp186

Beetle, darkling (*Tenebrionidae*)
AAL 435(bw); BNA cp7, 248-255(bw); MIS cp197; SGI
   cp143-145; WWD 72

Beetle, death-watch (Anobiidae)
BNA cp7, 198-203(bw); MAE 135

Beetle, dermestid (Derestidae)
BNA cp3, 193-195(bw)

Beetle, diving (Dytiscidae)
BNA 98(bw), 99(bw); MIS cp34(larva), cp44(larva), cp94-
   96

Beetle, diving (*Dytiscus marginalis*)
AAB 65, 107; AAL 428; AWR 26, 44(larva); PGT 162, 163

Beetle, dung See also Beetle, scarab

Beetle, dung (*Geotrupes* sp.)
BNA 142(bw); EOI 67

Beetle, dung (*Phanaeus* sp.)
BNA cp7; DIE cp1

Beetle, fire (*Pyrophorus* sp.)
MIS cp222

Beetle, fire-colored (*Dendroides* sp.)
BNA 261(bw); SGI cp148

Beetle, flower See also Beetle, June (green)

Beetle, flower (*Trichiotinus* sp.)
BNA 148(bw); SGI cp115

Beetle, flower (tumbling) (*Mordella* sp.)
BNA 265(bw); SGI cp149

Beetle, fuller (*Polyphylla fullo*)
DIE cp11

Beetle, glorious (*Plusiotis gloriosa*)
BNA cp8; SGI cp108

Beetle, glowworm (*Zarhipis* sp.)
BNA 183(bw); SGI cp127

Beetle, goldsmith (*Cotalpa lanigera*)
MIS cp234; SEF cp394

Beetle, goliath (*Goliathus druryi*)
EOI 66; WIW 26

Beetle, ground (*Carabidae*)
BNA cp1, 87(bw), 89(bw), 91(bw), 93(bw); MIS cp191,
   cp196, cp203, cp206; SEF cp380; SGI cp86

Beetle, Harlequin (*Acrocinus longimanus*)
AAL 438(bw); DIE cp17; EOI 66

Beetle, hercules or Rhinoceros (*Dynastes hercules*)
AAL 431(bw); DIE cp14; EOI 67; SGI cp112

Beetle, hercules (Eastern) (*Dynastes tityus*)
BNA cp8

Beetle, hister (Histeridae)
BNA cp1, 134(bw); MIS cp199

Beetle, Japanese (*Popilla japonica*)
BNA 146(bw); LBG cp386; MIS cp212; SGI cp105

Beetle, June (green) (*Cotinus* sp.)
BNA cp8; LBG cp387; MIS cp208; SGI cp113

Beetle, ladybird (Coccinellidae)
BNA cp5, 232-236(bw); MIS cp145-148

Beetle, ladybird (ash-gray) (*Olla abdominalis*)
BNA 236(bw); SGI cp139

Beetle, ladybird (convergent) (*Hippodamia convergens*)
BNA 236(bw); BOI 37; SGI cp140

Beetle, ladybird (nine-spotted) (*Coccinella novanotata*)
LBG cp380; SGI cp141

Beetle, ladybird (seven-spotted) (*Coccinella
   septempunctata*)
DIE cp19

Beetle, leaf (various genera)
BNA cp11, cp12, 290-303(bw); EOI 69; MIS cp149,
   cp150, cp151, cp155; WOM 52

Beetle, longhorn (various kinds)
BNA cp9-cp11, 282-288(bw); DIE cp30; SGI cp155-
   cp175; WIW 46

Beetle, longhorn (*Rosalia* sp.)
BNA cp9; DIE cp4

Beetle, longhorn (milkweed) (*Tetraopes* sp.)
BNA 288(bw); SGI cp177

Beetle, May or June (*Phyllophaga* sp.)
BNA 145(bw); MIS cp219

Beetle, net-winged (Lycidae)
AAL 436; BNA cp3; MIS cp160, cp163; SGI cp132

Beetle, pleasing fungus (*Cypherotylus californica*)
BNA cp5; SGI cp138

Beetle, pleasing fungus (other kinds)
BNA 226(bw); MIS cp139

Beetle, potato (*Lema trilineata*)
BNA 292(bw); MIS cp154

Beetle, rhinoceros (*Oryctes nasicornis*)
DIE cp22; SWF 46

Beetle, rove (Staphylinidae)
BNA cp2, 112-117(bw); MIS cp227, cp240; SGI cp93

Beetle, sawyer (*Monochamus* sp.)
BNA 286(bw); SGI cp173

Beetle, scarab (*Scarabaeidae*)
BNA cp7, cp8, 142-148(bw); BOI 33, 35; MAE 72; SGI
cp103; SWF 122; WWD 72

Beetle, scarab or shining leaf chafer (*Plusiotis* sp.)
BNA cp8; SGI cp107

Beetle, sexton (*Nicrophorus* sp.)
BNA 121(bw); SGI cp91

Beetle, skunk (desert) (*Eleodes armata*)
BNA 253(bw); SGI cp146

Beetle, snout See Weevil

Beetle, soldier (various genera)
BNA cp2; EOI 74-75; MIS cp162, cp168; SGI cp131

Beetle, stag or pinching bug (*Pseudolucanus capreolus*)
AAL 429; BNA 136(bw)

Beetle, stag (European) (*Lucanus cervus*)
AAL 429; BOI 28, 29; DIE cp29; EOI 66; SWF 46

Beetle, stag (giant) (*Lucanus elaphus*)
BNA cp7; MIS cp217; SEF cp383; SGI cp96

Beetle, tiger (*Cincindela* sp.)
DIE cp25; EOI 67; JAD cp378; SGI cp82

Beetle, tiger (other kinds)
BNA cp1; MIS cp195, cp207, cp238, cp239

Beetle, tortoise (various genera)
BNA cp12; BOI 26; WWW 92

Beetle, water (various genera)
PGT 165-168; SGI cp87

Beetle, water scavenger (*Hydrobius* sp.)
BNA 107(bw); MIS cp93; SGI cp89; WNW cp347

Beetle, whirligig (*Gyrinus* and other genera)
AAL 426(bw); BNA 103(bw); MIS cp91, cp92; PGT 170;
SGI cp88; WNW cp348

Beetle, wood-boring (metallic) (*Buprestidae*)
BNA cp4, 167-170(bw); MIS cp179, cp210, cp237; SGI
cp117

Beggar-tick See Shepherd's-needle

Belladonna See Nightshade, deadly

Bellbird, three-wattled (*Procnias tricarunculata*)
EOB 325

Bellflower, tall (*Campanula americana*)
NWE cp625

Bellwort, perfoliate (*Uvularia perfoliata*)
CWN 5

Beluga or belukha or white whale (*Delphinapterus leucas*)
BAA 47; CRM 209; EAL 313; GEM IV:325, 376-377, 379-382; HWD 31, 70; MEM 201, 203; MIA 117; NAF cp617, cp672; WDP 72; WOW 165; WPR 182-183

Beluga, Russian (fish) (*Huso huso*)
AAL 324-325; AWR 104-105; MIA 499

Bergamot, wild (*Monarda fistulosa*)
BCF 53; CWN 219; LBG cp257; NWE cp511; SEF cp483;
WAA 186, 189

Betony, wood See Lousewort, common

Bettong (*Bettongia penicillata*) See also Kangaroo, rat
CRM 29; GEM I:357

Bigeye (*Priacanthus arenatus*) See also Catalufa
ACF cp21; AGC cp339; KCR cp26; NAF cp386; RCK 83

Bigeye, short (*Pristigenys alta*)
ACF cp21; NAF cp387

Bilby or rabbit bandicoot (*Macrotis lagotis*)
CRM 21; GEM I:302

Bindweed, field (*Convolvulus arvensis*)
CWN 203; LBG cp121; SWW cp173; WTV 141

Bindweed, hedge (*Convolvulus sepium*)
BCF 21; CWN 203; NWE cp579

Binturong (*Arctictis binturong*)
GEM III:529; MEM 139; MIA 95

Birch, black or sweet or cherry (*Betula lenta*)
NTE cp176, cp379, cp488, cp613; SEF cp76, cp166; TSV
36, 37

Birch, bog (*Betula glandulosa*)
WFW cp113

Birch, gray (*Betula populifera*)
LBG cp428; NTE cp182, cp598

Birch, paper (*Betula papyrifera*)
LBG cp429; NTE cp179, cp615; NTW cp206; SEF cp67

Birch, river (*Betula nigra*)
NTE cp178; PMT cp[20]; SEF cp69; WNW cp471

Birch, silver or white (*Betula pendula*)
NTE cp181; NTW cp208; OET 137

Birch, Virginia roundleaf (*Betula niger*)
NTE cp202

Birch, water (*Betula occidentalis*)
NTE cp177; NTW cp205; WFW cp90

Birch, yellow (*Betula alleghaniensis*)
NTE cp180, cp487, cp617; SEF cp68

Birchir (*Polypterus ornatipinnis*)
AWR 209(bw); MIA 499

Bird of Paradise (*Paradisaea raggiana*)
EOB 441; NOA 54; SWF 109-111

Bird of Paradise, blue (*Paradisaea rudolphi*)
MIA 399; SAB 81; WGB 209

Bird of Paradise, King of Saxony (*Pteridophora alberti*)
MIA 399; WGB 209

Bird of Paradise, lesser (*Paradisaea minor*)
AAL 197

Bird of Paradise flower (*Strelitzia reginae*)
TSP 131

Bird's nest fungus (*Cyathus* sp.)
FGM p48; NAM cp632; SGM cp383, cp384

Birdsfoot trefoil (*Lotus corniculatus*)
CWN 167; LBG cp199; NWE cp356; SWW cp337; WAA
184; WTV 88

Bishopbird (*Euplectes* sp.)
AAL 88; EOB 425

Bishop's cap See Mitrewort

Bison (*Bison bison*)
ASM cp278; CMW 263(bw); CRM 193; GEM I:136-137,
V:360, 366, 398-403; LBG cp31; MIA 145; MNC 291(bw);
MNP cp36; MPS 307(bw); NAW 20, 21; SWM 160; WMC
238

Bison, European or wisent (*Bison bonasus*)
CRM 193; GEM V:404; MIA 145; NHU 111; SWF 72-73

Bistort (*Polygonum* sp.)
PGT 51; SWW cp114

Bitterling (*Rhodeus amarus*)
AAL 332; AWR 47, 216(bw); MIA 515

Bittern, American (*Botaurus lentiginosus*)
AGC cp567; EAB 518; FEB 130; FWB 129; HHH 269; MIA
209; NAB cp24; NAW 103; PBC 72; UAB cp12; WGB 19;
WNW cp532; WON 34

Bittern, Australian (*Botaurus poiciloptilus*)
HHH 279

Bittern, Eurasian (*Botaurus stellaris*)
EOB 71; HHH 273; RBB 17; WBW 71

Bittern, least (*Ixobrychus exilis*)
AAL 103; EAB 519; FEB 131; FWB 128; HHH 237; NAB
cp17; PBC 71; UAB cp11; WBW 67; WNW cp533; WOB
146; WOI 189

Bittern, South American (*Botaurus pinnatus*)
HHH 265

Bittern, sun (*Eurypyga helias*)
EOB 156; MIA 239; WBW 137(bw); WGB 49

Bittern (other *Ixobrychus* sp.)
HHH 235, 241, 247, 251, 253, 261

Bitterroot (*Lewisia rediviva*)
SWW cp548; WAA 36; WFW cp513

Bittersweet (*Celastrus scandens*)
BCF 51; NWE cp382; PPM cp25(berries)

Bittersweet (*Solanum* sp.) See Nightshade, bittersweet

Bitterweed See Sneezeweed

Black widow See under Spider

Blackbird, Brewer's (*Euphagus cyanocephalus*)
EAB 903; FWB 280; LBG cp549, cp550; NAB cp515,
cp570; UAB cp560, cp622

Blackbird, European (*Turdus merula*)
BAS 12; MIA 337; RBB 137; SPN 7; WGB 147

Blackbird, red-winged (*Agelaius phoeniceus*)
AGC cp609; AWR 148; EAB 902; FEB 357; FWB 279;
LBG cp545, cp546; NAB cp560, cp568; NAW 104; PBC
342; SPN 14, 146; UAB cp557, cp614; WNW cp580,
cp588; WOB 63; WON 168-170

Blackbird, rusty (*Euphagus carolinus*)
EAB 903; FEB 358; NAB cp569; SPN 145; UAB cp561,
cp623

Blackbird, tricolored (*Agelaius tricolor*)
EAB 903; UAB cp615; WNW cp587

Blackbird, yellow-headed (*Xanthocephalus
xanthocephalus*)
AAL 188; AWR 148; EAB 903; EOB 415; FEB 354; FWB
276; NAB cp518, cp583; NAW 105; PBC 341; SPN 147;
UAB cp428, cp556, cp612; WNW cp586

Blackbuck (*Antilope cervicapra*)
CRM 195; GEM V:477; MEM 577; MIA 153; WWD 94-95

Blackcap (*Sylvia atricapilla*)
MAE 96; MIA 345; RBB 149

Black-eyed Susan or Coneflower (*Rudbeckia hirta*)
BCF 99; LBG cp176; NWE cp281; WAA 10, 244

Black-eyed-Susan vine (*Thunbergia alata*)
TSP 97

Blackfish, Alaskan (*Dallia pectoralis*)
EAL 59; NAF cp194

Blacksmith (*Chromis punctipinnis*)
NAF cp326; PCF cp30

Bladdernut (*Staphylea trifolia*)
FWF 51; NTE cp353, cp425; TSV 70, 71

Bladderpod (*Isomeria arborea*)
SWW cp16, cp350

Bladderwort, greater (*Utricularia vulgaris*)
PGT 63; SWW cp283

Bladderwort, horned (*Utricularia cornuta*)
NWE cp313; WNW cp218

Bladderwort, swollen (*Utricularia inflata*)
NWE cp271; WNW cp213

Blazing-star (*Liatris* sp.)
CWN 289; LBG cp259-262; NWE cp543-546; WAA 188;
    WTV 200

Blazing-star (*Mentzelia laevicaulis*)
SWW cp212

Bleeding-heart, Western (*Dicentra formosa*)
SWW cp505; WFW cp486

Bleeding-heart, wild (*Dicentra eximia*)
CWN 113; NWE cp501; SEF cp477

Blenny (various genera)
ACF cp43, cp44; AGC cp428; NAF cp458, cp459, cp461,
    cp462, cp466; NNW 92; PCF cp39; RUP 132

Bloodroot (*Sanguinaria canadensis*)
BCF 159; CWN 75; NWE cp53; SEF cp440; WAA 159;
    WTV 7

Blue flag See under Flag

Blue tang (*Acanthurus coeruleus*)
ACF cp36; AGC cp366; KCR cp24; MIA 569; NAF cp328,
    cp337; WWF cp63

Bluebells, alpine (*Mertensia alpina*)
WAA 231

Bluebells, California See Desert Bell

Bluebells, mountain (*Mertensia ciliata*)
SWW cp648; WFW cp534; WNW cp300

Bluebells, Virginia (*Mertensia virginica*)
CWN 197; NWE cp649; SEF cp497; WNW cp299

Bluebells-of-Scotland See Harebell

Blueberry (*Vaccinium* sp.)
FWF 173; NWE cp232, cp657; SEF cp438, cp510; WNW
    cp259

Bluebird, eastern (*Sialia sialis*)
AAL 172-173; EAB 891; EOB 369; FEB 393; LBG cp524;
    MIA 335; NAB cp440; NAW 105; PBC 278, 279; SPN 83;
    WGB 145; WOB 210, 211; WON 150

Bluebird, mountain (*Sialia currucoides*)
EAB 891; FWB 427; LBG cp525; NAB cp441; SPN 84;
    UAB cp496, cp497; WFW cp276; WON 151

Bluebird, Western (*Sialia mexicana*)
EAB 890; FWB 428; JAD cp589; UAB cp500; WFW cp275

Blueblossom (*Ceanothus thyrsiflorus*)
NTW cp153, cp396; WFW cp222

Bluebonnet, Texas (*Lupinus texensis*)
LBG cp270; NWE cp639; WAA 122

Bluebuttons (*Knautia arvensis*)
CWN 251

Blue-curls (*Trichostema dichotomum*)
NWE cp612; WAA 165

Blue-eyed grass (*Sisyrinchium* sp.)
BCF 77; CWN 23; NWE cp594; SWW cp583; WAA 183;
    WFW cp519; WTV 151

Blue-eyed Mary (*Collinsia verna*)
NWE cp645

Blue-eyes, Arizona (*Evolvulus arizonicus*)
JAD cp69; SWW cp582

Bluefish (*Pomatomus saltatrix*)
AGC cp401; MIA 553; NAF cp556

Bluegill (*Lepomis macrochirus*)
NAF cp85; WNW cp48

Bluegrass, Kentucky (*Poa pratensis*)
LBG cp91; NWE cp29

Bluetail, red-flanked (*Tarsiger cyanurus*)
NHU 104

Bluethroat (*Luscinia svecica*)
AAL 172; EAB 891; UAB cp499

Bluets (*Houstonia caerulea*)
BCF 119; CWN 209; LBG cp288; NWE cp593; SEF cp490;
    WAA 182 (*H.serpyllifolia*); WTV 154

Blueweed or blue-devil (*Echium vulgare*)
CWN 195

Blusher (*Amanita rubescens*)
FGM cp26; GSM 151; NAM 132; NMT cp138; SGM cp11

Blusher, false or Panther (*Amanita pantherina*)
NAM cp130, cp131, cp679; PPC 111

Boa constrictor (*Boa constrictor*)
MIA 447; MSW cp16; LSW 19, 20; RNA 139

Boa, rainbow (*Epicrates cenchria*)
LSW 28-29, 35; MSW cp73

Boa, rosy (*Lichanura trivirgata*)
ARA cp508, cp525; JAD cp219, cp224; RNA 139; WRA
    cp36

Boa, rubber (*Charina bottae*)
ARA cp472; LSW 43; MIA 447; NAW 235; RNA 139; WFW cp599; WRA cp36

Boa, sand (*Eryx* sp.)
LSW 44, 46

Boa, tree (*Corallus* sp.)
LSW 32; MSW cp74

Boa, tree (emerald) (*Boa* or *Corallus caninus*)
AAL 240; LSW 30, 31; MIA 447; MSW cp23; SWF 189

Boa, water See Anaconda

Boa (other kinds)
AAL 238(bw); LSW 22, 23, 34, 47, 50, 52; MSW cp76

Boar, wild (*Sus scrofa*)
AAL 74(bw); ASM cp275; GEM V:24-27; MAE 27; MEM 501; MIA 129; NAW 22; SEF cp615; SWF 64-65; WMC 229

Bobcat (*Felis rufus*)
ASM cp 267, 269; AWC 81-85; CGW 157; CMW 247(bw); CRM 155; GEM III:611; GMP 306(bw) (head only); JAD cp517, cp520; MEM 19, 51; MIA 103; MNC 277(bw); MNP cp4a; MPS 302(bw); NAW 23, 25; NMM cp[28]; RFA 302-310 (bw); RMM cp[13]; SEF cp613; SWM 129; WFW cp362; WMC 204; WNW cp616; WWD 142-143

Bobolink (*Dolichonyx oryzivorus*)
EAB 905; FEB 356; FWB 278; LBG cp544; MIA 383; NAB cp555, cp567; SPN 148; UAB cp581, cp607; WGB 193

Bobwhite (*Colinus virginianus*)
EAB 704, 705; EOB 134; FEB 205; FWB 193; LBG cp503; MIA 233; NAB cp257; PBC 120; UAB cp280, cp282; WGB 43; WOB 212-213

Boletus See Mushroom, bolete

Bonefish (*Albula vulpes*)
AGC cp404; MIA 501; MPC cp279; NAF cp578

Boneset (*Eupatorium* sp.)
CWN 287; LBG cp150; NWE cp194; TSP 119

Boneset, climbing (*Mikania scandens*)
NWE cp203

Bongo (*Tragelaphus* or *Boocerus eurycerus*)
GEM V:353; MIA 141; NHA cp16; SWF 120-121; WOM 109

Bonito (*Sarda* sp.)
ACF cp48; PCF cp32

Bonito, oceanic See Tuna, skipjack

Bonobo See Chimpanzee, pygmy

Bontebok (*Damaliscus dorcas*)
CRM 191; MEM 563; MIA 149

Booby, blue-footed (*Sula nebouxii*)
AAL 98(bw); BWB 96-97; EAB 53; FWB 59; HSW 100; OBL 75, 166; UAB cp61; WOI 41, 42

Booby, brown (*Sula leucogaster*)
EAB 53; FEB 54; FWB 58; MIA 205; NAB cp77; OBL 35, 75, 235; UAB cp60; WGB 15

Booby, masked or blue-faced (*Sula dactylatra*)
EAB 52; HSW 101; NAB cp78; OBL 158, 191

Booby, Peruvian (*Sula variegata*)
EOB 53; OBL 74

Booby, red-footed (*Sula sula*)
AAL 7; BWB 98; EAB 53; EOB 59; HSW 102; OBL 38, 178, 179; UAB cp59; WOB 148; WOI 40

Boodie (*Bettongia lesueur*)
CRM 29; MEM 867, 870

Boomslang (*Dispholidus typus*)
LSW 218, 219; MIA 453

Borer (various kinds)
MIS cp157, cp164, cp169, cp180, cp181, cp184, cp188, cp194, cp220, cp232, cp241, cp244, cp245; SGI cp118-120, cp158, cp160, cp166, cp176, cp199; WNW cp352

Bottlebrush (*Callistemon citrinus*)
TSP 49

Bouncing Bet (*Saponaria officinalis*)
BCF 61; CWN 119; NWE cp163; PPC 100; SWW cp154

Bouto See River Dolphin, Amazon

Bowerbird, golden (*Prionodura newtoniana*)
EOB 437; WOI 131

Bowerbird, great (*Chlamydera nuchalis*)
AAL 196; WOI 130

Bowerbird, regent (*Sericulus chrysocephalus*)
MAE 45; NOA 127; WOI 131

Bowerbird, satin (*Ptilonorhynchus violaceus*)
AAL 196; BAS 40; EOB 436, 438-439; MIA 399; WGB 209

Bowfin (*Amia calva*)
AWR 219(bw); EAL 23; MIA 499; NAF cp44

Bowman's-root (*Gillenia trifoliata*)
CWN 85; NWE cp89

Box jelly or sea wasp (*Carybdea* sp.)
BMW 90-91

Boxelder (*Acer negundo*)
NTE cp335, cp494; NTW cp299, cp528; SEF cp133

Boxfish (*Ostracion tuberculatus*)
RUP 134

Bramble, dwarf (*Rubus lasiococcus*)
SWW cp42; WFW cp385

Brambling (*Fringilla montifringilla*)
RBB 178

Brant (*Branta bernica*)
AGC cp555; EAB 218, 219; FEB 114; FWB 116; MPC
   cp528; NAB cp172; PBC 80; RBB 25; UAB cp164; WPR
   110

Breadfruit (*Artocarpus altilis*) See also Jackfruit
TSP 13

Bream (*Abramis brama*)
AWR 46; MIA 515

Bristletail (various kinds)
EOI 22; MIS cp86; WIW 181(bw)

Brittlebrush (*Encelia farinosa*)
JAD cp136; SWW cp246

Brittlestar (various genera)
EAL 275, 280-281; KCR cp31, p33; MPC cp358-362,
   cp364; MSC cp565-571; SAS cpA30; SCS cp60; WPR
   137

Broadbill, green (*Calyptomena* sp.)
EOB 306

Brocket See under Deer

Brodiaea (*Brodiaea* sp.)
WAA 187, 230

Brooklime See Speedwell

Broom, Scotch (*Cytisus scoparius*)
SWW cp354; TSV 84, 85

Broomrape or cancer root (*Orobanche* sp.)
CWN 245; NWE cp68; SWW cp602, cp613

Broomsedge (*Andropogon virginicus*)
LBG cp98

Bryony, black (*Tamus communis*)
PPC 104(fruit)

Bryony, white (*Bryonia dioica*)
PPC 64(fruit)

Bryozoan (Ectoprocta family)
AGC cp276, cp287, cp288; EAL 190; MSC cp34, cp47-51,
   cp65, cp73, cp76, cp112-117; PGT 105; SAS cpA34

Buckbean (*Menyanthes trifoliata*)
CWN 187; PGT 40; SWW cp155; WNW cp258

Buckeye, California (*Aesculus californica*)
NTW cp302, cp382; SWW cp176; WFW cp105, cp148

Buckeye, Mexican (*Ungnadia speciosa*)
NTE cp318; NTW cp285, cp499

Buckeye, yellow (*Aesculus octandra*)
NTE cp358, cp384, cp531; SEF cp131, cp202

Buckeye (other *Aesculus* sp.)
NTE cp355-cp359, cp367, cp382, cp383, cp530; PMT
   cp[13]

Buckthorn, alder See Alder, black

Buckthorn, Carolina (*Rhamnus caroliniana*)
NTE cp145

Buckthorn, cascara (*Rhamnus purshiana*)
NTW cp193; WFW cp93

Buckthorn, European (*Rhamnus cathartica*)
LBG cp426; NTE cp201; PPC 96

Buckthorn, glossy (*Rhamnus frangula*)
LBG cp409, cp474; NTE cp75, cp545, cp610

Buckwheat, desert (*Eriogonum* sp.)
JAD cp359; SWW cp148

Buckwheat tree (*Cliftonia monophylla*)
NTE cp45; WNW cp446

Budgerigar (*Melopsittacus undulatus*)
EAB 695; EOB 227; MIA 261; WGB 71

Buffalo, African or Cape (*Synceros caffer*)
CRM 185; GEM V:345, 367, 376-386; MEM 546-47; MIA
   145; SAB 14

Buffalo, American See Bison

Buffalo burr (*Solanum rostratum*)
SWW cp14, cp210

Buffalo, smallmouth (fish) (*Ictiobus bubalus*)
MIA 517; NAF cp101

Buffalo, water (*Bubalus bubalis*)
CRM 185; GEM V:374; MIA 143

Buffalo, water (dwarf) See Anoa

Bufflehead See under Duck

Bugle (*Ajuga reptans*)
CWN 229

Bugler, scarlet (*Penstemon centranthifolius*)
SWW cp409; WFW cp476

Bugloss, viper's See Blueweed

Bulbul, black (*Hypsipetes madagascariensis*)
MIA 321; WGB 131

Bulbul, black-eyed (*Pycnonotus barbatus*)
MIA 321; WGB 131

Bulbul, red-whiskered (*Pycnonotus jocosus*)
EAB 54; FEB 375; MIA 321; NAB cp508; WGB 131

Bullfinch (*Pyrrhula pyrrhula*)
MIA 385; RBB 185; WGB 195

Bullfrog (*Rana catesbeiana*)
AAL 264, 285; ARA cp187, cp190; ARC 129; ARN 35(bw);
   CGW 29; FTW cp4; JAD cp288; MIA 483; NAW 281; PGT
   214; WNW cp137; WRA cp14

Bullfrog, South African (*Pyxicephalus adspersus*)
ERA 51; FTW p7

Bullhead (*Ictalurus* sp.)
MIA 519; NAF cp53, cp56, cp57; WNW cp86

Bumblebee (*Bombus* sp.)
AAL 480; BOI 227; LBG cp374; MIS cp505, cp506; SAB
   20; SEF cp405; SGI cp306; WFW cp578

Bumblebee, cuckoo (*Psithyrus vestalis*)
EOI 123

Bumelia (*Bumelia* sp.)
NTE cp42-44; NTW cp94

Bunchberry (*Cornus canadensis*)
NWE cp60, cp446; SEF cp445, cp506; SWW cp36; WAA
   161; WFW cp383

Bunting, black-headed (*Emberiza melanocephala*)
EOB 406

Bunting, corn (*Miliaria* or *Emberiza calandra*)
EOB 406; RBB 190

Bunting, indigo (*Passerina cyanea*)
EAB 298; FEB 395; FWB 307; NAB cp438, cp513; NAW
   108; PBC 361; SPN 125

Bunting, lapland See Longspur, lapland

Bunting, lark (*Calamospiza melanocorys*)
EAB 299; FEB 381; FWB 303; LBG cp535; NAB cp549,
   cp566; SPN 132; UABcp570, cp606

Bunting, lazuli (*Passerina amoena*)
EAB 298; FEB 391; FWB 304; NAB cp437; SPN 124; UAB
   cp501; WFW cp296

Bunting, McKay's (*Plectrophenax hyperboreus*)
EAB 299; UAB cp604

Bunting, painted (*Passerina ciris*)
EAB 298; EOB 402; FEB 390; FWB 306; MIA 373; NAB
   cp476, cp477; PBC 362; SEF cp320; UAB cp522, cp525;
   WGB 181

Bunting, reed (*Emberiza schoeniclus*)
EOB 403; MIA 369; RBB 189; SPN 8; WGB 179

Bunting, snow (*Plectrophenax nivalis*)
EAB 299; FEB 380; FWB 302; LBG cp543; MIA 369; NAB
   cp547; NAW 109; PBC 394; RBB 187; UAB cp579, cp605;
   WGB 179

Bunting, varied (*Passerina versicolor*)
EAB 299; FEB 392; FWB 305

Burbot (*Lota lota*)
AWR 217(bw); MIA 529; NAF cp43

Burdock, lesser (*Arctium minus*)
BCF 13; CWN 285; NWE cp487; WAA 50; WTV 193

Burnet, Canadian (*Sanguisorba canadensis*)
NWE cp141

Burningbush, eastern (*Euonymous atropurpureus*)
NTE cp172, cp555, cp581; SEF cp81, cp187

Bur-reed (*Sparganium* sp.)
CWN 49; NWE cp41; PGT 44; WNW cp315

Burrfish (*Chilomycterus* sp.)
ACF cp60; NAF cp306

Bush baby, dwarf (*Galago demidovii*)
MEM 334; NHP 104(bw); SWF front.

Bush baby, greater (*Otolemur crassicaudatus*)
GEM II:90; MIA 59; NHP 102 (bw)

Bush baby, lesser or Senegal (*Galago senegalensis*)
AAL 37; GEM II:78, 79, 87, 89; MEM 332; MIA 59; NHP
   58(bw), 99(bw); SWF 37

Bush baby, thick-tailed (*Galago crassicaudatus*)
MEM 318, 333, 335

Bush baby, Western needle-nailed (*Euoticus
   elegantulus*)
NHP 103 (bw)

Bushbuck (*Tragelaphus scriptus*) See also Kudu and
   Nyala
GEM V:344, 345; WOM 104

Bush-clover (*Lespedeza* sp.)
CWN 165; NWE cp538, cp573

Bushmaster (*Lachesis muta*)
MIA 459; MSW cp100; LSW 386, 387

Bushtit (*Psaltriparus* sp.)
EAB 897; EOB 382; FWB 357; UAB cp485; WFW cp264

Bustard, Australian (*Choriotis australis*)
WPP 128

Bustard, great (*Otis tarda*)
AAL 124(bw); MIA 239; WGB 49; WPP 48

Bustard, kori (*Ardeotis kori*)
BAS 128; EOB 153; WPP 85; WWD 78

Bustard, little (*Tetrax tetrax*)
EOB 155

Bustard, little black (*Afrotis atra*)
MIA 239; WGB 49

Butter-and-eggs (*Linaria vulgaris*)
BCF 55; CWN 239; LBG cp201; NWE cp331; SWW cp301;
WAA 184

Buttercup See also Celandine; Crowfoot

Buttercup, bulbous (*Ranunculus bulbosus*)
CWN 63; NWE cp262

Buttercup, celery-leaved (*Ranunculus sceleratus*)
PGT 37; PPC 95; WAA 242

Buttercup, common (*Ranunculus acris*)
BCF 87; LBG cp193; NWE cp264; PPC 94; WAA 54

Buttercup, creeping (*Ranunculus repens*)
CWN 65

Buttercup, early (*Ranunculus fascicularis*)
CWN 65

Buttercup, hispid (*Ranunculus hispidus*)
WTV 53

Buttercup, kidney-leaf (*Ranunculus abortivus*)
NWE cp259

Buttercup, sagebrush (*Ranunculus glaberrimus*)
SWW cp199

Buttercup, subalpine (*Ranunculus eschscholtzii*)
SWW cp200; WFW cp451

Buttercup, swamp (*Ranunculus septentrionalis*)
CWN 63; NWE cp263; WNW cp211

Buttercup, water (*Ranunculus aquatilis*)
NWE cp260 (*R.flabellaris*); SWW cp44; WNW cp250

Buttercup, Western (*Ranunculus occidentalis*)
WAA 249

Butterfish (*Paprilus* sp.)
ACF p50; PCF cp31

Butterfly, admiral (poplar) (*Limenitis populi*)
AAB 108; MIS cp618; PAB cp672; SGB cp87, cp135

Butterfly, admiral (red) (*Vanessa atalanta*)
MIS cp618; PAB cp672; SEF cp367; SGB cp135; SGI
cp234; WIW 162

Butterfly, admiral (white) (*Limenitis* or *Basilarchia arthemis*)
MIS cp624; PAB cp650; SGI cp226

Butterfly, alfafa or orange sulfur (*Colias eurytheme*)
PAB cp86, cp97, cp100, cp130; SGI cp209

Butterfly, Apollo (*Parnassius apollo*)
NHU 150; SGB cp107

Butterfly, baltimore (*Euphydryas phaeton*)
WNW cp391

Butterfly, birdwing (*Ornithoptera* sp.)
BOI 137; SGB cp282; WIW 22 (larva)

Butterfly, black witch (*Ascalapha odorata*)
MIS cp561; SGB cp212

Butterfly, blue (acmon) (*Plebejus acmon*)
LBG cp322, cp323; SGI cp224

Butterfly, blue (common) (*Polyommatus icarus*)
LBG cp329; PAB cp44(larva), cp467; SGB cp115

Butterfly, blue (Northern) (*Lycaeides argyrognomon*)
WFW cp542, cp545; WNW cp405

Butterfly, blue (western pigmy) (*Brephidium exilis*)
JAD cp413; SGI cp222

Butterfly, brimstone (*Gonepteryx rhamni*)
SGB cp14, cp71

Butterfly, brown (eyed) (*Lethe* or *Satyrodes eurydice*)
LBG cp346; MIS cp598; PAB cp701, cp704; SGB cp197;
WNW cp402, cp409

Butterfly, buckeye (*Junonia* or *Precis coenis*)
LBG cp345; MIS cp1(larva), cp607; PAB cp688, cp691;
SGI cp230

Butterfly, cabbage (European) (*Pieris rapae*)
MIS cp21(larva), cp582; SGB cp194

Butterfly, checkerspot (various genera)
JAD cp421; PAB cp556-590; SGI cp231, cp233; WFW
cp553, cp564

Butterfly, comma (*Polygonia c-album*)
PAB cp371, cp375; SGB cp114; WIW 95

Butterfly, copper (*Lycaena* sp.)
AAL 457; MIS cp616; SGB cp88; SGI cp219-220; WPP 46

Butterfly, copper (blue) (*Chalceria heteronea*)
JAD cp414, cp415; PAB cp466, cp492, cp516

Butterfly, copper (various genera)
PAB cp511-530, cp536-540; WFW cp543

Butterfly, crescent-spot (*Phyciodes* and other genera)
LBG cp340, cp341; JAD cp421, cp425; PAB cp627-632;
SGB cp192, cp193, cp246

Butterfly, diana (*Speyeria diana*)
PAB cp323, cp656, cp657; SGI cp235

Butterfly, dog face (*Colias* or *Zerene* sp.)
MIS cp579; PAB cp98, cp101; SGB cp208; SGI cp210

Butterfly, elfin (*Incisalia* sp.)
MIS cp605; PAB cp449-459; SGB cp178

Butterfly, hairstreak See Hairstreak butterfly

Butterfly, large white (*Pieris brassicae*)
SAB 132; SGB cp112; WIW 171 (larva)

Butterfly, marbled white (*Melanargia galathea*)
WIW 99; WPP 45

Butterfly, metalmark (*Calephelis* and other genera)
JAD cp427, cp428; LBG cp343; MIS cp613, cp626; PAB
  cp541-555; SGB cp152; WNW cp398

Butterfly, monarch (*Danaus plexippus*)
AAL 444, 445; EOI 99; JAD cp419; LBG cp335; MIS
  cp16(larva), cp620; PAB cp35(larva), cp596; SAB 57; SGB
  cp162; SGI cp29, cp239; WIW 163; WPP 19

Butterfly, morpho (*Morpho* sp.)
SGB cp238

Butterfly, mourning cloak (*Nymphalis antiopa*)
MIS cp2(larva), cp625; PAB cp24(larva), cp681; SEF
  cp369; SGB cp100; SGI cp236

Butterfly, orange tip (*Anthocaris* sp.)
JAD cp404; PAB cp37, cp57, cp74-76; SGB cp42; SGI
  cp214; WPP 45

Butterfly, painted lady (*Cynthia* or *Vanessa cardui*)
JAD cp420; LBG cp338; MIS cp11(larva), cp596; PAB
  cp18(larva); SGB cp136

Butterfly, painted lady (American) (*Cynthia* or *Vanessa
  virginiensis*)
LBG cp337; MIS cp619; PAB cp13(larva), cp669; SGB
  cp206

Butterfly, patch (*Chlosyne* sp.)
JAD cp429, cp430; PAB cp578, cp659, cp660, cp675,
  cp683; SGB cp156, cp218; SGI cp232

Butterfly, peacock (*Inachis io*)
DIE cp33; SGB cp81

Butterfly, peacock (white) (*Anartia jatrophae*)
MIS cp608; PAB cp690, cp693; WNW cp404, cp406

Butterfly, pearly eye (*Enodia* or *Lethe portlandia*)
MIS cp602; PAB cp694, cp695; SGB cp165; SGI cp237

Butterfly, phoebus (*Parnassius phoebus*)
JAD cp407; MIS cp591; PAB cp28(larva), cp71; WFW
  cp541

Butterfly, queen (*Danaus gilippus berenice*)
AAL 460; JAD cp418; MIS cp17(larva), cp622; PAB
  cp33(larva), cp594; SGB cp161

Butterfly, question mark (*Polygonia interrogationis*)
MIS cp7(larva), cp615; PAB cp15(larva), cp373, cp374

Butterfly, red-spotted purple (*Basilarchia* or *Limenitis
  astyanax*)
BOI 139; LBG cp321, cp330; PAB cp317, cp320; SGB
  cp150; SGI cp227

Butterfly, satyr (various genera)
PAB cp706-717

Butterfly, sister (western) (*Adelpha bredowii*)
PAB cp31(larva), cp652; SGI cp225; WFW cp555

Butterfly, skipper See Skipper

Butterfly, sleepy orange (*Eurema nicippe*)
PAB cp117, cp120; SGB cp170; SGI cp213

Butterfly, sulfur (*Colias* sp.)
LBG cp295, cp296; PAB cp85-97, cp100, cp103, cp110-
  114, cp130, cp131; SGI cp208; WFW cp546, cp558; WNW
  cp385, cp386

Butterfly, sulfur (*Phoebis* sp.)
MIS cp581; PAB cp108, cp115-116, cp118-119, cp121;
  SGI cp212

Butterfly, viceroy (*Limenitis archippus*)
AAL 459(bw); EOI 99; LBG cp334; MIS cp621; PAB
  cp26(larva), cp597; SGI cp228; WNW cp392

Butterfly, wood nymph (*Cercyonis pegala*)
LBG cp348; WFW cp550, cp551

Butterfly, zebra (*Heliconius charitonius*)
BOI 142-143; MIS cp585; PAB cp34(larva), cp644; SEF
  cp371; SGI cp238

Butterflyfish, banded (*Chaetodon striatus*)
ACF cp36; AGC cp369; RCK 33; NAF cp341; WWF cp32

Butterflyfish (other *Chaetodon* sp.)
ACF cp36, cp64; AGC cp367, cp368; EAL 113; KCR cp22;
  MIA 561; NAF cp338, cp339, cp340; PCF cp30; RCK 54,
  62, 146; RUP 74

Butterflyfish (other genera)
EAL 43; MIA 561; RUP 66

Butterflyweed (*Asclepias tuberosa*)
BCF 97; CWN 191; LBG cp228; NWE cp379; SWW cp371;
  WTV 104

Butternut (*Juglans cinerea*)
FWF 53; NTE cp330, cp533; SEF cp138, cp200

Butterwort (*Pinguicula vulgaris*)
NWE cp596; SWW cp585; WNW cp302

Butterwort, yellow (*Pinguicula lutea*)
NWE cp272; WNW cp208

Buttonbush (*Cephalanthus occidentalis*)
FWF 97; NTE 73, 402; NWE cp227; NTW 109, cp353,
  cp539; SEF cp60, cp163; TSV 52, 53; WNW cp237

Buttonwood See Mangrove, gray

Buzzard, auger or jackal (*Buteo rufofuscus*)
BOP 9, 71; EOB 113

Buzzard, common or steppe (*Buteo buteo*)
EOB 103; MIA 223; RBB 44; WGB 33

Buzzard, fishing (*Busarellus nigricollis*)
MIA 223; WGB 33

Buzzard, gray-faced eagle (*Butastur indicus*)
NHU 122

Buzzard, honey (*Pernis apivorus*)
EOB 110; MIA 219; NHU 97; WGB 29

Buzzard, long-legged (*Buteo rufinus*)
EOB 111; NHU 171

By-the-wind sailor (*Velella velella*)
AGC cp217; MPC cp390; MSC cp515, cp516; SCS cp30

C

Cabezon (*Scorpaenichthys marmoratus*)
MIA 547; MPC cp252; NAF cp420; PCF p16

Cachalot See Whale, sperm

Cacomistle See Ringtail, South American

Cactus, barrel (*Echinocactus grusonii*)
TSP 119

Cactus, barrel (*Ferocactus* sp.)
JAD cp122; SWW cp234; WAA 206

Cactus, beavertail (*Opuntia basilaris*)
JAD cp41; SWW cp473; WAA 207

Cactus, calico or lace or rainbow (*Echinocereus* sp.)
JAD cp123; SWW cp233; WAA 176, 209

Cactus, cereus See Cereus, night-blooming

Cactus, claret-cup (*Echinocereus triglochidiatus*)
JAD cp42; SWW cp423; WAA 169

Cactus, cushion (*Coryphantha vivipara*)
JAD cp401; SWW cp471

Cactus, fishhook (*Mamillaria* sp.)
JAD cp39; SWW cp470; WAA 209

Cactus, saguaro (*Cereus giganteus*)
JAD cp331; NTW cp312, cp375; OET 255; SWW cp183;
    WWD 48-49

Cactus, Simpson's hedgehog (*Pediocactus simpsonii*)
JAD cp38; SWW cp472

Caddisfly (various genera)
MIS cp335, cp337, cp338; PGT 196, 197; SGI cp196

Caecilian (*Gymnopis multiplicata* and other genera)
ERA 17; MIA 471

Caiman, smooth-fronted (*Paleosuchus palpebrosus*)
ERA 142

Caiman, spectacled (*Caiman crocodilus*)
AAL 215(bw); ARA cp257, cp260; MIA 415; RNA 209;
    WNW cp173

Calabash tree (*Crescentia cujete*)
PMT cp[42]

Calamus See Flag, sweet

Calla, wild See Arum, water

Callimico See Marmoset, Goeldi's

Calliopsis See Tickseed

Calthrops See Punctureweed

Camas, common (*Camassia quamash*)
SWW cp630; WFW cp524

Camel, bactrian (*Camelus bactrianus*)
CRM 173; GEM V:95, 604; MEM 512; MIA 133; WWD 98-
    99

Camel, dromedary (*Camelus dromedarius*)
GEM V:86-94; MIA 133

Camellia (*Camellia japonica*)
NTW cp164, cp393

Camphor tree (*Cinnamomum camphora*)
NTE cp70; NTW cp95

Camphorweed (*Heterotheca subaxillaris*)
LBG cp168; NWE cp291

Campion, bladder (*Silene cucubalus*)
BCF 31; CWN 115; LBG cp140; NWE cp71; SWW cp33;
    WTV 16

Campion, brilliant (*Lychnis fulgens*)
NHU 118

Campion, moss (*Silene acaulis*)
NWE cp457

Campion, red (*Lychnis dioica*)
CWN 117

Campion, starry (*Silene stellata*)
CWN 115; NWE cp70

Campion, white (*Lychnis alba*)
CWN 115; LBG cp141; NWE cp69; SWW cp34; WTV 23

Canary, yellow (*Serinus canaria*)
MIA 385; WGB 195

Cancer root See Broomrape

Candelabra, fairy or rock-jasmine (*Androsace septentrionalis*)
WFW cp394

Candlelabra tree (*Euphorbia candelabrum*)
OET 267

Candystick (*Allotropa virgata*)
SWW cp403; WFW cp478

Candyweed (*Polygala lutea*)
CWN 249; WTV 102

Cane (*Arundinaria gigantea*)
TSV 42, 43

Cannonball tree (*Couroupita guianensis*)
TSP 25

Canvasback See under Duck

Caper bush (*Capparis spinosa*)
TSP 49

Capercaillie (*Tetrao urogallus*)
EOB 135; NHU 98; RBB 54; SAB 91; SWF 71; WOM 70

Capuchin See under Monkey

Capybara (*Hydrochoerus hydrochaeris*)
AAL 49; AWR 173; GEM III:336, 337; MEM 696-699; MIA 185; WWW 100-101

Caracal (*Felis caracal*)
AWC 113-115; CRM 155; GEM III:596, 597; MIA 101

Caracara, Audubon's or common (*Caracara cheriway*)
BOP 50, 60; EAB 295; NAB cp312; UAB cp305; WOB 52, 53; WWD 156

Caracara, crested (*Polyborus plancus*)
EOB 107; FEB 219; FGH cp1; FWB 213; JAD cp541; LBG cp495; MIA 225; WGB 35

Caracara, yellow-headed (*Milvago chimachino*)
BOP 63

Cardinal (*Cardinalis cardinalis*)
EAB 300; EOB 409; FEB 389; FWB 431; JAD cp600, cp601; MIA 373; NAB cp406, cp407; NAW 110; PBC 357; SEF cp319; SPN 120, 121; UAB cp448, cp467; WGB 183; WOM 158; WOB 142(head only)

Cardinalfish (*Apogon* sp.)
ACF cp28; PCF cp30

Cardinalflower (*Lobelia cardinalis*)
BCF 109; CWN 247; NWE cp430; SWW cp419; WAA 42; WFW cp474; WNW cp277

Caribou (*Rangifer tarandus*)
ASM cp290; BAA 41, 103; CRM 181; GEM V:242-253; MAF 81; MEM 521, 523; MIA 137; MNC 283(bw); MNP cp1a; MPS 305(bw); NAW 9; SEF cp618; SWM 153; WFW cp377; WPR 74-75

Carob (*Ceratonia siliqua*)
NTW cp273

Carp (*Cyprinus carpio*)
AAL 320; MIA 513; NAF cp118, cp119; PCF p6; PGT 205; WNW cp37, cp38

Carpetshark or Wobbegong (*Orectolobus* sp.)
RUP 186; SJS 23, 59; SSW 118

Carpetshark, collared or catshark (*Parascyllium collare*)
EAL 136; SJS 23, 44

Carrionflower (*Smilax herbacea*)
CWN 19; FWF 191; NWE cp40, cp655

Cashew (*Anacardium occidentale*)
OET 269

Cassowary (*Casuarius* sp.)
AAL 91(bw); BAS 68; EOB 18, 26; MIA 197; NOA 130; SWF 108; WGB 7

Castor bean (*Ricinus communis*)
PPC 43; PPM cp9; TSP 75

Casuarina, river-oak (*Casuarina cunninghamiana*)
NTW cp1

Cat, bay or Bornean red (*Felis badia*)
CRM 157; ROW cp2;

Cat, black-footed (*Felis nigripes*)
GEM III:617; MEM 53;

Cat, desert (Turkestan) (*Felis thinobia*)
GEM III:616

Cat, fishing (*Felis viverrina*)
AWC 208-211; MEM 53

Cat, flat-headed (*Felis planiceps*)
CRM 157; ROW cp2

Cat, Geoffroy's (*Felis geoffroyi*)
AWC 189; CRM 153

Cat, golden (*Felis temmincki*)
CRM 157; GEM III:598(*Profelis aurata*); MEM 53; MIA 101

Cat, Iriomote (*Felis iriomotensis*)
CRM 157; GEM III:589

Cat, jaguarundi See Jaguarundi

Cat, jungle (*Felis chaus*)
AWC 201-204; MEM 53;

Cat, leopard (*Felis bengalensis*)
AWC 196; CRM 157; GEM III:590, 592-593; MEM 53; MIA 101; ROW cp2;

Cat, marbled (*Pardofelis* or *Felis marmorata*)
AWC 206; GEM III:595; ROW cp2

Cat, margay See Margay

Cat, marsupial (*Satanellus* sp.)
MEM 839

Cat, Pallas' (*Felis* or *Otocolobus manul*)
GEM III:619(head only); MIA 103

Cat, pampas (*Lynchailurus pajeros*)
GEM III:627; MIA 101

Cat, ringtail See Ringtail

Cat, rust (*Prionailurus rubiginosus*)
GEM III:591

Cat, sand (*Felis margarita*)
GEM III:617; MEM 53

Cat, serval See Serval

Cat, tiger or little spotted (*Felis tigrinus*) See also Quoll
CRM 157; MEM 52

Cat, wild (African) (*Felis silvestris* or *libyca*)
AWC 102, 103; GEM III:614, 615; MEM 52; MIA 103

Cat, wild (European or Scottish) (*Felis sylvestris*)
AWC 96-107; GEM III:577; MEM 51; NNW 63; WOM 72, 73

Catalpa, Northern (*Catalpa speciosa*)
FWF 19; NTE cp97, cp441; NTW cp126, cp377; SEF cp61, cp149; TSV 46, 47

Catalpa, Southern (*Catalpa bignoides*)
LBG cp410; NTE cp94; NTW cp125, cp378; SEF cp62

Catalufa (*Priacanthus* or *Pseudopriacanthus* sp.)
AAL 353; PCF cp30

Catamount See Mountain lion

Catbird, gray (*Dumetella carolinensis*)
EAB 576; FEB 324; FWB 426; MIA 331; NAB cp420; PBC 269; SPN 94; UAB cp477; WGB 141; WOB 24-25

Catclaw (*Acacia* sp.)
JAD cp303, cp317; LBG cp443, cp462; NTE cp295, cp296, cp397, cp502; NTWcp255, cp262, cp329, cp524

Caterpillar, tent (*Malacosoma* sp.)
MIS cp546; SGB cp183; SGI cp242; WIW 157(bw)

Caterpillar, woolly bear (*Isia isabella*)
LBG cp301(moth); MIS cp5, cp541(moth)

Catfish, electric (*Malapterurus electricus*)
AWR 208(bw); MIA 521

Catfish, gafftopsail (*Bagre marinus*)
ACF p13; MIA 521; NAF cp501

Catfish, hardhead (*Arius felis*)
ACF p13; MIA 521; NAF cp500

Catfish, Pungas (*Pangasius pangasius*)
AWR 214(bw); MIA 521

Catfish, walking (*Clarias batrachus*)
NAF cp45; WNW cp83

Catfish (other kinds)
AAL 337lbw); AWR 215(bw); NAF cp47, cp54, cp55

Catnip (*Nepeta cataria*)
CWN 227; NWE cp183

Cat's ear (*Hypochoeris radicata*)
CWN 297; WTV 80

Cat's ears, elegant (*Calochortus elegans*)
SWW cp47

Catshark (*Stegostoma fasciatum* or *tigrinum*)
RCK 191; RUP 187

Catshark, false (*Pseudotriakis microdon*)
EAL 137; SJS 30

Catshark (other kinds)
EAL 131; SJS 29

Cattail (*Typha latifolia*)
BCF 141; CWN 49; FWF 139; NWE cp309; WAA 201; WNW cp220

Cavefish, southern (*Typhlichthys subterraneus*)
NAF cp216

Cavy, Patagonian See Mara

Cavy, rock (*Kerodon rupestris*)
GEM III:333; MEM 691; MIA 185

Cedar, Alaska yellow (*Chamaecyparis nootkatensis*)
NTW cp57, cp451; WFW cp56

Cedar, blue Atlas (*Cedrus atlantica*)
NTW cp46, cp429; PMT cp[26]

Cedar, deodar (*Cedrus deodara*)
NTW cp47; PMT cp[27]

Cedar, incense (*Libocedrus* or *Calocedrus decurrens*)
NTW cp58, cp453; PMT cp[22]; WFW cp57

Cedar of Lebanon (*Cedrus libani*)
NTE cp16; NTW cp48; OET 81

Cedar, Port Orford (*Chamaecyparis lawsoniana*)
NTW cp55; WFW cp58

Cedar, red (Eastern) (*Juniperus virginiana*)
LBG cp400; NTE cp35; SEF cp38; TSV 18, 19

Cedar, red (Southern) (*Juniperus silicicola*)
NTE cp36

Cedar, red (Western) or giant (*Thuja plicata*)
NTW cp60, cp452; PMT cp[136]; WFW cp59, cp142; WNW cp427

Cedar, stinking See Torreya, Florida

Cedar, white (Atlantic) (*Chamaecyparis thyoides*)
NTE cp34, cp486; SEF cp39; WNW cp429

Cedar, white (Northern) (*Thuja occidentalis*)
NTE cp32, cp485; SEF cp37; TSV 16, 17; WNW cp428

Celandine, greater (*Chelidonium major*)
BCF 67; CWN 75; NWE cp254; PPC 68

Celandine, lesser (*Ranunculus ficaria*)
CWN 61; NNW 17

Centaury (*Centaurium calycosum*)
LBG cp242; SWW cp451

Century plant (*Agave* sp.)
JAD cp339; SWW cp356; TSP 107; WAA 173; WWD 44

Cercocarpus (*Cercocarpus* sp.) See also Mahoney, mountain
LBG cp414, cp486; NTW cp185, cp186

Cereus, night-blooming (*Cereus undatus* or *greggii*)
JAD cp87; SWW cp181; TSP 87

Cero (*Scomberomorus regalis*)
ACF cp48; NAF cp562

Chachlalaca (*Ortalis* sp.)
AAL 115; EAB 134; EOB 139; FEB 199; MIA 227; NAB cp272; WGB 37

Chafer (various kinds)
EOI 75; SGI cp101-cp104

Chaffinch (*Fringilla coelebs*)
EOB 416; MAE 29; MIA 385; RBB 177; SAB 67; WGB 195

Chameleon (*Chamaeleo* sp.)
AAL 221; ERA ii, 71, 96-97; MIA 423; SAB 25, 38(head only), 78; SWF 129; WWW 148

Chameleon, common or flap-necked (*Chamaeleo dilepsis*)
ERA 86, 95; SWF 129

Chamois (*Rupicapra rupicapra*)
CRM 199; GEM V:295, 298, 496-499; MEM 585, 590, 591; MIA 155; WOM 56-57

Chanterelle, false See Jack o' Lantern

Chanterelle, yellow (*Cantharellus cibarius*)
FGM cp7; GSM 64; NAM cp308, cp427; NMT cp113; SGM cp154

Charr (*Salvelinus* sp.)
EAL 59; MIA 507

Chat, crimson (*Ephthianura tricolor*)
MIA 349; WGB 159; WWD 184

Chat, orange (*Ephthianura aurifrons*)
EOB 398

Chat, yellow-breasted (*Icteria virens*)
AAL 185; EAB 983; FEB 464; FWB 369; NAB cp359; PBC 331; SEF cp294; SPN 25(bw), 118; UAB cp413

Checker mallow or checkers (*Sidalcea* sp.)
SWW cp125, cp509; WAA 190; WNW cp290

Checkerberry (*Gaultheria procumbens*)
CWN 175; NWE cp229; WAA 160; WTV 45

Cheetah (*Acinonyx jubatus*)
AAB 81; AAL 55(bw); AWC 108, 124, 129-147; CRM 155; GEM I:180-181, III:363-365, 582-588; MEM 8-9, 40-43; MIA 103; NHU 187; SAB 22, 36, 101; WPP 91, 93; WWD 62

Cherry, bitter (*Prunus emarginata*)
NTW cp148, cp363; WFW cp186

Cherry, black (*Prunus serotina*)
FWF 141; LBG cp419, cp458; NTE cp136, cp447, cp548; NTW cp146, cp356, cp464; SEF cp91, cp153; TSV 64, 65

Cherry, Catalina (*Prunus lyonii*)
NTW cp119, cp160, cp354, cp467

Cherry, hollyleaf (*Prunus ilicifolia*)
NTW cp176, cp355, cp468; WFW cp147

Cherry, Mahaleb (*Prunus mahaleb*)
NTW cp161, cp366

Cherry, pin (*Prunus pensylvanica*)
LBG cp417, cp456, cp473; NTE cp124, cp423, cp550, cp584; NTW cp138, cp362, cp469; SEF cp84, cp152, cp189

Cherry, sour (*Prunus cerasus*)
NTE cp126; NTW cp145

Cherry, sweet or Mazzard (*Prunus avium*)
NTE cp129, cp547; NTW cp194, cp358, cp465

Cherubfish (*Centropyge argi*)
ACF cp37; NAF cp321

Chestnut, American (*Castanea dentata*)
NTE cp150, cp526, cp609; SEF cp85; TSV 32, 33

Chestnut, dwarf or Allegheny chinkapin (*Castanea pumila*)
FWF 161; NTE cp148, cp403; SEF cp90; TSV 132, 133

Chia (*Salvia columbariae*)
JAD cp64; SWW cp654

Chickadee, black-capped (*Parus atricapillus*)
EAB 896; FEB 278; FWB 353; MIA 357; NAB cp428; SEF cp333; SPN 66; UAB cp487; WGB 167; WON 146

Chickadee, boreal (*Parus hudsonicus*)
EAB 897; FEB 277; FWB 354; NAB cp511; SPN 67; UAB cp544

Chickadee, Carolina (*Parus carolinensis*)
EAB 896; NAB cp427; PBC 251; SEF cp332

Chickadee, chestnut-backed (*Parus rufescens*)
EAB 899; FWB 355; UAB cp545; WFW cp262

Chickadee, mountain (*Parus gambeli*)
EAB 899; FWB 352; NAW 111; UAB cp488; WFW cp261

Chickaree See Squirrel, Douglas or Squirrel, red

Chicken-of-the-woods fungus (*Polyporus sulphureus*)
NMT cp111

Chickweed (*Stellaria media*)
LBG cp136; NWE cp48

Chickweed, field (*Cerastium arvense*)
BCF 7; CWN 119; SWW cp18

Chickweed, mouse-ear (*Cerastium vulgatum*)
CWN 119; LBG cp138; NWE cp49

Chicory (*Chichorium intybus*)
BCF 105; CWN 295; LBG cp291; NWE cp605; SWW cp598; WTV 160

Chicory, desert (*Rafinesquia neomexicana*)
JAD cp86; SWW cp76

Chiff-chaff (*Phylloscopus collybita*)
RBB 151

Chimaera or ratfish (*Hydrolagus colliei*)
EAL 143; MIA 497; MPC cp258; NAF cp362; PCF p6

Chimpanzee (*Pan troglodytes*)
CRM 87; GEM I:48, II:27, 361, 464-485; MEM 422-25; MIA 77; NHP 60(bw), 168(bw), 170(bw); SWF 116, 117

Chimpanzee, pygmy (*Pan paniscus*)
CRM 87; GEM II:361; MEM 423; MIA 77; NHP 169(bw)

China rose (*Hibiscus rosa-sinensis*)
TSP 61

Chinaberry or china tree (*Melia azedarach*) See also Soapberry, Western
NTE cp341, cp469; NTW cp295, cp394; TSP 35; TSV 120, 121

Chinchilla, long-tailed (*Chinchilla lanigera*)
CRM 129; MIA 187; WOM 186

Chinchilla (other kinds)
GEM III:321, 322; MEM 702

Chinchweed (*Pectis papposa*)
JAD cp133; SWW cp235

Chinese houses, purple (*Collinsia heterophylla*)
SWW cp521; WFW cp496

Chinkapin or chinquapin See also Chestnut, dwarf

Chinkapin, bush (*Castanopsis sempervirens*)
WFW cp169

Chinkapin, Florida (*Castanea alnifolia*)
NTE cp151, cp527

Chinkapin, giant (*Castanopsis chrysophylla*)
NTW cp92, cp501; WFW cp78, cp178

Chinkapin, Ozark (*Castanea ozarkensis*)
NTE cp149; SEF cp87

Chipmunk (*Eutamias* sp.)
ASM cp2, cp3, cp7, cp13, cp14; MNP cp26; WFW cp332

Chipmunk, Eastern (*Tamias striatus*)
ASM cp8; CGW 169; GEM III:66-71; GMP 137(bw), cp4; LBG cp90; MIA 161; MNC 157(bw); MPS 131(bw); SEF cp585; WMC 117

Chipmunk, least (*Eutamias minimus*)
ASM cp4; CMW 88 (bw); JAD cp497; MNC 155(bw); MPS 130(bw); NAW 24; NMM cp[7]; RMM cp[4]; WFW cp333

Chipmunk, Siberian or striped squirrel (*Tamias sibiricus*)
GEM III:65

Chipmunk, Uinta or Colorado (*Eutamias umbrinus*)
ASM cp1, cp5; CMW 94 (bw); WFW cp331

Chipmunk, yellow-pine (*Eutamias amoenus*)
ASM cp6; CMW 90 (bw); SWM 24; WFW cp334

Chiru See Antelope, Tibetan

Chital See Deer, axis

Chiton (several genera)
AAL 489; AGC cp79-81; BCS 51; EAL 250; MPC cp176-186; MSC cp370-381; NNW 90; SCS cp16

Chlamydomonas (*Chlamydomonas* sp.)
PGT 72

Chokeberry, red (*Pyrus arbutifolia*)
NWE cp218; TSV 66, 67

Chokecherry (*Prunus virginiana*)
LBG cp421, cp457, cp472; NTE cp134, cp446, cp549, cp605; NTW cp162, cp357, cp466; WFW cp156

Cholla, jumping (*Opuntia fulgida*)
JAD cp325; NTW cp311, cp391

Cholla, teddybear or jumping (*Opuntia bigelovii*)
JAD cp332; SWW cp6

Cholla, tree (*Opuntia imbricata*)
JAD cp333; SWW cp474; WAA 66

Chough (*Pyrrhocorax pyrrhocorax*)
EOB 445; RBB 169; WOM 89(*P.graculus*)

Christmas-berry See Toyon

Christmas cactus, desert (*Opuntia leptocaulis*)
JAD cp162; SWW cp429

Chub (*Hybopsis* sp.)
NAF cp126, cp135, cp143, cp162, cp164, cp170; WNW cp66

Chub, Bermuda (*Kyphosus sectatrix*)
ACF cp36; MIA 561; MPC cp274; NAF cp526

Chub, Tui (*Gila bicolor*)
JAD cp298; NAF cp127

Chub (other kinds)
ACF cp36; AWR 46; NAF cp112-cp116, cp139, cp166, cp186, cp187

Chubsucker, creek (*Erimyzon oblongus*)
NAF cp104

Chuckwalla (*Sauromalus obesus*)
ARA cp331; JAD cp189; MIA 417; NAW 256(bw); RNA 107; WRA p20

Chuck-will's-widow (*Coprimulgus carolinensis*)
EAB 578, 579; FEB 257; NAB cp278; SEF cp251

Chukar (*Alectoris chukar* or *graeca*)
EAB 704, 705; EOB 135; FWB 192; JAD cp544; UAB cp281; WOM 127

Chuparosa (*Beloperone californica*)
JAD cp160; SWW cp407

Cicada, dogday See Harvestfly, dogday

Cicada, grand Western (*Tibicen dorsata*)
BOI 53-55; MIS cp 289

Cicada, seventeen year or periodical (*Magicicada* sp.)
MIS cp291; SEF cp416; SGI cp66

Cicada (other kinds)
DIE cp7, cp13; EOI 59; WIW 105(bw)

Cicada-killer See under Wasp

Cicely, sweet (*Osmorhiza claytoni*)
NWE cp169

Cichlid (*Cichlasoma* sp.)
NAF cp79; WNW cp52

Cigar plant See Waxweed

Cigar tree See Catalpa, Northern

Cinquefoil (*Potentilla* sp.)
BCF 37; CWN 77, 79; LBG cp186, cp191; NWE cp247, cp250, cp252; SWW cp215, cp351, cp396; WAA 190; WFW cp203; WTV 54

Cisco (*Coregonus autumnalis*)
AWR 221(bw)

Civet, African (*Civettictis civetta*)
MEM 137; MIA 93

Civet, African palm (*Nandinia binotata*)
GEM III:530; MIA 93

Civet, banded palm (*Hemigalus derbyanus*)
CRM 147; GEM III:532; MEM 139; MIA 95; ROW cp3

Civet, Indian (*Viverricula indica*)
GEM III:519, 521

Civet, malagasy (*Fossa fossa*)
CRM 147

Civet, masked palm (*Paguma larvata*)
GEM III:529; MIA 93

Civet, Oriental or Malay (*Viverra tangalunga* or *megaspila*)
CRM 145; GEM III:519, 521; MEM 139; ROW cp3

Civet, otter (*Cynogale bennetti*)
CRM 145; GEM III:534; MIA 95

Civet, palm (*Arctogalidia* sp.)
GEM III:531; MEM 137

Civet (other kinds)
CRM 147; MEM 139; ROW cp3

Clam, giant (*Tridacna* sp.)
BMW 114; EAL 267; NOA 97; WWF cp54; WWW 155

Clam, razor (*Ensis directus*)
AAL 494(bw); MSC cp308; SAS cpB27

Clam, razor (*Solen* sp.)
EAL 250; SAS cpB27

Clammyweed (*Polanisia dodecandra*)
JAD cp81; SWW cp166

Clarkia (*Clarkia* sp.) See also Farewell-to-Spring
SWW cp485, cp531

Cleavers (*Galium aparine*)
NWE cp171

Cliffrose (*Cowania mexicana*)
NTW cp250, cp323; WFW cp104, cp201

Clingfish (*Gobiesox* and other genera)
ACF p14; PCF p21; RUP 99

Clintonia See also Lily, bluebeard and Lily, wood (Speckled)

Clintonia, red (*Clintonia andrewiana*)
SWW cp422, cp561; WFW cp473

Clotburr See Cockleburr

Clover, crimson (*Trifolium incarnatum*)
CWN 157

Clover, hop (*Trifolium agrarium*)
NWE cp316

Clover, jackass (*Wislizenia refracta*)
JAD cp101; SWW cp324

Clover, mustang (*Linanthus montanus*)
SWW cp437; WFW cp502

Clover, prairie (*Petalostemon* sp.)
LBG cp253; NWE cp515; SWW cp93, cp496

Clover, rabbit's foot (*Trifolium arvense*)
BCF 39; CWN 157; NWE cp518; WTV 138

Clover, red (*Trifolium pratense*)
BCF 41; CWN 159; LBG cp229; NWE cp516; SWW cp499;
WAA 52

Clover, sweet (*Melilotus* sp.)
CWN 161; LBG cp151, cp218; NWE cp135, cp326; SWW
cp119

Clover, white (*Trifolium repens*)
CWN 159; LBG cp157; NWE cp120; SWW cp89

Clownfish (*Amphiprion* sp.)
EAL 115; NOA 107; RCK 93, 119, 158; RUP 104, 109,
110; WWF cp42; WWW 142

Club-rush (*Scirpus lacustris*)
PGT 45

Coatimundi, ring-tailed (*Nasua nasua*)
ASM cp205; GEM III:461; JAD cp513; MEM 102; MIA 85;
NAW 26; WFW cp357; WWD 138-140

Cobia (*Rachycentron canadum*)
AAL 355(bw); ACF p29; AGC cp408; MIA 553; NAF cp586

Cobra, black-necked spitting (*Naja nigrocollis
nigrocollis*)
ERA 127; LSW 251, 252

Cobra, Egyptian (*Naja haje*)
LSW 257

Cobra, Indian (*Naja naja*)
AAL 252; ERA 124; LSW 247, 248, 253; MIA 455; MSW
cp92; WPP 58-59

Cobra, king (*Ophiophagus hannah*)
AAL 200; LSW 263; MIA 455; WPP 58

Cobra plant (*Darlingtonia californica*)
SWW cp10; WAA 194

Cockatiel (*Nymphicus hollandicus*)
MIA 259; WGB 69

Cockatoo, roseate See Galah

Cockatoo, sulfur-crested (*Cacatua galerita*)
EOB 223; MIA 259; WGB 69

Cockatoo (other kinds)
BAS 125; NOA 163; WWD 181

Cockle, giant Atlantic (*Dinocardium robustum*)
AGC cp130; SAS cpB26

Cockleburr (*Xanthium strumarium*)
FWF 153; NWE cp44

Cock-of-the-rock, Peruvian (*Rupicola peruviana*)
AAL 156; EOB 327; MIA 303; WGB 113

Cockroach, American or waterbug (*Periplaneta
americana*)
MIS cp125

Cockroach, German or Croton bug (*Blattella germanica*)
MIS cp124; SGI cp37

Cockroah, Oriental or Asiatic (*Blatta orientalis*)
MIS cp126

Cod, Atlantic (*Gadus morhua*)
AAL 338(bw); ACF p15; AGC cp389; MIA 527; NAF cp492

Coelacanth (*Latimeria chalumnae*)
AAL 322(bw); EAL 125; MIA 497; WWW 169

Coendu See Porcupine, prehensile-tailed

Coffee (*Coffea* sp.)
OET 266, 267

Coffee, wild (*Triosteum perfoliatum*)
NWE cp427

Coffee tree, Kentucky (*Gymnocladus dioica*)
NTE cp311, cp503; OET 211; PMT cp[58]

Cohosh, black (*Cimicifuga racemosa*)
NWE cp134

Cohosh, blue (*Caulophyllum thalictroides*)
CWN 99; FWF 179; NWE cp388

Colic-root (*Aletris farinosa*)
NWE cp127

Colobus monkey, black (*Colobus satanas*)
CRM 83

Colobus monkey, black and white or guereza (*Colobus
guereza*)
GEM II:319; MEM 402-403; MIA 73; NHP 150(bw)

Colobus monkey, Kirk's or Zanzibar red (*Colobus kirki*)
CRM 83

Colobus monkey, Olive (*Colobus verus* or *Procolobus
verus*)
CRM 83; MIA 73

Colobus monkey, red (*Colobus badius*)
CRM 83; MIA 73; NHP 148

Colobus monkey, Western or Western black and white
(*Colobus polykomos*)
CRM 83; GEM II:322; WOM 107

Coltsfoot (*Tussilago farfara*)
CWN 259; NWE cp297

Colugo, Philippine (*Cynocephalus volans*)
MEM 446; ROW 13(bw)

Columbine (*Aquilegia vulgaris*)
PPC 59

Columbine, blue (*Aquilegia caerulea*)
SWW cp605; WAA 226, 227

Columbine, red or Western (*Aquilegia formosa*)
SWW cp400; WFW cp483

Columbine, wild or Eastern (*Aquilegia canadensis*)
BCF 123; CWN 65; FWF 3; NWE cp426; SEF cp474; WAA
79, 159; WTV 109

Columbine, yellow (*Aquilegia chrysantha*)
SWW cp289; WAA 226, 227; WFW cp445

Comb jelly (*Pleurobrachia* and other genera) See also
Sea gooseberry
AAL 508-509; AGC cp227; BCS 47; BMW front., 89; EAL
182, 183; MPC cp382; MSC cp491-493

Combfish (*Zaniolepis* sp.)
PCF cp22

Comfrey (*Symphytum officinale*)
CWN 199

Compass plant (*Silphium laciniatum*)
LBG cp171; NWE cp293

Conch (*Strombus* sp.)
AGC cp121; SCS cp50

Condalia or blueweed (*Condalia* sp.)
NTE cp41; NTW cp84

Condor, Andean (*Vultur gryphus*)
BOP 29(head only), 37, 40; EOB 107; WOM 183; WWW
10

Condor, California (*Gymnogyps californianus*)
BOP 39; EAB 917; FGH cp2; FWB 205; MIA 217; NAW
112; WGB 27; WOB 247-249; WWW 38(head only)

Coneflower, prairie (*Ratibida pinnata*)
NWE cp280

Coneflower, three-lobed (*Rudbeckia triloba*)
CWN 265; WTV 94

Coney (*Epinephelus fulvus*)
ACF cp26; MIA 549

Coney, Caribbean (*Cephalopholis fulva*)
AAL 352; KCR cp23; RCK 24; RUP 147, 150

Cony See Pika

Cooloola monster (*Cooloola propator*)
EOI 42 (bw)

Coot, American (*Fulica americana*)
AAL 122; AGC cp601; BWB 217-219; EAB 722, 723; FEB
122; FWB 121; MIA 237; NAB cp134; PBC 134; UAB
cp111; WGB 47; WNW cp535

Coot, crested (*Fulica cristata*)
EOB 145

Coot, European (*Fulica atra*)
EOB 151; PGT 225; RBB 60

Cooter See Turtle, Florida cooter; Turtle, red-eared;
Turtle, river cooter

Copepod (*Cyclops* sp.)
AAL 399(bw), 400; PGT 131; WWW 51

Copperhead (*Agkistrodon contortrix*)
ARA cp649-cp652, cp655; ARC 240, 241; ARN 72(bw);
CGW 97; JAD cp244; LSW 351-353; NAW 238; RNA 199;
SEF cp530

Coquina (*Donax variabilis*)
AGC cp131; MSC cp321

Coral bells (*Heuchera sanguinea*)
SWW cp420

Coral, boulder (*Montastrea* sp.)
KCR p8, cp18; WWF cp8

Coral, brain (*Diploria* sp. and other genera)
AAB 133; AGC cp279; BMW 112-113(close-up); KCR p9;
MSC cp2-5

Coral, elkhorn (*Acropora palmata*)
AGC cp272; KCR p7; MSC cp32; WWF cp1

Coral, finger (*Porites porites*)
KCR p7, cp17

Coral, fire (*Millepora* sp.)
AGC cp274; KCR p11, cp17; MSC cp25

Coral, flower (*Mussa angulosa*)
KCR p10; MSC cp1

Coral, lettuce-leaf (*Agaricia* sp.)
KCR p11; MSC cp33

Coral, orange cup (*Balanophyllia elegans*)
BCS 73; MPC cp442, cp443

Coral, pillar (*Dendrogyra cylindrus*)
KCR p7; MSC cp6

Coral, porous (*Porites astreoides*)
KCR p8; MSC cp9

Coral, precious (*Corallium rubrum*)
AAB 9

Coral, red or orange (*Tubastrea coccinea*)
KCR cp30; WWF cp6

Coral, soft (red) (*Gersemia rubiformis*)
AGC cp255; MPC cp446, cp451; MSC cp37, cp39

Coral, staghorn (*Acropora cervicornis*)
AAL 527(bw); AGC cp273; KCR p7; MSC cp55; WWF cp5

Coral, star (*Montastrea* and other genera)
AGC cp285; KCR p8, p10; MSC cp10, cp11, cp174

Coral, starlet (stoke's) (*Dichocoenia stokesi*)
AGC cp286; KCR p10; MSC cp7

Coral (other kinds)
KCR p7-p12; RUP 13; WWF cp2, cp4

Coralbean or coral tree (*Erythrina* sp.)
NTE cp354, cp368; NWE cp432; SEF cp127, cp145; SWW
    cp425; TSP 29; WAA 165

Coralbean, Southeastern (*Erythrina herbacea*)
NTE cp354, cp368; SEF cp127, cp145

Coralberry (*Symphoricarpos orbiculatus*)
TSV 162, 163

Coral-root See under Orchid

Cordgrass, prairie (*Spartina pectinata*)
LBG cp105

Cordgrass, saltmeadow (*Spartina patens*)
AGC cp464

Cordgrass, saltmarsh (*Spartina alternifolia*)
AGC cp466

Cordon bleu (*Uraeginthus* sp.)
EOB 422

Corepsis See Tickseed

Cormorant See also Shag

Cormorant, black-faced (*Phalacrocorax fuscescens*)
OBL 34

Cormorant, blue-eyed (*Phalacrocorax atriceps*)
BWB 93; EOB 63

Cormorant, brandt's (*Phalacrocorax penicillatus*)
EAB 119; FWB 73; MPC cp534; UAB cp77

Cormorant, common or great (*Phalacrocorax carbo*)
EAB 121; FEB 68; NAB cp102; OBL 26, 202; RBB 15;
    WGB 17

Cormorant, double-crested (*Phalacrocorax auritus*)
AGC cp575; EAB 120; FEB 69; FWB 70; MPC cp533;
    NAB cp99; NAW 113; OBL 46, 159; PBC 56; UAB cp76;
    WNW cp520; WON 30, 31

Cormorant, Galapagos or flightless (*Nannopterum harrisi*)
AAL 99(bw); BWB 92; EOB 63; MIA 207; OBL 47; WGB
    17

Cormorant, king or rock (*Phalacrocorax albiventor*)
HSW 109

Cormorant, little black (*Phalacrocorax sulcirostris*)
HSW 35

Cormorant, olivaceous (*Phalacrocorax olivaceus*)
EAB 121; FEB 70; FWB 71; NAB cp101; WNW cp519

Cormorant, pelagic (*Phalacrocorax pelagicus*)
EAB 120; FWB 72; HSW 14-15; MPC cp532; UAB cp74

Cormorant, pied (*Phalacrocorax varius*)
HSW 107

Cormorant, pygmy (*Halietor pygmaeus*)
OBL 43

Cormorant, red-faced (*Phalacrocorax urile*)
EAB 121; UAB cp75

Cormorant, red-footed (*Phalacrocorax gaimardi*)
BMW 197; OBL 150

Cormorant, reed (*Phalacrocorax* or *Halietor africanus*)
EOB 53; HSW 108; MIA 207; WGB 17

Cormorant, spotted (*Phalacrocorax punctatus*)
HSW 110

Corn cockle (*Agrostemma githago*)
CWN 117; LBG cp237; NWE cp469; PPC 57; SWW
    cp462; WTV 118

Corn snake See Ratsnake, red

Cornetfish, red (*Fistularia petimba*)
ACF p17; NAF cp436

Corydalis See Fitweed

Cottonmouth or Water moccasin (*Agkistrodon piscivorus*)
AAL 258(bw), 259(bw); AGC cp442; ARA cp654, cp656,
    cp657; ARC 242; LSW 355-357; MIA 459; NAW 240; RNA
    201; WNW cp175, cp177, cp194

Cottontail, desert (*Sylvilagus audubonii*)
ASM cp242; CMW 80(bw); GEM IV:300; JAD cp508; MEM
    717; MIA 193; MPS 116(bw); NMM cp[5]

Cottontail, Eastern (*Sylvilagus floridanus*)
ASM cp238; CGW 197; CMW 77(bw); GEM IV:302-303;
    GMP 122(bw); LBG cp48; MEM 715, 717; MNC 141(bw);
    MPS 116(bw); NAW 27; SEF cp594; WMC 106, 107

Cottontail, New England (*Sylvilagus transitionalis*)
ASM cp245; CGW 197; GMP 126(bw); SEF cp593; WMC 109

Cottontail, Nuttall's or Mountain (*Sylvilagus nuttalli*)
ASM cp237; MPS 117(bw); NAW 28; RMM cp[11]; SWM 17

Cottonwood, black (*Populus trichocarpa*)
NTW cp207; WFW cp94

Cottonwood, eastern (*Populus deltoides*)
LBG cp432; NTE cp185, cp186, cp512, cp611; SEF cp65, cp181; TSV 156, 157; WNW cp472

Cottonwood, Fremont (*Populus fremontii*)
NTW cp209, cp540; WFW cp96, cp167

Cottonwood, narrowleaf (*Populus angustifolia*)
NTW cp135; WFW 80

Cottonwood, swamp (*Populus heterophylla*)
NTE cp190; WNW cp470

Coucal (*Centropus* sp.)
EOB 231, 233

Cougar See Mountain lion

Courser, cream-colored (*Cursorius cursor*)
EOB 182; MIA 247; WGB 57

Cowbird, bronzed (*Molothrus aeneus*)
EAB 905; FEB 362; FWB 284; LBG cp551; NAB cp517, cp574; UAB cp559, cp621

Cowbird, brown-headed (*Molothrus ater*)
EAB 905; FEB 363; FWB 285; JAD cp613; MIA 383; NAB cp516, cp571; PBC 351; UAB cp558, cp620; WOB 102

Cowfish (*Lactophrys* sp.)
ACF cp60; NAF cp315

Cowrie, chestnut (*Cypraea* or *Zonaria spadicea*)
BCS 87; MSC cp444

Cowrie, tiger (*Cypraea tigris*)
RUP 55

Cowslip See Marsh marigold

Coyote (*Canis latrans*)
ASM cp263, 264; CGW 157; CMW 191(bw); GEM IV:104-106; GMP 246(bw); JAD cp523, cp524; LBG cp37, cp38; MEM 62, 63; MIA 79; MNC 237(bw); MNP 229(bw); MPS 249(bw); NAW 4; NMM cp[20]; RFA 54-61(bw); RMM cp[14]; SWM 74, 76; WFW cp370, cp371; WMC 178; WPP 142; WWD 118

Coypu See Nutria

Crab (various genera)
MSC cp628-690

Crab, arrow (*Stenorhynchus seticornis*)
KCR cp30; MSC cp574; SCS cp55; WWF cp39

Crab, blue (*Callinectes sapidus*)
AGC cp147; MSC cp657; SAS cpC27; SCS cp53

Crab, coral (*Carpilius corallinus*)
AGC cp141; MSC cp648

Crab, fiddler (*Annulipes* sp.)
AAB 125; EAL 241

Crab, fiddler (*Uca* sp.)
AGC cp142, cp143; MPC cp343; MSC cp628-630; SAS cpC34; SCS cp57

Crab, ghost (*Ocypoda* sp.)
AAL 403(bw); AGC cp144; BMW 38-39; EAL 238-239; MSC cp631; SCS cp58; WOI 181; WWW 194

Crab, green (*Carcinus maenas*)
AGC cp145

Crab, hermit (various genera)
AGC cp156, cp159; BCS 94; EAL 151; KCR cp30; MPC cp339; MSC cp676, cp677, cp680-687; SAB 116; SAS cpC21, cpC22; SCS cp52

Crab, horseshoe (*Limulus polyphemus*)
AAL 395(bw); AGC cp160; BCS 52, 53, 63; EAL 245; MSC cp666; WWW 176

Crab, Jonah (*Cancer borealis*)
AGC cp139; SAS cpC26

Crab, kelp (*Pugettia producta*)
MPC cp334

Crab, King (*Paralithodes* or *Lopholithodes* sp.)
AAL 403; BCS 106

Crab, lady (*Ovalipes* sp.)
AGC cp135; SAS cpC28

Crab, land (*Cardisoma* sp.)
EAL 240; MSC cp632; SCS cp58

Crab, land (*Gecarcinus* sp.)
SCS cp58; WOI 181

Crab, marsh (*Sesarma* sp.)
SAS cpC32; SCS cp57

Crab, mole (*Emerita* sp.)
AGC cp154; MPC cp342; MSC cp689, cp690

Crab, mud (various genera)
AGC cp133; MPC cp327; SAS cpC31; SCS cp56

Crab, Oregon (*Cancer oregonensis*)
MPC cp335

Crab, porcelain (*Petrolisthes* sp.)
MPC cp328; SAS cpC23; SCS cp52; WWW 150

Crab, red (*Cancer productus*)
MPC cp336

Crab, rock (*Cancer irroratus* or *antennarius*)
AGC cp140; BCS 40; MPC cp338; SAS cpC26

Crab, sally light-foot (*Grapsus grapsus*)
BWB 32; HSG cp40; MSC cp649; SCS cp56; WOI 39

Crab, sargassum (*Portunus sayi*)
AGC cp136

Crab, shore (*Carcinus* or *Pachygrapsus* sp.)
EAL 233; HSG cp39; MPC cp331; SCS cp56

Crab, shore (*Hemigrapsus* sp.)
MPC cp326, cp329, cp332

Crab, soldier (*Mictyris* or *Coenobita* sp.)
AAB 74; BMW 41

Crab, spider (*Libinia* or *Mithrax* sp.)
AGC cp138; SCS cp54

Crab, sponge (*Dromidia* sp.)
AGC cp151; KCR cp30; MSC cp669; SAS cpC24

Crab, stone (Florida) (*Menippe mercenaria*)
AGC cp134; SAS cpC30

Crab, swimming (*Liocarcinus* or *Portunus* sp.)
HSG cp42; KCR cp30; SAS cpC29; SCS cp53

Crab, wharf (*Sesarma cinereum*)
AGC cp146; SAS cpC32

Crab (other kinds)
AGC cp137; BCS 75; BMW 75; MAE 102; MPC cp325,
cp330, cp337, cp340, cp341; MSC cp628-cp690; SCS
cp53, cp56; SWF 170-171

Crabgrass, smooth (*Digitaria ischaemum*)
NWE cp31

Crake, African black (*Limnocorax flavirostra*)
AAL 122; AWR 95; EOB 150

Crake, corn (*Crex crex*)
EOB 150; MIA 237; WGB 47

Cranberry (*Vaccinium*)
FWF 111; NWE cp78, cp452, cp567; WNW cp266; WTV
139

Crane, Australian (*Grus rubicundus*)
COW cp11, 140(bw)

Crane, black-crowned (including West African)
(*Balearica pavonina*)
COW cp2, p1, 76(bw); EOB 145-147; LOW 76(bw)

Crane, blacknecked (*Grus nigricollis*)
COW cp20, cp21, p24, 216(bw)

Crane, blue (*Anthropoides paradisea*)
COW cp4, cp5, 86(bw), p4; EOB 143; WBW 131

Crane, crowned (including South African) (*Balearica
regulorum*)
COW cp1, cp3, p2, p3; MIA 235; WBW 135; WGB 45;
WPP 80, 81

Crane, demoiselle (*Anthropoides virgo*)
COW cp6, 94(bw), p5; EOB 144; NHU 170; WBW 133

Crane, Eurasian (*Grus grus*)
COW cp23, 226(bw)

Crane, hooded (*Grus monachus*)
COW cp22, p20-p23, 206(bw)

Crane, Japanese or red-crowned (*Grus japonensis*)
BAS 20; BWB 36, 37; COW cp18, cp19, p17-p19,
196(bw); EOB 145; WBW 127

Crane, sandhill (*Grus canadensis*)
COW cp14, cp15, p12-p14, 170(bw); EAB 122, 123; FEB
148; FWB 137; LBG cp505; NAB cp30; PBC 124; UAB
cp16; WBW 123; WNW cp531; WON 81, 82; WPP 138-
139

Crane, sarus (*Grus antigone*)
AWR 110; COW cp12, 150(bw); WBW 121

Crane, wattled (*Bugeranus carunculatus*)
AAL 121(bw); COW cp7, cp8, p6, p7, 120(bw)

Crane, white (Siberian, Asiatic, or great) (*Grus* or
Bugeranus leucogeramus)
AWR 108-109; COW cp9, cp10, p8, 130(bw); NHU 50, 80

Crane, white-naped (*Grus vipio*)
COW cp13, p9, p10, p11, 160(bw)

Crane, whooping (*Grus americana*)
BAS 144; COW cp16, cp17, 184(bw); EAB 122, 123; FEB
149; FWB 136; MIA 235; NAB cp29; WGB 45; WOB 223,
251-255; WON 80

Cranesbill, dovesfoot (*Geranium molle*) See also
Geranium, wild
CWN 139; WTV 132

Crape myrtle (*Lagerstroemia* sp.)
NTW cp118, cp392; PMT cp[71]; TSP 63

Crappie, black (*Pomoxis nigromaculatus*)
NAF cp80; WNW cp50

Crayfish, freshwater (*Astacus pallipes*)
MAE 95 (head only)

Crayfish, painted (*Panulirus versicolor*)
RUP 38

Crazyweed See Locoweed, purple

Cream cup (*Platystemon californicus*)
LBG cp189; SWW cp208

Creambush (*Holodiscus discolor*)
WFW cp192

Creeper, brown or tree (*Certhia familiaris*)
EAB 124; EOB 392; FEB 314; FWB 341; NAB cp355; PBC 259; RBB 165; SEF cp348; SPN 72; UAB cp388; WFW cp268

Creole-fish (*Paranthias furcifer*)
ACF cp26; NAF cp504

Creosote bush (*Larrea tridentata*)
JAD cp342; LBG cp114

Cress (*Cardamine* sp.)
CWN 105; NWE cp144; WTV 9

Cress, bitter (*Cardamine cordifolia*)
SWW cp133; WNW cp246

Cress, hoary (*Cardaria draba*)
LBG cp147; SWW cp159

Cress, Winter (*Barbarea vulgaris*)
BCF 71; CWN 101; LBG cp206; NWE cp345

Crestfish (*Lophotus lacepedei*)
ACF cp22; PCF cp46

Cricket, bush (various kinds)
BOI 96; SAB 12-13; WIW 114, 150

Cricket, cave (*Ceuthophilus secretus*)
MIS cp249

Cricket, field (*Gryllus* sp.)
AAL 411(bw); AGC cp491; LBG cp350; MIS cp252; SGI cp29

Cricket, house (*Acheta domestica*)
MIS cp251

Cricket, Jerusalem (*Stenopelmatus fuscus*)
JAD cp375; MIS cp247; SGI cp27

Cricket, mole (European) (*Gryllotalpa vinae*)
EOI 45; WIW 104 (bw)

Cricket, mole (Northern) (*Gryllotalpa hexadactyla*)
MIS cp250; SGI cp30

Cricket, Mormon (*Anabrus simplex*)
JAD cp374; LBG cp351; MIS cp257

Cricket, tree (*Oecanthus fultoni*)
SGI cp28

Crimson rosella See under Parrot

Crinoid (sea lilies and feather stars) (*Crinozoa*)
RCK 205; RUP 29, 99; WWF cp47, cp48

Croaker, Atlantic (*Micropogonias undulatus*)
AGC cp384; NAF cp479

Croaker, yellowfin (*Umbrina roncador*)
MPC cp242; NAF cp480; PCF p33

Croaker (other kinds)
NAF cp477; PCF p33

Crocodile, African (*Crocodylus* sp.)
AAL 215(bw); AWR 90, 91; MIA 415

Crocodile, American (*Crocodylus acutus*)
AGC cp438; ARA cp258, cp261; NAW 253; RNA 209; WNW cp174

Crocodilefish (*Cociella* sp.)
RCK 52; RUP 120-121

Crocus, Autumn (*Colchicum autumnale*)
PPC 1; PPM cp8

Crombec (*Sylvietta* sp.)
EOB 369

Crossbill, common or red (*Loxia curviostra*)
EAB 301; EOB 418, 420; FEB 385; FWB 473; MAE 56; MIA 385; NAB cp412; RBB 184; SEF cp312; SPN 142; SWF 48; UAB cp452, cp553; WFW cp310; WGB 195

Crossbill, parrot (*Loxia pytopsittacus*)
EOB 420

Crossbill, white-winged (*Loxia leucoptera*)
EAB 301; FEB 383; FWB 472; NAB cp413; SEF cp315; UAB cp453, cp554

Crossvine (*Bignonia capreolata*)
WTV 111

Croton (*Codiaeum variegatum*)
TSP 52

Crow, American or common (*Corvus brachyrhynchos*)
EAB 125; EOB 232; FEB 365; FWB 288; MIA 401; MPC cp555; NAB cp579; NAW 115; PBC 247; SEF cp266; SPN 65; UAB cp562, cp625; WGB 211

Crow, carrion (*Corvus corone*)
EOB 443; RBB 172

Crow, fish (*Corvus ossifragus*)
AGC cp608; EAB 125; FEB 364; NAB cp580

Crow, northwestern (*Corvus caurinus*)
FWB 289; UAB cp624

Crow, pied (*Corvus albus*)
EOB 444

Crowfoot (*Ranunculus* sp.)
AWR 43; CWN 61; PGT 49

Crownbeard See Daisy, cowpen

Crown-of-thorns (*Euphorbia milii*)
TSP 57

Crown-of-thorns (starfish) (*Acanthaster planci*)
RCK 97; WWF cp11

Crucifixion thorn (*Canotia holacantha*)
JAD cp314; NTW cp79

Cuckoo, black-billed (*Coccyzus erythropthalmus*)
EAB 131; FEB 269; FWB 263; NAB cp521; SEF cp268;
   WON 103

Cuckoo, channel-billed (*Scythrops novaehollandiae*)
MIA 265; WGB 75

Cuckoo, European (*Cuculus canorus*)
BAS 48; EOB 233; MIA 265; RBB 103; WGB 75

Cuckoo, great spotted (*Clamator glandarius*)
EOB 231

Cuckoo, mangrove (*Coccyzus minor*)
AGC cp611; NAB cp523

Cuckoo, yellow-billed (*Coccyzus americanus*)
EAL 130, 131; EOB 232; FEB 268; FWB 262; MIA 265;
   NAB cp522; PBC 194; UAB cp562; WGB 75

Cuckoo-roller (*Leptosomus discolor*)
EOB 279; MIA 285; WGB 95

Cuckoo-shrike (*Campephaga* or *Coracina* sps.)
EOB 349, 352

Cucumber tree (*Magnolia acuminata*)
NTE cp93, cp389; PMT cp[76]; SEF cp58; TSV 22, 23

Cucumber, wild or burr (*Echinocystis lobata*)
BCF 65; NWE cp43, cp205

Cucumber-root, Indian (*Medeola virginiana*)
CWN 5; FWF 159; NWE cp3; SEF cp454; WTV 76

Culver's root (*Veronicastrum virginicum*)
NWE cp133

Cunner (*Tautogolabris adspersus*)
ACF cp39; AGC cp341; NAF cp357

Curassow, great (*Crax rubra*)
EOB 134; MIA 227; WGB 37

Curassow, nocturnal (*Nothocrax urumutum*)
MIA 227; WGB 37

Curlew, bristle-thighed (*Numenius tahitiensis*)
EAB 791

Curlew, bush stone (*Burhinus magnirostris*)
BAS 69

Curlew, Eurasian (*Numenius arquata*)
EOB 167; RBB 78; WBW 207

Curlew, Hudsonian See Whimbrel

Curlew, long-billed (*Numenius americanus*)
AAL 126; BWB 182; EAB 792; FEB 150; FWB 143; LBG
   cp508; NAB cp246; NAW 101; PBC 145; UAB cp217;
   WBW 205; WPP 24

Curlew, short-billed See Whimbrel

Curlew, stone (*Burhinus oedicneus*)
AAL 128(bw); EOB 179; MIA 245; RBB 63; WBW 319;
   WGB 55

Currawong (*Strepera* sp.)
EOB 435

Cuscus, gray (*Phalanger orientalus*)
CRM 27; GEM I:309; MEM 855

Cuscus, spotted (*Phalanger maculatus*)
AAL 27(bw), 29; GEM I:305-307; MEM 851-853; NOA 52;
   WOI 143; WWW 16-17

Cusk (*Brosme brosme*)
ACF p15

Cusk-eel (various genera)
ACF p16; MPC cp301; NAF cp448-cp450; PCF cp20

Cushion star See under Starfish

Custard apple (*Annona reticulata*)
TSP 11

Cutlassfish, Atlantic (*Trichiurus lepturus*)
NAF cp433

Cuttlefish (*Sepia* sp.)
AAB 89; EAL 256, 272; RCK 121; RUP 113-119; WWF
   cp27-cp30

Cycad (*Cycas* and other genera)
OET 58, 59

Cypress, Arizona (*Cupressus arizonica*)
NTW cp67, cp447; WFW cp60, cp143

Cypress, Baker (*Cupressus bakeri*)
NTW cp68, cp442

Cypress, bald (*Taxodium distichum*)
NTE cp27; PMT cp[135]; SEF cp36; TSV 12, 13; WNW
   cp433

Cypress, false (*Chamaecyparis* sp.)
NTE cp33; NTW cp56

Cypress, gowen (*Cupressus goveniana*)
NTW cp65, cp454

Cypress, MacNab (*Cupressus macnabiana*)
NTW cp66, cp443; WFW cp55

Cypress, Monterey (*Cupressus macrocarpa*)
NTW cp63, cp444; OET 91

Cypress, Sargent (*Cupressus sargentii*)
NTW cp64, cp445

Cypress, tecate (*Cupressus guadalupensis*)
NTW cp69, cp446

Cyrilla See Titi (plant)

D

Dab (*Limanda limanda*)
MIA 577

Dace (various genera)
JAD cp296; MIA 515; NAF cp142, cp156, cp158, cp159, cp176, cp184, cp185, cp192

Daddy-long-legs, brown (*Phalangium opilio*)
AAL 398(bw); LBG cp391; MIS cp673

Daddy-long-legs, eastern (*Leiobunum* sp.)
MIS cp674; SEF cp421

Dagger pod (*Phoenicaulis cheiranthoides*)
SWW cp541

Dahoon (*Ilex cassina* or *myrtifolia*)
NTE cp52, cp54, cp556; SEF cp50; WNW cp448

Daisy, blackfoot or desert (*Melampodium leucanthum*)
JAD cp92; LBG cp130; SWW cp65; WAA 172

Daisy, cowpen (*Verbesina encelioides*)
LBG cp179; SWW cp251

Daisy, English (*Bellis perennis*)
LBG cp131; SWW cp66

Daisy, Michaelmas See Aster, New York

Daisy, oxeye (*Chrysanthemum leucanthemum*)
BCF 101; CWN 279; LBG cp132; NWE cp91; SWW cp68

Daisy, seaside (*Erigeron glaucosus*)
MPC cp614; SWW cp477

Daisy, showy (*Erigeron speciosus*)
SWW cp480; WFW cp505

Daisy, stemless (*Townsendia excapa*)
SWW cp73; WFW cp380

Daisy, Tahoka (*Machaeranthera tanacetifolia*)
JAD cp62; LBG cp292; SWW cp595

Daisy, woolly (*Eriophyllum wallacei*)
JAD cp127; SWW cp241

Daisy, yellow spiny (*Haplopappus spinulosus*)
JAD cp131; SWW cp236

Dame's rocket (*Hesperis matronalis*)
BCF 73; CWN 109; WAA 183, 234

Damselfish (*Pomacentrus* and other genera) See also Beaugregory
ACF cp38, cp64; NAF cp324; RCK 158; WWF cp34

Damselfish, yellowtail (*Microspathodon chrysurus*)
ACF cp38; AGC cp337; NAF cp322

Damselfly See also Dragonfly

Damselfly (*Enallagma* sp.)
MIS cp353-355; PGT 140; SGI cp12

Damselfly, narrow-winged (Coenagrionidae)
BOI 173-175; MIS cp35(naiad), cp39(naiad)

Damselfly (Other kinds)
AAB [5], 65(larva); EOI 26; KCR cp22; MIS cp350, cp357, cp366, cp378; WNW cp367

Dandelion (*Taraxacum officinale*)
BCF 17; CWN 299; NWE cp235, cp305; SWW cp267

Dandelion, desert (*Malacothrix glabrata*)
JAD cp125; SWW cp257

Dandelion, dwarf (*Krigia virginica*)
CWN 297

Darner See Dragonfly

Darter (fish) (*Etheostroma* sp.)
AWR 219(bw); NAF cp220-cp250; WNW cp61, 62

Darter (fish) (*Percina* sp.)
NAF cp226, cp228, cp234-237, cp242

Darter, American (bird) See Anhinga

Darter, crystal or sand (fish) (*Ammocrypta* sp.)
NAF cp251-cp255

Darter, snail (fish) (*Percina tanasi*)
NAF cp241; WNW cp60

Dayflower (*Commelina* sp.)
CWN 27; NWE cp611; SWW cp600, cp601; WFW cp529

Dead man's fingers (coral) (*Alcyonium* and other genera)
AAL 518(bw); BCS 44; EAL 173; KCR p12

Dead man's fingers (fungus) (*Xylaria polymorpha*)
FGM cp1; NAM cp697; SGM cp345

Dead nettle See also Henbit

Dead nettle, purple (*Lamium purpureum*)
CWN 233; WTV 170

Death camus (*Zigadenus* sp.)
LBG cp144; NWE cp137; WFW cp399

Death cap (*Amanita phalloides*)
FGM cp27; NAM cp113, cp673; NMT cp59; PPC 112;
 SGM cp4

Death cap, false (*Amanita citrina*)
FGM cp27; GSM 153; NAM cp125; NMT cp95; SGM cp5

Deer, axis (*Axis* or *Cervus axis*)
GEM V:147, 150; MEM 524; WPP 8-9

Deer, black-tailed See Deer, mule

Deer, brocket (*Mazama americana*)
GEM V:227; MEM 523

Deer, brow-antlered (*Cervus eldi*)
CRM 177

Deer, fallow (*Dama dama*)
ASM cp288; CRM 175; GEM V:124-125, 131, 152-160;
 MEM 531; NAW 30(bw)

Deer, Florida key (*Odocoileus virginianus clavium*)
ASM cp286

Deer, hog (*Axis* or *Cervus porcinus*)
CRM 177; GEM V:149; SWF 81

Deer, hog (Calamian) (*Axis calamianensis*)
GEM V:148

Deer, marsh (*Blastocerus dichotomus*)
CRM 179; MEM 522

Deer, mouse or Chevrotain (*Tragulus* sp.)
GEM V:116-117, 120(bw), 122(bw); MEM 516; MIA 135

Deer, mule or black-tailed (*Odocoileus hemionus*)
ASM cp289; CMW 252(bw); CRM 179; GEM V:200, 213,
 216; JAD cp528; LBG cp35; MAE 49; MNC 285(bw); MNP
 cp7a, 285(bw); MPS 303(bw); NAW 30, 31; NMM cp[30];
 RMM cp[16]; SWM 145, 147; WFW cp373; WPP 145

Deer, musk (*Moschus* sp.)
AAL 77; CRM 175; GEM V:135, 136; MEM 518; MIA 135;
 WOM 138

Deer, pampas (*Ozotoceros bezoarticus*)
CRM 179; GEM V:218; MEM 522; MIA 139; WPP 177

Deer, Pere David's (*Elaphurus davidiensis*)
CRM 179; GEM V:161-163; MEM 525; MIA 137

Deer, red or elk or wapiti (*Cervus elaphus*) See also Elk,
 American
ASM cp287; CMW 249(bw); CRM 177; GEM V:144, 176-
 188, 195(Altai and Kashmir); LBG cp32; MAE 26, 55; MEM
 10; MNC 283(bw); MNP 283(bw); MPS 302(bw); NHU 49
 (Bukhara variety); NMM cp[31]; NNW 51, 138; SAB 63;
 SWF 53-55; SWM 141; WFW cp376; WMC 231; WOM
 59-61, 170

Deer, roe (*Capreolus capreolus*)
GEM V:201-211; MEM 524, 525; MIA 137; NNW 37; SWF
 56

Deer, rusa (Prince Albert's spotted) (*Cervus alfredi*)
ROW front., 57(bw)

Deer, sika (*Cervus nippon*)
CRM 177; GEM V:174; MEM 525; MNC 300(bw); WMC
 232, 233

Deer, swamp See Barasingha

Deer, Thorold's (*Cervus albirostris*)
CRM 177; GEM V:197

Deer, tufted (*Elaphodus cephalophus*)
GEM V:138; MEM 525; MIA 135

Deer, water (*Hydropotes inermis*)
GEM V:198; MEM 525; MIA 135; SAB 74

Deer, white-tailed (*Odocoileus virginianus*)
ASM cp285; CGW 205; CMW 255(bw); CRM 179; GEM
 V:212, 214; GMP 317(bw); LBG cp33; MEM 523; MIA 137;
 MNC 287(bw); MNP 287(bw); MPS 303(bw); NAW 32, 33;
 SAB 69; SEF cp616; SWF 147-149; SWM 144; WFW
 cp374; WMC 234-236; WNW cp617; WPP 146

Deerberry (*Vaccinium stamineum*)
TSV 74, 75

Deer-brush (*Ceanothus integerrimus*)
WFW cp198, cp221

Deer-weed (*Lotus scoparis*)
SWW cp284; WFW cp438

Deer's-tongue See Monument plant

Degu (*Octodon degus*)
GEM III:343; MEM 703; MIA 187

Desert bell (*Phacelia campanularia*)
JAD cp70; SWW cp652

Desert calico (*Langloisia matthewsii*)
JAD cp51; SWW cp488

Desert candle (*Caulanthus inflatus*)
JAD cp164; SWW cp391

Desert gold (*Liananthus aureus*)
JAD cp145; SWW cp218

Desert plume See Prince's plume

Desert star, Mohave (*Monoptilon bellioides*)
JAD cp90; SWW cp67

Desert trumpet (*Eriogonum inflatum*)
JAD cp108; SWW cp275

Desert velvet (*Psathyrotes ramosissima*)
JAD cp118; SWW cp277

Desman, Pyrenean (*Galemys pyrenaicus*)
CRM 43; GEM I:503; MEM 767

Desman, Russian (*Desmana moschata*)
CRM 43; GEM I:510; MIA 37; NHU 113

Destroying angel (*Amanita virosa*)
NAM cp123, cp124, cp672; NMT cp11; PPC 114; PPM
cp27, cp28; SGM cp8

Devil's bit (*Chamaelirium luteum*)
CWN 13; NWE cp124; SEF cp431; WTV 18

Devil's claw (*Proboscidea althaeafolia*)
JAD cp117, cp165; LBG cp118, cp197; SWW cp13, cp288

Devil's club (*Oplopanax horridum*)
WFW cp160

Devil's urn fungus (*Urnula craterium*)
FGM cp1; NAM cp613

Devil's walking-stick (*Aralia spinosa*)
NTE cp322, 538; TSV 170, 171

Dewberry, swamp (*Rubus hispidus*)
NWE cp57; WNW cp252

Dhole See Dog, dhole

Dibatag (*Ammodorcas clarkei*)
CRM 97; MEM 577; MIA 153

Dickcissel (*Spiza americana*)
AAL 183(bw); FEB 341; FWB 464; LBG cp529; MIA 373;
NAB cp545; SPN 126; WGB 183

Dikdik (*Madoqua* sp.)
GEM V:318, 331; WPP 84

Dikdik, Kirk's (*Madoqua kirkii*)
MEM 575; MIA 151; NHA 108(bw), 111(bw) (head only)

Dikkop See Thick-knee, water

Dingo (*Canis dingo*)
GEM IV:100-103; MAE 6, 7; MEM 84, 85; MIA 79; NOA
239

Dipper, American (*Cinclus mexicanus*)
EAB 199; EOB 357; FWB 349; MIA 331; NAW 180; SPN
80; UAB cp492; WFW cp271; WGB 141; WNW cp579

Dipper, Eurasian (*Cinclus cinclus*)
RBB 126

Dittany (*Cunila origanoides*)
CWN 217; WTV 148

Diver See Loon

Dobsonfly (Corydalidae)
BOI 189; MIS cp45(larva), cp332, cp333; WNW cp359,
cp360

Dock (*Rumex* sp.)
BCF 25; JAD cp46; LBG cp220, cp222; NWE cp439; PGT
40; SWW cp374

Dodder (*Cuscuta gronovii*)
CWN 205; NWE cp202

Dog, bush (*Speothos venaticus*)
CRM 135; GEM IV:149; MEM 85; MIA 81

Dog, dhole or red (*Cuon alpinus*)
CRM 135; GEM IV:141; MEM 80, 81; MIA 81

Dog, hunting or painted (*Lycaon pictus*)
AAL 58(bw); CRM 135; GEM IV:134-139; MEM 76-79;
MIA 81

Dog, prairie See Prairie dog

Dog, raccoon (*Nyctereutes procyonoides*)
GEM IV:143-145; MEM 85; MIA 81

Dog, wild See Dog, dhole or Dog, hunting

Dogbane, common See Indian hemp

Dogbane, spreading (*Apocynum androsaemifolium*)
CWN 189; LBG cp234; NWE cp557; SWW cp537; WFW
cp509

Dogfish, piked or spiny (*Squalus acanthias*)
AGC cp414; MIA 493; MPC cp323; NAF cp589; SJS 20

Dogfish, sandy or spotted (*Scyliorhinus* sp.)
MIA 491; SSW 126, 127(bw)

Dogwood (*Cornus florida*)
NTE cp76, cp454, cp554, cp579; SEF cp52, cp160, cp190;
TSV 44, 45

Dogwood, alternate-leaf or pagoda (*Cornus alternifolia*)
NTE cp77, cp409; PMT cp[35]; TSV 160, 161

Dogwood, Kousa (*Cornus kousa*)
PMT cp[36]

Dogwood, Pacific (*Cornus nuttallii*)
NTW cp111, cp384; PMT cp[37]; SWW cp180; WFW cp69

Dogwood, red-osier or Western (*Cornus stolonifera*)
NTE cp78; NTW cp110; NWE 237; SEF cp51; WFW cp70;
WNW cp325, cp444

Doll's eyes (*Actaea pachypoda*)
CWN 67; FWF 41; NWE cp162, cp238; SEF cp428, cp501

Dolphin, Amazon See River dolphin, Amazon

Dolphin, Atlantic spotted (*Stenella plagiodon*)
NAF cp640; WDP 97; WMC 215

Dolphin, blind See River dolphin, Ganges

Dolphin, bottle-nosed (*Tursiops truncatus*)
AGC cp321; EAL 306; GEM IV:396, 408, 411; HSG cp55;
HWD 6-7, 37; MEM 100; MIA 115, MPC cp204, cp205;
NAF cp639, cp642; NAW 34; WDP 90; WMC 212; WOW
275

Dolphin, bottle-nosed (Pacific) (*Tursiops gilli*)
BMW 161

Dolphin, brindled (*Stenella attenuata*)
NAF cp641; WOW 266

Dolphin, Chilean or black (*Cephalorhynchus eutropia*)
WOW 250

Dolphin, clymene or Helmut (*Stenella clymene*)
NAF cp645; WDP 99; WOW 263

Dolphin, Commerson's or piebald (*Cephalorhynchus commersonii*)
EAL 307; HWD 39, 90; MEM 183; WDP 116, 130(bw), 133; WOW 248

Dolphin, common (*Delphinus delphis*)
AAL 53; AGC cp322; CRM 207; EAB 107; EAL 305, 307; HWD 35, 78; MEM 183; MIA 115; MPC cp203, cp206; NAF cp633, cp648; WDP 90; WOW 271

Dolphin, dusky (*Lagenorhynchus obscurus*)
EAL 307; GEM IV:403; HWD 89; WDP 125; WOW 237

Dolphin, estuarine or little bay (*Sotalia fluviatilis*)
WDP 82; WOW 188

Dolphin, Fraser's or shortsnout (*Lagenodelphis hosei*)
CRM 207; HWD 37; WDP 150; WOW 242

Dolphin, Heaviside's or Benguela (*Cephalorhynchus heavisidii*)
CRM 207; WDP 133; WOW 251

Dolphin, Hector's or New Zealand (*Cephalorhynchus hectori*)
HWD 38; WDP 133; WOW 253

Dolphin, hourglass (*Lagenorhynchus cruciger*)
HWD 37; WDP 125; WOW 238

Dolphin, humpbacked (Atlantic) (*Sousa teuszii*)
EAL 307; MEM 183; WDP 82, 86; WOW 186

Dolphin, humpbacked (Indo-Pacific or Chinese white) (*Sousa chinensis*)
HWD 35, 43; MIA 115; WOW 184

Dolphin, Irrawaddy or snubfin (*Orcaella brevirostris*)
CRM 205; GEM IV:406; HWD 31; WDP 134; WOW 256

Dolphin, northern right whale (*Lissodelphis borealis*)
MEM 183; NAF cp673; WDP 150; WOW 246

Dolphin, Peale's or blackchin (*Lagenorhynchus australis*)
WDP 125; WOW 240

Dolphin, Risso's or grey (*Grampus griseus*)
AGC cp319; EAL 308; HWD 38; MEM 184; MIA 115; MPC cp212; NAF cp654; WDP 90; WOW 254

Dolphin, rough-toothed (*Steno bredanensis*)
EAL 306; MEM 182; NAF cp643; WDP 82; WOW 181

Dolphin, spinner (*Stenella longirostris*)
GEM IV:407; HWD 39, 76, 92; MEM 192; NAF cp638, cp644; WDP 99; WOW 261

Dolphin, spotted (*Senella attenuata*)
CRM 207; EAL 307; GEM IV:406; HWD 10-11, 38, 116; MEM 183; WDP 97; WOW 268

Dolphin, striped (*Stenella coerulevalba*)
HWD 93; MIA 115; MPC cp207; NAF cp646; WDP 90; WOW 265

Dolphin, white-beaked (*Lagenorhynchus albirostris*)
HWD 36; NAF cp650; WDP 123; WOW 230

Dolphin, white-sided (Atlantic) (*Lagenorhynchus acutus*)
EAL 306; HSG cp56; MEM 182; NAF cp649; WDP 123; WOW 232

Dolphin, white-sided (Pacific) (*Lagenorhynchus obliquidens*)
EAL 291; GEM I:142-143, IV:402, 409; HWD 36; MPC cp208; NAF cp632, cp647; WDP 116, 123; WOW 234

Dolphinfish (*Coryphaena* sp.)
ACF p30; HSG cp49; MIA 555; NAF cp582; PCF cp46

Doris See Nudibranch

Dormouse, common or hazel (*Muscardinus avellanarius*)
GEM III:280-283; MAE 31; MEM 678, 679; NNW 141; SWF 66

Dormouse, edible or fat (*Glis glis*)
GEM III:277, 278; MEM 678; MIA 181; SWF 67

Dormouse, forest (*Dryomys nitedula*)
GEM III:288; NHU 159

Dormouse, garden or orchard (*Eliomys quercinus*)
GEM III:284-285, 287; MEM 680-681; SWF 34

Dotterel (*Eudromias morinellus*)
RBB 66; WBW 201

Douroucouli See Monkey, night

Dove, collared (*Streptopelia decaocto*)
MIA 255; RBB 101; WGB 65

Dove, diamond (*Geopelia cuneata*)
MIA 255; WGB 65

Dove, ground (*Columbina passerina*)
EAB 710; FEB 190; FWB 176; JAD cp553; LBG cp510; MIA 253; NAB cp325; PBC 189; UAB cp344; WGB 63

Dove, Inca (*Columbina inca*)
EAB 711; FEB 192; FWB 177; JAD cp552; NAB cp326; UAB cp345

Dove, mourning (*Zenaida macroura*)
AAL 133(bw); EAB 710; EOB 219; FEB 191; FWB 179; JAD cp551; LBG cp509; MIA 253; NAB cp322; PBC 189; SEF cp270; UAB cp349; WGB 63; WON 101

Dove, quail (*Geotrygon* sp.)
EAB 711

Dove, quail (blue-headed) (*Starnoenas cyanocephala*)
MIA 253; WGB 63

Dove, rock (*Columba livia*)
EAB 712; FEB 196; FWB 182; MIA 253; NAB cp326; UAB cp346; WGB 63

Dove, spotted (*Streptopelia chinensis*)
EAB 712; EOB 216; FWB 181; UAB cp351

Dove, stock (*Columba oenas*)
RBB 99

Dove, superb fruit or purple-crowned pigeon
(*Ptilinopus superbus*)
MIA 255; WGB 65

Dove tree (*Davidia involucrata*)
PMT cp[45]

Dove, turtle See Turtledove

Dove, white-winged (*Zenaida asiatica*)
EAB 713; FEB 195; FWB 183; JAD cp550; NAB cp321; PBC 189; UAB cp348

Dove, white-tipped or white-fronted (*Leptotila verreauxi*)
EAB 712; FEB 194; NAB cp324

Dovekie (*Alle alle*)
AGC cp523; AOG cp1; DBN cp12; EAB 47; FEB 63; MIA 251; NAB cp96, cp98; WGB 61

Dowitcher, longbilled (*Limnodromus scolopaceus*)
AGC cp595; BWB 175; FEB 160; FWB 149; NAB cp213; UAB cp213, cp239; WBW 263

Dowitcher, shortbilled (*Limnodromus griseus*)
AGC cp594; BAA 127; FEB 161; FWB 148; MPC cp565; NAB cp201, cp212; UAB cp212; WBW 261(bw)

Dracaena, giant (*Cordyline australis*)
NTW cp305

Dragon, flying (*Draco* sp.)
AAB 108; ERA 94; MIA 421

Dragon tree or dragon blood tree (*Dracaena draco*)
TSP 27

Dragonfish See Angler-fish

Dragonfly See also Damselfly

Dragonfly, darner (brown) (*Boyeria vinosa*)
MIS cp377; WNW cp365

Dragonfly, darner (green) (*Anax junius*)
BOI 176; MIS cp343, cp346; WNW cp366

Dragonfly, darter (*Libellula* and other genera)
EOI 28; PGT 145-147; WIW 86

Dragonfly, skimmer (*Libellula* and other genera)
BOI 177; MIS cp347-349, 359-361, 365, 370, 371, 373, 375, 376; SGI cp8; WIW 15

Dragonfly (other kinds)
EOI title page, 31; MIS cp345, cp362, 363; SGI cp6, cp7, cp9-cp11; WFW cp569; WIW 35; WNW cp364

Dragonhead See Obediant plant

Drill (*Papio leucophacus*)
CRM 77; MEM 372; MIA 69

Drongo (*Dicrurus* sp.)
AAL 194(bw); EOB 430

Dropwort, water (*Oenanthe crocata*)
PPC 87

Drum, black (*Pogonias cromis*)
AGC cp380; MIA 559; NAF cp475

Drum, freshwater (*Aplodinotus grunniens*)
MIA 559; NAF cp96

Drum, red (*Sciaenops ocellatus*)
ACF p35; AGC cp385; NAF cp478

Duck, black (*Anas rubripes*)
AGC cp533; DNA 54; EAB 201; FEB 111; NAB cp133; NAW 118; PBC 85; WNW cp489

Duck, bufflehead (*Bucephala albeola*)
AGC cp542; DNA 138, 139; EAB 200; FEB 86; FWB 103; MPC 524; NAB cp127, cp151; UAB cp103, cp143; WNW cp508

Duck, canvasback (*Aythya valisineria*)
AAL 107; AGC cp530; DNA 81, 82; EAB 200; FEB 101; FWB 85; NAB cp110, cp154; NAW 109; NNW cp499; PBC 93; UAB cp95, cp137

Duck, eider See Eider

Duck, ferruginous (*Aythya nyroca*)
DNA 95, 97

Duck, gadwall (*Anas strepera*)
AGC cp533; DNA 38, 39; EAB 208; FEB 103; FWB 90; NAB cp135, cp139; PBC 85; RBB 28; UAB cp125, cp147; WNW cp496

Duck, goldeneye See Goldeneye

Duck, harlequin (*Histrionicus histrionicus*)
AGC cp538; DNA 121, 123; EAB 201; FEB 97; FWB 96; MPC cp525; NAB cp120, cp160; UAB cp115; WNW cp504

Duck, mallard (*Anas platyrhynchos*)
AGC cp544; BWB 70-71; DNA 48, 49, 51; EAB 209; FEB
93; FWB 98; MIA 215; NAB cp107, cp137; NAW 171; PBC
83, 84; RBB 30; UAB cp97, cp146; WGB 25; WNW cp492;
WON 49, 50

Duck, mandarin (*Aix galericulata*)
AWR 119-121; BWB 74; DNA 27; EOB 94; MIA 215;
WGB 25

Duck, masked (*Oxyura dominica*)
DNA 166, 167; FEB 109

Duck, Mexican (*Anas diazi*)
EAB 202; UAB cp123

Duck, mottled (*Anas fulvigula*)
AGC cp532; FEB 110

Duck, muscovy (*Cairina moschata*)
MIA 215; WGB 25

Duck, oldsquaw (*Clangula hyemalis*)
AGC cp552; DNA 125-127; EAB 210, 211; FEB 87; FWB
106; MPC cp526; NAB cp116, cp152; NAW 177; UAB
cp98, 99, 139

Duck, pintail (*Anas acuta*)
AAL 107; AGC cp529; DNA 58-60; EAB 212, 213; EOB
96; FEB 105; FWB 88; MPC cp527; NAB cp115, cp138;
PBC 86; RBB 31; UAB cp124, cp148; WNW cp498

Duck, redhead (*Aythya americana*)
AGC cp531; BWB 73; DNA 87, 88; EAB 213; FEB 100;
FWB 86; NAB cp109, cp153; PBC 91; UAB cp94, cp136;
WNW cp500

Duck, ring-necked (*Aythya collaris*)
AGC cp546; DNA 89, 91; EAB 202; FEB 82; FWB 102;
NAB cp121, cp156; PBC 92; UAB cp90, cp134; WNW
cp501

Duck, ruddy (*Oxyura jamaicensis*)
AGC cp539; DNA 168-171; EAB 203; EOB 95; FEB 99;
FWB 95; MIA 215; NAB cp111, cp159; RBB 39; UAB
cp116, cp127; WGB 25; WNW cp510

Duck, spotbill (*Anas peocilorhyncha*)
DNA 56, 57

Duck, torrent (*Merganetta armata*)
AWR 163

Duck, tufted (*Aythya fuligula*)
DNA 98, 99; EAB 202; EOB 97; FEB 83; MAE 97; NAB
cp122, cp155; RBB 34; UAB cp91

Duck, white-headed (*Oxyura leucocephala*)
DNA 172, 173

Duck, wood (*Aix sponsa*)
AAL 109; BWB 74; DNA 25; EAB 204, 205; FEB 96; FWB
97; NAB cp119, cp164; NAW 120; PBC 90; SEF cp262;
SWF 145; UAB cp114, cp135; WNW cp491; WOB 120,
121

Duckbill platypus See Platypus

Duckweed (*Lemna* sp.)
PGT 66; WNW cp313

Duckweed, rootless (*Wolfia arrhiza*)
PGT 66

Dugong (*Dugong dugon*) See also Manatee
CRM 167; EAL 341, 346; GEM IV:530-532; MEM 293,
298; MIA 111; NOA 89

Duiker (*Cephalophus* sp.)
CRM 189; GEM V:326-328; MEM 556, 557; MIA 145;
NHA cp9; SWF 121

Duiker, common or gray (*Sylvicapra grimmia*)
GEM V:325; MEM 556, 559; MIA 145

Dung roller or tumblebug (*Canthon* sp.)
BNA 140(bw); LBG cp379; SGI cp98

Dunlin or Red-backed sandpiper (*Calidris alpina*)
AGC cp585; EAB 793; FEB 167; FWB 162; MPC cp578;
NAB cp192, cp209; PBC 161; RBB 72; UAB cp188, cp231;
WBW 271

Dunnart (*Sminthopsis* sp.)
CRM 19; GEM I:267, 270; MEM 839, 842; MIA 17; NOA
170; WWD 178-179

Dunnock (*Prunella modalaris*)
EOB 363; MIA 331; RBB 128; WGB 141

Durgon, black (*Melichthys niger*)
NAF cp325; PCF cp46

Dusty miller (*Artemisia stelleriana*)
AGC cp458

Dutchman's breeches (*Dicentra cucullaria*)
BCF 171; CWN 113; NWE cp114; SEF cp450

Dutchman's pipe (*Aristolochia durion*)
NWE cp387; SEF cp457

E

Eagle, African fishing (*Haliaeetus vocifer*)
AWR 100-101; BWB 236, 241, 242

Eagle, bald (*Haliaectus leucocephalus*)
AGC cp606; BOP 100, 101, 106, 108; BWB 244-247; EAB
503, 504; EOB 110, 123; FEB 216; FGH cp18-cp20; FWB
209; MIA 221; MPC cp553; NAB cp305, cp307; NAW
121(head only), 122; PBC 113(head only); UAB cp333,
cp335; WGB 31; WNW cp562; WOB 8-9(head only), 28-
29, 206, 207; WON 60, 61

Eagle, bateleur (*Terathopius ecaudatus*)
BOP 103; MIA 221; WGB 31

Eagle, black or Verreaux's (*Aquila verreauxii*)
BOP 102

Eagle, black-breasted snake (*Circaetus gallicus pectoralis*)
EOB 111

Eagle, booted (*Hieraaetus pennatus*)
EOB 117

Eagle, crested serpent (*Spilornis cheela*)
MIA 221; WGB 31

Eagle, golden (*Aquila chrysaetos*)
BOP 11, 104, 105; EAB 505; FEB 217; FGH cp18-cp20; FWB 208; JAD cp540; LBG cp494; MIA 223; NAB cp308; NHU 152; PBC 112; RBB 45; UAB cp332; WGB 33; WOM 74; WWD 123

Eagle, harpy (*Harpia harpyja*)
MIA 223; WGB 33

Eagle, martial (*Poleamaetus bellicosus*)
BOP 97

Eagle, monkey-eating (*Pithecophaga* sp.)
WOI 112(head only); WWW 32(head only)

Eagle, sea or white-tailed (*Haliaeetus* sp.)
EOB 110, 118, 119; FGH p25

Eagle, short-toed (*Circaetus gallicus*)
MAE 26

Eagle, Spanish imperial (*Aquila heliaca adalberti*)
MAE 26

Eagle, steppe or tawny (*Aquila rapax*)
NHU 48

Earth-ball fungus (*Scheroderma citrinum*)
FGM p48; NAM cp654; PPC 142; SGM cp376

Earthcreeper, scale-throated (*Upucerthia dumentaria*)
EOB 312

Earth-scale fungus (*Agrocybe* sp.)
FGM cp38; NAM cp225

Earthstar fungus (*Geastrum* sp.)
FGM p47; GSM 249; NAM cp634-639; SGM cp361-363

Earthtongue fungus (*Microglossum* and other genera)
FGM cp1; NAM cp682, cp685, cp686; SGM cp343; WNW cp338

Earthworm (*Lumbricus terrestris*)
AAL 488; EAL 206; NNW 137

Earwig (*Forficula auricularia* and other genera)
FOI 10; LDQ cp377; MIS cp88-90; SGI cp33; WIW 98(bw)

Earworm, corn (*Heliothis zea*)
SGI cp263

Echidna, long-beaked (*Zaglossus bruijni*)
AAL 26(bw); CRM 17; GEM I:195, 206; MEM 819

Echidna, short-beaked (*Tachyglossus aculeatus*)
AAB 18; GEM I:41, 195, 202; MEM 818, 820; NOA 23-25

Edelweiss (*Leontopodium alpinum*)
WOM 31

Eel, American (*Anguilla rostrata*)
ACF cp10; NAF cp39, cp445; WNW cp84

Eel, conger (*Conger* sp.)
ACF p9; EAL 30; MIA 505

Eel, cusk See Cusk-eel

Eel, electric (*Electrophorus electricus*)
AWR 207(bw); MIA 513

Eel, European (*Anguilla anguilla*)
AAL 328; AWR 49; EAL 26; MIA 505; NNW 78; SAB 53(elver)

Eel, gulper (*Eurypharynx pelecanoides*)
EAL 37; MIA 505

Eel, moray (*Gymnothorax* sp.)
AGC cp431; BMW 129; EAL 32; KCR cp25; MPC cp304; NAF cp453; WWF cp38

Eel, moray (other genera)
AAL 328; ACF cp10; EAL 29; MIA 505; MPC cp304; NAF cp451, cp452; PCF p6; RUP 54, 106

Eel, snake (*Ophichthus* sp.)
ACF p11; PCF p6

Eel, spiny (*Mastacembelus congicus*)
AWR 209(bw); MIA 573

Eel, wolf (*Anarrhichthys ocellatus*)
MPC cp303; NAF cp455; PCF cp40

Eel (other kinds)
ACF p63; EAL 30, 33, 36, 37; NAF cp446; WWW 198

Eelgrass (*Zostera marina*)
MPC cp508

Eelpout (*Lycodes* sp.)
ACF p45; PCF p11

Egret, cattle (*Ardeola* or *Bubulcus ibis*)
AAL 103; AWR 35; BWB 42; EAB 520, 521; FEB 136; FWB 132; HHH 29, 143; LBG cp487; MAE 93; MIA 209; NAB cp4, cp8; PBC 64; SAB 14; UAB cp5, cp6; WBW 39; WGB 19

Egret, great or American white (*Egretta alba*)
BMW 52, 58-59; EAB 520; HHH 27, 87; MIA 209; NAW 124; WBW 41; WGB 19; WOB 33

Egret, great or common (*Casmerodius albus*)
AGC cp562; BWB 44, 45; FEB 135; FWB 134; NAB cp2, cp6; PBC 66; UAB cp3, cp4; WNW cp526

Egret, intermediate (*Egretta intermedia*)
HHH 29, 111; WBW 47

Egret, little (*Egretta garzetta*)
HHH 31, 127, 133; WBW 45

Egret, reddish (*Dichroanassa* or *Egretta rufescens*)
AGC cp566; BWB 43; EAB 521; FEB 138; HHH 33, 91; NAB cp3, cp15; PBC 65

Egret, reef See Heron, reef

Egret, snowy (*Egretta thula*)
AGC cp563; BAS 104; BWB 44; EAB 521; FEB 134; FWB 133; HHH 33, 123; NAB cp1, cp5; PBC 67; UAB cp1, cp2; WBW 50(bw); WNW cp527; WOB 134

Egret, Swinhoe's or Chinese (*Egretta eulophotes*)
HHH 33, 135

Eider, common (*Somateria mollissima*)
AGC cp547; BAA 124, 125, 133(bw); DNA 108, 109, 111; EAB 207; EOB 97; FEB 81; FWB 111; MIA 215; NAB cp117, cp145; NAW 126; RBB 35; UAB cp109, cp145; WGB 25

Eider, king (*Somateria spectabilis*)
AGC cp548; DNA 112, 113; EAB 206; FEB 80; NAB cp118, cp146; NHU 79; UAB cp106, cp144

Eider, spectacled (*Somateria fischeri*)
DNA 115, 116; EAB 206; UAB cp107, cp142

Eider, Steller's (*Polysticta stelleri*)
DNA 118, 119; EAB 207; UAB cp108

Eland, Common or cape (*Taurotragus oryx*)
MEM 549; MIA 141; NHA cp2

Eland, giant or Lord Derby's (*Taurotragus derbiamus*)
CRM 183; MEM 551; NHA cp1

Elder or elderberry (*Sambucus canadensis* or *nigra*)
FWF 73; NTE cp336, cp413, cp539; NWE cp217; PPC 99; TSV 172, 173; WNW cp242, cp438

Elder, blue (*Sambucus cerulea*)
NTW cp281, cp348, cp460; WFW cp162

Elder, Mexican (*Sambucus mexicana*)
NTW cp280, cp347

Elder, red (*Sambucus callicarpa*)
NTW cp282, cp346, cp487; WFW cp157, cp191

Elecampane (*Inula helenium*)
CWN 259; LBG cp166; NWE cp296

Elephant, African (*Loxodonta africana*)
AAB 38; AAL 68-69; CRM 163; GEM I:22, IV:459, 463, 475-479, 504-520; MEM 453-61; MIA 123; SWF 132-133; WPP 72-73, 75

Elephant, Asian (*Elephas maximus*)
CRM 163; GEM IV:463, 468, 496-501; MEM 460; MIA 123

Elephant-ear fungus (*Gyromitra fastigiata*)
GSM 29

Elephant head (*Pedicularis groenlandica*)
SWW cp534; WFW cp492

Elephant-snout (*Campylomormyrus* and other genera)
AWR 208(bw); EAL 42; MIA 501

Elephant-tree (*Bursera microphylla*)
JAD cp312, cp315; NTW cp257, cp462

Elf cup fungus, scarlet (*Sarcoscypha coccinea*)
FGM cp1; NAM cp605; NMT cp7; SGM cp415

Elk, American or Wapiti (*Cervus canadensis*) See also Deer, red
GEM V:190-193; MEM 527; MIA 137; NAW 35, 36; SWF 160

Elliottia or Southern-plume (*Elliottia racemosa*)
NTE cp65, cp443

Elm, American (*Ulmus americana*)
NTE cp162, cp373, cp498; NTW cp152, cp321, cp531; SEF cp86; TSV 196, 197; WNW cp465

Elm, Chinese (*Ulmus parvifolia*)
NTE cp168; NTW cp156; PMT cp[141]

Elm, English (*Ulmus procera*)
NTE cp163; NTW cp154

Elm, Siberian (*Ulmus pumila*)
LBG cp416; NTE cp166; NTW cp151

Elm, slippery (*Ulmus rubra*)
NTE cp160, cp500; SEF cp94, cp177

Elm, water (*Planera aquatica*)
NTE cp167; WNW cp464

Elm, winged (*Ulmus alata*)
NTE cp165, cp497

Elm (other *Ulmus* sp.)
NTE cp161, cp164, cp205, cp499; PMT cp[140]

Emu (*Dromaius novaehollandiae*)
EOB 19, 24; MIA 197; NOA 201; WOI 132; WGB 7; WPP 127(head only); WWW 78-79

Epaulette tree (*Pterostyrax hispida*)
PMT cp[111]

Ergot (*Claviceps purpurea*)
SGM cp346

Ermine (*Mustela erminea*)
AAL 62(bw); ASM cp208, cp210; CGW 153; CMW
　219(bw); GEM III:388, 395-397; GMP 273(bw); LBG cp52;
　MEM 110, 111; MIA 87; MNC 255(bw); MPS 274(bw);
　NAW 37; NNW 56; RFA 190, 198(bw); RMM cp[10]; SEF
　cp602, cp604; WFW cp344; WPR 80, 81

Escolar (*Lepidocybium flavobrunneum*)
ACF cp48; PCF p47

Eucalyptus (*Eucalyptus* sp.)
NOA 197; NTW cp86, cp100, cp352, cp450, cp449; OET
　215, 217; SWF 101; TSP 29

Euglena (*Euglena* sp.)
PGT 87

Evening primrose (*Oenothera* sp.)
BCF 47; CWN 145; LBG cp190, cp242; WTV 72

Evening primrose, birdcage (*Oenothera deltoides*)
JAD cp93; SWW cp61; WAA 212, 213

Evening primrose, common (*Oenothera biennis*)
CWN 145; LBG cp205; NWE cp354; WAA 38

Evening primrose, Hooker's (*Oenothera hookeri*)
SWW cp219; WFW cp448

Evening primrose, pink (*Oenothera speciosa*)
WAA 22, 25

Evening primrose, showy (*Oenothera speciosa*)
LBG cp235; NWE cp468

Everlasting, pearly (*Anaphalis margaritacea*)
CWN 281; NWE cp196; SWW cp145; WFW cp422

Everlasting, sweet or catfoot (*Gnaphalium obtusifolium*)
CWN 281; NWE cp197

Eyebright (*Euphrasia americana*)
NWE cp502

F

Fairy duster (*Calliandra eriophylla*)
JAD cp48; SWW cp567

Fairy slipper See Orchid, calypso

Fairybell, wartberry (*Disporum trachycarpum*)
SWW cp81; WFW cp402

Falanouc (*Eupleres goudoti*)
CRM 147; GEM III:534; MIA 95

Falcon, African pygmy (*Polihierax semitorquatus*)
EOB 107

Falcon, aplomado (*Falco femoralis*)
BOP 46; EAB 297; EOB 107; FGH cp21; JAD cp543; UAB
　cp322

Falcon, barred forest (*Micrastur ruficollis*)
EOB 106; MIA 225; WGB 35

Falcon, brown (*Falco berigora*)
MIA 225; WGB 35

Falcon, Eleonora's (*Falco eleonorae*)
WOI 174-175

Falcon, laughing (*Herpetotheres cachinnans*)
EOB 106

Falcon, little (*Falco longipennis*)
BOP 47

Falcon, peregrine (*Falco peregrinus*)
AAL 114; AGC cp603; BAA 121(bw); BOP 48, 56, 57; EAB
　297; EOB 123; FEB 231; FGH cp23; FWB 230; MIA 225;
　NAB cp315; NHU 55; PBC 116; RBB 50; UAB cp321,
　cp323; WGB 35; WOB 261-263

Falcon, prairie (*Falco mexicanus*)
BOP 51, 61, 62; EAB 297; FEB 233; FGH cp22; FWB 231;
　UAB cp320

Falcon, red-footed (*Falco vespertinus*)
SAB 55

Falconet, collared or red-legged (*Microhierax
　caerulescens*)
MIA 225; WGB 35

Fantail (*Rhipidura* sp.)
AAL 176; EOB 378

Fanwort (*Cabomba caroliniana*)
CWN 59

Farewell-to-Spring (*Clarkia amoena*)
LBG cp240; SWW cp466; WFW cp511

Farkleberry See Sparkleberry

Feather bells (*Stenanthium gramineum*)
NWE cp142

Feather plume (*Dalea formosa*)
LBG cp231; SWW cp491

Featherduster See under Worm

Featherfoil (*Hottonia inflata* or *palustris*)
CWN 179; NWE cp110; PGT 58-59; WNW cp314

Felwort (*Swertia perennis*)
SWW cp592; WFW cp525

Fennel (*Foeniculum vulgare*)
LBG cp207; SWW cp323; WTV 85

Fer-de-lance (*Bothrops atrox*)
LSW 383; MIA 459; MSW p98

Fer-de-lance, false (*Xenodon rabdocephalus*)
LSW 192

Fern, cinnamon (*Osmunda cinnamomea*)
WNW cp304

Fern, marsh (*Thelypteris palustris*)
WNW cp305

Fern, royal (*Osmunda regalis*)
FWF 125; WNW cp306

Fern, tree (various genera)
OET 56, 57

Fern, water (*Azolla filiculoides*)
PGT 61

Fern, water or water velvet (*Salvinia natans*)
PGT 62

Ferret, black-footed (*Mustella nigripes*)
ASM cp212; CMW 225(bw); CRM 139; GEM III:406; MEM
112; MIA 87; MNP 259(bw); MPS 275(bw)

Fever-tree See Pinckneya

Fiddleneck (*Amsinckia retrorsa*)
LBG cp160; SWW cp372

Fieldfare (*Turdus pilaris*)
RBB 138; SPN 16

Fig, strangler (*Ficus destruens*)
NOA 124

Filaree or Storksbill (*Erodium cicutarium*)
CWN 141; JAD cp57; NWE cp462; SWW cp452; WAA 233

File fish (*Aluterus* and *Monacanthus* sp.)
ACF cp59; AGC cp348; MIA 581; NAF cp316-cp318; RUP
133; WWW 22-23

File fish, orange-spotted or taillight (*Cantherhines
pullus*)
KCR cp24; RCK 40

File shell (*Lima* sp.)
MSC cp350; RCK 44; RUP 26; WWW 136

Finch, Cassin's (*Carpodacus cassinii*)
EAB 303; FWB 471; UAB cp462, cp584; WFW cp309

Finch, house (*Carpodacus mexicanus*)
AAL 190(bw); EAB 302; FEB 405; FWB 467; JAD cp617,
cp618; NAB cp410, cp559; PBC 368, 369; SPN 140; UAB
cp460, cp585

Finch, large ground (*Geospixa magnirostris*)
WGB 181

Finch, plush-capped (*Catamblyrhynchus diadema*)
EOB 407

Finch, purple (*Carpodacus purpureus*)
EAB 304; FEB 404; FWB 470; MIA 385; NAB cp409,
cp558; PBC 366; SEF cp313; SPN 144; UAB cp461,
cp583; WGB 193

Finch, rosy (*Leucosticte* sp.)
EAB 302; FWB 466; UAB cp459, cp465

Finch, snow (*Montifringilla nivalis*)
MIA 391; WGB 201

Finch, warbler (*Certhidea olivacea*)
WGB 181

Finch, woodpecker (*Cactospiza pallida*)
WOI 27

Finch, zebra (*Poephila guttata*)
WWD 182-183

Finfoot, American See Sungrebe

Finfoot, African (*Podica senegalensis*)
AAL 124(bw); MIA 239; WGB 49

Fir, balsam (*Abies balsamea*)
NTE cp22, cp480; SEF cp33, cp172

Fir, bristlecone (*Abies bracteata*)
NTW cp36, cp427; WFW cp135

Fir, Douglas (*Pseudotsuga menziesii*)
NTW cp41, cp426; WFW cp134

Fir, Douglas (bigcone) (*Pseudotsuga macrocarpa*)
NTW cp42, cp425; WFW cp132

Fir, Fraser (*Abies fraseri*)
NTE cp21; SEF cp32

Fir, grand (*Abies grandis*)
NTW cp39; WFW cp48

Fir, Korean (*Abies koreana*)
PMT cp[2]

Fir, noble (*Abies procera*)
NTW cp35, cp428; WFW cp133

Fir, Nordman or Caucasian (*Abies nordmanniana*)
PMT cp[3]

Fir, Pacific silver or Cascades (*Abies amabilis*)
NTW cp38; WFW cp47

Fir, red (*Abies magnifica*)
NTW cp34; OET 72; WFW cp43

Fir, subalpine or Rocky Mountain (*Abies lasiocarpa*)
NTW cp37; WFW cp46

Fir, white or Colorado (*Abies concolor*)
NTW cp40; PMT cp[1]; WFW cp45

Fire tree or bush (*Embothrium coccineum*)
TSP 27

Firebrat (*Thermobia domestica*)
MIS cp85

Firecracker flower (*Dichelostemma ida-maia*)
SWW cp412

Firefly (Lampyridae)
BNA cp3, 189-191(bw); MIS cp171-173

Firefly, eastern (*Photinus pyralis*)
SGI cp129

Firefly, western (*Ellychnia californica*)
SGI cp128

Firefly, woods (*Photuris pennsylvanica*)
LBG cp383; SEF cp392; SGI cp130

Fireweed (*Epilobium angustifolium*)
CWN 147; NWE cp541; SEF cp480; SWW cp536; WAA 159; WFW cp493

Fireweed, dwarf (*Epilobium latifolium*)
WAA 248

Fisher (*Martes pennanti*)
ASM cp213; CGW 145; CMW 217(bw); MAE 49; MNC 253(bw); MPS 274(bw); RFA 178-184(bw); SEF cp603; SWF 156; SWM 102; WMC 189

Fitweed, Case's (*Corydalis caseana*)
SWW cp118; WFW cp427

Five-spot, desert (*Malvastrum rotundifolium*)
JAD cp43; SWW cp465; WAA 210

Flag, blue (*Iris versicolor*)
BCF 151; CWN 21; NWE cp620; WNW cp297

Flag, sweet (*Acorus calamus*)
CWN 45; NWE cp310; WNW cp221

Flag, yellow (*Iris pseudacorus*)
AWR 43; CWN 21; NWE cp314; PGT 43; WNW cp219

Flagfish (*Jordanella floridae*)
NAF cp196; WNW cp100

Flame of the woods (*Ixora fulgens*)
TSP 63

Flame or fire vine (*Senecio confuscus*)
TSP 93

Flamefish (*Apogon maculatus*)
KCR cp26; NAF cp378

Flameflower (*Pyrostegia venusta*)
TSP 91

Flamingo, Andean (*Phoenicopterus andinus*)
BWB 39

Flamingo, greater (including American) (*Phoenicopterus ruber*)
AAB 98-99; AWR 97; BAS 88; BMW 62-63; EAB 326; EOB xvi-xvii, 84-87; FEB 142; MAE 93; MIA 211; NAB cp12; NAW 128, 129; WBW 119; WGB 21

Flamingo, lesser (*Phoenicopterus minor*)
AWR 96; EOB 89; SAB 131

Flamingo tongue (*Cyphoma gibbosum*)
AGC cp128; MSC cp449; SCS cp50

Flashlight fish (various genera)
ACF p62; BMW 188-189; NNW 126; RUP 166; WWW 119

Flatpod (*Idahoa scapigera*)
LBG cp115; SWW cp15

Flax (*Linium* sp.)
CWN 129; JAD cp149; LBG cp287; NWE cp600; SWW cp229, cp589; WFW cp520; WTV 86

Flea, beach (*Talorchestia* or *Orchestoidea* sp.)
MPC cp350; MSC cp586, cp587

Flea, cat (*Ctenocephalides felis*)
MIS cp76; SGI cp195

Flea, human (*Pulex irritans*)
MIS cp68

Flea, snow (*Achorutes nivicola*)
MIS cp77

Flea, water (*Daphnia* sp.)
AAB 135; AAL 399(bw); EAL 223; NNW 84-85; PGT 129

Fleabane, common (*Erigeron philadelphicus*)
WAA 180

Fleabane, daisy (*Erigeron strigosus* or *annuus*)
BCF 115; CWN 277; NWE cp93, cp484; WAA 180; WTV 24

Fleabane, marsh (*Pluchea purpurascens*)
NWE cp563; WTV 197

Fleabane, spreading (*Erigeron divergens*)
JAD cp89; LBG cp133; SWW cp69

Flicker, common, gilded, northern, or yellow-shafted (*Colaptes auratus*)
AAB 17; AAL 152; EAB 1036, 1037; EOB 296; FEB 308; FWB 326; JAD cp568; MIA 293; NAB cp348; PBC 214; SEF cp277; SWF 150; UAB cp370-373; WFW cp248; WGB 103; WOB 92, 93; WON 135, 136; WWD 161

Flier (*Centrarchus macropterus*)
NAF cp83

Floating hearts (*Nymphoides aquatica*) See also Water lily, fringed
NWE cp81; WNW cp254

Flounder (*Bothus* sp.)
ACF p55; AGC cp358; NAF cp283

Flounder (*Paralichthys* sp.)
ACF p55; NAF cp71, cp285

Flounder, blackback or winter (*Pseudopleuronectes americanus*)
AGC cp357; BCS 31, 32; MIA 579; NAF cp275

Flounder, peacock (*Bothus lunatus*)
AAL 373; ACF p55; BMW 133; MIA 577; RUP 116; WWF cp19

Flounder, spotfin (*Cyclopsetta fimbriata*)
NAF cp286

Flounder, starry (*Paralichthys stellatus*)
MIA 577; MPC cp307; NAF cp72, cp278, cp279; PCF p43

Flounder, three-eye (*Ancylopsetta dilecta*)
NAF cp287

Flower-of-an-hour (*Hibiscus trionum*)
CWN 87; SWW cp230; WTV 87

Flower-piercer (*Diglossa cyanea*)
BAS 108

Fly agaric (*Amanita muscaria*)
FGM cp26; GSM 155; NAM cp143, cp144, cp680; NMT cp2; PPC 110; PPM cp29; SGM cp2, cp3; SWF 61

Fly, alder (*Sialis* sp.)
EOI 63 (larva); MIS cp331; PGT 159

Fly, bee (*Bombylius* sp.)
AAL 464; LBG cp370, cp371; SGI cp328

Fly, bee (other kinds)
AGC cp485; MIS cp495, cp496, cp512, cp514-516; SGI cp327

Fly, black (*Simulium* or *Bibio* sp.)
MIS cp416; SEF cp406; SGI cp315; WFW cp575; WNW cp379

Fly, blow (*Lucilia illustris*)
AAL 469; SGI cp342

Fly, bluebottle (*Calliphora vomitoria*)
MAE 135; MIS cp431

Fly, crane (*Tipula* sp.)
BOI 206-207; DIE cp28; EOI 81; PGT 175; MIS cp406, cp407; SGI cp309; WNW cp373

Fly, deer (*Chrysops* sp.)
AGC cp483; MIS cp424; SEF cp407; SGI cp321; WFW cp576

Fly, drone or hover (*Eristalis* or *Eristalomyia tenax*)
AAL 464; BOI 210-211; EOI 82; MIS cp508; PGT 191; SGI cp331

Fly, dung (common yellow) (*Scatophaga stercoraria*)
EOI 86; MIS cp440; SAB 85; SGI cp340

Fly, empid or dance (*Empis livida*)
WIW 111

Fly, flesh (*Sarcophaga* sp.)
MIS cp425; SGI cp343

Fly, fruit (*Drosophila melanogaster*)
SGI cp338; WNW cp381

Fly, greenbottle (*Phaenicia sericata*)
MIS cp431

Fly, horse (*Tabanus* sp.)
AGC cp484; MIS cp427-429; SGI cp319; WNW cp378

Fly, house (*Musca domestica*)
AAL 467(bw); MIS cp421; SGI cp341

Fly, hover or flower (*Metasyrphus americanus* or *Syrphus* sp.)
LBG cp365; MIS cp487; SGI cp332; WIW 39, 143

Fly, mantid or false mantid (*Mantispa* or *Climaciella* sp.)
MIS cp303-306

Fly, marsh (*Tetanocera* and other genera)
SGI cp336; WNW cp380

Fly, may (various kinds)
MIS cp36(naiad), cp37(naiad), cp38(naiad), cp391-396; SGI cp5; WNW cp375

Fly, may (Eurasian) (*Ephemera danica*)
AAB 139; EOI 24; PGT 149; WIW 134

Fly, robber (various genera) See also Bee killer
AAL 464; BOI 209; EOI 85; LBG cp361; MIS cp398-400; SGI cp326

Fly, scorpion (*Panorpa* sp.)
BOI 193; MIS cp409; SGI cp194; WIW 69(bw)

Fly, screw-worm (*Cochliomyia hominivorax*)
MIS cp432

Fly, shore (*Ephydra cinerea*)
SGI cp337

Fly, stone (various kinds)
AAL 415(bw); EOI 41; MIS cp40(naiad), cp41(naiad), cp339, cp340; SGI cp13

Fly, tachinid (various genera)
MIS cp422, cp423, cp437-439, cp485, cp500; SGI cp347-349

Fly, white (greenhouse) (*Trialeurodes vaporariorum*)
MIS cp 58; SGI cp74

Flycatcher, acadian (*Empidonax virescens*)
EAB 359; NAB cp461; SEF cp327; UAB cp519

Flycatcher, alder (*Empidonax alnorum*)
EAB 361; FWB 390

Flycatcher, ash-throated (*Myiarchus cinerascens*)
EAB 360; FEB 286; FWB 403; JAD cp573; NAB cp472;
  UAB cp548; WGW cp253

Flycatcher, boat-billed (*Megarhynchus pitangua*)
EOB 318

Flycatcher, brown-crested (*Myiarchus tyrannulus*)
FEB 287; FWB 404; JAD cp574

Flycatcher, buff-breasted (*Empidonax fulvifrons*)
FWB 397

Flycatcher, Coue's (*Contopus pertinax*)
EAB 361

Flycatcher, dusky (*Empidonax oberholseri*)
EAB 361; EOB 319; FWB 405; UAB cp546

Flycatcher, gray (*Empidonax wrightii*)
EAB 363; FWB 393; JAD cp569; UAB cp490

Flycatcher, great crested (*Myiarchus crinitus*)
EAB 362; FEB 289; NAB cp469; PBC 227; SPN 56

Flycatcher, Hammond's (*Empidonax hammondii*)
EAB 363; UAB cp517

Flycatcher, kiskadee See Kiskadee, great

Flycatcher, least (*Empidonax minimus*)
EAB 362; FEB 283; NAB cp464

Flycatcher, olive-sided (*Contopus borealis*)
EAB 364; FEB 280; FWB 394; NAB cp470; SEF cp282;
  SPN 57; UAB cp547; WFW cp250

Flycatcher, paradise (*Terpsiphone* sp.)
EOB 379; NHU 129

Flycatcher, pied (*Ficedula hypoleuca*)
EOB 368; RBB 155

Flycatcher, royal (*Onychorhynchus cornatus*)
EOB 321; MIA 307; WGB 117

Flycatcher, scissor-tailed (*Muscivora forficata*)
EAB 365; EOB 321; FEB 294; FWB 406; LBG cp519; MIA
  305; NAB cp418; PBC 226; UAB cp469; WGB 115

Flycatcher, short-tailed pygmy (*Myironis ecaudata*)
EOB 320

Flycatcher, spotted (*Muscicapa striata*)
MIA 351; RBB 154; WGB 161

Flycatcher, sulphur-bellied (*Myiodynastes luteiventris*)
FWB 411

Flycatcher, vermilion (*Pyrocephalus rubinus*)
EAB 365; EOB 321; FEB 297; FWB 407; JAD cp572; MIA
  309; NAB cp415, cp512; SPN 58; UAB cp451, cp551;
  WGB 119

Flycatcher, western (*Empidonax difficilis*)
EAB 367; FWB 391; UAB cp549; WGW cp252

Flycatcher, Wied's crested (*Myiarchus tyrannulus*)
EAB 367; UAB cp520

Flycatcher, willow or traill's (*Empidonax trailli*)
EAB 367; FEB 285; NAB cp462; PBC 231; SPN 59; UAB
  cp518

Flycatcher, yellow-bellied (*Empidonax flaviventris*)
EAB 366; FEB 284; NAB cp463

Flying fish (*Cypselurus* sp.)
AAB 108; ACF p18; MIA 532; PCF p12

Flying fox, Indian (*Pteropus giganteus*)
ANB ii; CRM 45; GEM I:547(head only); LOB 46(bw),
  47(bw)

Flying fox (other kinds)
ANB 6; NOA 215; SWF 92

Flying squirrel, giant (*Petaurista petaurista*)
CRM 99; GEM III:101(head only), 102; MIA 161

Flying squirrel, Northern (*Glaucomys sabrinus*)
ASM cp184; CGW 169; CMW 118(bw); CRM 97; GEM
  III:98; GMP 161(bw), cp9; MIA 161; MNC 173(bw); MPS
  138(bw); SEF cp591; SWM 39; WFW cp336; WMC 132

Flying squirrel, Southern (*Glaucomys volans*)
ASM cp181, cp183; CGW 169; GEM III:98; GMP 157(bw),
  cp8; MEM 614, 616; MNC 175(bw); MPS 175(bw); NAW
  89; SEF cp589, cp590; WMC 129; WOM 159

Flying squirrel, Zenker's (*Idiurus zenkeri*)
CRM 103; MIA 165

Foamflower (*Tiarella cordifolia*)
CWN 127; NWE cp123; SEF cp430; WAA 161

Forget-me-not (*Myosotis scorpoides*)
CWN 197; NWE cp642; WNW cp301

Fossa (*Cryptoprocta ferox*)
CRM 145; GEM III:376(head only), 538; MIA 97

Four o'clock, desert or Colorado (*Mirabilis multiflora*)
JAD cp52; SWW cp439; WFW cp514

Four o'clock, trailing (*Allionia incarnata*)
JAD cp59; SWW cp458

Fowl, jungle See Junglefowl

Fowl, mallee See Mallee fowl

Fox, arctic (*Alopex lagopus*)
ASM cp251, cp252, cp265; BAA 14, 65, 87(bw), 111; GEM IV;53, 117; MAE 80; MEM 72, 73; MIA 79; NAW 37; NNW 29; RFA 74-78(bw); SWM 81, 82; WOI 49

Fox, bat-eared (*Otocyon megalotis*)
GEM IV:60, 157, 158; MEM 72-73; WWD 55

Fox, Blandford's (*Vulpes cana*)
CRM 131

Fox, Cape or silver-backed (*Vulpes chama*)
GEM IV:130

Fox, Corsac (*Vulpes corsac*)
CRM 131

Fox, crab-eating or savannah (*Dusicyon thous*)
GEM IV:147; MEM 75; MIA 81

Fox, fennec (*Vulpes zerba* or *Fennecus zerba*)
CRM 131; GEM I:129, IV:52, 132; MEM 70; MIA 79; WWW 40

Fox, gray (*Urocyon cinereoargenteus*)
ASM cp250, cp257; CGW 157; CMW 202(bw); GEM IV:146; GMP cp16; JAD cp522; LBG cp43; MEM 71; MNC 245(bw); MPS 252(bw); NAW 38; NMM cp[23]; RFA 94-99 (bw); SEF cp611; WFW cp366, cp368; WMC 184

Fox, hoary (*Dusicyon vetulus*)
GEM IV:153

Fox, kit or swift (*Vulpes velox*)
ASM cp258-260; CMW 199(bw); CRM 131; GEM IV:129, 130; JAD cp521; LBG cp39, cp40; MEM 25, 71; MNC 241(bw); MPS 251(bw); NMM cp[22]; WWD 117(head only)

Fox, pale (*Vulpes pallida*)
GEM IV:130

Fox, pampas (*Dusicyon gymnocercus*)
GEM IV:152

Fox, red (*Vulpes vulpes*)
AAL 57; ASM cp249, cp253-255; CGW 157; CMW 197(bw); GEM IV:118-126, 129; GMP cp15; LBG cp41, cp42, cp44, cp45; MAE 53, 127; MEM xii-xiii, 69; MIA 79; MNC 243(bw); MNP cp56; MPS 251(bw); NAW 39; NHU 106; NMM cp[21]; NNW 66; RFA 81-93(bw); SEF cp612; SWF 68; SWM 85-87; WFW cp364, cp365, cp367; WMC 182

Franciscana See River dolphin, La Plata

Francolin, black (*Francolinus francolinus*)
FEB 204

Francolin, red-necked (*Francolinus afer*)
MIA 229; WGB 39

Frangipani (*Plumeria* sp.)
OET 269; TSP 37

Franklin tree (*Franklinia alatamaha*)
NTE cp214, cp437; PMT cp[51]

Fremontia, California (*Fremontodendron californicum*)
NTW cp237, cp324; SWW cp358; WFW cp202

Friarbird, noisy (*Philemon corniculatus*)
EOB 400-401

Frigatebird, great (*Fregata minor*)
AAL 88; BWB 166-167; EOB 53, 64; HSW 97; OBL 10, 78, 167

Frigatebird, magnificent (*Fregata magnificens*)
AAL 100; AGC cp494; EAB 372-374; FEB 52; HSW 96; MIA 207; NAB cp85, cp86; OBL 42, 158, 162; UAB cp57; WGB 17; WOB 27; WOI 95-99

Fringe cups (*Tellina grandiflora*)
SWW cp106; WFW cp410

Fringe tree (*Chionanthus virginicus* or *retusus*)
NTE cp84, cp419; PMT cp[32], cp[33]; TSV 54, 55

Fringehead (*Neoclinus* sp.)
NAF cp465; PCF cp38

Fritillary, great spangled (*Speyeria cybele*)
MIS cp611; PAB cp609; WGW cp562, cp563

Fritillary, gulf (*Agraulis* or *Dione vanillae*)
AAB 48; AAL 460; MIS cp3(larva), cp614; PAB cp19(larva), cp593

Fritillary, regal (*Speyeria idalia*)
LBG cp336; WPP 168

Fritillary, scarlet (*Fritillaria recurva*)
SWW cp395; WFW cp482

Fritillary (various genera)
LBG cp342, cp344; PAB cp604-626, cp633-642; WNW cp396

Frog, African clawed (*Xenopus laevis*)
AAL 279(bw), 281; ARA cp222; FTW cp78, p73; MIA 473; WRA fig.14

Frog, barking (*Hylactophryne augusti*)
ARA cp154; WRA p12

Frog, bull See Bullfrog

Frog, carpenter (*Rana virgatipes*)
AAL 288-289; ARA cp205; ARC 137; SEF cp536; WNW cp139

Frog, cascades (*Rana cascadae*)
ARA cp199; WFW cp613; WRA p13

Frog, Casque-headed (*Hemiphractus proboscideus*)
ERA 44

Frog, chirping (*Syrrhophus* sp.)
ARA cp167-cp168; ART p[5]

Frog, chorus (Brimley's) (*Pseudacris brimleyi*)
ARA cp175; ARC 124; WNW cp148

Frog, chorus (Mountain) (*Pseudacris brachyphona*)
ARA cp163; ARC 123

Frog, chorus (ornate) (*Pseudacris ornata*)
AAL 308-309; ARA cp176, cp177; ARC 126

Frog, chorus (Southern) (*Pseudacris nigrita*)
ARA cp183; ARC 125; SEF cp541

Frog, chorus (striped or Western) (*Pseudacris triseriata*)
ARA cp179; ARC 127; CGW 25; LBG cp571; WNW cp132; WRA cp16

Frog, common or European (*Rana temporaria*)
AAL 291(bw); AWR 52; FTW p2; MAE 97; MIA 483; PGT 209

Frog, crawfish (*Rana areolata*)
ARA cp194, cp195, cp198; ARC 128

Frog, cricket (Northern) (*Acris crepitans*)
AAL 309; ARA cp153; ARC 111; CGW 25; MIA 481; WNW cp133; WRA cp16

Frog, cricket (Southern) (*Acris gryllus*)
ARA cp161, cp162; ARC 112; WNW cp141, cp142

Frog, Darwin's (*Rhinoderma darwini*)
ERA 14; FTW fig.19; MIA 477

Frog, edible (*Rana esculenta*)
AAL 287(bw); AWR 52; ERA 50; FTW p71; PGT 210

Frog, flying (*Agalychnis spurrelli*)
AAB 109

Frog, foothill yellow-legged (*Rana boylii*)
ARA cp209; WFW cp614; WRA cp14

Frog, goliath (*Rana goliath*)
AAL 289

Frog, grass (little) (*Limnaoedus ocularis*)
ARA cp172; ARC 122

Frog, green or bronze (*Rana clamitans*)
AAL 288-289; ARA cp189, cp213; ARC 130; ARN 36 (bw); CGW 29; WNW cp134

Frog, holy cross (*Notaden bennetti*)
WWD 186; WWW 149

Frog, horned (*Ceratophyrys* or *Megophrys* sp.)
ERA 37, 47; FTW cp22, cp23, cp31, p84; SAB 33

Frog, leaf or leaf-folding (*Phyllomedusa* or *Pachymedusa* sp.)
ERA 3, 47; FTW cp24; SWF 168; WWW 141

Frog, leopard See Leopard frog

Frog, marsh (*Rana ridibunda*)
MIA 483

Frog, marsupial (*Gastrotheca* or *Assa* sp.)
AAL 316; FTW cp57; MIA 481; NOA 129; SAB 88

Frog, mink (*Rana septentrionalis*)
ARA cp200; ARN 37 (bw); CGW 29

Frog, mountain yellow-legged (*Rana muscosa*)
ARA cp207; WFW cp610; WRA p13, cp14

Frog, narrowmouthed (*Gastrophryne* sp.)
AAL 298(bw); ARA cp219, cp221; ARC 138; FTW cp100; JAD cp282; LBG cp565; MIA 473; WRA cp16

Frog, pickerel (*Rana palustris*)
AAL 288; ARA cp201; ARC 133; ARN 40(bw); CGW 29; NAW 285; SEF cp543; WNW cp144

Frog, pig (*Rana grylio*)
ARA cp188; ARC 131; ART p[8]

Frog, poison arrow or poison dart (*Dendrobates* sp.)
AAL 264, 292; ERA 15, 49; FTW cp27, cp93; MIA 481; SAB 107

Frog, poison arrow (Turquoise) (*Dendrobates auratus*)
ERA 13; FTW cp92; WWW 145

Frog, red-legged (*Rana aurora*)
ARA cp208, cp215; WFW cp611; WRA p13, cp14

Frog, reed or rush (*Hyperolius* sp.)
AAL 293; AWR 81; ERA 59; FTW cp98

Frog, river (*Rana heckscheris*)
ARA cp210; ARC 132; WNW cp135

Frog, Seychelles (*Sooglossus seychellensis*)
ERA 51; MIA 485

Frog, sheep (*Hypopachus variolosus*)
ARA cp218; MIA 473

Frog, spotted (*Rana pretiosa*)
ARA cp206; ERA 42; WFW cp612; WRA p13

Frog, tailed (*Ascaphus truei*)
AAL 280; ARA cp165; ERA 42; FTW fig.12; MIA 475; WFW cp615; WRA cp16

Frog, Tarahumara (*Rana tarahumarae*)
ARA cp193; WRA p13

Frog, tree See Treefrog

Frog, wood (*Rana sylvatica*)
AAL 289; ARA cp211, cp212, cp214, cp216; ARC 136; ARN 38(bw); CGW 29; NAW 278-279; SEF cp542; WNW cp140, cp149; WRA cp14

Frogbit (*Hydrocharis morsus-ranae*)
PGT 64

Frogfish (*Antennarius* sp.)
ACF p14; PCF p13; "Long-lure or orange" (*A.multicellata*)
EAL 91; KCR cp26; NAF cp296, cp298; RCK 32

Frogmouth (*Podargus* sp.)
EOB 248, 253; MIA 273; SAB 32; WGB 83; WWW 157

Frostweed (*Helianthemum canadense*)
CWN 97; NWE cp258

Fuchsia, California (*Zauschneria californica*)
SWW cp405; WGW cp475

Fulmar, northern (*Furmarus glacialis*)
AAL 95(bw); AGC cp501; BAS 145; EAB 839; EOB 48;
FEB 56; FWB 55; HSW 81; MIA 201; MPC cp600; NAB
cp40; OBL 51, 58, 166; RBB 11; UAB cp25, cp73; WGB
11; WPR 150

Fumitory (*Fumaria officinalis*)
WTV 135

G

Gadwall See under Duck

Gaillardia See Indian blanket

Galago See Bush baby

Galah (*Cacatua roseicapilla*)
EOB 224-225; NOA 198

Galingale See Rush, sweet

Gallinule See also Reedhen

Gallinule, common or Florida or moorhen (*Gallinula chloropus*)
BWB 220; EAB 722; FEB 121; FWB 120; MAE 97; MIA
237; NAB cp247; PBC 133; PGT 223; RBB 59; UAB
cp246; WGB 47; WNW cp536

Gallinule, purple (*Porphyrula martinica*)
AAL 122; AGC cp600; BWB 220; EAB 723; EOB 144;FEB
120; NAB cp248; PBC 132; PGT 225; WNW cp537; WOB
149

Gallito, crested (*Rhinocrypta lanceolata*)
EOB 328

Galliwasp (*Diploglossus lessorae*)
MIA 439

Gannet, Cape (*Morus capensis*)
EOB 53, 58, 61; OBL 74

Gannet, northern (*Morus bassanus*)
AAL 98(bw); AGC cp517; BWB 94; EAB 51; FEB 55; HSW
99; MIA 205; NAB cp76; NAW 132; OBL 11, 34, 206-207;
PBC 54, 55; RBB 14; WGB 15; WOI 63; WON 23

Gar (*Atractosteus* or *Lepisosteus* sp.)
AAL 326(bw); AWR 218(bw); EAL 22; MIA 499; NAF cp25-
cp28; WNW cp78

Garfish (*Belone belone*)
MIA 532

Garganey (*Anas querquedula*)
DNA 62-64

Garibaldi (*Hypsypops rubicunda*)
AAL 320; BCS 72; MPC cp259, cp260; NAF cp319, cp320;
PCF cp30

Garland flower (*Daphne cneorum*)
PPC 4

Garlic, wild (*Allium* sp.)
CWN 9; NWE cp553; WTV 177

Garter snake, black-nosed or blackneck (*Thamnophis cyrtopsis*)
ARA cp536; JAD cp220; RNA 151; WRA cp42

Garter snake, Butler's (*Thamnophis butleri*)
ARA cp529; RNA 147

Garter snake, checkered (*Thamnophis marcianus*)
ARA cp515; JAD cp216; LBG cp594; LSW 171; RNA 151;
WRA cp42

Garter snake, eastern or common (*Thamnophis sirtalis*)
ARA cp530-545, cp576; ARC 236; ARN 61(bw), 62(bw);
CGW 93; LBG cp592; LSW 170; MIA 451; MSW p65;
NAW 241; RNA 149; SEF cp526, cp527; WFW cp594

Garter snake, Mexican (*Thamnophis eques*)
ARA cp528; RNA 147; WRA cp42

Garter snake, narrow headed (*Thamnophis rufipunctatus*)
ARA cp548; RNA 153; WRA cp43

Garter snake, Northwestern (*Thamnophis ordinoides*)
AAL 243(bw); ARA cp512; RNA 151; WRA cp42

Garter snake, plains (*Thamnophis radix*)
ARA cp534, cp543; ART p[24]; LBG cp593; RNA 147;
WRA cp42

Garter snake, shorthead (*Thamnophis brachystoma*)
CGW 93; RNA 147

Garter snake, two-striped (*Thamnophis hammondii*)
ARA cp504; WRA p48

Garter snake, Western aquatic or Santa Cruz (*Thamnophis couchii*)
RNA 151; WRA cp43

Garter snake, Western terrestrial (*Thamnophis elegans*)
ARA cp511; JAD cp215; LBG cp597; RNA 149; WFW
   cp595; WRA cp43

Gaur or gayal (*Bos gaurus* or *B.g.frontalis*)
CRM 187; GEM V:361, 387-89; MIA 143

Gaura, scarlet (*Gaura coccinea*)
SWW cp532; WFW cp494

Gavial (*Gavialis gangeticus*)
MIA 414

Gayfeather (*Liatris punctata*) See also Blazing-star
SWW cp533; WAA 140, 184

Gaywings (*Polygala paucifolia*)
CWN 249; NWE cp496; WAA 161

Gazelle, Clark's See Dibatag

Gazelle, Dama (*Gazella dama*)
GEM V:462; MEM 577

Gazelle, Dorcas (*Gazella dorcas*)
GEM V:469

Gazelle, giraffe See Gerenuk

Gazelle, goitered (*Gazella subgutturosa*)
GEM V:474; MEM 577; NHU 186

Gazelle, Grant's (*Gazella granti*)
AAL 85; GEM V:463; NHA 9(bw), 10(bw), 121(bw)

Gazelle, Indian or Edmi or mountain (*Gazella gazella*)
GEM V:467, 468

Gazelle, slender-horned (*Gazella leptoceros*)
GEM V:466

Gazelle, Thompson's (*Gazella thomsoni*)
GEM V:300, 470, 471; MEM 578, 579; MIA 153; NHA
   115(bw); SAB 23; WPP 90

Gecko, ashy (*Sphaerodactylus cinereus* or *elegans*)
ARA cp396, cp398; ERA 67; LOW 111; RNA 73

Gecko, banded (Texas) (*Coleonyx brevis*)
ARA cp393, cp394; JAD cp177, cp178; RNA 67

Gecko, banded (Western) (*Coleonyx variegatus*)
ARA cp392, cp395; JAD cp176, cp179; NAW 258; RNA
   67; WRA cp35; WWD 58(head only)

Gecko, Indo-Pacific or fox (*Hemidactylus garnoti*)
ARA cp401; RNA 69

Gecko, leaf-tailed (*Uroplatus finibriatus*)
MIA 427; WOI 160-161

Gecko, leaf-toed (*Phyllodactylus xanti*)
ARA cp391; JAD cp175; RNA 71; WRA cp35

Gecko, Madagascan day (*Phelsuma laticauda*)
ERA 100

Gecko, Mediterranean (*Hemidactylus turcicus*)
ARA cp397; RNA 71; WRA cp35

Gecko, reef (*Sphaerodactylus notatus*)
ARA cp399, cp400; RNA 73

Gecko, reticulated (*Coleonyx reticulatus*)
ART p[11]; RNA 67

Gecko, spiny-tailed (*Diplodactylus ciliaris*)
WWD 186

Gecko, web-footed (*Palmatogecko rangei*)
MIA 425; WWD 67, 68

Gecko, yellow-headed (*Gonatodes albogularis*)
ARA cp402; RNA 71

Gecko (other kinds)
AAL 216; ERA 88, 101; MIA 425; RNA 69, 71; WRA cp35,
   cp47; WWW 154

Gemsbok (*Oryx gazella*)
AAB 112-113; MEM 563, 569; WPP 87; WWD front., 87

Genet, common or small spotted (*Genetta genetta*)
AAB 49; CRM 145; MEM 135, 141, 142; MIA 93

Genet, large-spotted (*Genetta tigrina*)
AAL 56; GEM III:513, 525; WPP 96

Genet, panther (*Genetta maculata*)
GEM III:525

Gentian, closed or bottle (*Gentiana andrewsii*)
CWN 185; NWE cp650

Gentian, explorer's or blue (*Gentiana calycosa*)
SWW cp609; WFW cp533

Gentian, fringed (*Gentiana crinita*)
CWN 185; NWE cp617; WAA 183

Gentian, fringed (Western) (*Gentiana dentosa*)
SWW cp610; WAA 231

Gentian, great yellow (*Gentiana lutea*)
PPC 79

Gentian, Northern or rose (*Gentiana amarella*)
SWW cp603

Gentian, prairie (*Eustoma grandiflorum*)
LBG cp282; SWW cp612

Gentian, seaside or catchfly (*Eustoma exaltatium*)
NWE cp522; WAA 166

Geranium, Richardson's (*Geranium richarsonii*)
SWW cp43; WFW cp386

Geranium, sticky (*Geranium viscosissimum*)
SWW cp456

Geranium, wild (*Geranium maculatum*)
BCF 93; CWN 139; NWE cp466; WAA 156

Gerardia, purple (*Agalinis purpurea*)
NWE cp520

Gerbil, great (*Rhombomys opimus*)
MEM 676; MIA 175; NHU 190

Gerbil, Lybian or Mongolian (*Meriones* sp.)
MEM 675-677

Gerbil, South African or Namib (*Gerbillurus* sp.)
GEM III:255; MAE 19; MEM 676; MIA 175

Gerenuk (*Litocranius walleri*)
GEM V:479; MEM 582, 583; MIA 153; NHA cp5; WPP 86

Gharial, false (*Tomistoma schlegelii*)
ERA 142, 143

Ghost-flower (*Mohavea confertiflora*)
JAD cp147; SWW cp228

Gibbon, crested, black, white-cheeked or concolor
(*Hylobates concolor*)
CRM 85; GEM II:338-339, 352, 353; MIA 75; NHP
161(bw)

Gibbon, dark-handed or agile (*Hylobates lar agilis*)
GEM II:329

Gibbon, gray or Muller's (*Hylobates lar mulleri*)
GEM II:336-337; ROW 31(bw)

Gibbon, Hoolock or white-browed (*Hylobates hoolock*)
CRM 85; GEM II:352, 353; MIA 75

Gibbon, Kloss's or Dwarf (*Hylobates klossi*)
CRM 85; MEM 420; MIA 75

Gibbon, Moloch or silvery (*Hylobates lar moloch*)
GEM II:352, 354; MEM 418; SAB 91

Gibbon, pileated or capped (*Hylobates lar pileatus*)
GEM II:353; MIA 75

Gibbon, siamang See Siamang

Gibbon, white-handed or Lar (*Hylobates lar*)
CRM 85; GEM II:333-335, 352; MAE 44; MEM 415; MIA
75; NHP 162(bw); SWF 88, 89

Gila monster (*Heloderma suspectum*)
ARA cp332; ERA 104; JAD cp180; LOW 178; MIA 441;
NAW 260(bw); RNA 93; WRA p20

Gilia (*Ipomopsis* sp.) See also Skyrocket
NWE cp179, cp428; SWW cp137, cp413, cp442, cp590;
WAA 165

Gill-over-the-ground or creeping charlie (*Glechoma
hederacea*)
BCF 27; CWN 233; NWE cp648; SWW cp618

Ginger, wild (*Asarum canadense*)
BCF 127; CWN 51; NWE cp386; SEF cp459

Ginger, wild (long-tailed) or Western (*Asarum caudatum*)
SWW cp383; WFW cp463

Ginkgo (*Ginkgo biloba*)
NTE cp235, cp619; NTW cp235; OET 60, 61

Ginseng (*Panax ginseng*)
NHU 118

Ginseng, American (*Panax quinquefolium*)
FWF 105; NWE cp1, cp441; WTV 114

Ginseng, dwarf (*Panax trifolius*)
CWN 131; NWE cp160

Giraffe (*Giraffa camelopardalis*)
AAL 76; CRM 183; EAB 18; GEM V:258-259, 267-277;
MEM 13, 534-540; MIA 139; SAB 84; WPP front., 67-69

Glasswort (*Salicornia* sp.)
AGC cp469, cp470; NWE cp34, cp437; SCS cp18

Glider, greater (*Petauroides* or *Schoinobates volans*)
AAL 28; GEM I:314; MEM 858; MIA 21; SWF 104

Glider, pygmy or feathertail (*Acrobates pygmaeus*)
GEM I:326; MEM 21, 860

Glider, squirrel (*Petaurus norfolcensis*)
MEM 856, 857; NOA 134

Glider, sugar (*Petaurus breviceps*)
AAB 108; GEM I:226, 318, 319; MEM 859; MIA 21; NOA
134, 135; SAB 45

Glider, yellow-bellied (*Petaurus australis*)
GEM I:322; MEM 859

Glider (other *Petaurus* sp.)
CRM 27; SWF 105

Globeflower (*Trollius laxus*)
NWE cp255; SWW cp57; WAA 230

Globemallow, Coulter's (*Sphaeralcea coulteri*)
JAD cp157; SWW cp365

Globemallow, desert (*Sphaeralcea ambigua*)
JAD cp158; SWW cp398

Globemallow, mountain or stream (*Iliamna rivularis*)
SWW cp510; WFW cp499; WNW cp289

Globemallow, scarlet (*Sphaeralcea coccinea*)
SWW cp366; WFW cp454

Glowworm (Phengodidae)
BNA 183(bw); NNW 124

Gnatcatcher, black-tailed (*Polioptila melanura*)
EAB 1031; FWB 380; JAD cp588; UAB cp494

Gnatcatcher, blue-gray (*Polioptila caerulea*)
EAB 1033; FEB 456; FWB 381; MIA 345; NAB cp443;
   PBC 281; SEF cp325; SPN 82; UAB cp495; WFW cp274;
   WGB 155

Gnateater, rufous (*Conopophaga lineata*)
EOB 329

Gnu, brindled or blue or Wildebeest (*Connochaetes taurinus*)
AAB 77(head only); AAL 84; CRM 193; GEM V:424-431;
   MEM 563; MIA 149; WPP 64-65

Gnu, white-tailed (*Connochaetes gnou*)
CRM 193

Goanna See Monitor lizard, Gould's

Goat, mountain (*Oreamnos americanus*)
ASM cp280; CMW 266(bw); CRM 199; GEM V:500-502;
   MEM 585, 587; MIA 155; MNP cp3a; MPS 308(bw); NAW
   40, 41; SWM 163, 164; WOM 36; WWW 24-25

Goat, wild (*Capra aegagrus*)
AAL 86(bw); CRM 201; GEM I:171(head only), V:535(bw),
   536; MEM 585; WOI 166

Goatfish (*Mulloidichthys* and other genera)
ACF cp28; PCF cp35; RCK 165

Goatfish, red (*Mullus auratus*)
ACF cp28; NAF cp415

Goatsbeard (*Aruncus* sp.)
NWE cp153; SWW cp117; WFW cp414

Goatsbeard (*Tragopogon* sp.)
FWF 13; LBG cp165; NWE cp295

Goats-rue (*Tephrosia virginiana*)
NWE cp524

Go-away bird (*Corythaixoides* sp.)
EOB 232, 234; MIA 267; WGB 77

Goby (several genera)
ACF cp46, cp47; MPC cp300; NAF cp468-cp470; PCF
   p19, cp20; RUP 98; WWW 73

Godwit, bar-tailed (*Limosa lapponica*)
EAB 793

Godwit, black-tailed (*Limosa limosa*)
RBB 76; WBW 217

Godwit, Hudsonian (*Limosa haemastica*)
AGC cp596; EAB 793; FEB 153; NAB cp232

Godwit, marbled (*Limosa fedoa*)
AGC cp597; BWB 182; EAB 793; EOB 168; FEB 152;
   FWB 144; MPC cp564; NAB cp231; UAB cp210; WBW
   213

Goldcrest (*Regulus regulus*)
EOB 371; RBB 153

Golden club (*Orontium aquaticum*)
CWN 45; NWE cp308; WNW cp222

Golden rain or shower (*Cassia fistula*)
TSP 19

Golden rain tree (*Koelreuteria paniculata*)
PMT cp[69]

Golden stars (*Bloomeria crocea*)
SWW cp319

Goldeneye, Barrow's (*Bucephala islandica*)
DNA 141, 143; EAB 208; FEB 84; FWB 105; NAB cp125,
   cp149; UAB cp104, cp130; WNW cp505

Goldeneye, common or American (*Bucephala clangula*)
DNA 145, 147; EAB 208; FEB 85; FWB 104; NAB cp126,
   cp150; PBC 94, 95; RBB 36; UAB cp105, cp131; WNW
   cp506; WON 53

Goldenrod (*Solidago* sp.) See also Silverrod
CWN 251-255; LBG cp208, cp213-216; MPC cp603;
   NWE cp337-344; SWW cp308, cp309; WFW cp433

Goldenrod, early (*Solidago juncea*)
BCF 113; WAA 42; WTV 95

Goldenseal (*Hydrastis canadensis*)
FWF 65; NWE cp117

Goldfields (*Lasthenia chrysostoma*)
LBG cp178; SWW cp263

Goldfinch, American (*Carduelis tristis*)
AAL 190(bw); EAB 305; FEB 397; FWB 438; LBG cp552;
   NAB cp385, cp386; NAW 133; PBC 372; SPN 141; UAB
   cp430; WGB 195

Goldfinch, European (*Carduelis carduelis*)
MIA 385; RBB 180

Goldfinch, Lawrence's (*Carduelis lawrencei*)
EAB 304; FWB 440; UAB cp418, cp431

Goldfinch, lesser (*Carduelis psaltria*)
EAB 305; FEB 398; FWB 437; UAB cp417

Goldfish (*Carassius auratus*)
MIA 513; NAF cp120, cp121

Goldthread (*Coptis groenlandica*)
CWN 73; NWE cp62; SEF cp435; WNW cp256

Goosander See Merganser, common

Goose, bar-headed (*Anser indicus*)
EOB 93

Goose, barnacle (*Branta leucopsis*)
RBB 24

Goose, brant See Brant

Goose, Canada (*Branta canadensis*)
AGC cp554; BWB 58-59; EAB 220, 221; EOB 95; FEB 115; FWB 118, 119; MIA 213; NAB cp170; NAW 135; PBC 79; RBB 23; UAB cp162; WGB 23; WNW cp512; WON 46

Goose, Egyptian (*Alopochen aegyptiacus*)
DNA 15

Goose, emperor (*Chen* or *Anser canagicus*)
EAB 222; EOB 92; MPC cp529; WWW 87; SAB ii; UAB cp160

Goose, graylag (*Anser anser*)
MIA 213; RBB 22; WGB 23

Goose, Hawaiian (*Branta sandvicensis*)
EOB 93

Goose, magpie (*Anseranas semipalmata*)
EOB 92; MIA 213; WGB 23

Goose, pink-footed (*Anser fabalis brachyrhyncus*)
EOB 93; RBB 20

Goose, red-breasted (*Rufibrenta ruficollis*)
EOB 92; NHU 79

Goose, Ross' (*Chen rossii*)
EAB 223; FEB 117; FWB 114; UAB cp158; WNW cp515

Goose, snow or blue (*Chen* or *Anser caerulescens*)
AGC cp553, cp555; BWB 48; EAB 224, 225; EOB 100; FEB 116; FWB 115; NAB cp171; NAW 136; PBC 81; UAB cp159, cp161; WNW cp513, cp514; WPR 110

Goose, white-fronted (*Anser albifrons*)
EAB 223; FEB 113; FWB 117; NAB cp169; RBB 21; UAB cp163, cp165; WNW cp511

Gooseberry (*Ribes rotundifolium* or *roezlii*)
FWF 175; TSV 26, 27; WFW cp153

Goosefish, American (*Lophius americanus*)
ACF p14; AGC cp355; BCS 33; NAF cp292; WWW 128(head only)

Gopher, pocket See Pocket gopher

Goral (*Nemorhaedus goral*)
CRM 199; GEM V:505; MEM 584; MIA 155

Gorilla (*Gorilla gorilla*)
CRM 87; GEM I:48, II:21, 360, 424-461; MEM 307, 313, 432-39; MIA 77; NHP 172(bw)

Gorilla, mountain (*Gorilla gorilla beringei*)
MAE 32; NHP 173, 174(bw); SWF 118; WOM 110-113

Goshawk, chanting (dark) (*Melierax metabates*)
MIA 221; WGB 31

Goshawk, chanting (pale) (*Melierax poliopterus*)
BOP 77; EOB 110

Goshawk, northern (*Accipiter gentilis*)
AAB 48; BOP 8, 82; EAB 506; FEB 223; FGH cp6, cp7; FWB 233; MAE 59; MIA 223; NAB cp296; NHU 96; SEF cp255; UAB cp326; WFW cp223; WGB 33

Gourami (*Osphronemus goramy*)
MIA 573

Gourd, buffalo or wild (*Cucurbita foetidissima*)
JAD cp140; LBG cp196; SWW cp344

Grackle, boat-tailed (*Quiscalus major*)
AGC cp607; EAB 905; FEB 361; NAB cp519, cp575; PBC 350

Grackle, common or purple (*Quiscalus quiscula*)
EAB 905; EOB 414; FEB 360; FWB 283; MIA 383; NAB cp573; NAW 137; PBC 350; UAB cp618; WGB 193

Grackle, great-tailed (*Quiscalus mexicanus*)
EAB 905; FWB 282; NAB cp576; UAB cp619

Grass, bluestem (*Andropogon* sp.)
LBG cp97, cp101; NWE cp400

Grass, bottlebrush (*Hystrix patula*)
NWE cp404

Grass, cord See Cordgrass

Grass, crab See Crabgrass

Grass, cotton (*Eriophorum* sp.)
NHU 77; NWE cp233; SWW cp96; WNW cp238; WTV 108

Grass, fluff (*Erioneuron pulchellum*)
JAD cp361

Grass, gama (*Tripsacum dactyloides*)
LBG cp108

Grass, gramma or mesquite (*Bouteloua gracilis*)
NWE cp398

Grass, Indian (*Sorghastrum nutans*)
LBG cp94; NWE cp401

Grass of Parnassus (*Parnassia* sp.)
CWN 127; NWE cp73; SWW cp24; WFW cp391; WTV 50

Grass, orchard (*Dactylis glomerata*)
LBG cp104

Grass pink See under Orchid

Grass, redtip (*Agrostis alba*)
LBG cp93; NWE cp403

Grass, squirreltail (*Hordeum jubatum*)
LBG cp107; WPP 14

Grass, sweet vernal (*Anthoxanthum odoratum*)
LBG cp96; NWE cp399

Grass, timothy (*Phleum pratense*)
LBG cp106; NWE cp32

Grass, velvet (*Holcus lanatus*)
LBG cp92

Grass widow (*Sisyrinchium douglasii*)
SWW cp464; WAA 244

Grass, wool (*Scirpus cyperinus*)
NWE cp30; WNW cp322

Grasshopper (various kinds)
BOI 105-119; DIE cp3, cp9, cp20, cp27; EOI 4, 44; MIS
   cp259-282; SAB 34-35; SGI cp16-cp21; WWW 163

Grasshopper, horse-lubber (*Taeniopoda eques*)
JAD cp371; SGI cp15

Grasshopper, lubber (*Brachystola* or *Dactylotum* sp.)
BOI 108; MIS cp262; WWD 154-155

Grasshopper, meadow (*Chorthippus parallelus*)
AAL 411(bw); MAE 29

Grayling, Arctic (*Thymallus arcticus*)
AAL 331(bw); AWR 221(bw); NAF cp67

Grayling, European (*Thymallus thymallus*)
EAL 54; MIA 507

Graysby (*Epinephelus cruentatus*)
AAL 352; ACF cp26

Greasewood (*Sarcobatus vermiculatus*)
JAD cp345

Grebe, eared or black-necked (*Podiceps nigricollis*)
DBN cp8; EAB 407; FWB 81; NAB cp181, cp183; NAW
   139; RBB 10; UAB cp178, cp180; WNW cp486; WON 16

Grebe, great crested (*Podiceps cristatus*)
AAL 93; BWB 208; EOB 43; MIA 199; RBB 8; WGB 9

Grebe, horned or Slavonian (*Podiceps auritus*)
AGC cp524; BAS 36; BWB 205; DBN cp9; EAB 408; EOB
   43; FEB 75; FWB 80; MPC cp516; NAB cp182, cp184;
   NAW 138; PBC 41; RBB 9; UAB cp179, cp181; WNW
   cp485; WON 17

Grebe, least (*Tachybaptus dominicus*)
DBN cp6; EAB 409; FEB 77; NAB cp177, cp179

Grebe, little (*Tachybaptus ruficollis*)
AAL 93; BAS 84; MIA 199; RBB 7; WGB 9

Grebe, pied-billed (*Podilymbus podiceps*)
AGC cp526; DBN cp5; EAB 409; FEB 76; MPC cp518;
   NAB cp178, cp180; NAW 138; UAB cp176, cp177; WNW
   cp483; WOB 79; WON 15

Grebe, red-necked (*Podiceps grisegena*)
AGC cp527; DWB 210; DBN cp7; EAB 409; FEB 74; FWB
   79; MPC cp513; NAB cp185; UAB cp175; WNW cp484

Grebe, western (*Aechmophorus occidentalis*)
BWB 206, 207; DBN cp10; EAB 408; FWB 78; MPC
   cp514; SAB 109; UAB cp174; WNW cp482; WOB 76-79

Green dragon (*Arisaema dracontium*)
CWN 47; NWE cp12; WNW cp311

Greenfinch (*Carduelis chloris*)
RBB 179

Greenling (*Hexagrammos* sp.)
MIA 547; MPC cp246, cp250; NAF cp400, cp411; PCF
   cp22

Greenshank (*Tringa nebularia*)
RBB 80; WBW 237

Grenadier, Pacific (*Coryphaenoides acrolepis*)
PCF p48

Grison (*Galictis cuja*)
GEM III:411; MEM 112; MIA 87

Gromwell or Puccoon (*Lithospermum* sp.)
NWE cp375; SWW cp209, cp331

Grosbeak, black-headed (*Pheucticus melanocephalus*)
EAB 306; FWB 439; NAB cp399, cp562; PBC 359; SPN
   123; UAB cp443, cp582; WFW 295

Grosbeak, blue (*Guiraca caerulea*)
EAB 306; FEB 394; FWB 429; JAD cp604; NAB cp439;
   PBC 360; UAB cp498, cp550

Grosbeak, evening (*Hesperiphona vespertina*)
EAB 307; FEB 396; FWB 436; NAB cp384; PBC 364; SEF
   cp305; UAB cp426, cp429; WFW cp311

Grosbeak, pine (*Pinicola enucleator*)
EAB 307; EOB 419; FEB 384; FWB 434; MIA 385; NAB
   cp414, cp424; PBC 370; SEF cp314; UAB cp466, cp481;
   WFW cp308; WGB 195

Grosbeak, rose-breasted (*Pheucticus ludovicianus*)
EAB 307; EOB 407; FEB 382; MIA 373; NAB cp408,
   cp561; PBC 359; SEF cp316; SPN 122; WGB 183

Grosbeak, scarlet See Rosefinch

Ground cone, California (*Boschniakia strobilacea*)
SWW cp386; WFW cp470

Ground ivy See Gill-over-the-ground

Ground squirrel, African or Cape (*Xerus inauris*)
GEM III:74; MAE 18, 65; MEM 599; SAB 130

Ground squirrel, African (Western) or striped (*Xerus
   erythropus*)
AAL 43(bw); GEM III:73; MEM 617; MIA 161; WWD 84-85

Ground squirrel, antelope (*Spermophilus leucurus*)
WWD 50

Ground squirrel, Arctic (*Spermophilus parryii* or *undulatus*)
ASM cp30; BAA 61(bw); SWM 3; WPR 72

Ground squirrel, Belding's (*Spermophilus beldingi*)
ASM cp27; GEM III:62; LBG cp85; MEM 624, 625

Ground squirrel, California (*Spermophilus beecheyi*)
ASM cp17; GEM III:58; MNP cp8a

Ground squirrel, Columbian (*Spermophilus columbianus*)
ASM cp29; GEM III:64; LBG cp81; NAW 90; SWM 32

Ground squirrel, European (*Citellus* or *Spermophilus citellus*)
CRM 95; GEM III:57; NHU 46

Ground squirrel, Franklin's (*Spermophilus franklinii*)
ASM cp26; LBG cp84; MNC 161(bw); MPS 133(bw)

Ground squirrel, golden mantled (*Spermophilus lateralis*)
ASM cp9; CMW 106(bw); GEM III:60-61; NMM cp[9]; RMM cp[3]; SWM 35; WFW cp335

Ground squirrel, Mexican (*Spermophilus mexicanus*)
ASM cp16; JAD cp502

Ground squirrel, Richardson's (*Spermophilus richardsonii*)
ASM cp22; GEM III:64; LBG cp83; MNC 163(bw); MPS 134(bw); SWM 33; WPP 151

Ground squirrel, round-tailed (*Spermophilus tereticaudus*)
ASM cp19; GEM III:59; JAD cp504

Ground squirrel, spotted (*Spermophilus spilosoma*)
ASM cp18; CMW 104(bw); JAD cp503; LBG cp88; MPS 134(bw); NMM cp[11]

Ground squirrel, thirteen lined (*Spermophilus tridecemlineatus*)
ASM cp15; CMW 102(bw); LBG cp89; MIA 161; MNC 165(bw); MNP 200(bw); MPS 135(bw); SWM 34

Ground squirrel, Townsend's (*Spermophilus townsendii*)
ASM cp25; JAD cp501

Ground squirrel, Uinta (*Spermophilus armatus*)
ASM cp28; CMW 100 (bw); LBG cp82

Ground squirrel, Wyoming (*Spermophilus elegans*)
CMW 98 (bw); RMM cp[5]

Groundcherry (*Physalis heterophylla*)
CWN 201; NWE cp273; WTV 79

Groundcherry, purple (*Physalis lobata*)
JAD cp67; LBG cp285; SWW cp581

Groundnut or Indian potato (*Apios americana*)
CWN 155; NWE cp412; WTV 189

Groundsel (*Senecio* sp.)
CWN 257; SWW cp239, cp242, cp313; WFW cp444; WNW cp205; WPR 55

Groundsel tree (*Baccharis halimifolia*)
NWE cp220; SEF cp99; TSV 198, 199

Grouper (*Cephalopholis miniatus*)
RCK 56; RUP 102

Grouper (*Mycteroperca* sp.)
ACF cp25; KCR cp23; PCF cp29; RCK 139

Grouper, black (*Mycteroperca bonaci*)
MIA 549; NAF cp511

Grouper, Nassau (*Epinephelus striatus*)
AGC cp382; NAF cp516; RCK 40

Grouper, yellow-mouth (*Mycteroperca interstitialis*)
ACF cp25; RCK 31

Grouper (other kinds)
AAL 352; NAF cp509; RCK 58, 141, 147

Grouse, black (*Lyrurus tetrix*)
MIA 233; RBB 53; WGB 43

Grouse, blue (*Dendragapus obscurus*)
AAL 116; EAB 410; FWB 201; UAB cp260, cp261; WFW cp224

Grouse, hazel or hazelhen (*Tetrastes bonasia*)
NHU 99

Grouse, red See Ptarmigan, willow

Grouse, ruffled (*Bonasa umbellus*)
AAL 117; EAB 411; FEB 212; FWB 202; NAB cp268; NAW 140, 142; SEF cp247; UAB cp263; WFW cp225; WOB 72-73; WON 75-77

Grouse, sage (*Centrocercus urophasianus*)
BAS 28; EAB 412, 413; EOB 133; FWB 200; JAD cp545; LBG cp499; NAW 141; SAB 83; UAB cp253, cp257; WOB 64-67; WPP 162-163

Grouse, sharp-tailed (*Tympanuchus phasianellus*)
EAB 415; FEB 208; FWB 198; LBG cp502; NAB cp259, cp260; NAW 141; UAB cp252, cp256; WON 79

Grouse, spruce (*Dendeagapus canadensis*)
EAB 413; FEB 209; FWB 195; NAB cp270; SEF cp248; UAB cp262

Grunion (*Leuresthes tenuis*)
EAL 99; MIA 537; MPC cp298; NAF cp569; PCF p12

Grunt, french (*Haemulon flavolineatum*)
AAL 356; ACF cp32; KCR cp23; RCK 37

Grunt, white (*Haemulon plumieri*)
ACF cp32; NAF cp532

Grysbok (*Raphicerus melanotis*)
CRM 195; MIA 151

Guan, crested (*Penelope purpurascens*)
MIA 227; WGB 37

Guan, horned (*Oreophasis derbianus*)
EOB 134

Guanaco (*Lama guanicoe* or *huanacos*)
AAL 75(bw); CRM 173; GEM I:126, V:97, 103, 105, 106;
MEM 515; MIA 133; WPP 190; WWD 126

Guava (*Psidium guajava*)
PFH 148-149; TSP 37

Guava, pineapple (*Feijoa sellowiana*)
TSP 31

Gudgeon (*Gobio gobio*)
MIA 515

Guereza See Colobus, black and white

Guillemot See also Murre, common

Guillemot, black or Tystie (*Cepphus grylle*)
AOG cp5; DBN cp16; EAB 46; EOB 211; FEB 62; NAB
cp92; OBL 127; RBB 97

Guillemot, pigeon (*Cepphus columba*)
AAL 132(bw); AOG cp6; BWB 107; DBN cp17, cp18; EAB
46; FWB 60; MPC cp536; UAB cp80

Guinea fowl, helmeted (*Numidia meleagris*)
MIA 233; WGB 43

Guinea fowl, vulturine (*Acryllium vulturinum*)
EOB 135; WPP 97

Guinea pig (*Cavia porcellus*)
MEM 693

Guitarfish (*Rhinobatus* sp.)
MIA 495; PCF p3

Gulfweed See Sargassum

Gull, Andean (*Larus serranus*)
OBL 86

Gull, black-backed or black-headed (*Larus ridibundus*)
EAB 419; FEB 34; FWB 31; HSW 132, 133; MIA 249; NAB
cp46; RBB 85; SAB 23, 82

Gull, black-backed (great) See Gull, great

Gull, black-backed (lesser) (*Larus fuscus*)
EAB 424; FEB 38; NAB cp41; OBL 18; RBB 87; SAB 111

Gull, black-headed (Patagonian) (*Larus maculipennis*)
OBL 86

Gull, black-backed (Southern) or kelp (*Larus
dominicanus*)
AAL 131; HSW 130; OBL 15

Gull, Bonaparte's (*Larus philadelphia*)
AGC cp506; EAB 418, 419; FEB 33; FWB 34; HSW 38-
39; MPC cp590; NAB cp48, cp53, cp55; PBC 173; UAB
cp29, cp32; WNW cp554

Gull, brown-headed (*Larus brunnicephalus*)
OBL 86-87

Gull, Californian (*Larus californicus*)
EAB 419; FWB 27; HSW 128; MPC cp595; NAW 147;
UAB cp19, cp37; WNW cp556; WOB 119

Gull, common or mew (*Larus canus*)
EAB 425; FWB 23; HSW 132; MPC cp598; RBB 86; UAB
cp23, cp35

Gull, dolphin or Magellan (*Larus* or *Gabianus scoresbyi*)
AAL 131; EOB 190; HSW 29

Gull, Franklin's (*Larus pipixcan*)
EAB 420; FEB 37; FWB 32; NAB cp45; OBL 119; UAB
cp31, cp34; WNW cp553

Gull, glaucous (*Larus hyperboreus*)
EAB 421; FEB 44; FWB 21; HSW 9; NAB cp33, cp36;
PBC 166; UAB cp20

Gull, glaucous-winged (*Larus glaucescens*)
EAB 421; FWB 20; HSW 135; MPC cp594; OBL 230; UAB
cp21, cp36

Gull, great or great black-backed (*Larus marinus*)
AGC cp504; BWB 154-155; EAB 421; EOB 188; FEB 39;
HSW 131; NAB cp42, cp51; OBL 122; PBC 168; RBB 89;
SAB 104

Gull, grey (*Larus modestus*)
OBL 35

Gull, grey-headed (*Larus cirrocephalus*)
OBL 42, 222

Gull, Heermann's (*Larus heermanni*)
EAB 421; FWB 30; HSW 26; MPC cp591; NAW 146; UAB
cp28, cp40

Gull, herring (*Larus argentatus*)
AAL 131; AGC cp503; BAA 115; BWB 146-151; EAB 422,
423; FEB 40; FWB 24; HSW 129; MIA 249; MPC cp597;
NAB cp37, cp50; NAW 144; OBL 18, 227; PBC 169, 170;
RBB 88; SAB 23; UAB cp18, cp38; WGB 59; WNW cp557;
WON 93

Gull, Iceland (*Larus glaudoides*)
EAB 425; FEB 43; NAB cp31, cp34; OBL 27; PBC 167

Gull, ivory (*Pagophila eburnea*)
BAA 123; EAB 424; EOB 189; FEB 46; MIA 249; NAB
cp32, cp49; WGB 59; WPR 144-145

Gull, laughing (*Larus atricilla*)
AGC cp505; EAB 424; FEB 36; FWB 33; HSW 133; NAB cp43, cp56; OBL 47; PBC 172

Gull, lava or dusky (*Larus fuliginosus*)
HSW 137; OBL 19

Gull, little (*Larus minutus*)
EAB 425; EOB 189; FEB 32; HSW 137; NAB cp47; OBL 107

Gull, mew See Gull, common

Gull, Pacific (*Larus pacificus*)
HSW 30; OBL 82

Gull, ring-billed (*Larus delawarensis*)
AGC cp502; EAB 427; FEB 41; FWB 26; HSW 133; MPC cp599; NAB cp38, cp52; NAW 146; OBL 83; PBC 171; UAB cp24; WNW cp555

Gull, Ross' (*Rhodostethia rosea*)
EOB 189; NHU 83

Gull, Sabine's (*Xema sabini*)
EAB 426; EOB 189; FEB 35; FWB 35; HSW 27(bw); MPC cp589; NAB cp44; OBL 107; PBC 175; UAB cp30

Gull, silver (*Larus novaehollandiae*)
BWB 154; HSW 19, 131; OBL 87

Gull, simeon (*Larus belcheri*)
OBL 219

Gull, sooty or Aden (*Larus hemprichi*)
HSW 136; OBL 82

Gull, swallow-tailed (*Creagus* or *Larus furcatus*)
BWB 145; HSW 137; OBL 127, 159, 183

Gull, Thayer's (*Larus thayeri*)
EAB 426; FEB 45; FWB 25; MPC cp592; NAB cp35; UAB cp17

Gull, Western (*Larus occidentalis*)
EAB 426; FWB 29; HSW 135; MPC cp596; UAB cp22, cp39

Gull, white-eyed (*Larus leucophthalmus*)
OBL 83

Gull, yellow-footed (*Larus livens*)
FWB 28

Gum, black See Tupelo

Gum, sweet See Sweet gum

Gum tree See Eucalyptus

Gumweed (*Grindelia squarrosa*)
LBG cp182; NWE cp298; SWW cp259

Gundi (*Ctenodactylus* and other genera)
GEM III:297(bw); MEM 707; MIA 189

Gunnel (*Pholis* sp.)
ACF p45; PCF cp39

Gunnel, penpoint (*Apodichthys flavidus*)
MPC cp302; NAF cp447

Gurnard, flying (*Dactylopterus volitans*)
ACF p52; KCR cp22; MIA 545; WWW 139

Gymnure See Moonrat

Gyrfalcon (*Falco rusticolus*)
BOP 4-5, 41; EAB 296; FEB 225; FGH cp24; FWB 217; MIA 225; NAB cp316; NHU 85; UAB cp309, cp318; WGB 35; WPR 90, 91

H

Hackberry (*Celtis occidentalis*) See also Sugarberry
NTE cp169, cp551; SEF cp77, cp188

Hackberry, netleaf (*Celtis reticulata*)
NTE cp170; NTW cp201, cp474

Haddock (*Melanogrammus aeglefinus*)
ACF p15; AGC cp388; EAL 87; MIA 527; NAF cp491

Hagfish (various genera)
ACF p9; EAL 17; MIA 489; NAF cp444

Haha (*Cyanea* sp.)
PFH 98-101

Hairstreak butterfly, black (*Strymondia pruni*)
SGB cp129

Hairstreak butterfly, brown (*Thecla butulae*)
SGB cp30(larva), 130

Hairstreak butterfly, early (*Erora laeta*)
MIS cp588; PAB cp409

Hairstreak butterfly, gray (*Strymon melinus*)
LBG cp319; MIS cp587; PAB cp386; SGB cp203; SGI cp218

Hairstreak butterfly, green (*Callophrys rubi*)
EOI 106; SGB cp54

Hairstreak butterfly (other kinds)
AAL 457; BOI 140, 141; JAD cp440, 441; PAB cp382-448; SGI cp217; WFW cp544

Hake (*Merluccius* sp.)
EAL 86; MIA 529; MPC cp290; NAF 487, cp488; PCF p11

Hake (*Urophycis* sp.)
ACF p15; MIA 527; NAF cp473, cp474

Halfbeak (*Hyporhamphus unifasciatus*)
NAF cp434

Halfmoon (*Medialuna californiensis*)
MPC cp275; NAF cp541; PCF cp34

Halibut, Atlantic (*Hippoglossus hippoglossus*)
ACF p57 ; MIA 577

Halibut, California (*Paralichthys californicus*)
BMW 48-49; MIA 577; MPC cp312; NAF cp282; PCF p43

Halibut, Pacific (*Hippoglossus stenolepis*)
MPC cp305; NAF cp280; PCF p45

Halosaur (*Halosaurus* sp.)
EAL 37

Hammercop See Stork, hammerhead

Hammerhead See Shark, hammerhead

Hamster, common (*Cricetus cricetus*)
CRM 109; GEM III:206; MEM 673; MIA 169

Hamster, golden (*Mesocricetus auratus*)
GEM III:211; MIA 169

Hardhack See Steeplebush

Hare, Arctic (*Lepus arcticus*)
AAB 69; ASM cp233, cp236; GEM IV:249, 258; NHU 107; SWM 19; WOI 46-47; WPR 59, 61

Hare, bushman (*Bunolagus monticularis*)
CRM 93; MEM 715

Hare, calling See Pika

Hare, Cape (*Lepus capensis*)
ASM cp243; GEM IV:286; MIA 191

Hare, European or field (*Lepus europaeus*)
CGW 197; GEM IV:246, 248, 251, 264-9, 276, 281-283; MEM 713, 715; WPP 38, 39

Hare, hispid (*Caprolagus hispidus*)
CRM 95; MEM 715; MIA 191

Hare, Japanese (*Lepus brachyurus*)
GEM IV:286

Hare, Northern (*Lepus timidus*)
ASM cp234

Hare, Pampas See Mara

Hare, rock (greater red) (*Pronolagus crassicaudatus*)
MEM 715

Hare, snowshoe or varying (*Lepus americanus*)
AAL 42(bw); ASM cp231, cp235; CGW 197; CMW 82(bw); GEM IV:266-267; GMP 130(bw), 131(bw); MEM 722, 723; MIA 191; MNC 143(bw); MPS 117(bw); NAW 42; NNW 44; SEF cp592; SWM 20; WFW cp350; WMC 112

Harebell (*Campanula rotundifolia*)
CWN 205; NWE cp618; SEF cp498; SWW cp611; WAA 230; WFW cp532

Harrier, northern or hen (*Circus cyaneus*)
AAL 112-113; AGC cp604; BOP 84(head only), 87; kEAB 510; FEB 235; FGH cp3; FWB 215; JAD cp531; LBG cp489; MIA 221; NAB cp309, cp310; RBB 42; UAB cp317; WGB 31; WNW cp561; WON 63; WPP 160-161

Harrier, pied (*Circus melanoleucus*)
EOB 111

Harrier, spotted (*Circus assimilis*)
BOP 86

Hartebeest, Coke's or Kongoni (*Alchephalus busephalus*)
CRM 191; GEM V:419, 604; MEM 563; MIA 149; NHA cp11; WPP 83

Harvest fish (*Peprilus alepidotus*)
AGC cp364; NAF cp546

Harvestfly, dogday (*Tibicen canicularis*)
AAL 421; MIS cp290; SEF cp415; SGI cp65

Harvestman See Daddy-long-legs

Hatchet fish (*Argyropelecus* sp.)
AAL 331(bw), 332; MAE 111; MIA 509; PCF p48; WWW 186

Haw, black (*Viburnum prunifolium*) See also Witherod
FWF 199; NTE cp203; SEF cp103; TSV 80, 81

Hawfinch (*Coccothraustes coccothraustes*)
BAS 108; RBB 186

Hawk, black (*Buteogallus anthracinus*)
EAB 507; FGH cp16; FWB 210; UAB cp339

Hawk, broad-winged (*Buteo platypterus*)
BOP 70; EAB 507; FEB 239; FGH cp10, cp13; FWB 222; NAB cp297; SEF cp257; WOB 106

Hawk, Cooper's (*Accipiter cooperii*)
BOP 80, 81; EAB 507; FEB 241; FGH cp6, cp7; FWB 221; JAD cp532; MIA 223; NAB cp293; SEF cp254; UAB cp327; WGB 33

Hawk, duck See Falcon, peregrine

Hawk, ferruginous (*Buteo regalis*)
AAL 111; BOP 76, 116-117; EAB 508, 509; FGH cp12, cp15; FWB 228; JAD cp538; LBG cp493; NAW 101(head); UAB cp315

Hawk, fish See Osprey

Hawk, Galapagos (*Buteo galapagoensis*)
WOI 28-29, 30

Hawk, gray (*Buteo nitidus*)
EAB 509; FGH cp8; FWB 224; UAB cp340

Hawk, great black (*Buteogallus urubitinga*)
BOP 68

Hawk, harrier (African) (*Polyboroides* sp.)
EOB 111; MIA 221; WGB 31

Hawk, Harris' (*Parabuteo unicinctus*)
BOP 83; EAB 509; FEB 218; FGH cp17; FWB 212; JAD
cp533; NAB cp311; UAB cp311

Hawk, marsh See Harrier, Northern Hawk, pigeon See
Merlin

Hawk, red-shouldered (*Buteo lineatus*)
EAB 511; FEB 239; FGH cp9; FWB 226; NAB cp298; PBC
110; SEF cp256; UAB cp312, cp313; WNW cp564

Hawk, red-tailed (*Buteo jamaicensis*)
BOP 65, 69(head only), 84; EAB 512, 513; FEB 237; FGH
cp11, cp14; FWB 211; JAD cp536, cp537; LBG cp491,
cp492; MIA 223; NAB cp300; NAW 149; SEF cp258; UAB
cp314, cp342; WGB 33; WON 66, 67

Hawk, roadside (*Buteo magnirostris*)
FGH cp8

Hawk, rough-legged (*Buteo lagopus*)
BOP 79; EAB 513; FEB 236; FGH cp10, cp15; FWB 229;
JAD cp539; NAB cp295; NAW 150; UAB cp310, cp316

Hawk, sharp-shinned (*Accipiter striatus*)
BOP 72, 73; EAB 513; FEB 240; FGH cp6, cp7; FWB 220;
NAB cp294; NAW 152; PBC 107; SEF cp253; UAB cp324,
cp325; WON 65

Hawk, short-tailed (*Buteo brachyurus*)
EAB 513; FEB 228; FGH cp4, cp13; NAB cp319; WNW
cp560

Hawk, sparrow (*Accipiter nisus*)
NAW 152; NHU 43; RBB 43; SWF 50-51

Hawk, sparrow (black-mantled) (*Accipiter
melanchoryphus*)
EOB 111

Hawk, Swainson (*Buteo swainsonii*)
BOP 88; EAB 514; FEB 234; FGH cp12, cp13; FWB 216;
JAD cp534, cp535; LBG cp490; NAB cp299; SAB 54; UAB
cp308, cp343

Hawk, white-tailed (*Buteo albicaudatus*)
EAB 515; FEB 229; FGH cp17; NAB cp301

Hawk, zone-tailed (*Buteo albonotatus*)
BOP 75; EAB 515; FGH cp16; FWB 207; UAB cp341

Hawk-eagle, ornate (*Spizaetus ornatus*)
BOP 85; EOB 111

Hawkfish (various genera)
AAL 7; AWF cp38; BMW 150-151; RCK 137; RUP 149

Hawkmoth, death's head (*Acherontia atropos*)
BOI 156(larva); EOI 101(larva); SGB cp36; WIW 31(larva)

Hawkmoth, elephant (*Deilephila* sp.)
NNW 141; SGB cp62; WWW 141

Hawkmoth, eyed (*Smerinthus ocellata*)
NNW 167; SGB cp127

Hawkmoth, hummingbird (*Macroglossum stellatarum*)
AAB 110; SGB cp91

Hawkmoth (other kinds)
EOI 98; SGB cp80; WWW 151

Hawk's beard (*Crepis acuminata*)
SWW cp272

Hawkweed, orange (*Hieracium aurantiacum*)
LBG cp227; NWE cp374

Hawkweed, yellow See King devil

Hawthorne, Biltmore (*Crateagus intricata*)
LBG cp434, cp454; NTE cp250, cp450

Hawthorne, black (*Crateagus douglassii*)
NTW cp219; WFW cp97

Hawthorne, cockspur (*Crateagus crus-galli*)
NTE cp219, cp412; SEF cp102

Hawthorne, Columbia (*Crateagus columbiana*)
NTW cp217, cp473

Hawthorne, downy (*Crateagus mollis*)
NTE cp220, cp452; SEF cp74, cp155

Hawthorne, fanleaf (*Crateagus flabellata*)
LBG cp435; NTE cp245

Hawthorne, fireberry (*Crateagus chrysocarpa*)
NTE cp247; NTW cp215, cp472

Hawthorne, fleshy (*Crateagus succulenta*)
NTE cp226; NTW cp184

Hawthorne, green (*Crateagus viridis*)
NTE cp221, cp453; PMT cp[41]; SEF cp73

Hawthorne, oneseed or common (*Crateagus monogyna*)
NTW cp216, cp471

Hawthorne, Washington (*Crateagus phaenopyrum*)
NTE cp243, cp448, cp552; NTW cp214, cp360, cp475

Hawthorne (other *Crateagus* sp.)
LBG cp427; NTE cp210-cp218, cp225, cp241-cp252,
cp451; NTW cp181, cp470

Hazel, Turkish (*Corylus colurna*)
PMT cp[39]

Hazelnut, American (*Corylus americana*)
TSV 190, 191

Hazelnut, beaked (*Corylus cornuta*)
FWF 49; WFW cp168

Heal-all (*Prunella vulgaris*)
BCF 9; CWN 225; NWE cp646

Heath, beach (*Hudsonia tomentosa*)
AGC cp455; NWE cp363

Heather, mountain (*Cassiope* sp.)
SWW cp174; WAA 248; WFW cp193

Hedgehog, European (*Erinaceus europaeus*)
AAL 32; GEM I:414-415, 449-464; MEM 739, 754-757;
    MIA 33; NNW 29, 115; SWF 60

Hedgehog (other kinds)
GEM I:465; MEM 751, 752; MIA 33; NHU 178

Hedge-nettle, great (*Stachys cooleyae*)
SWW cp518; WFW cp490; WNW cp284

Heliotrope, sweet-scented (*Heliotropium
    convolvulaceum*)
JAD cp101; SWW cp51

Heliozoan (Actinophryidae)
EAL 161; PGT 89

Hellbender (*Cryptobranchus alleganiensus*)
AAL 267(bw); ARA cp24; ARC 43; CGW 45; ERA 24, 26;
    MIA 463; WNW cp105

Hellebore, false (*Veratrum viride*)
See also Lily, corn (California)
CWN 17; NWE cp27; PPC 107 (*V.album*), 108; WNW
    cp320

Helmet brid (*Euryceros prevostii*)
EOB 353

Hemipode, Andalusian or little buttonquail (*Turnix
    sylvatica*)
EOB 159; MIA 235; WGB 45

Hemlock, Carolina (*Tsuga caroliniana*)
NTE cp19

Hemlock, Eastern (*Tsuga canadensis*)
NTE cp20, cp483; SEF cp34; TSV 14, 15

Hemlock, mountain (*Tsuga mertensiana*)
NTW cp49, cp433; WFW cp49, cp139

Hemlock, poison (*Conium maculatum*)
CWN 137; PPC 70; PPM cp22

Hemlock, water (*Cicuta douglasii* or *maculata*)
NWE cp187; SWW cp167; WNW cp240

Hemlock, Western (*Tsuga heterophylla*)
NTW cp52, cp434; OET 74; WFW cp140

Hempweed, climbing See Boneset, climbing

Hen of the woods (*Grifolia frondosus*)
FGM cp14; NAM cp474, cp476

Henbane (*Hyoscyamus niger*)
PPC 80

Henbit (*Lamium amplexicaule*)
CWN 233; NWE cp571; SWW cp487; WTV 171

Hepatica, round-lobed or liverwort (*Hepatica
    americana*)
BCF 125; CWN 73; NWE cp464, cp592; SEB cp487; WAA
    157; WTV 164

Herb-Christopher (*Actaea spicata*)
PPC 55

Herb-Paris (*Paris quadrifolia*)
PPC 90

Herb-Robert (*Geranium robertianum*)
CWN 141; NWE cp461

Hercules club (*Zanthoxylum clava-herculis*)
NTE cp323

Heron, Agami (*Agamia agami*)
HHH 181

Heron, black (*Egretta ardesiaca*)
BAS 117; EOB 70; HHH 103; MIA 209; WGB 19

Heron, black-headed (*Ardeola melanocephala*)
HHH 65

Heron, boat-billed (*Cochlearius cochlearius*)
AWR 165; HHH 211; MIA 209; WBW 55; WGB 19

Heron, goliath (*Ardea goliath*)
AWR 74; HHH 79; WBW 21

Heron, gray (*Ardea cinerea*)
AWR 66; BAS 149; EOB 68-69; HHH 45; MIA 209; RBB
    18; WBW 19; WGB 19

Heron, great blue (*Ardea herodias*)
AAL 103; AGC cp565; BMW 60, 61; EAB 522, 523; FEB
    139; FWB 135; HHH 27, 52, 53; MPC cp558; NAB cp10,
    cp14; NAW 154; PBC 61; UAB cp15; WBW 23; WNW
    cp528; WON 4, 37, 39

Heron, green or green-backed (*Butorides striatus*)
AAL 103; AGC cp571; BAS 120; BMW 57; BWB 31-33;
    EAB 525; FEB 133; FWB 130; HHH 17, 175; NAB cp18;
    NAW 156; PBC 62; UAB cp7, cp9; WBW 29, 31; WNW
    cp534; WOB 49-51

Heron, little blue (*Florida* or *Egretta caerula*)
AGC cp564, cp570; EAB 524, 525; FEB 137; HHH 33,
    119; NAB cp7, cp16; NAW 155; PBC 63; WNW cp521,
    cp525

Heron, Louisiana or tricolored (*Egretta* or *Hydranassa
    tricolor ruficollis*)
EAB 527; FEB 140; NAB cp13; PBC 68; PGT 226; WOB
    135

Heron, night See Night heron

Heron, pied (*Egretta picata*)
HHH 95

Heron, pond (Chinese) (*Ardeola bacchus*)
HHH 157; MAE 121

Heron, pond (Indian) (*Ardeola grayii*)
HHH 153; WBW 33

Heron, pond (other *Ardeola* sp.)
HHH 29, 161, 165

Heron, purple (*Ardea purpurea*)
EOB 67; HHH 83; WBW 27

Heron, reef (*Egretta gularis* or *sacra*)
HHH 33, 139; OBL 138

Heron, Squacco (*Ardeola ralloides*)
HHH 149; WBW 35

Heron, Sumatran (*Ardea sumatrana*)
HHH 75

Heron, tiger (*Tigrisoma* sp.)
AWR 165; HHH 215, 219, 221, 225, 229

Heron, tricolored (*Egretta tricolor*)
AGC cp568; BWB 28-29; HHH 107

Heron, whistling (*Syrigma sibilatrix*)
HHH 37

Heron, white-faced (*Egretta novaehollandiae*)
HHH 115

Heron, white-necked (*Ardea pacifica*)
HHH 61

Heron (other kinds)
HHH 41, 57, 71, 169, 231

Herring, Atlantic or sea (*Clupea harengus*)
AAL 326(bw); ACF p12; MIA 503

Herring, lake See Cisco

Herring, ox-eye (*Magalops cyprinoides*)
AWR 210(bw)

Herring, Pacific (*Clupea harengus pallasi*)
MPC cp289; NAF cp574; PCF p7

Hibiscus (*Hibiscus coccineus*) See also China rose,
Rose-mallow
NWE cp420; WNW cp275

Hickory, bitternut (*Carya cordiformis*)
NTE cp332; SEF cp135

Hickory, black (*Carya texana*)
NTE cp333

Hickory, mockernut (*Carya tomentosa*)
NTE cp348; OET 244; SEF cp137; TSV 34, 35

Hickory, nutmeg (*Carya myristiciformis*)
NTE cp347; WNW cp440

Hickory, pignut (*Carya glabra*)
NTE cp343, cp377, cp534, cp627; SEF cp136, cp199

Hickory, shagbark (*Carya ovata*)
NTE cp344, cp537, cp608; PMT cp[25]; SEF cp132

Hickory, shellbark (*Carya laciniosa*)
NTE cp345, cp536, cp626

Hickory, water (*Carya aquatica*)
NTE cp324; WNW cp439

Hind, red (*Epinephelus guttatus*)
ACF cp26; KCR cp23; NAF cp502

Hind, rock (*Epinephelus advensionis*)
ACF cp26; NAF cp505; RUP 144

Hind, speckled (*Epinephelus drummondhayi*)
ACF cp26; NAF cp503

Hippopotamus (*Hippopotamus amphibius*)
AWR 84-87; CRM 173; GEM V:66-79; MAE 95; MEM
507-511; MIA 131

Hippopotamus, pygmy (*Choeropsis liberiensis*)
CRM 173; GEM V:7, 62-64; MEM 506; MIA 131; WWW
191

Hitch (*Lavinia exilicauda*)
NAF cp125

Hoatzin (*Opisthocomus hoatzin*)
BAS 108; EOB 232; MIA 227; WGB 37; WWW 172

Hobblebush (*Viburnum alnifolium*)
NWE cp215

Hobby, northern (*Falco subbuteo*)
FGH p25; MIA 225; RBB 49; WGB 35

Hog, giant forest (*Hylochoerus meinertzhageni*)
CRM 171; GEM V:38; MEM 500; MIA 129

Hog, pygmy (*Sus salvanius*)
CRM 171; GEM V:31

Hogchoker (*Trinectes maculatus*)
AGC cp359; NAF cp70, cp290

Hogfish (*Lachnolaimus maximus*)
ACF cp39; MIA 565; RUP 71

Hogfish, Spanish (*Bodianus rufus*)
AAL 320; ACF cp39; AGC cp342; NAF cp369; RCK 33;
RUP 66; WWF cp21

Hognose snake, Eastern (*Heterodon platyrhinos*)
AAL 246(bw); AGC cp444; ARA cp485, cp563, cp565;
ARC 206, 207; ARN 65(bw); CGW 85; LBG cp606; LSW
189; RNA 167

Hognose snake, Southern (*Heterodon simus*)
ARA cp585; ARC 209; ERA 124; LSW 192; RNA 167

Hognose snake, Western (*Heterodon nasicus*)
ARA cp572; JAD cp261; LBG cp604; LSW 191; RNA 167;
  WPP 159; WRA p40

Holly, American (*Ilex opaca*)
NTE cp212, cp559; SEF cp105, cp193; TSV 158, 159

Holly, European or English (*Ilex aquifolium*)
NTE cp213; NTW cp180, cp489; OET 226

Holly, Sarvis or serviceberry (*Ilex amelanchier*)
NTE cp115

Holly (other *Ilex* sp.)
NTE cp197, cp558; PMT cp[66]

Honesty (*Lunaria annua*)
FWF 33; SWW cp441

Honey creeper, Hawaiian (*Himatione sanguinea*)
WOI 88-89

Honey creeper, blue or red-legged (*Cyanerpes cyaneus*)
EOB 407

Honeyeater (various genera)
BAS 124; EOB 399; MIA 365; NOA 220; WGB 175, 177

Honeyguide (*Indicator indicator*)
AAL 149(bw); MIA 289; WGB 99

Honeysuckle, bush (*Diervilla lonicera*)
NWE cp358; TSV 88, 89

Honeysuckle, Japanese (*Lonicera japonica*)
NWE cp20, cp360; WTV 15

Honeysuckle, swamp (*Rhododendron viscosum*)
NWE cp208; TSV 56, 57; WNW cp229

Honeysuckle, trumpet (*Lonicera sempervirens*)
NWE cp431

Hoopoe (*Upupa epops*)
AAL 149(bw); EOB 276, 278; MIA 287; WGB 97; WPP 42-43

Hoopoe, wood (*Phoeniculus purpurens*)
EOB 279

Hop tree (*Ptelea trifoliata*)
NTE cp352, cp417, cp501; NTW cp301, cp343, cp532;
  OET 240; PMT cp[110]; TSV 194, 195

Horn of plenty fungus (*Craterellus* sp.)
FGM cp7; GSM 61; SGM cp156

Hornbeam (*Carpinus caroliniana*)
NTW cp147, cp596; PMT cp[24]; SEF cp89; TSV 192, 193

Hornbeam, European (*Carpinus betulus*)
PMT cp[23]

Hornbeam, hop (*Ostrya virginiana*)
FWF 37; NTE cp222, cp371, cp491, cp628; PMT cp[91];
  SEF cp75, cp167, cp176

Hornbeam, hop (chisos) (*Ostrya chisosensis*)
NTW cp199

Hornbean, hop (Knowlton) (*Ostrya knowltonii*)
NTW cp200, cp533

Hornbill, great Indian (*Buceros bicornis*)
MIA 287; SWF 84; WGB 97

Hornbill, ground (*Bucorvus* sp.)
EOB 282; MIA 287; WGB 97; WPP 78

Hornbill, helmeted (*Rhinoplex vigil*)
MIA 287; WGB 97

Hornbill, Northern pied (*Anthracoceros albirostris*)
EOB 281

Hornbill, red-billed (*Tockus erythrorhynchus*)
EOB 284; MIA 287; WGB 97; WPP 78

Hornbill, yellow-billed (*Tockus flavirostris*)
AAL 150; BAS 112; EOB 283; WPP 78

Hornbill (other kinds)
EOB 280; WOI 114-115; WWW 131

Horned lizard, coast (*Phrynosoma coronatum*)
LOW 59; RNA 125; WRA p21

Horned lizard, desert (*Phrynosoma platyrhinos*)
ARA cp339, cp341; RNA 127; WRA p21

Horned lizard, flat-tailed (*Phrynosoa mcalli*)
ARA cp334; JAD cp194; RNA 127; WRA p21

Horned lizard, regal (*Phrynosoma solare*)
ARA cp335; JAD cp195; RNA 125; WRA p21

Horned lizard, round-tailed (*Phrynosoma modestum*)
ARA cp336; JAD cp193; LOW 15; RNA 127; WRA p21

Horned lizard, short-horned (*Phrynosoma douglassii*)
AAL 223(bw); ARA cp337, cp338; ART p[12]; JAD cp197;
  RNA 127; WFW cp583; WRA p21

Horned lizard, Texas or California (*Phrynosoma cornutum*)
ARA cp340, cp342; ARC 177; ERA 94; JAD cp196; LBG
  cp586; MIA 417; NAW 261; RNA 125; WRA p21

Hornero, rufous (*Furnarius rufus*)
EOB 312, 315; MIA 297; WGB 107

Hornet, bald-faced (*Vespula maculata*)
MIS cp483

Hornet, worker or European (*Vespa crabro*)
BOI 212; WIW 125(bw)

Hornet, worker (giant) (*Vespa crabro germana*)
LBG cp367; MIS cp444; SEF cp403

Hornworm, tobacco (*Manduca sexta*)
MIS cp25, cp558(moth); SGI cp253

Hornworm, tomato (*Manduca quinquemaculata*)
MIS cp26

Horse balm (*Collinsonia canadensis*)
NWE cp329

Horse, Przewalski's (*Equus przewalskii*)
CRM 169; GEM IV:553(head), 586-588 (Camarque), 589;
   MEM 484; MIA 125

Horsechestnut (*Aesculus hippocastanum* or *carnea*)
NTE cp360, cp407, cp625; NTW cp303, cp380, cp500;
   PMT cp[12]

Horsenettle (*Solanum* sp.)
JAD cp73; LBG cp123; NWE cp80; SWW cp584

Horsetail or scouring rush (*Equisetum* sp.)
FWF 135; PGT 36, 49

Horseweed (*Erigeron canadensis*)
NWE cp156

Houndfish (*Tylosurus crocodilus*)
ACF p17; MIA 532

Hound's-tongue, Western (*Cynoglossum grande*)
SWW cp644; WFW cp517

Howler monkey, black (*Alouatta caraya*)
GEM II:150; MEM 354; MIA 65; SAB 91

Howler monkey, mantled (*Alouatta palliata*)
CRM 69; GEM II:144; MEM 368, 369; NHP 123 (bw)

Howler monkey, red (*Alouatta seniculus*)
AAB 96; GEM II:144; MIA 65

Howler monkey (other *Alouatta* sp.)
CRM 69; GEM II:149

Huckleberry (*Vaccinium* sp.)
WFW cp150, cp163

Huemul or guemal (*Hippocamelus* sp.)
CRM 181; GEM V:220-222; MEM 522

Huisache (*Acacia farnesiana*)
LBG cp446, cp463; NTE cp297, cp396; NTW cp264,
   cp331; SEF cp164

Hummingbird, Allen's (*Selasphorus sasin*)
EAB 528; EOB 258; FWB 320; HLB 6, 79, 90; UAB cp390,
   cp392

Hummingbird, Anna's (*Calypte anna*)
AAL 142; EAB 529; EOB 259; FWB 317; HLB 8, 51, 58,
   64, 68, 75-78 +; NAW 159; UAB cp399; WFW cp236

Hummingbird, bee (*Mellisuga helenae*)
EOB 261; MIA 279; WGB 89

Hummingbird, Berylline (*Amazilia beryllina*)
HLB 10, 55

Hummingbird, black-chinned (*Archilochus alexandri*)
EAB 529; FWB 312; HLB 12, 72, 73, 96; JAD cp563; UAB
   cp398; WFW cp235; WOB 141

Hummingbird, blue-throated (*Lampornis clemenciae*)
EAB 529; FWB 311; HLB 14, 50, 70, 71, 144, 145; UAB
   cp396, 403

Hummingbird, broad-billed (*Cynanthus latirostris*)
EAB 531; EOB 263; FWB 310; HLB 16, 55, 69, 74, 82;
   UAB cp394, cp405

Hummingbird, broad-tailed (*Selasphorus platycercus*)
EAB 531; FWB 318; HLB 18, 65, 91, 132; UAB cp393,
   cp402; WFW cp238

Hummingbird, buff-bellied (*Amazilia yucatanensis*)
EAB 531; FEB 273; HLB 20; NAB cp481

Hummingbird, calliope (*Stellula calliope*)
EAB 530; FWB 316; HLB 22, 93; NAW 157; UAB cp400,
   cp407; WFW cp237; WOB 91

Hummingbird, costa (*Calypte costae*)
AAB 110; EAB 532; FWB 314; HLB xii, 24, 80, 98; JAD
   cp564; NAW 160; UAB cp401

Hummingbird, giant (*Patagona gigas*)
EOB 260; MIA 279; WGB 89

Hummingbird, Lucifer (*Calothorax lucifer*)
FWB 315; HLB 26, 39, 163

Hummingbird, magnificent or Rivoli (*Eugenes fulgens*)
EAB 533; FWB 309; HLB 28, 51, 97; UAB cp395, cp404

Hummingbird, ruby-throated (*Archilochus colubris*)
AAL 142; EAB 533; FEB 275; FWB 319; HLB 30, 88, 89,
   142, 153; MIA 279; NAB cp479; NAW 157; SEF cp321;
   WGB 89; WON 126

Hummingbird, rufous (*Selasphorus rufus*)
BAS 108; EAB 533; FEB 274; FWB 321; HLB 32, 66, 83,
   86, 139; NAB cp480; NAW 158, 161; UAB cp391, cp406;
   WFW cp239

Hummingbird, sword-billed (*Ensifera ensifera*)
EOB 261

Hummingbird, violet-chested (*Sternocylta cyanopectus*)
AAL 88

Hummingbird, violet-crowned (*Amazilia violiceps*)
EAB 533; FWB 313; HLB 34, 40; UAB cp397

Hummingbird, white-eared (*Hylocharis leucotis*)
EAB 534; FWB 308; HLB 36, 69, 81, 84, 94, 162

Humphead or nurseryfish (*Kurtus* sp.)
AWR 211(bw); EAL 118

Hutia (*Capromys* sp.)
CRM 125; GEM III:351; MEM 703; MIA 187

Hyacinth, wild (*Camassia scilloides*)
NWE cp621

Hydra (*Hydra* sp.) See also Hydroid
AAB 130; EAL 161; PGT 95

Hydrangea, wild (*Hydrangea arborescens*)
TSV 76, 77

Hydroid or hydrozoan (various kinds)
AAL 172, 515(bw); AGC cp256-263; BMW 96-97; MPC
cp481-486; MSC cp38, cp67, cp69-89; SAS cpA9, cpA10

Hydromedusa (various kinds)
AGC cp224; MPC cp385, cp388; MSC cp494, cp498-501,
cp505

Hyena, brown (*Hyaena brunnea*)
CRM 149; GEM III:566; MIA 99

Hyena, spotted (*Crocuta crocuta*)
AAL 24, 62(bw); CRM 149; GEM III:557-561, 568, 569;
MEM 154, 157; MIA 99; WPP 103

Hyena, striped (*Hyaena hyaena*)
CRM 149; GEM III:562-563; MIA 99

Hyrax, bush or yellow spotted (*Heterohyrax* sp.)
GEM IV:540, 543; MEM 464, 465; MIA 123

Hyrax, rock (*Procavia* sp.)
AAL 70(bw); GEM IV:535, 538, 540, 541, 544; MEM 463,
464; MIA 123; WOM 114

Hyrax, tree (*Dendrohyrax* sp.)
GEM IV:546; MIA 123

Hyssop (*Hyssopus officinalis*)
CWN 229

I

Ibex (*Capra ibex*)
CRM 201; GEM V:304, 516-528; MEM 585; MIA 155;
SAB 75; WOM 36, 44, 45

Ibis, glossy (*Plegadis falcinellus*)
AGC cp500; EAB 567; EOB 75; FEB 144; MIA 211; NAB
cp27; PBC 74; WBW 97; WGB 21

Ibis, olive or green (*Bostrychia* or *Lampribis olivacea*)
WBW 105

Ibis, sacred (*Threskiornis aethiopica*)
EOB 75; WBW 97

Ibis, scarlet (*Eudocimus ruber*)
AWR 168; BAS 117; BMW 66-67; EOB 78; MAE 89

Ibis, white (*Eudocimus albus*)
AGC cp561; EAB 568, 569; FEB 145; NAB cp26; WOB 36

Ibis, white-faced (*Plegadis chihi*)
EAB 570; FWB 138; NAB cp28; UAB cp14; WNW cp523

Ibis, wood See Stork, wood

Ibisbill (*Ibidorhyncha struthersii*)
MIA 245; NHU 65, 154; WGB 55

Icefish (*Pagetopsis macropterus*)
MIA 567; WPR 181

Iceplant, common (*Mesembryanthemum crystallinum*)
JAD cp88; MPC cp613; SWW cp78

Iguana, common (*Iguana iguana*)
ARA cp388; LOW 30; MIA 417; SWF 166-167

Iguana, crested (*Brachylophus vitiensis*)
ERA 93

Iguana, desert (*Dipsosaurus dorsalis*)
ARA cp345; JAD cp188; NAW 262; RNA 105; WRA cp24

Iguana, Galapagos (*Conolophus subscristatus*)
MIA 419; WOI 37; WWW 20-21

Iguana, marine (*Amblyrhynchus cristatus*)
AAL 200; ERA 98-99; MIA 419; RCK 143; WOI front., 38-
39; WWW 173-175

Iguana, rhinoceros (*Cyclura cornuta*)
LOW 51

Iguana, spiny-tailed (*Ctenosaura* sp.)
ARA cp333; MIA 419; RNA 105; WRA cp47

Impala (*Aepyceros melampus*)
EAB 105; GEM V: 312-313, 456-459; MEM 563, 568;
MIA 153; NHA cp19, cp20; WPP 86

Indian apple (*Datura discolor*)
WAA 210

Indian blanket (*Gaillardia pulchella*)
JAD cp60; LBG cp226; NWE cp419; SWW cp394; WAA 64

Indian hemp (*Apocynum cannabinum*)
CWN 189

Indian paintbrush (*Castilleia coccinea*)
CWN 241; LBG cp225; NWE cp433

Indian paintbrush, desert (*Castilleja chromosa*)
JAD cp159; SWW cp417

Indian paintbrush, giant red (*Castilleja miniata*)
SWW cp418; WFW cp472

Indian paintbrush, sulfur (*Castilleja sulphurea*)
SWW cp310; WFW cp434

Indian pipe (*Monotropa uniflora*)
BCF 157; CWN 151; NWE cp112; SEF cp449; SWW cp92;
WAA 160; WFW cp428

Indigo, false (prairie) (*Baptisia leucantha*)
LBG cp154; NWE cp146

Indigo, wild (*Baptisia tinctoria*)
NWE cp328

Indigobush (*Amorpha fruticosa*)
NWE cp631

Indri (*Indri indri*)
CRM 63; GEM II:71; MIA 57; NHP 92(bw)

Ink cap, common (*Coprinus atramentarius*)
FGM p34; GSM 233; NAM cp19; NMT cp28; PPC 122;
SGM cp223

Ink cap, desert (*Podaxis pistillaris*)
JAD cp166

Ink cap, mica or glistening (*Coprinus micaceus*)
FGM p34; GSM 231; NAM cp42; SGM cp226

Ink cap, shaggy (*Coprinus comatus*)
FGM p34; NAM cp704; NMT cp67; SGM cp224; SWF 61

Innocence See Bluets

Inside-out flower, Northern (*Vancouveria hexandra*)
SWW cp103; WFW cp400

Iora, common (*Aegithina tiphia*)
EOB 351

Irbis See Leopard, snow

Iris, Douglas' (*Iris douglasiana*)
LBG cp279; SWW cp608; WFW cp527

Iris, dwarf crested (*Iris cristata*)
NWE cp619; SEF cp495

Iris, red (*Iris fulva*)
NWE cp385; WNW cp276

Iris, tough-leaved (*Iris tenax*)
LBG cp278; SWW cp607; WFW cp528

Iris, wild See Flag, blue or yellow

Iris (other *Iris* sp.)
SWW cp606; WAA 151; WTV 165

Irish lord (*Hemilepidotus* sp.)
MPC cp254; NAF cp413; PCF p16

Irish moss (*Chondrus crispus*)
AGC cp477

Ironweed (*Vernonia noveboracensis*)
CWN 289; LBG cp247; NWE cp565; WTV 191

Ironweed, desert (*Olneya tesota*)
JAD cp319; NTW cp261

Ironweed, tall (*Vernonia altissima*)
LBG cp294; NWE cp607

Isopod (various genera)
AAL 345; AGC cp178; KCR cp26; MAE 102; MPC cp351,
cp354; MSC cp583-585, cp594; SAS cpC14; WPR 136,
137; WWW 140

J

Jabiru (*Ephippiorhynchus asiaticus*)
NOA 213

Jacamar, rufous-tailed (*Galbula ruficauda*)
EOB 295; MIA 289; WGB 99

Jacana, African (*Actophilornis africanas*)
AAL 126; AWR 95; BWB 220; EOB 180

Jacana, Australian (*Irediparra gallinacea*)
BWB 221-223

Jacana, Northern (*Jacana spinosa*)
EAB 571; EOB 179, 187; FEB 123; MIA 241; NAB cp250;
WGB 51

Jacana, pheasant-tailed (*Hydrophasianus chirungus*)
BWB 220

Jacaranda (*Jacaranda mimosaefolia*)
TSP 33

Jack (*Caranx* sp.)
ACF p30; KCR cp24; NAF cp551, cp552, cp555; PCF
cp31; RCK 41

Jackal, black-backed or silverbacked (*Canis mesomelas*)
GEM IV:108-114; MEM 64; SAB 60-61; WPP 103

Jackal (other *Canis* sp.)
CRM 133; GEM IV:107, 108, 115(head only); WPP 60-61

Jackdaw (*Corvus monedula*)
EOB 445; RBB 170

Jackfruit (*Artocarpus scortechinii*)
OET 265

Jack-in-the-pulpit (*Arisaema triphyllum*)
BCF 147; CWN 47; FWF 79; NWE cp9, cp390, cp443;
   PPM cp10; SEF cp456, cp503; WAA 160; WNW cp309,
   cp327; WTV 168

Jackrabbit, antelope (*Lepus alleni*)
ASM cp239; GEM IV:291; JAD cp510; MAE 61

Jackrabbit, black-tailed (*Lepus californicus*)
ASM cp240; CMW 86(bw); JAD cp509; LBG cp46; MEM
   716, 718; MIA 191; MNP cp5a; MPS 118(bw); NAW 43;
   WMC 113; WWD 54

Jackrabbit, white-tailed (*Lepus townsendii*)
ASM cp232, cp241; CMW 84(bw); GEM IV:285; LBG cp47;
   MNC 145(bw); MPS 118(bw); NAW 44; SWM 21; WPP
   150; WWD 114-115

Jackknife fish (*Equetus lanceolatus*)
ACF p34; KCR cp26; NAF cp343

Jack-o-Lantern fungus (*Omphalotus* sp.)
FGM cp17; NAM cp310, cp426, cp483; PPM cp32

Jacksmelt (*Atherinopsis californiensis*)
PCF p12

Jacob's ladder (*Polemonium* sp.)
NWE cp626; WAA 253

Jaeger, long-tailed (*Stercorarius longicaudus*)
AAL 129(bw); EAB 847; FEB 51; FWB 48; NAB cp87; OBL
   79; UAB cp54

Jaeger, parasitic or arctic skua (*Stercorarius parasiticus*)
AGC cp493; EAB 846; EOB 199; FEB 50; FWB 46; HSW
   40, 125; MPC cp550; NAB cp89; OBL 151; RBB 84; UAB
   cp53

Jaeger, pomarine (*Stercorarius pomarinus*)
EAB 847; EOB 199; FEB 49; FWB 47; MPC cp551; NAB
   cp88; OBL 39; UAB cp55

Jaguar (*Panthera onca*)
ASM cp274; AWC 168-183; CRM 153; GEM I:27, IV:20-
   23; MEM 48; MIA 105; NAW 45

Jaguarundi (*Felis yagouaroundi*)
ASM cp272; AWC 184; CRM 153; GEM III:623; MEM 52

Jasmine, blue (*Clematis crispa*)
WTV 157

Jawfish (*Opistognathus* sp.)
ACF cp43

Jay (*Garrulus glandarius*)
EOB 445, 446; MIA 401; RBB 167; WGB 211

Jay, blue (*Cyanocitta cristata*)
EAB 126; FEB 373; MIA 401; NAB cp435; NAW 163; PBC
   216; SEF cp324; SPN 64; WGB 211; WOB 44; WON 142,
   144

Jay, brown (*Cyanocorax morio*)
FEB 370

Jay, gray or Canadian (*Perisoreus canadensis*)
EAB 126; EOB 447; FEB 371; FWB 292; NAB cp425; SEF
   cp330; SPN 63; UAB cp471; WFW cp255

Jay, green (*Cyanocorax yncas*)
EAB 127; FEB 374; MIA 401; NAB cp483; WGB 211

Jay, Mexican (*Aphelocoma ultramarina*)
EAB 127; UAB cp506

Jay, pinyon (*Gymnorhinus cyanocephalus*)
EAB 127; FWB 294; UAB cp507; WFW cp258

Jay, scrub (*Aphelocoma coerulescens*)
EAB 127; FEB 372; FWB 293; NAB cp436; SEF cp323;
   UAB cp505; WFW cp257

Jay, Steller's (*Cyanocitta stelleri*)
EAB 127; EOB 445; FWB 295; NAW 163; UAB cp502;
   WFW cp256

Jelly mold (various genera)
NAM cp563-567; NOA 122

Jellyfish (various genera) See also Sea nettle
BCS 104; BMW 94, 99, 178-179; EAL cp4, 150, 172;
   MSC cp495, cp497, cp502-514; WPR 180; WWW 214,
   215, 219, 220, 222-223

Jellyfish, blue button (*Porpita porpita* or *linneana*)
KCR p16; MSC cp504; SAS cpA11; SCS cp30; WWW 221

Jellyfish, cannonball (*Stonolophus meleagris*)
AGC cp222; SAS cpA8; MSC cp507, cp514

Jellyfish, comb (*Mnemiopsis maccradyi*)
WWW 216-217

Jellyfish, compass (*Chrysaora hydrostatica* or
   hysoscella)
EAB 9; EAL 167; HSG cp3, cp32

Jellyfish, lion's mane (*Cyanea* sp.)
AGC cp219; BCS 46; HSG cp2; MPC cp389; SAS cpA7;
   WWW 220

Jellyfish, moon or common (*Aurelia aurita*)
AAL 516-517; AGC cp223; EAL 172; KCR p16; MAE 106;
   MPC cp384; MSC cp502; SCS cp30; WWW 213

Jellyfish, upside-down (*Cassiopeia xamachana*)
AGC cp221; KCR p16; MSC cp509; SCS cp30

Jerboa (*Jaculus* or *Allactaga* sp.)
GEM III:146; MEM 683; MIA 181; NHU 190; WWD 104

Jerusalem artichoke (*Helianthus tuberosus*)
BCF 15; CWN 263; LBG cp172; NWE cp283

Jerusalem thorn (*Parkinsonia aculeata*)
JAD cp307, cp322; NTW cp267, cp327, cp520; PMT
   cp[93]; TSP 35

Jessamine, yellow (*Gelsemium sempervirens*)
NWE cp357; WTV 55

Jewelflower, Arizona (*Streptanthus arizonicus*)
JAD cp107; SWW cp315

Jewelflower, mountain (*Streptanthus tortuosus*)
SWW cp316, cp382

Jewelweed See Touch-me-not, orange

Jewfish (*Epinephelus itajara*)
ACF cp25; AGC cp377; MIA 549; NAF cp508

Jimsonweed (*Datura stramonium*)
CWN 201; FWF 149; NWE cp86; PPC 72; PPM cp4

Jimsonweed or thorn apple, Southwestern (*Datura wrighti*)
JAD cp100; SWW cp52

Joe-pye-weed (*Eupatorium* sp.)
BCF 173; CWN 285; LBG cp248; NWE cp564; WAA 201; WNW cp287

John Dorie (*Zeus* sp.)
AAL 346(bw); EAL 105; MIA 539

Jojoba (*Simmondsia chinensis*)
JAD cp347

Joshua tree (*Yucca brevifolia*)
JAD cp326; NTW cp310; OET 260

Judas tree (*Cercis siliquastrum*) See also Redbud
OET 205; TSP 21

Jumping mouse, meadow (*Zapus hudsonius*)
ASM cp93; CMW 184(bw); GMP cp13; LBG cp67; MEM 682; MIA 181; MNC 225(bw); MPS 248(bw); WMC 168; WNW cp599

Jumping mouse, Pacific (*Zapus trinotatus*)
ASM cp91

Jumping mouse, Western (*Zapus princeps*)
ASM cp94; GEM III:141; LBG cp66; NMM cp[18]; WFW cp328

Jumping mouse, woodland or Northern (*Napaeozapus insignis*)
ASM cp92; CGW 185; GEM III:144; GMP cp14; MNC 227(bw); SEF cp584; WMC 170

Junco, dark-eyed or slate-colored (*Junco hyemalis*)
AAL 183(bw); EAB 308, 309; FEB 401; FWB 477; JAD cp612; MIA 369; NAB cp429; PBC 383; SEF cp267; SPN 139; UAB cp483; WFW cp305, cp306; WGB 179

Junco, gray-headed (*Junco caniceps*)
EAB 309; UAB cp480

Junco, yellow-eyed (*Junco phaeonotus*)
EAB 309; FWB 476; UAB cp479

June bug or June beetle See Beetle, June

Junglefowl, red (*Gallus gallus*)
MIA 231; WGB 41

Juniper, alligator (*Juniperus deppeana*)
LBG cp397, cp477; NTW cp74, cp457; WFW cp62

Juniper, California (*Juniperus californica*)
NTW cp72, cp456; WFW cp61

Juniper, common (*Juniperus communis*)
LBG cp401, cp475; NTE cp39, cp543

Juniper, oneseed (*Juniperus monosperma*)
LBG cp399; NTW cp75, cp461; WFW cp66

Juniper, pinchot (*Juniperus pinchotii*)
NTE cp38; NTW cp77

Juniper, redberry (*Juniperus erythrocarpa*)
NTW cp76, cp482

Juniper, Rocky Mountain (*Juniperus scopulorum*)
NTW cp70, cp455; WFW cp64

Juniper, Utah (*Juniperus osteosperma*)
LBG cp398, cp476; NTW cp78, cp458; WFW cp65

Juniper, Western (*Juniperus occidentalis*)
NTW cp73, cp459; WFW cp63

K

Kagu (*Rhynocheto jubatus*)
EOB 157; MIA 239; WGB 49

Kakapo or owl parrot (*Strigops habroptilus*)
BAS 28; MIA 261

Kangaroo See also Wallaroo

Kangaroo, gray (Eastern) (*Macropus giganteus*)
AAB 19; AAL 31(bw); CRM 33; GEM I:382, 383; MEM 868, 869; NOA 60; WOI 120-123; WPP 114-115

Kangaroo, gray (Western) (*Macrotus fuliginosus*)
GEM I:381

Kangaroo, rat See also Bettong

Kangaroo, rat (desert) (*Caloprymnus capestris*)
CRM 29; MEM 867

Kangaroo, rat (long-nosed) (*Potorous tridactylus*)
WOI 125

Kangaroo, rat (musky) (*Hypsiprymnodon moschatus*)
MEM 866; NOA 38

Kangaroo, rat (rufous) (*Aepyprymnus rufescens*)
MEM 865, 867; MIA 23; NOA 40

Kangaroo rat See under Rat

Kangaroo, red (*Macropus rufus*)
AAL 30(bw); GEM I:213(head only), 229, 346-354; MAE
   120; MEM 863; MIA 25; NOA 57; WPP 112

Kangaroo, tree (*Dendrolagus* sp.)
AAL 30(bw); CRM 25; GEM I:388-391; MEM 866, 871;
   MIA 25; NOA 55; SWF 104

Kapok tree See Silk cotton tree

Karung See Snake, wart

Katsura tree (*Cercidiphyllum japonicum*)
PMT cp[28]

Katydid (various genera)
BOI 110-115; EOI 44; MIS cp254, cp255, cp258, cp272,
   cp283-288; NNW 117; SEF cp413; SGI cp22-26

Kea (*Nestor notabilis*)
EOB 222; MIA 259; WGB 69; WOM 145

Kelp, giant or bull (*Macrocystis* and *Nereocystis* sp.)
BMW 68; MPC cp490, cp491, cp493; NOA 73

Kelp, winged (*Alaria* sp.)
MPC cp489

Kelpfish (various genera)
MIA 567; MPC cp266; NAF cp467, cp472; PCF cp38

Kestrel, American (*Falco sparverius*)
BOP 10, 44, 52-54, 59; EAB 297; FEB 230; FGH cp21;
   FWB 227; JAD cp542; LBG cp496; NAB cp314; NNW 13;
   PBC 118; UAB cp330, cp331; WOM 156; WON 70, 71

Kestrel, common or Eurasian (*Falco tinnunculus*)
EOB 107, 108; FGH cp21; GEM I:156; MAE 29, 122; MIA
   225; RBB 47; WGB 35

Kestrel, greater (*Falco rupicoloides*)
BOP 45

Kiang (*Equus kiang*)
CRM 169; MIA 125

Killdeer (*Charidrius vociferus*)
AAL 124(bw); EAB 716; EOB 166; FEB 170; FWB 156;
   LBG cp506; MPC cp581; NAB cp235, cp238; NAW 166;
   PBC 140; UAB cp184, cp187; WBW 197; WOB 104, 105;
   WON 7, 85-87

Killer whale (*Orcinus orca*)
AGC cp323; CRM 209; EAL 308; GEM IV:387-394; HWD
   34, 80; MEM 184; MIA 115; MPC cp209; NAF cp651;
   NAW 94; SWM 62; WDP 134, 136; WMC 216; WOW 212

Killer whale, false (*Pseudorca crassidens*)
EAL 307; HWD 34; MEM 183; MPC cp213; NAF cp637,
   cp655; WDP 134; WOW 215

Killer whale, pygmy (*Feresa attenuata*)
NAF cp656; WDP 116; WOW 218

Killifish (*Fundulus* and other genera)
ACF p19; AGC cp393-394; MIA 537; NAF cp198-cp214,
   cp346-cp353; PCF p12; WNW cp64, cp101

King devil (*Hieracium pratense*)
BCF 107; CWN 301; LBG cp164; NWE cp306; WTV 78

Kingbird, Cassin's (*Tyrannus vociferans*)
FWB 401; LBG cp516

Kingbird, eastern (*Tyrannus tyrannus*)
EAB 369; FEB 293; FWB 395; LBG cp518; MIA 305; NAB
   cp423; PBC 224; SPN 52; UAB cp474; WGB 115

Kingbird, gray (*Tyrannus dominicensis*)
EAB 369; FEB 292; NAB cp426; SEF cp280

Kingbird, thick-billed (*Tyrannus crassirostris*)
EAB 368; FWB 402

Kingbird, tropical (*Tyrannus melancholicus*)
EAB 369; FWB 400; NAB cp468; UAB cp 420

Kingbird, Western or Arkansas (*Tyrannus verticalis*)
EAB 369; FEB 288; FWB 399; JAD cp575; LBG cp517;
   NAB cp467; PBC 225;UAB cp421

Kingfish (*Menticirrhus* sp.)
ACF p35; AGC cp386; NAF cp481

Kingfisher, Amazon (*Chloroceryle amazona*)
EOB 267

Kingfisher, belted (*Ceryle alcyon*)
AAL 145(bw); AGC cp602; EAB 572; EOB 267; FEB 270;
   FWB 264; MIA 283; MPC cp557; NAB cp433; NAW 167;
   UAB cp503, cp504; WGB 93; WNW cp568; WOB 98, 99;
   WON 129

Kingfisher, blue-breasted (*Halcyon malimbica*)
EOB 267

Kingfisher, eurasian (*Alcedo atthis*)
AWR 32; BWB 224, 232-233; EOB 269; MIA 283; PGT
   227; RBB 111; WGB 93

Kingfisher, giant (*Ceryle maxima*)
BWB 235

Kingfisher, green (*Chloroceryle americana*)
EAB 572; FEB 271; NAB cp482; UAB cp523

Kingfisher, malachite (*Corythornis cristatas*)
AWR 32-33(head only); WWW 143

Kingfisher, pied (*Ceryle rudis*)
EOB 266

Kingfisher, pygmy (*Ceyx pictus*)
BWB 231

Kingfisher, ringed (*Megaceryle torquata*)
EAB 572; NAB cp434

Kingfisher, sacred (*Halcyon sancta*)
AAL 145(bw)

Kingfisher, shovel-billed (*Clytoceyx rex*)
MIA 283; WGB 93

Kingfisher, stork-billed (*Pelargopsis capensis*)
BWB 231

Kingfisher, white-collared or mangrove (*Halcyon chloris*)
MIA 283; WGB 93

Kinglet, golden-crowned (*Regulus satraps*)
EAB 1033; FEB 432; MIA 347; NAB cp458; SEF cp337; UAB cp509; WFW cp272; WGB 157

Kinglet, ruby-crowned (*Regulus calendula*)
EAB 1033; FEB 477; FWB 358; NAB cp459; PBC 283; SEF cp326; SPN 81; UAB cp510; WFW cp273

Kingsnake, California mountain or Coral (*Lampropeltis zonata*)
ARA cp599; LSW 131, 132; MSW cp83; RNA 183; WFW cp591; WRA cp37

Kingsnake, common (*Lampropeltis getulus* varieties)
ARA cp483, cp522, cp560, cp561, cp590, cp592, cp594; ARC 211; JAD cp234, cp246; LSW 114-121; MIA 453; MSW p31, p32; NAW 243; RNA 181; SEF cp525; WFW cp593; WRA cp37; WWD 149

Kingsnake, gray-banded or Mexican (*Lampropeltis mexicana*)
ARA cp601-cp603; JAD cp242; LSW 122, 123; RNA 183

Kingsnake, mole or prairie (*Lampropeltis calligaster*)
ARA cp557, cp569; ARC 210; LBG cp607; LSW 116; RNA 181

Kingsnake, scarlet (*Lampropeltis triangulum*)
ARC 212, 213; LSW 130; SAB 37

Kingsnake, Sonoran or Huachuca mountain (*Lampropeltis pyromelana*)
ARA cp598; LSW 124, 125; RNA 183; WFW cp592; WRA cp37

Kinkajou (*Potos flavus*)
GEM III:464; MEM 107; MIA 85

Kinnikinick See Bearberry or Dogwood, red-osier

Kiskadee, great (*Pitangus sulphuratus*)
EAB 363; EOB 320; FEB 291; MIA 305; NAB cp390; WGB 115

Kite, black-shouldered (*Elanus caerulus*)
FEB 226; FGH cp5; FWB 219

Kite, brahminy (*Haliastur indus*)
MIA 219; WGB 29

Kite, hook-billed (*Chondrohierax uncinatus*)
FGH cp8

Kite, Mississippi (*Ictinia mississippiensis*)
BOP 89; EAB 517; FEB 221; FGH cp5; FWB 218; NAB cp304; SEF cp260

Kite, red (*Milvus milvus*)
EOB 111; MIA 219; RBB 41; WGB 29

Kite, snail or Everglades (*Rostrhamus sociabilis*)
BOP 89; EAB 516; FEB 220; FGH cp4; MIA 219; NAB cp320; WGB 29; WNW cp558; WOB 132, 133

Kite, swallow-tailed (*Elanoides forficatus*)
EAB 517; FEB 227; FGH cp4; NAB cp302; SEF cp259; WNW cp559

Kite, white-tailed (*Elanus leucurus*)
EAB 517; NAB cp303; UAB cp319

Kittiwake, black-legged (*Rissa tridactyla*)
AGC cp500; BMW 198-199; EAB 427; EOB 189; FEB 42; FWB 22; HSW 126, 127; MIA 249; MPC cp593; NAB cp39, cp54, cp57; NAW 168-169; OBL 114, 170, 191, 226; PBC 174; RBB 90; SAB 111; UAB cp26, cp33; WGB 59; WOB 88

Kittiwake, red-legged (*Rissabrevirostris*)
EAB 427; UAB cp27

Kiwi (*Apteryx australis*)
EOB 18, 27; MIA 197; SWF 108; WGB 7

Klamath weed See St.John's-wort, common

Klipspringer (*Oreotragus oreotragus*)
GEM V:332, 334; MEM 576; MIA 151; NHA cp3; SAB 74; WPP 87

Knapweed, spotted (*Centaurea maculosa*)
CWN 293; NWE cp486; WAA 188

Knifefish, banded (*Gymnotus carapo*)
AWR 207(bw); MIA 513

Knot, red (*Calidris canutus*)
EAB 795; FEB 178; FWB 172; MPC cp566; NAB cp191, cp211; PBC 156; RBB 70; UAB cp223, cp230

Koala (*Phascolarctos cinereus*)
AAB 38; AAL 28; CRM 25; GEM I:331-338; MEM 872-875; MIA 19; NOA 149; SWF 107; WWW 105

Kob (*Kobus kob*)
AAL 83(bw); MEM 562, 565; MIA 147; NHA cp13, cp14; WPP 86-87

Koel (*Eudynamys scolopacea*)
EOB 232; MIA 265; WGB 75

Kokako (*Callaeas cinerea*)
EOB 432

Komodo dragon (*Varanus komodoensis*)
AAL 237; ERA 107; MIA 441; WOI 116; WWW 177

Kongoni See Hartebeest, coke's

Kookaburra, laughing (*Dacelo* sp.)
AAL 145(bw); EOB 268; MIA 283; NOA 144; WGB 93

Kouprey (*Bos sauveli*)
CRM 187; GEM V:389

Kowari (*Dasyuroides byrnei*)
MEM 845; MIA 17; NOA 167

Krait (*Bungarus* sp.) See also Seasnake
LSW 266-268

Kreuzotter See Adder (*Vipera berus*)
Krill (*Euphausia superba*)
HSG cp28

Kudu, greater (*Tragelaphus strepsiceros*)
GEM I:171(head), V:349, 351; MIA 141

Kudu, lesser (*Tragelaphus imberbis*)
GEM V:348; NHA cp17

Kudzu (*Pueraria lobata*)
NWE cp576

Kufah or Habu (*Trimeresurus* sp.)
LSW 363, 366

Kultarr (*Antechinomys laniger*)
MEM 838

Kusu See Opossum, brush-tailed

L

Labrador tea (*Ledum glandulosum*)
NWE cp209; SWW cp177; WFW cp185; WNW cp234

Lacebark (*Hoheria populnea*)
PMT cp[60]

Lacerta See Lizard, green and Lizard, wall

Lacewing, brown (*Hemerobius* sp.)
MIS cp328; SGI cp80

Lacewing, giant (*Polystoechotes* sp.)
MIS cp330; NNW 34

Lacewing, green (*Chrysopa* sp.)
AAL 424; ROI 180; LRG cp358; MIO cp330, SGI cp78;
    WIW 95

Ladybug See Beetle, ladybird

Ladyfish (*Elops saurus*)
NAF cp580

Lady's slipper, California (*Cypripedium californicum*)
ONA 25; SWW cp86

Lady's slipper, clustered (*Cypripedium fasciculatum*)
ONA 27; SWW cp376; WFW cp468

Lady's slipper, mountain (*Cypripedium montanum*)
ONA 29; SWW cp85; WAA 142

Lady's slipper, pink (*Cypripedium acaule*)
CWN 29; GWO 74; ONA 21; NWE cp493; SEF cp470;
    WAA 220, 221; WTV 133

Lady's slipper, ram's head (*Cypripedium arietinum*)
ONA 21; WAA 221

Lady's slipper, showy or green (*Cypripedium reginae*)
CWN 29; GWO 82; NWE cp494; ONA 31; WAA 220;
    WNW cp264

Lady's slipper, white (small) (*Cypripedium candidum*)
ONA 25; WAA 220

Lady's slipper, yellow (*Cypripedium calceolus*)
BCF 163; CWN 31; GWO 52; NWE cp320; ONA 23; SEF
    cp468; WAA 220; WTV 65

Lady's slipper (other) NHU 118; ONA 27, 29

Lady's thumb (*Polygonum persicaria*)
NWE cp528; PGT 39

Lamb's quarters (*Chenopodium album*)
NWE cp19

Lammergeier (*Gypaetus barbatus*)
MIA 219; WGB 29; WOM 74

Lampern (*Lampetra fluviatilis*)
MIA 489

Lamprey, brook (*Lampetra* sp.)
AWR 218(bw); EAL 14, 16; NAF cp37

Lamprey, Pacific (*Lampetra tridentata*)
MPC cp297; NAF cp42, cp442

Lamprey, sea (*Petromyzon marinus*)
ACF p9; AGC cp411; MIA 489; NAF cp38, cp443

Lampshell (*Terebratulina* sp.)
EAL 189; MSC cp358

Lancelet (*Branchiostoma lanceolatum*)
EAL 287

Langur, common or Hanuman (*Presbytis entellus*)
CRM 79; GEM II:296-307; NHP 154 (bw)

Langur, douc (*Pygathrix nemaeus*)
CRM 81; NHP 156

Langur (other genera) See also Monkey, leaf
CRM 79, 81; MEM 401, 402, 409-411

Lanternfish (*Myctophum* sp.)
HSG cp46; NNW 103

Lapwing (*Vanellus vanellus*)
EAB 717; EOB 163; MIA 241; RBB 69; WBW 169; WGB
51

Lapwing, sociable (or Sociable Plover) (*Chettusia gregaria*)
NHU 170

Larch See also Tamarack

Larch, European (*Larix decidua*)
NTE cp17, cp630; PMT cp[72]

Larch, golden (*Pseudolarix kaempferi*)
PMT cp[109]

Larch, Japanese (*Larix kaempferi*)
PMT cp[73]

Larch, subalpine (*Larix lyallii*)
NTW cp45; WFW cp50

Larch, Western (*Larix occidentalis*)
NTW cp44, cp424; WFW cp51, cp131

Lark, crested (*Galerida cristata*)
AAL 163; MAE 27

Lark, desert (*Ammomanes deserti*)
MIA 313; WGB 123

Lark, finch (*Eremopterix* sp.)
EOB 337

Lark, horned or shore (*Eremophila alpestris*)
AGC cp612; EAB 573; EOB 337; FEB 344; FWB 270; JAD
cp576; LBG cp520; MIA 313; NAB cp556; PBC 236; SPN
60; UAB cp603; WGB 123

Lark, short-toed or red-capped (*Calandrella cinerea*)
MIA 313; WGB 123

Lark, thick-billed (*Ramphocoris clotbey*)
MIA 313; WGB 123

Larkspur, Nuttall's (*Delphinium nuttallianum*)
JAD cp63; SWW cp634

Larkspur, prairie (*Delphinium virescens*)
LBG cp155, cp156; NWE cp152

Larkspur, rocket (*Delphinium ajacis*)
PPC 5; WTV 158

Larkspur, spring (*Delphinium tricorne*)
NWE cp622; SEF cp493

Laurel, alpine (*Kalmia microphylla*)
SWW cp560; WFW cp213

Laurel, California (*Umbellularia californica*)
NTW cp93, cp333, cp498; WFW cp77

Laurel, great See Rhododendron, white

Laurel, sheep (*Kalmia angustifolia*)
NWE cp585; WNW cp288

Laurel, swamp (*Kalmia polifolia*)
CWN 177

Laurelcherry, Carolina (*Prunus caroliniana*)
NTE cp138, cp408

Lavender, sea (*Limonium carolinianum*)
AGC cp451; CWN 177; NWE cp651

Laver, purple (*Porphyra umbilicalis*)
AGC cp476

Leadplant (*Amorpha canescens*)
LBG cp275; NWE cp632

Leaf insect (*Phyllium* sp.)
EOI 49; WIW 147

Leafbird (*Chloropsis* sp.)
AAL 166; EOB 353

Leafhopper, redbanded or scarlet and green (*Graphocephala coccinea*)
AAL 421; BOI 57; MIS cp112; SGI cp71

Leaflove (*Phyllastrephus scandens*)
MIA 321; WGB 131

Leafscraper, gray-throated (*Sclerurus albigularis*)
EOB 314

Leatherback See under Turtle

Leatherleaf (*Chamaedaphne calyculata*)
NWE cp231; WNW cp260

Leatherwood See Titi (plant)

Lechwe (*Kobus leche*)
CRM 189; GEM V:452; MAE title; MIA 147

Leech, fish (*Piscicola geometra*)
PGT 117

Leech, horse (*Haemopsis sanguisuga*)
PGT 119

Leek, wild or ramp (*Allium tricoccum*)
CWN 9; NWE cp157

Lemming, brown (*Lemmus sibiricus*)
ASM cp63

Lemming, collared or arctic (*Dicrostonyx torquatus*)
ASM cp78; GEM III:232; MEM 651, 653

Lemming, Greenland (*Dicrostonyx groenlandicus*)
GEM III:230-231

Lemming, Norway (*Lemmus lemmus*)
GEM III:227; MEM 653, 657; MIA 173

Lemming, Southern bog (*Synaptomys cooperi*)
ASM cp65; CGW 185; CRM 111; GMP 207(bw); LBG
cp55; MIA 173; MNC 223(bw); MPS 247(bw); WMC 161;
WNW cp600

Lemming, steppe (*Lagurus lagurus*)
NHU 173

Lemur, black (*Lemur macaco*)
AAL 37; CRM 61; GEM II:35, 57; NHP 101; WOI 153

Lemur, brown (*Lemur fulvus*)
CRM 61; GEM II:58-60; MEM 321; WOI 150

Lemur, dwarf (*Cheirogaleus* sp.)
GEM II:47; MEM 326; NHP 85(bw)

Lemur, four-marked (*Phaner furcifer*)
CRM 61; NHP 86(bw)

Lemur, gliding (*Cynocephalus variegatus*)
WOI 102-104

Lemur, gray gentle (*Hapalemur griseus*)
CRM 63; GEM II:49, 50; MEM 321; NHP 89 (bw)

Lemur, mongoose (*Lemur mongoz*)
CRM 61; MEM 322

Lemur, mouse (*Microcebus* sp.)
AAL 24; CRM 59; GEM II:43; MEM 327; NHP 84(bw); WOI
157

Lemur, ring-tailed (*Lemur catta*)
GEM II:18, 23, 53, 56, V:610-611; MEM 323-325; MIA
57; NHP 87 (bw), 100; WOI 155

Lemur, ruffed or variegated (*Varecia variegata*)
CRM 63; GEM II:58; MEM 321; MIA 57; NHP 88 (bw)

Lemur, sifaka (*Propithecus* sp.)
AAL 36; MEM 329; WOI front., 154

Lemur, sifaka (Verreaux's) (*Propithecus verreauxi*)
CRM 63; GEM II:65-69; MIA 57; NHP 91(bw)

Lemur, sportive or weasel (*Lepilemur mustelinus*)
CRM 63; MEM 321; NHP 90 (bw)

Lemur, woolly (*Avahi laniger*)
MIA 57

Leopard (*Panthera pardus*)
AAL 55(bw), AWC 148, 154-167; CRM 151; GEM IV:25-
29; MEM 44-47; MIA 105; WPP 94-95

Leopard, American See Jaguar

Leopard, clouded (*Neofelis nebulosa*)
CRM 157; GEM IV:3; MIA 103

Leopard frog, Northern (*Rana pipiens*)
AAL 285; ARA cp192, cp203; ARN 39(bw); CGW 29; JAD
cp286; MIA 483; NAW 283; WFW cp616; WNW cp130,
cp143; WRA cp15

Leopard frog, Plains (*Rana blairi*)
ARA cp197, cp202; LBG cp572; WRA cp15

Leopard frog, Rio Grande (*Rana berlandieri*)
ARA cp196; JAD cp285; WRA cp15

Leopard frog, Southern (*Rana sphenocaphala*)
ARA cp191, cp204; ARC 134; CGW 29; SEF cp537

Leopard, snow (*Panthera uncia*)
AWC 153; CRM 151; GEM IV:1-3; MEM 49; MIA 105

Lettuce, miner's (*Montia perfoliata*)
SWW cp35; WFW cp396

Leucaena, little-leaf (*Leucana retusa*)
NTW cp270, cp332, cp514

Lightningbug See Firefly

Lignum vitae (*Guaiacum officinale*)
PMT cp[57]

Lignum vitae, Texas (*Guaiacum angustifolium*)
NTE cp298, cp470, cp553; NTW cp256, cp399, cp484

Lilac, California See Blueblossum

Lily, adobe (*Fritillaria pluriflora*)
SWW cp502

Lily, alpine (*Lloydia serotina*)
SWW cp50

Lily, amber (*Anthericum torreyi*)
SWW cp368; WFW cp457

Lily, Atamasco or zephyr (*Zephranthes atamasco*)
NWE cp85; WTV 13

Lily, avalanche (*Erythronium montanum*)
SWW cp55; WFW cp390

Lily, blackberry (*Belamcanda chinensis*)
NWE cp369; WTV 103

Lily, bluebead or corn (*Clintonia borealis*)
CWN 9; NWE cp6, cp658; SEF cp455, cp509

Lily, calla (*Zantedeschia aethiopica*)
TSP 133

Lily, Canada or yellow meadow (*Lilium canadense*)
BCF 81; CWN 3; LBG cp224; NWE cp277, cp368; WAA
186

Lily, climbing or glory (*Gloriosa rothschildiana*)
TSP 85

Lily, corn (California) (*Veratrum californicum*)
SWW cp115; WFW cp413

Lily, desert (*Hesperocallis undulata*)
JAD cp98; SWW cp54

Lily, glacier or yellow fawn (*Erythronium grandiflorum*)
SWW cp280; WAA 230; WFW cp446

Lily, leopard (*Lilium catesbaei*)
WAA 164

Lily, Mariposa See also Cat's ears, elegant

Lily, Mariposa or globe (*Calochortus* sp.)
JAD cp152, cp154; LBG cp194; SWW cp87, cp225, cp226,
   cp279, cp362, cp363, cp461; WAA 231; WFW cp401

Lily, May See Lily-of-the-valley, false

Lily, pinewoods (*Nemastylis purpurea*)
NWE cp615

Lily, rain (*Zephyranthes longifolia*)
JAD cp153; LBG cp195; SWW cp213

Lily, sand (*Leucocrinum montanum*)
SWW cp53

Lily, sego (*Calochortus nuttalli*)
JAD cp99; LBG cp122; SWW cp48; WAA 205; WFW
   cp388

Lily, spider (*Hymenocallis* sp.)
NWE cp104; SCS cp18; WNW cp228

Lily, swamp (*Crinum americanum*)
NWE cp90; WAA 165, 201; WNW cp227

Lily, tiger (*Lilium columbianum*)
SWW cp359; WFW cp458

Lily, trout See Trout lily

Lily, turk's cap (*Lilium superbum*)
CWN 3; NWE cp367; WNW cp278; WTV 106

Lily, Washington or Cascade (*Lilium washingtonianum*)
SWW cp56; WFW cp389

Lily, wood or Rocky Mountain (*Lilium philadelphicum*)
BCF 79; CWN 3; NWE cp371; SEF cp462; SWW cp399;
   WFW cp481

Lily, wood (speckled) (*Clintonia umbellulata*)
FWF 197

Lily, zephyr (*Zephyranthes candida*) See also Lily,
   Atamasco and rain
TSP 133

Lily-of-the-valley (*Convallaria majalis*)
PPC 2; PPM cp6

Lily-of-the-valley, false and wild (*Maianthemum* sp.)
CWN 15; PCC 83; SWW cp109; WFW cp408

Limpet (*Patella vulgata*)
EAL 268; MSC cp382-391

Limpkin (*Aramus guarauna*)
AAL 121(bw); BWB 30; EAB 573; EOB 144, 149; FEB 143;
   MIA 235; NAB cp23; PBC 125; WBW 136(bw); WGB 45;
   WNW cp522

Linden See Basswood, white

Linden, european (*Tilia x europaea*)
NTE cp156; NTW cp197

Linden, silver (*Tilia tomentosa*)
PMT cp[137]

Ling, European (*Molva molva*)
ACF p15

Lingcod (*Ophiodon elongatus*)
MIA 527; MPC cp247; NAF cp428; PCF cp22

Linnet (*Acanthis cannabina*)
RBB 182

Linsang, African (*Poiana richardsoni*)
MEM 139; MIA 93

Linsang, banded (*Prionodon linsang*)
MIA 93; ROW cp3

Linsang, spotted (*Prionodon pardicolor*)
CRM 147; GEM III:527; MEM 140

Lion (*Panthera leo*)
AAB 128-129; AAL 24; AWC 16-54; CRM 151; GEM
   III:578-579, 581, IV:31-48; MEM 28-35; MIA 105; WPP
   23, 98-101

Lion, mountain See Mountain lion

Lionfish (*Pterois* or *Pteropterus* sp.)
AAL 130(bw); BMW 134-135; EAL 103; MIA 545; RCK 52;
   RUP 171; WWF cp24

Lionfish (*Scorpaenidae* sp.)
AAL 320; NOA 96

Live-forever (*Sedum purpureum*)
NWE cp552

Lizard, alligator (King or Arizona) (*Gerrhonotus kingi*)
ARA cp446; RNA 87; WRA cp29

Lizard, alligator (Northern) (*Gerrhonotus coeruleus*)
AAL 235(bw); ARA cp448; RNA 89; WFW cp585; WRA
   cp29

Lizard, alligator (Panamint) (*Gerrhonotus panamintinus*)
ARA cp447; RNA 87; WRA cp29

Lizard, alligator (Southern) (*Gerrhonotus multicarinatus*)
ARA cp445, cp449; ERA 104; MIA 439; RNA 89; WRA
cp29

Lizard, alligator (Texas) (*Gerrhonotus liocephalus*)
LOW 168; RNA 89

Lizard, armadillo (*Cordylus cataphractus*)
AAL 229

Lizard, blue-tongued (*Tiliqua scincoides*)
WWD 189

Lizard, brush (*Urosaurus* sp.)
ARA cp380; JAD cp190; RNA 113; WRA cp47, p25

Lizard, bunch-grass (*Sceloporus scalaris*)
ARA cp367, cp378; RNA 117; WRA cp26

Lizard, canyon (*Sceloporus merriami*)
ARA cp368; RNA 117

Lizard, collared (black or desert) (*Crotaphytus insularis*)
ARA cp359; JAD cp186; RNA 109

Lizard, collared (common) (*Crotaphytus collaris*)
AAL 7; ARA cp355, cp356; JAD cp200, cp201; LBG
cp591; LOW 55; MAE 61; MIA 417; RNA 107; WRA cp24;
WWD 57

Lizard, collared (reticulate) (*Crotaphytus reticularis*)
ARA cp360; RNA 109

Lizard, curltail (*Leiocephalus carinatus*)
ARA cp364; RNA 133

Lizard, earless (greater) (*Cophosaurus texamus*)
AAL 224; ARA cp361; JAD cp204; LOW 78; RNA 131;
WRA cp22

Lizard, earless (keeled) (*Holbrookia propingua*)
ARA cp365; RNA 129

Lizard, earless (lesser, Northern, or speckled) (*Holbrookia
maculata*)
ARA cp346, cp366, cp370; JAD cp192; LBG cp585; RNA
129; WRA cp22

Lizard, earless (spot-tailed) (*Holbrookia lacerata*)
ARA cp348; LBG cp587; RNA 131

Lizard, European (*Lacerta vivipara*)
LOW 80(bw)

Lizard, fence (Eastern) (*Sceloporus undulatus*)
ARA cp375; ARC 178; CGW 97; LBG cp589; MIA 417;
RNA 123

Lizard, fence (Western) (*Sceloporus occidentalis*)
ARA cp379; LOW 54; RNA 123; WFW cp584; WRA cp26

Lizard, Florida scrub (*Sceloporus woodi*)
ARA cp376; RNA 123

Lizard, frilled or frilled dragon (*Chlamydosaurus kingii*)
AAL 218(bw); ERA 90; LOW 67; MIA 421; NOA 230; WOI
137; WWD 170-171; WWW 88-89

Lizard, fringetoed (*Acanthodactylus* sp.)
WWD 66

Lizard, fringetoed (Coachella Valley) (*Uma inornata*)
RNA 133; WRA cp23

Lizard, fringetoed (Colorado desert) (*Uma notata*)
ARA cp343; ERA 94; JAD cp182; RNA 133; WRA cp23

Lizard, fringetoed (Mojave) (*Uma scoparia*)
ARA cp344; JAD cp181; RNA 131; WRA cp23

Lizard, glass (Eastern) (*Ophisaurus ventralis*)
AAL 235(bw); ARA cp453, cp456; ARC 191; RNA 91

Lizard, glass (Island) (*Ophisaurus compressus*)
ARA cp454; ARC 190; RNA 91

Lizard, glass (slender) (*Ophisaurus attenuatus*)
ARA cp455; ARC 189; LBG cp581; NAW 260; RNA 91

Lizard, green (*Lacerta* sp.)
AAL 233; ARA cp386; ERA 101; MIA 435; RNA 83; WOM
46

Lizard, horned (*Phrynosoma* sp.) See Horned lizard

Lizard, legless (*Anniella* sp.)
ARA cp452; MIA 439; RNA 91; WRA cp34, cp47

Lizard, leopard (blunt-nosed) (*Gambelia silus*)
ARA cp347; JAD cp187; LBG cp584; RNA 109; WRA cp24

Lizard, leopard (long-nosed) (*Gambelia wislizenii*)
ARA cp357; JAD cp191; RNA 109; WRA cp24

Lizard, mesquite (*Sceloporus grammicus*)
ARA cp371; ART p[13]; RNA 117

Lizard, night (common or desert) (*Xantusia virgilis*)
AAL 226(bw); ARA cp403, cp406; JAD cp183, cp184; MIA
429; RNA 85; WRA p30

Lizard, night (granite) (*Xantusia henshawi*)
ARA cp404, cp405, cp407; ERA 100; JAD cp185; LOW
151; RNA 85; WRA p30

Lizard, night (Island) (*Xantusia riversiana*)
ARA cp408; NAW 254; WRA p30

Lizard, rock (*Petrosaurus* sp.)
ARA cp358; RNA 111; WRA cp25, cp46

Lizard, rosebelly (*Sceloporus variabilis*)
ARA cp373; RNA 115

Lizard, ruin (*Podarcis sicula*)
ARA cp387

Lizard, sagebrush (*Sceloporus graciosus*)
ARA cp377; JAD cp212; RNA 121; WRA cp26

Lizard, shingleback (*Trachydosaurus rugosus*)
MAE 66; WWD 189; WWW 74

Lizard, side-blotched (*Uta stansburiana*)
ARA cp363; JAD cp202; LBG cp590; RNA 111; WRA p25

Lizard, six-lined racerunner (*Cnemidophorus
    sexlineatus*)
ARA cp411; ARC 188; LBG cp580; RNA 95

Lizard, small-scaled (*Urosaurus microscutatus* or
    U.nigricaudus)
ARA cp382; RNA 113; WRA p25

Lizard, spiny (blue) (*Sceloporus cyanogenys*)
ARA cp352; RNA 119

Lizard, spiny (Clark) (*Sceloporus clarkii*)
AAL 225; ARA cp349, cp372; RNA 121; WRA cp27

Lizard, spiny (crevice) (*Sceloporus poinsetti*)
ARA cp354; JAD cp198; RNA 119; WRA cp27

Lizard, spiny (desert) (*Sceloporus magister*)
ARA cp350; JAD cp199; RNA 121; WRA cp27

Lizard, spiny (granite) (*Sceloporus orcutti*)
ARA cp351; RNA 121; WRA cp27

Lizard, spiny (Texas) (*Sceloporus olivaceus*)
ARA cp381; RNA 119

Lizard, spiny (Yarrow's) (*Sceloporus jarrovi*)
ARA cp353; LOW 126; NAW 264; RNA 119

Lizard, striped plateau (*Sceloporus virgatus*)
ARA cp374; RNA 123; WRA cp26

Lizard, toad-headed agamid (*Phrynocephalus* sp.)
ERA 95; NHU 181; WWD 102-103

Lizard, tree (*Urosaurus ornatus*)
ARA cp369; JAD cp205; RNA 113; WRA p25

Lizard, wall (*Lacerta muralis*)
AAL 231(bw); MIA 435

Lizard, wall (*Podarcis* sp.)
LOW 144(bw), 146; RNA 83

Lizard, whiptail See Whiptail lizard

Lizard, worm See Worm-lizard

Lizard, zebra-tailed (*Callisaurus draconoides*)
ARA cp362; JAD cp203; RNA 131; WRA cp22

Lizardfish (*Snyodus* sp.)
ACF p13; EAL 68, 69; KCR cp25; NAF cp430; RUP 148

Lizard's tail (*Saururus cernuus*)
CWN 49; NWE cp121; WAA 201; WNW cp236; WTV 32

Llama (*Lama guanicoe f. glama*)
GEM V:107, 589(bw); WOM 180-181

Loach (*Botia* sp.)
EAL 78

Lobelia, great (*Lobelia siphilitica*)
LBG cp276; NWE cp636; WAA 42, 51; WTV 120, 162

Lobelia, spiked (*Lobelia spicata*)
NWE cp634

Lobster claw (*Heliconia rostrata*)
TSP 123

Lobster, northern (*Homarus americanus*)
AGC cp162; MSC cp624

Lobster, rock (California) (*Panulirus interruptus*)
MPC cp346; MSC cp623

Lobster, slipper (*Scyllarides* sp.)
MSC cp626, cp627; RCK 45; RUP 45

Lobster, spiny (*Palinurus vulgaris* or *argus*)
AAL 404; AGC cp161; EAL 232; KCR cp29; SAS cpC20;
    SCS cp51

Lobster, squat (*Galathea strigosa*)
EAL 237; NNW 97

Lobster (other kinds)
BCS 41, 62; NNW 99; RCK 45, 159; RUP 40-41, 45

Loco, melon (*Apodanthera undulata*)
JAD cp141; SWW cp345

Locoweed, purple (*Oxytropis lambertii*)
LBG cp189; NWE cp540; SWW cp516; WFW cp489

Locoweed, showy (*Oxytropis splendens*)
LBG cp265; NWE cp526; SWW cp515

Locoweed, white (*Oxytropis sericea*)
SWW cp124

Locoweed, woolly (*Astragalus mollissimus*)
LBG cp268; NWE cp531

Locust (tree) (*Robinia* sp.)
NTE cp305, cp467; TSV 104, 105

Locust, black (tree) (*Robinia pseudo-acacia*)
FWF 93; LBG cp444, cp453, cp484; NTE cp304, cp442,
    cp507; NTW cp268, cp373, cp519; SEF cp129, cp157;
    TSV 58, 59

Locust, desert (insect) (*Schistocerca gregaria*)
EOI 13, 47

Locust, honey (tree) (*Gleditsia triacanthos*)
FWF 155; NTE cp306, cp506; NTW cp269, cp518; OET
    206; WNW cp434

Locust, New Mexico (tree) (*Robinia neomexicana*)
NTW cp271, cp385, cp515; WFW cp219

Locust, seventeen-year See Cicada, periodical

Loggerhead See under Turtle

Logrunner, northern (*Orthonyx spaldingi*)
EOB 379

Longclaw (*Macronyx* sp.)
EOB 337, 339

Longspur, chestnut-collared (*Calcarius ornatus*)
EAB 310, 311; FEB 343; FWB 268; LBG cp542; NAB
cp552; UAB cp597

Longspur, lapland (*Calcarius lapponicus*)
EAB 310, 311; EOB 403; FEB 342; FWB 269; LBG cp540;
NAB cp551; PBC 393; UAB cp576

Longspur, McCown's (*Calcarius mccownii*)
EAB 310; FEB 340; FWB 267; LBG cp539; NAB cp553;
UAB cp595

Longspur, Smith's (*Calcarius pictus*)
EAB 310; FEB 345; LBG cp541; NAB cp554; UAB cp596

Lookdown fish (*Selene vomer*)
ACF p29; AWR 158; MIA 553; NAF cp545

Loon, Arctic or black-throat (*Gavia arctica*)
DBN cp1; EAB 574; EOB 41(head only); MPC cp511; NAB
cp187; RBB 6; UAB cp167, cp170

Loon, common (*Gavia immer*)
AAL 92; AGC cp525; BWB 208, 209; DBN cp4; EAB 575;
FEB 72; FWB 77; MPC cp517; NAB cp188; NAW 170;
SEF cp263; UAB cp169, cp173; WNW cp481; WON 12, 13

Loon, Pacific (*Gavia pacifica*)
FWB 75

Loon, red-throated (*Gavia stellata*)
AGC cp528; BAA 128; BWB 200; DBN cp2; EAB 574; FEB
73; FWB 74; MIA 199; MPC cp512; NAB cp186; RBB 5;
UAB cp166, cp171; WGB 9; WPR 109

Loon, yellow-billed (*Gavia adamsii*)
DBN cp3; EAB 574; FWB 76; UAB cp168, cp172

Loosejaw (*Malacosteus niger*)
EAL 61(bw); MAE 111

Loosestrife, purple (*Lythrum salicaria*)
BCF 45; CWN 143; NWE cp542; PGT 38; SWW cp535;
WAA 57; WNW cp293; WTV 187

Loosestrife (other kinds) (*Lysimachia* sp.)
CWN 179, 181; NWE cp321, cp322, cp353, cp562; SWW
cp294-296; WNW cp200, cp203, cp286; WTV 83

Lopseed (*Phryma leptostachya*)
NWE cp145

Loquat (*Eriobotrya japonica*)
TSP 57

Lords-and-ladies (*Arum maculatum*)
PPC 61, 62

Lorikeet, rainbow (*Trichoglossus haematodus*)
EOB 222, 229; MIA 259; SWF 98-99; WGB 69

Loris, slender (*Loris tardigradus*)
CRM 59; GEM II:36-38, 77, 84; MEM 333, 335; MIA 59;
NHP 95 (bw)

Loris, slow (*Nycticebus coucang*)
AAL 37; CRM 59; GEM II:77, 81, 85; MEM 335; MIA 59;
NHP 95(bw); ROW 27(bw); WOI 110

Lotus, American (*Nelumbo lutea*)
CWN 59; NWE cp245; WNW cp215

Lotus, sacred (*Nelumbo nucifera*)
TSP 103

Lotusbird See Jacana, Australian

Louse, book (Psocoptera)
NNW 72; WIW 195 (bw)

Louse, fish (*Argulus* sp.)
PGT 126, 127

Louse, head or body (*Pediculus humanus* or *capitis*)
EOI 53; MIS cp70; SGI cp44

Louse, sucking (Siphunculata)
WIW 195 (bw)

Louse, water (*Asellus* sp.)
PGT 133

Lousewort, bracted (*Pedicularis bracteosa*)
WFW cp431

Lousewort, common (*Pedicularis canadensis*)
NWE cp436; SEF cp478

Lousewort, woolly (*Pedicularis lanata*)
WAA 252

Lovebird, Fischer's (*Agapornis fischeri*)
EOB 222

Lovebug (*Plecia nearctica*)
SGI cp316

Luina, silvery (*Luina hypoleuca*)
SWW cp271; WFW cp435

Lumpfish or Atlantic lumpsucker (*Cyclopterus lumpus*)
ACF p52; AGC cp346; MIA 547; NAF cp297, 299

Lungfish, African (*Protopterus* sp.)
AAL 322(bw); AWR 209(bw); EAL 127; MIA 497

Lungfish, Australian (*Neoceratodus forsteri*)
AWR 128; EAL 126; MIA 497

Lungwort See Bluebells, alpine

Lupine (*Lupinus* sp.)
WOI 144; WPR 88

Lupine, blue-pod (*Lupinus polyphyllus*)
SWW cp638; WFW cp538

Lupine, Coulter's (*Lupinus sparsiflorus*)
JAD cp72; SWW cp637

Lupine, miniature (*Lupinus bicolor*)
LBG cp273; SWW cp636

Lupine, tree (*Lupinus arboreus*)
MPC cp601

Lupine, wild (*Lupinus perennis*)
CWN 157; LBG cp269; NWE cp640; WTV 155

Lychnis, evening See Campion, white

Lynx (*Felis lynx*)
AAL 55(bw); ASM cp268; AWC 76, 87-89, 92-95; CGW 157; CMW 245 (bw); CRM 155; GEM II:608; MAE 54; MEM 51; MIA 101; MPS 301(bw); NAW 47; NNW 63; RFA 293-299 (bw); SEF cp614; SWM 128; WFW cp361; WOM 71

Lynx, Pardel or Spanish (*Felix lynx pardina*)
AWC 90-91; GEM III:610; WOM 71

Lyonia or maleberry (*Lyonia ferruginea*)
NTE cp102; NWE cp228

Lyontree (*Lyonothamnus floribundus*)
NTW cp88, cp296, cp351

Lyrebird, Superb (*Menura novaehollandiae* or *superba*)
AAL 160, 161; EOB 308, 309, 310; MIA 311; SWF 102-103; WGB 121

M

Macaque monkey, bonnet (*Macaca radiata*)
GEM II:219; MEM 374, 378; MIA 67; NHP 43(bw); WPP 52-53

Macaque monkey, crab-eating (*Macaca fascicularis*)
CRM 75; MEM 375, 377; WOI 107

Macaque monkey, Japanese or red-faced (*Macaca fuscata*)
CRM 75; GEM II:209, 286-295; MAE 51; MEM 386; MIA 67; WOI 73, 74

Macaque monkey, lion-tailed (*Macaca silenus*)
CRM 75; GEM II:232(head only), 233; NHP 129(bw)

Macaque monkey, pig-tailed (*Macaca nemestrina*)
CRM 75; MEM 375; NHP 129(bw); WWW 39(head only)

Macaque monkey, rhesus (*Macaca mulatta*)
GEM II:214-217; MEM 317, 375, 387

Macaque monkey (other kinds)
GEM II:209-213, 218, 223; MEM 374, 375, 389; MIA 67; NHP 43(bw), 129(bw); ROW 29(bw)

Macaw, blue and yellow (*Ara ararauna*)
BAS 129; EOB 224

Macaw, hyacinth (*Anodorhynchus hyacinthinus*)
AAL 134; BAS 222

Macaw, military or red-blue-and-green (*Ara chloroptera*)
BAS 117; WWW 135

Macaw, scarlet (*Ara macao*)
EOB 227; MIA 263; SWF 34; WGB 73; WWW 135

Mackerel (*Scomber* sp.)
ACF cp48; AGC cp399, cp400; MPC cp291; NAF cp564-565; PCF cp32

Mackerel, Jack (*Trachurus symmetricus*)
MPC cp293; NAF cp560; PCF cp31

Mackerel, Spanish (*Scomberomorus maculatus*)
ACF cp48; NAF cp563

Madder, wild (*Galium mollugo*)
LBG cp152; NWE cp181

Madrone, Pacific (*Arbutus menziesii*)
NTW cp114, cp371, cp488; WFW cp76, cp158

Madrone, Texas (*Arbutus texana*)
NTE cp120; NTW cp159, cp372

Madtom or stonecat (*Noturus* sp.)
NAF cp46-cp52

Magnolia See also Catalpa, Northern, and Sweet bay

Magnolia, Ashe or sandhill (*Magnolia ashei*)
NTE cp89

Magnolia, bigleaf (*Magnolia macrophylla*)
NTE cp90, cp430; SEF cp53

Magnolia, Fraser (*Magnolia fraseri*)
NTE cp91, cp434; SEF cp55

Magnolia, pyramid or mountain (*Magnolia pyramidata*)
NTE cp92

Magnolia, saucer (*Magnolia* x *soulangiana*)
NTE cp86; NTW cp113

Magnolia, Southern (*Magnolia grandifolia*)
NTE cp58, cp432, cp565; NTW cp96, cp374, cp485; SEF cp40, cp161

Magnolia, umbrella (*Magnolia tripetela*)
NTE cp88, cp431; SEF cp54

Magpie, azure-winged (*Cyanopica cyana*)
EOB 442; MAE 26; NHU 131

Magpie, black-billed (*Pica pica*)
EAB 129; EOB 445; FEB 369; FWB 290; JAD cp580; LBG cp523; MIA 401; NAB cp584; RBB 168; UAB cp617; WGB 211; WOB 209

Magpie, green (*Cissa chinensis*)
EOB 445

Magpie, red-billed blue (*Urocissa erythrorhyncha*)
EOB 445

Magpie, yellow-billed (*Pica nuttalli*)
EAB 128; UAB cp616

Mahoney, mountain (*Cercocarpus* sp.)
NTW cp82; WFW cp72, cp92

Maidenhair tree See Ginkgo

Mako (*Isurus* sp.)
ACF p62; MIA 491; SJS 28

Mallard See under Duck

Mallee fowl (*Leipoa ocellata*)
EOB 134; MAE 63; MIA 227; WGB 37; WOI 135

Mallow See also Globemallow; Rosemallow

Mallow, Carolina (*Modiola caroliniana*)
NWE cp372

Mallow, musk (*Malva moschata*)
CWN 89; NWE cp467

Mallow, prickly (*Sida spinosa*)
CWN 89

Mallow, seashore (*Kosteletzkya virginica*)
NWE cp521; WAA 196; WTV 145

Mamba, black or green (*Dendroaspis* sp.)
LSW 261; MIA 455; MSW cp93

Manakin, red-capped (*Pipra mentalis*)
EOB 325

Manatee See also Dugong

Manatee (*Trichechus manatus*)
AGC cp331; EAL 341-345; CRM 167; GEM IV;521(head only), 527, 528; MEM 294-297; MIA 111; MNP 278(bw); NAW 48; WMC 225

Manatee, Amazonian (*Trichechus inunguis*)
AAL 70(bw); EAL 341; MEM 293, 302; WWW 177

Mandarin, wild (*Streptopus amplexifolius*)
CWN 11

Mandrake (*Mandragora officinarum*)
PPC 84

Mandrake (*Podophyllum* sp.) See Mayapple, common

Mandrill (*Papio sphinx*) See also Drill
AAL 24(head only); CRM 77; GEM II:22, 254-255(head only), 256; MEM 340, 372; MIA 69; NHP 137; SAB 82

Mangabey, agile or crested (*Cercocebus galeritus*)
CRM 73; GEM II:263; MIA 67

Mangabey, black (*Cercocebus aterrimus*)
GEM II:263

Mangabey, gray or white-cheeked (*Cercocebus albigena*)
GEM II:263; MEM 380, 391; MIA 67

Mangabey, white or collared (*Cercocebus torquatus*)
CRM 73; GEM II:263; MEM 370(head only);NHP 131(bw)

Mango (*Mangifera indica*)
OET 264; TSP 33

Mangrove, black (*Avicennia germinans*)
AGC cp446; NTE cp56, cp435; SCS cp24

Mangrove, gray (Buttonwood) (*Conocarpus erectus*)
SCS cp24

Mangrove, red (*Rhizophora mangle*)
AGC cp447; SCS cp22; TSP 105

Mangrove, white (*Laguncularia racemosa*)
AGC cp445; SCS cp24

Mannikin See Munia, white-backed

Manroot or man-of-the-earth (*Ipomoea pandurata*)
WTV 34

Manta ray (*Manta* sp.)
EAL 140; HSG cp13; MIA 495; NAF cp259, cp260; PCF p5; SJS 63; WWW 114

Mantis (various kinds)
BOI 90, 91; EOI 38, 39; LBG cp357; MIS cp298, cp301, cp302; SEF cp417; SGI cp34; WIW 110

Mantis, praying (*Mantis religiosa*)
LBG cp356; MIS cp299, cp300; WWW 94-95(*Stenovates pantherina*)
Manzanita (*Arctostaphylos manzanita*)
NTW cp105, cp370, cp478; WFW cp74, cp194

Maple, bigleaf (*Acer macrophyllum*)
NTW cp225; WFW cp102

Maple, black (*Acer nigrum*)
NTE cp260, cp622

Maple, Florida (*Acer barbatum*)
NTE cp262

Maple, hedge or field (*Acer campestre*)
PMT cp[5]

Maple, mountain (*Acer spicatum*)
NTE cp254; TSV 186, 187

Maple, Norway (*Acer platanoides*)
NTE cp259, cp388; NTW cp228, cp335

Maple, paperbark (*Acer griseum*)
PMT cp[8]

Maple, planetree or sycamore (*Acer pseudoplatanus*)
NTE cp253; NTW cp221

Maple, red (*Acer rubrum*)
NTE cp261, cp366+; NTW cp227, cp317-318, cp526;
    SEF cp120, cp180; TSV 182, 183; WNW cp479

Maple, Rocky Mountain or dwarf (*Acer glabrum*)
NTW cp230, cp300; WFW cp100

Maple, silver (*Acer saccharinum*)
NTE cp264, cp493, cp600; NTW cp229, cp316, cp527;
    SEF cp119, cp179; WNW cp478

Maple, striped (*Acer pensylvanicum*)
NTE cp255, cp623; TSV 184, 185

Maple, sugar (*Acer saccharum*)
NTE cp258, cp374, cp592; SEF cp121; TSV 180, 181

Maple, vine (*Acer circinatum*)
NTW cp231; WFW cp98

Maple (other *Acer* sp.)
NTE cp263; NTW cp226; PMT cp[4], cp[6], cp[7], cp[9]-
    cp[11]; TSV 178, 179; WFW cp99

Mara or Cavy, Patagonian (*Dolichotis* sp.)
CRM 127; GEM III:333, 334; MEM 694, 695; MIA 185

Mare's-tail (*Hippuris vulgaris*)
PGT 51

Margate (*Haemulon album*)
ACF cp32; KCR cp22; MIA 555

Margate, black (*Anisotremus surinamensis*)
ACF cp32; MIA 555

Margay or Tree ocelot (*Felis wiedi*)
AWC 190-195; CRM 153; GEM III:624, 625; MEM 52

Marguerite See Daisy, oxeye

Marigold, desert (*Baileya multiradiata*)
JAD cp128; SWW cp260

Marigold, marsh See Marsh marigold

Marijuana (*Cannabis sativa*)
LBG cp117; NWE cp20; PPC 67; PPM cp2

Mariposa See under Lily

Markhor (*Capra fulconeri*)
CRM 201; GEM V:532-534; WOM 139

Marlin (*Tetrapterus* sp.)
ACF cp49; HSG cp17; MIA 571; PCF cp32

Marlin, blue (*Makaira nigricans*)
ACF cp49; EAL 118; MIA 571

Marmoset, buffy-headed (*Callithrix flaviceps*)
GEM II:188; MEM 344

Marmoset, common or tufted (*Callithrix jacchus*)
CRM 65; GEM II:189; NHP 107(bw)

Marmoset, Goeldi's or callimico (*Callimico goeldii*)
CRM 67; GEM II:178(head only)-182; MEM 343; MIA 61;
    NHP 113(bw)

Marmoset, lion-headed (*Leontideus rosalia*)
SWF 186

Marmoset, pygmy (*Cebuella pygmaea*)
GEM II:184-186; MEM 307, 343; MIA 61; NHP 109(bw)

Marmoset, Santorem or tassel-ear (*Callithrix
    humeralifer*)
CRM 65; MEM 343

Marmoset, silvery or black-tailed (*Callithrix argentata*)
CRM 65; GEM II:190; MEM 343; MIA 61

Marmoset, white-eared (*Callithrix aurita*)
GEM II:188

Marmoset, white-fronted (*Callithrix geoffroyi*)
GEM II:191(bw)

Marmot, yellow-bellied (*Marmota flaviventris*)
ASM cp228; CMW 96(bw); GEM III:33, 49; MNP cp6a;
    MPS 131(bw): NAW 49; RMM cp[7]; SWM 30; WFW
    cp351

Marmot (other *Marmota* species)
ASM cp225, cp226, cp230; CRM 95; GEM III:35, 37, 40-
    44, 52; MEM 613, 615, 616; NHU 46, 158, 173; SWM 29;
    WFW cp352; WOM 84-85

Marsh marigold (*Caltha palustris*)
AWR 43; BCF 155; CWN 71; NWE cp276; PGT 37; PPC
    66; WAA 198; WNW cp209, cp253; WPR 54; WTV 63

Marsh marigold, Western (*Caltha leptosepala*)
SWW cp59; WFW cp381

Marsupial mouse or antechinus (*Antechinus* sp.)
CRM 15; GEM I:261, 264; MIA 17; NOA 139; WOI 125;
    WPP 119

Marsupial mouse (*Murexia* sp.)
CRM 15; MEM 839

Marsupial mouse, brush-tailed or phascogale
(*Phascogale* sp.)
CRM 19; GEM I:257, 258; MEM 839; NOA 39

Marsupial mouse (other genera)
CRM 15; GEM I:256; MEM 839, 843; WPP 119

Marten, American (*Martes americana*)
ASM cp211; CGW 145; CMW 215(bw); GEM III:417; MAE
   reverse title pg.; MEM 118; MIA 87; MNC 251(bw); MPS
   273(bw); RFA 166-171(bw); SEF cp599; SWM 101; WFW
   cp343; WOM 83

Marten (other *Martes* sp.)
AAL 62(bw); CRM 118, 141; GEM III:412, 413; MEM 118;
   NNW 42

Martin, house (*Delichon urbica*)
RBB 119

Martin, purple (*Progne subis*)
EAB 852; FEB 262; FWB 261; JAD cp577, cp578; MIA
   315; NAB cp332, cp336; SPN 61; UAB cp358, cp359;
   WGB 125

Massasauga See under Rattlesnake

Matamata (*Chelus fimbriatus*)
AAL 200; AWR 179; ERA 80; MIA 413; OTT 96; TOW cp3;
   TTW 51

Mayapple, common (*Podophyllum peltatum*)
BCF 161; CWN 99; NWE cp87; PPC 93; PPM cp26; SEF
   cp439; WTV 8

Mayflower See Arbutus, trailing

Mayflower, Canada (*Maianthemum canadense*)
NWE cp125; SEF cp452

Maypop See Passionflower

Meadow beauty (*Rhexia virginica*)
CWN 143; NWE cp477; WNW cp271; WTV 140

Meadow foam, Douglas (*Limnanthese douglasii*)
LBG cp128, cp187; SWW cp19, cp196

Meadowlark, eastern (*Sturnella magna*)
EAB 904; FEB 347; FWB 272; LBG cp547; MIA 383; NAB
   cp392; PBC 340; SPN 6, 149; UAB cp424; WGB 193;
   WON 173

Meadowlark, Western (*Sturnella neglecta*)
EAB 904; FEB 346; FWB 273; LBG cp548; NAB cp391;
   NAW 173; UAB cp423; WON 172

Meadow-rue (*Thalictrum* sp.)
BCF 89; CWN 89; NWE cp168, cp200; SEF cp384, cp434;
   SWW cp384; WFW cp460; WNW cp257

Meadow-sweet (*Spiraea latifolia*)
CWN 87; LBG cp149; NWE cp219

Mealy bug (*Pseudococcus* sp.)
MIS cp54; SGI cp77

Medick, black See Nonesuch

Medlar (*Mespilus germanica*)
OET 183

Medusafish (*Icichthys lockingtoni*)
PCF cp46

Meerkat, gray (*Suricata suricatta* or *Paracynictis selousi*)
GEM III:506-507, 515, 517, 554; MEM 24, 146; MIA 95;
   SAB 135; WWD 82

Meerkat, red (*Cynictis penicillata*)
GEM III:511-514, 548, 549

Menhaden (*Brevoortia* sp.)
ACF p12; AGC cp397; MIA 503; NAF cp576

Mercury, dog's or herb (*Mercurialis perennis*)
PPC 85

Merganser, Chinese (*Mergus squamatus*)
DNA 161, 162

Merganser, common (*Mergus merganser*)
AAL 109; DNA 163, 164; EAB 210, 211; FEB 90; FWB 84;
   NAB cp114, cp161; RBB 38; UAB cp101, cp128; WNW
   cp507

Merganser, hooded (*Lophodytes cucullatus*)
AAL 109; AGC cp540; BWB 76, 77; DNA 149-151; EAB
   210, 211; FEB 98; FWB 82; NAB cp128, cp163; NAW 174;
   UAB cp102, cp126; WNW cp509

Merganser, red-breasted (*Mergus serrator*)
AGC cp541; DNA 157, 159; EAB 211; EOB 95; FEB 91;
   FWB 83; MIA 215; NAB cp113, cp162; RBB 37; UAB
   cp100, cp129; WGB 25

Merlin or Pigeon hawk (*Falco columbarius*)
BOP 55, 58; EAB 297; FEB 232; FGH cp22; FWB 225;
   NAB cp313; PBC 117; RBB 48; UAB cp328, cp329

Mescalbean (*Sophora secundiflora*)
NTE cp340, cp471, cp504; NTW cp272, cp397, cp522;
   PMT cp[124

Mesite (*Mesitornis* sp.)
EOB 158; WGB 45

Mesquite, honey (*Prosopis glandulosa*)
JAD cp308, cp321; LBG cp445, cp461, cp485; NTE cp299,
   cp398, cp505; NTWcp259, cp517

Mesquite, screwbean (*Prosopis pubescens*)
JAD cp302, cp311, cp320; NTW cp260, cp330, cp525

Mesquite, velvet (*Prosopis velutina*)
JAD cp310; NTW cp263, cp523

Mezereon (*Daphne mezereum*)
PPC 3; PPM cp20

Midge (*Chironomus* sp.)
PGT 189; SGI cp314

Midge, green (*Tanytarsus* sp.)
LBG cp359; MIS cp379

Midshipman (*Porichthys* sp.)
ACF p14; EAL 90; MIA 531; MPC cp299; NAF cp463; PCF
cp34

Milk cap fungus (*Lactarius* sp.)
FGM cp40-42; GSM 116-131; NAM cp240, cp244,
cp247+; NMT cp19, cp20, cp52, cp132, cp148; PPC 134;
SGM cp118-133; WNW cp336

Milkfish (*Chanos* sp.)
EAL 70; MIA 509

Milkvetch (*Astragalus* sp.)
JAD cp50; LBG cp230; SWW cp318, cp430, cp490, cp557;
WFW cp471

Milkweed bug, large (*Oncopeltus fasciatus*)
MIS cp115; SGI cp27

Milkweed bug, small (*Lygaeus kalmii*)
MIS cp116; SGI cp57

Milkweed, climbing (*Sarcostemma cynanchoides*)
JAD cp83; SWW cp182

Milkweed, common (*Asclepias syriaca*)
BCF 49; CWN 191; FWF 23; LBG cp241; NWE cp234,
cp547; PPM cp24; WTV 179

Milkweed, orange See Butterflyweed

Milkweed, poison (*Asclepias subverticillata*)
JAD cp80; SWW cp164

Milkweed, purple (*Asclepias purpurascens*)
CWN 191

Milkweed, red (*Asclepias lanceolata*)
WTV 116

Milkweed, showy (*Asclepias speciosa*)
SWW cp549

Milkweed, swamp (*Asclepias incarnata*)
NWE cp548; WNW cp267

Milkweed, white (*Asclepias variegator* or *albicans*)
JAD cp82; NWE cp190; SWW cp163; WTV 22

Milkwort (*Polygala* sp.)
CWN 249; NWE cp377, cp348, cp517; SWW cp120

Milkwort, cross (*Polygala cruciata*)
CWN 249; WTV 184

Milkwort, fringed See Gaywings

Milkwort, shrubby (*Polygala chamaebuxus*)
WOM 50-51

Milkwort, yellow See Candyweed

Millipede (various genera)
AAB 124; AAL 405; EOI 2, 8, 10; JAD cp393

Mimosa, prairie (*Desmanthus illinoensis*)
LBG cp158; NWE cp119

Miner, bell (*Manorina melanophrys*)
NOA 146

Minivet, ashy (*Pericrotus divaricatus*)
NHU 128

Mink (*Mustela vison*)
ASM cp214; CGW 153; CMW 228(bw); CRM 139
(*M.lutreola*); GEM III:407; GMP 284(bw); MEM 116; MNC
261(bw); MPS 276(bw); NAW 49; NNW 82; PGT 233; RFA
205-210(bw); SWM 108; WFW cp345; WMC 194; WNW
cp613

Minnow (various genera)
NAF cp147-cp182; WNW cp71, cp72

Minnow, sheepshead (*Cyprinodon variegatus*)
ACF p19; AGC cp395; MIA 535; NAF cp197, cp349

Mint, coyote (*Monardella odoratissima*)
SWW cp616; WFW cp535

Mint, field (*Mentha arvensis*)
CWN 221; NWE cp186, cp561; SWW cp653

Mint, mountain (*Pycnanthemum* sp.)
CWN 223; NWE cp185

Mint, water (*Mentha aquatica*)
PGT 41

Mission bells (*Fritillaria lanceolata*)
SWW cp378; WFW cp467

Mistletoe, American (*Phoradendron* sp.)
FWF 57; NWE cp239; PPC 91; SEF cp499

Mite, velvet (Trombidiidae)
EOI 133; MIS cp629; WWD 153

Mite, water (*Acarina* or *Hydrachna* sp.)
AWR 25; PGT 125; WWW 52-53

Mitrewort or Bishop's cap (*Mitella* sp.)
CWN 127; NWE cp129; SWW cp9; WAA 160; WFW cp404

Mitrewort, false (*Tiarella anifoliata*)
SWW cp99; WFW cp397

Moccasin, Mexican or cantil (*Agkistrodon bilineatus*)
LSW 347, 348

Moccasin, water See Cottonmouth

Moccasin-flower See Lady's-slipper, pink

Mockernut See Hickory, mockernut

Mockingbird (*Mimus polyglottos*)
EAB 576; EOB 361; FEB 325; FWB 425; JAD cp590; MIA 331; NAB cp419; PBC 267; SPN 13, 92, 93; UAB cp475; WGB 141

Mockorange See Syringa, Lewis'

Mojarra (*Eucinostomus* sp.)
ACF p34; NAF cp542; PCF cp34

Mojarra, yellowfin (*Gerres cinereus*)
ACF p34; NAF cp518

Mole, broad-footed (*Scalopus latimanus*)
ASM cp40

Mole, coast (*Scalopus orarius*)
ASM cp42; MIA 37; MNP 169(bw)

Mole, eastern (*Scalopus aquaticus*)
AAL 32; ASM cp37, cp38; CGW 125; CMW 36(bw); GMP 73(bw); LBG cp77; MNC 99(bw); MPS 62(bw); NAW 50(bw); NNW 57, 153; WMC 60

Mole, European (*Talpa europea*)
GEM I:508-509; MEM 766; MIA 37

Mole, golden (Chrysochloridae)
CRM 39; GEM I:472, 473; MEM 765; MIA 33

Mole, hairy tailed (*Parascalops breweri*)
ASM cp39; CGW 125; GEM I:512; GMP 70(bw); MIA 37; MNC 97(bw); WMC 58

Mole, marsupial (*Notoryctes typhlops*)
GEM I:213, 297; MAE 68; MEM 842; MIA 19; NOA 172

Mole, shrew (*Neurotrichus gibbsii*)
ASM cp53; MAE 49; MEM 767; NAW 51

Mole, star-nosed (*Condylura cristata*)
ASM cp41; CGW 125; GEM I:512; GMP 76(bw); MEM 767; MIA 37; MNC 101(bw); NAW 51; WMC 62; WNW cp595

Mole-rat (*Tachyorcyctes* and other genera)
GEM III:149(bw), 312-313; MEM 667, 670, 709; MIA 169, 189; WWW 57

Mole-vole, Southern (*Ellobius fuscocapillus*)
GEM III:253; MEM 652; MIA 173

Molly Miller (*Scartella cristata*)
ACF cp43; NAF cp464

Molly, sailfin (*Poecilia latipinna*)
ACF p19; AGC cp396; NAF cp193, cp350; WNW cp56

Mollymawk See Albatross

Moloch See Thorny devil

Monal See Pheasant

Moneywort (*Lysimachia nummularia*)
CWN 181; NWE cp261; WTV 84

Mongoose, banded or zebra (*Mungos mungo*)
GEM III:536, 544; MEM 148; MIA 97

Mongoose, bushy-tailed (*Bdeogale crassicauda*)
MEM 149; MIA 97

Mongoose, crab (*Herpestes urva*)
GEM III:543

Mongoose, dwarf (*Helogale parvula*)
GEM III:556; MEM 134, 147, 149, 152-53

Mongoose, Egyptian (*Herpestes ichneumon*)
MEM 149

Mongoose, Indian gray (*Herpestes* sp.)
GEM III:368, 540; MIA 97

Mongoose, marsh or water (*Atilax paludinosus*)
GEM III:548; MEM 149; MIA 97

Mongoose, narrow-striped (*Mungotictis decemlineata*)
GEM III:537; MEM 149

Mongoose, ring-tailed (*Galidia elegans*)
MEM 149

Mongoose, slender (*Herpestes sanguineus*)
GEM III:539

Mongoose, white-tailed (*Ichneumia albicauda*)
GEM III:546; MEM 149; MIA 97

Mongoose, yellow See Meerkat, red

Monitor, dragon See Komodo dragon

Monitor lizard, Gould's or sand goanna (*Varanus gouldii*)
ERA 107; MIA 441; WOI 136; WWD 191

Monitor, Komodo See Komodo dragon

Monitor lizard, lace (*Varanus varius*)
ERA 71; MAE 6

Monitor lizard (other kinds)
AAB 79; AAL 236; AWR 124-125; ERA 105, 106; LOW 58; MIA 441; NHU 182; WWW 72, 195

Monkey, Allen's swamp (*Allenopithecus nigroviridis*)
MEM 380; MIA 71; NHP 145(bw)

Monkey, blue (*Cercopithecus mitis*)
CRM 73; GEM II:274

Monkey, capuchin (black capped or brown) (*Cebus apella*)
GEM II:159; MEM 355; NHP 115 (bw)

Monkey, capuchin (white-fronted) (*Cebus albifrons*)
GEM II:152; MIA 63

Monkey, colobus See Colobus

Monkey, De Brazza's (*Cercopithecus neglectus*)
GEM II:274; MIA 71; NHP 140(bw); WOM 101

Monkey, Diana or Rolaway (*Cercopithecus diana*)
CRM 73; GEM II:268; MIA 71; NHP 142(bw)

Monkey, golden (*Rhinopithecus roxellana*)
NHP 157

Monkey, green (*Cercopithecus sabaeus*)
GEM II:264-267

Monkey, howler See Howler monkey

Monkey, leaf (*Presbytis* sp.) See also Langur
GEM II:308, 311; NHP 153; NHP 155(bw); WOI 108-109

Monkey, leaf (*Semnopithecus* sp.)
MEM 400, 404; NHP 152

Monkey, lesser white-nosed (*Cercopithecus petaurista*)
GEM II:270-271

Monkey, L'Hoest's (*Cercopithecus lhoesti*)
CRM 73; GEM II:274

Monkey, macaque See Macaque

Monkey, mangabey See Mangabey

Monkey, marmoset See Marmoset

Monkey, mona (*Cercopithecus mona*)
GEM II:269

Monkey, moustached (*Cercopithecus cephus*)
GEM II:272-273(head only); MEM 380

Monkey, night (*Aotus trivirgatus*)
GEM II:119, 123; MEM 355, 365; MIA 63; NHP 115(bw);
SWF 186

Monkey, owl-faced (*Cercopithecus hamlyni*)
GEM II:274

Monkey, patas (*Erythrocebus patas*)
AAL 38(bw); GEM II:282-285; MEM 377, 381; MIA 71;
NHP 146(bw)

Monkey, proboscis (*Nasalis larvatus*)
CRM 81; GEM II:314; MAE 88; MEM 399; MIA 73; NHP
160; WOI 106; WWW 124(head only)

Monkey, red-bellied (*Cercopithecus erythrogaster*)
GEM II:270; MIA 71

Monkey, red-eared (*Cercopithecus erythrotis*)
CRM 73

Monkey, Schmidt's (*Cercopithecus ascanius schmidti*)
NHP 143(bw)

Monkey, snub-nosed (golden) (*Pygathrix roxellanae*)
CRM 81; MIA 73

Monkey, spider (*Ateles* sp.) See also Muriqui
CRM 71; GEM II:161(head only), 162(head only), 166,
168; MEM 354: MIA 65

Monkey, squirrel (*Saimiri* sp.)
CRM 67; GEM II:125-129; MEM 311, 354

Monkey, tantalus (*Cercopithecus aethiops tantalus*)
NHP 143(bw)

Monkey, vervet or grivet (*Cercopithecus aethiops*)
AAL 38(bw); GEM II:275-278; MEM 370, 377, 378, 383;
MIA 71; SEF 114-115

Monkey, woolly (*Lagothrix lagothricha*)
CRM 71; GEM II:162, 164, 169, 171; MEM 354, 359; MIA
65; NHP 127(bw); SWF 187

Monkeyflower (*Mimulus* sp.)
CWN 241; NWE cp616; SWW cp286, cp357, cp435,
cp545, cp564; WFW cp204, cp439, cp497; WNW cp217,
cp298; WTV 183

Monkfish See Shark, angel

Monkshood (*Aconitum* sp.)
NWE cp613; SWW cp635; WAA 230; WFW cp530

Montia, broad-leaved (*Montia cordifolia*)
SWW cp21; WFW cp397

Monument plant (*Frasera speciosa*)
SWW cp4; WFW cp411

Moon jelly See Jellyfish, moon

Moonfish (*Selene* sp.) See also Opah
NAF cp544; PCF cp31

Moonrat, greater or gymnure (*Echinosorex gymnurus*)
GEM I:446; MEM 751, 752; MIA 33; ROW 8(bw)

Moonrat, lesser (*Hylomys suillus*)
GEM I:448; MEM 752; ROW 9(bw)

Moonrat, Mindano (*Podogymnura truei*)
CRM 35; MEM 752; MIA 33

Moonseed (*Menispermum canadense*)
NWE cp204

Moorhen See Gallinule, common

Moose (*Alces alces*)
AAL 77; ASM cp291; CGW 205; GEM V:230-241; MAE
11; MEM 524, 532, 533; MIA 137; MNC 289(bw); MNP
289(bw); MPS 304(bw); NAW 2-3, 52, 53; NHU 61; SEF
cp617; SWF 158-159; SWM 149-151; WFW cp378;
WNW cp618; WPR 112-113

Moray See under Eel

Morel (*Morchella* sp.)
FGM cp2; GSM 33-35; NAM cp710, cp711, cp713; NMT cp46, cp47; SGM cp286

Morel, false (*Gyromitra* sp.)
FGM cp3; NAM cp714-720; PPC 126; PPM cp30

Morning-glory (*Ipomoea* sp.)
CWN 203; NWE cp580, cp654; PPM cp3; SWW cp467, cp573; TSP 87

Morning-glory, beach, or railroad vine (*Ipomoea pes-caprae*)
NWE cp578; SCS cp18; SWW cp468; TSP 103

Morning-glory, ivy-leaved (*Ipomoea hederacea*)
LBG cp283; NWE cp654; WTV 192

Morning-glory, red (*Ipomoea* or *Quamoclit coccinea*)
NWE cp418; WTV 122

Mosquito, house (*Culex pipiens*)
LBG cp362; MIS cp419; PGT 181-183; WNW cp376

Mosquito, malaria (*Anopheles gambiae*)
EOI 91; LBG cp360; MIS cp384; SGI cp313

Mosquito, salt marsh (*Aedes* sp.)
AGC cp487; MIS cp380-381; SGI cp311

Mosquito (other kinds)
AAB 43; MAE 134; MIS cp386-389; PGT 185; SGI cp312

Mosquitofish (*Gambusia affinis*)
ACF p19; AGC cp392; EAL 101; JAD cp294; NAF cp208, cp351; WNW cp97

Moss animal See Bryozoan

Moss, fairy See Fern, water

Moss, Spanish See Spanish moss

Moth, acraea (*Estigmene acraea*)
LBG cp303; MIS cp4(larva), cp527

Moth, Asian atlas (*Attacus atlas*)
AAL 448, 449; SGB cp270; SWF 87; WOI 100(head only); WWW 138

Moth, Atlas (*Coscinocera hercules*)
SAB 69; SGB cp272

Moth, burnet (*Zygaena* sp.)
EOI 100

Moth, cabbage (*Mamestra brassicae*)
SGB cp92

Moth, ceanothus (*Hyalophora euryalus*)
SGI cp246

Moth, cecropia (*Hyalophora cecropia*)
MIS cp20(larva), cp564; SGI cp248

Moth, cinnabar (*Callimorpha jacobaeae*)
AAB 40(larva); AAL 453

Moth, clio (*Ectypia clio*)
SGI cp254

Moth, codling (*Laspeyresia pomonella*)
MIS cp542

Moth, cynthia (*Samia cynthia*)
BOI 162-163; MIS cp12(larva), cp562; SGB cp122

Moth, Emperor or giant peacock (*Saturnia pyri*)
SGB cp124; WIW 101(bw)

Moth, Emperor gum (*Antheraea eucalypti*)
EOI 109(larva)

Moth, gypsy (*Lymantria dispar*)
MIS cp528; SEF cp354; SGB cp182

Moth, hawk See Hawk moth

Moth, imperial (*Eacles imperialis*)
MIS cp19(larva), cp550

Moth, io or bull's-eye (*Automeris io*)
MIS 27(larva), cp566; SGB cp148

Moth, lappet (*Gastropacha quercifolia*)
SGB cp70; WIW 94(bw)

Moth, luna or moon (*Actias luna*)
AAL 448; MIS cp23(larva), cp573; NNW 112; SEF cp349; SGB cp140; SGI cp243; SWF 87

Moth, oak beauty (*Biston strataria*)
WIW 154

Moth, polyphemus (*Antheraea polyphemus*)
MIS cp23(larva), cp567; SEF cp360; SGB cp143; SGI cp250

Moth, prometheus (*Callosamia promethea*)
MIS cp563; SEF cp358

Moth, regal (*Citheronia regalis*)
MIS cp28(larva), cp568; SGB cp157

Moth, rosy maple (*Dryocampa rubicunda*)
SEF cp351

Moth, satin (*Stilpnotia salicis*)
SGB cp202

Moth, satyr (Carolina) (*Hermeuptychia sasybius*)
SEF cp359

Moth, silk (*Automeris* sp.)
SAB 39(larva)

Moth, sphinx (pandora) (*Eumorpha pandorus*)
MIS cp547; SGI cp251

Moth, sphinx (willow) (*Smerinthus cerisyi*)
MIS cp565; SGI cp252

Moth, sphinx (other kinds)
JAD cp448; MIS cp548, cp549, cp554; SGB cp186, cp198,
   cp201

Moth, tiger (milkweed) (*Euchaetias egle*)
LBG cp306; MIS cp6(larva), cp529

Moth, tiger (ornate) (*Apantesis ornata*)
LBG cp302; MIS cp555

Moth, tiger (southern) (*Apantesis phyllira*)
SGI cp257

Moth, tiger (spotted) (*Halisidota maculata*)
MIS cp540; SEF cp355

Moth, tussock (white-marked) (*Orgyia leucostigma*)
MIS cp13(larva), cp545

Moth, underwing (various genera)
MIS cp556, cp557; SGB cp55; WIW 55(larva)

Motherwort (*Leonurus cardiaca*)
CWN 233; NWE cp559; WTV 180

Motmot, blue-crowned (*Momotus momota*)
EOB 271; MIA 285; WGB 95

Mouflon (*Ovis ammon* or *orientalis*)
AAL 86(bw); CRM 203; GEM V:544-548; MIA 157; WOM
   62

Mountain ash (*Pyrus americana*)
FWF 117; TSV 168, 169

Mountain ash, American (*Sorbus americana*)
NTE cp320, cp416, cp564; SEF cp130, cp194

Mountain ash, European or Rowan (*Sorbus aucuparia*)
NTE cp321; NTW cp289, cp477; OET 185

Mountain ash (other *Sorbus* sp.)
NTE cp319, cp415; NTW cp290; WFW cp108

Mountain lion (*Felis concolor*)
ASM cp270, cp271; AWC 56-75; CGW 125; CMW
   243(bw); CRM 153; GEM III:599, 601-607; JAD cp518,
   cp519; MAE 5, 61; MEM 55; MIA 101; MNC 273(bw); MNP
   cp4b; MPS 300(bw); NAW 46; RFA 280-289(bw); SWF
   173; SWM 127; WFW cp363; WMC 203; WOM 154; WPP
   177; WWD 141

Mountain pride (*Penstemon newberryi*)
SWW cp492

Mountain-laurel (*Kalmia latifolia*)
BCF 135; NTE cp57, cp440, cp462; NWE cp212, cp586;
   PPM cp18; SEF cp44, cp148, cp436, cp484; TSV 102,
   103; WAA 10

Mouse, African climbing (*Dendromus* sp.)
GEM III:270-273; MEM 670

Mouse, Australian (*Pseudomys* sp.)
GEM III:154

Mouse, beach See Mouse, oldfield

Mouse, birch (*Sicista betulina*)
GEM III:142-143; MIA 81; SWF 66

Mouse, brush (*Peromyscus boylii*)
ASM cp125; JAD cp490

Mouse, cactus (*Peromyscus eremicus*)
ASM cp133; JAD cp492

Mouse, California (*Peromyscus californicus*)
ASM cp120

Mouse, canyon (*Peromyscus crinitus*)
CMW 151(bw)

Mouse, chinchilla (*Chinchillula sahamae*)
CRM 107

Mouse, cotton (*Peromyscus gossypinus*)
ASM cp126; MNC 191(bw); MPS 212(bw); SEF cp581;
   WMC 145; WNW cp597

Mouse, deer (*Peromyscus maniculatus*)
AAB 143; ASM cp124, cp128; CGW 177; CMW 147(bw);
   GEM III:214; GMP 178(bw); JAD cp489; LBG cp62; MEM
   641; MIA 167; MNC 195(bw); MPS 213(bw); NAW 54; SEF
   cp580; SWM 46, 47; WFW cp329; WMC 141

Mouse, Florida (*Peromyscus floridanus*)
ASM cp132

Mouse, golden (*Ochrotomys nuttalli*)
ASM cp129; MIA 167; MNC 199(bw); MPS 215(bw); WMC
   147; WNW cp598

Mouse, grasshopper (Northern) (*Onychomys
leucogaster*)
ASM cp99; CMW 155(bw); JAD cp491; LBG cp70; MEM
   649; MIA 167; MNC 201(bw); MPS 215(bw); NMM cp[16]

Mouse, grasshopper (Southern) (*Onychomys torridus*)
ASM cp98; GEM III:4, 218; JAD cp488

Mouse, harsh-furred (*Lophuromys sikapusi*)
MEM 660

Mouse, harvest (*Micromys minutus*)
GEM III:126-127, 183, 184; MAE 121; MIA 179

Mouse, harvest (*Reithrodontomys* sp.)
ASM cp113, cp116, cp118, cp122; CMW 143(bw),
   145(bw); GEM II:219; JAD cp486, cp487; LBG cp63, cp64,
   cp65; MEM 13, 648; MIA 167; MNC 187(bw), 189(bw);
   MPS 181(bw), 182(bw), 211(bw); NAW 55, 56; WMC 138

Mouse, hopping (*Notomys* sp.)
CRM 117; MEM 661; MIA 179; NOA 185; WWD 52

Mouse, house (*Mus musculus*)
ASM cp123; CGW 185; CMW 182(bw); GEM I:152-153, III:177, 178; GMP 221(bw); JAD cp495; LBG cp61; MIA 179; MNC 297(bw); WMC 166

Mouse, jumping See Jumping mouse

Mouse, kangaroo (*Microdipodops* sp.)
ASM cp102, cp103; JAD cp478, cp479; MIA 163

Mouse, leaf-eared (*Phyllotis* sp.)
MEM 647, 648

Mouse, marsupial See Marsupial mouse

Mouse, meadow See Vole, meadow

Mouse, mole (*Notiomys* sp.)
MEM 646

Mouse, oldfield (*Peromyscus polionotus*)
ASM cp112; LBG cp71; WMC 140

Mouse, pinyon (*Peromyscus truei*)
ASM cp130; MPS 214(bw); NMM cp[14]

Mouse, pocket (Plains) (*Perognathus flavescens*)
ASM cp108; CMW 129(bw); LBG cp75; MNC 181(bw); MPS 178(bw); NAW 56

Mouse, pocket (silky) (*Perognathus flavus*)
ASM cp97; CMW 131(bw); JAD cp471; MIA 163; MPS 178(bw); NMM cp[15]

Mouse, pocket (Mexican spiny) (*Liomys irroratus*)
ASM cp115; MIA 163

Mouse, pocket (other *Perognathus* sp.)
ASM cp95-111; GEM III:139; JAD cp472-477; LBG cp68, cp69; MPS 177(bw), 179(bw)

Mouse, pygmy (*Baiomys* sp.)
ASM cp114; LBG cp74; MEM 644

Mouse, rock (*Petromyscus* sp.)
MAE 19; MPS 212(bw)

Mouse, spiny (*Acomys* sp.)
GEM III:200, 201(bw); MEM 661

Mouse, Texas (*Peromyscus attwateri*)
MPS 211(bw)

Mouse, white-ankled (*Peromyscus pectoralis*)
MPS 214(bw)

Mouse, white-footed (*Peromyscus leucopus*)
AAL 44; ASM cp127; CGW 177; CMW 149(bw); GMP 183(bw); MNC 193(bw); MPS 213(bw); NAW 57; SEF cp579; WMC 143

Mouse, wood (*Apodemys sylvaticus*)
GEM III:188; MEM 639; MIA 179; SAB 16

Mouse, yellow-necked (*Apodemus flavicollis*)
GEM III:181, 186-187; WPP 41

Mousebird (*Colius* sp.)
EOB 265

Mudminnow (*Umbra* sp.)
EAL 45; NAF cp195

Mudpuppy or water dog (*Necturus* sp.)
AAB 56; AAL 274(bw); ARA cp16, cp19-cp21; ARC 49-51; ARN 14(bw); CGW 45; ERA 27; MIA 467; NAW 291; WNW cp104

Mudskipper (*Periophthalmus* sp.) See also Goby
EAL 38; MAE 89

Mulberry, paper (*Broussonetia papyrifera*)
NTE cp236

Mulberry, red (*Morus rubra*)
NTE cp157; NTW cp203; SEF cp70; TSV 148, 149

Mulberry, Texas (*Morus microphylla*)
NTE cp183; NTW cp202, cp251

Mulberry, white (*Morus alba*)
FWF 129; NTE cp154, cp237, cp546, cp607; NTW cp204, cp252, cp463; SEF cp125

Mule's ears (*Wyethia amplexicaulis*)
LBG cp169; SWW cp249

Mulgara (*Dasycercus cristicauda*)
CRM 17; MEM 845; MIA 17

Mullein (*Verbascum thapsus*)
BCF 59; CWN 235; LBG cp219; NWE cp336; SWW cp298; WAA 61

Mullein, moth (*Verbascum blattaria*)
CWN 235; LBG cp192; NWE cp150, cp335; SWW cp297; WAA 184

Mullet (*Mugil* sp.)
ACF p41; AGC cp405; MIA 559, 565; MPC cp277, cp287; NAF cp484, cp485; PCF cp35

Munia, white-backed (*Lonchura striata*)
EOB 424

Muntjac (*Muntiacus* sp.)
CRM 175; GEM V:137; MEM 524; MIA 135; NNW 167

Muriqui (*Brachyteles arachnoides*)
CRM 71; GEM II:176, 177; MEM 357; MIA 65

Murre, common or guillemot (*Uria aalge*)
AGC cp520; AOG cp3; BAA 95(bw); DBN cp14, cp18; EAB 46; EOB 207; FEB 60; FWR 61; HSW 151; MIA 251; MPC cp535; NAB cp94; OBL 39, 207; RBB 95; UAB cp79, cp81; WGB 61

Murre, thick-billed (*Uria lomvia*)
AOG cp4; DBN cp15; EAB 47; FEB 61; NAB cp93; OBL 142; PBC 187; UAB cp78; WOB 88; WOI 67

Murrelet, ancient (*Synthliboramphus antiquua*)
EAB 47; FWB 64; MIA 47; MPC cp538

Murrelet, Kittlitz (*Brachyramphus brevirostre*)
DBN cp20; EAB 47; MIA 47

Murrelet, marbled (*Brachyramphus marmoratus*)
FWB 62; MIA 251; WGB 61

Murrelet, Xantus (*Synthliboramphus hypoleucus*)
MPC cp537

Mushroom, amanita (*Amanita* sp.)
FGM cp25-28; GSM 146-162; NAM cp112-149, cp164-167; NMT cp8-cp11, cp24, cp25, cp59+; SGM cp1, cp5-16

Mushroom, amanita (Caesar's) (*Amanita caesarea*)
FGM cp25; NAM cp142, cp681; NMT cp1; SGM cp1

Mushroom, amanita (spring or fool's) (*Amanita verna*)
FGM cp27; NMT cp10; PPC 113; SGM cp7

Mushroom, bolete (*Boletus* sp.)
FGM cp13; GSM 98-113; NAM cp370, cp382-385, cp395-399+; NMT cp41, cp42, cp43, cp54, cp55+; PPC 115-118; SGM cp239-257

Mushroom, coral (various genera)
NAM cp733-752

Mushroom, cultivated (*Agaricus bisporus*)
SGM cp206

Mushroom, cultivated (*Psalliota hortensis*)
NMT cp121

Mushroom, fairy helmet (*Mycena* sp.)
FGM cp19

Mushroom, fairy-ring (*Marasinius oreades*)
NAM cp17

Mushroom, fly See Fly agaric

Mushroom, fried chicken (*Lyophyllum* sp.)
FGM cp18; NAM cp266; NMT cp34

Mushroom, funnel-cap (*Leucopaxillus* sp.)
NAM cp239, cp277

Mushroom, funnel-cup (*Clitocybe infundibuliformis*)
NMT cp129

Mushroom, hedgehog (*Hericium erinaceus*)
GSM 52; NAM cp547

Mushroom, honey (*Armillaria mellea*)
FGM cp15; GSM 185; NMT cp105; SGM cp85; SWF 61

Mushroom, horse (*Psalliota arvensis*)
FGM cp31; NMT cp13; SGM cp204

Mushroom, meadow (*Agaricus campestris*)
NAM cp153, cp154

Mushroom, oyster (*Pleurotus ostreatus*)
FGM cp20; GSM 182; NAM cp484, cp497; NMT cp38; SGM cp113

Mushroom, parasol (*Lepiota* sp.)
FGM cp29; GSM 164-168; NAM cp169-180; NMT cp12, cp116-cp119, cp139; PPC 135; SGM cp20-cp27

Mushroom, pig's-ears (*Discina perlata*)
NAM cp618

Mushroom, pig's-ears (*Gomphus clavatus*)
FGM cp7

Mushroom, poison-pie (*Hebeloma crustuliniforme*)
NAM cp259; PPC 127

Mushroom, russula or brittlegill (*Russula* sp.)
FGM cp39, cp40; GSM 132-137; NAM cp249, cp252, cp286+; NMT cp4, cp21, cp39, cp61+; SGM cp134-150; WNW cp335

Mushroom, scale-cap (*Pholiota* sp.)
FGM cp33; NAM cp181-192+

Mushroom, thimble-cap (*Verpa* sp.)
NAM cp708, cp712

Mushroom, velvet-foot (*Flammulina velutipes*)
GSM 178; NAM cp63

Mushroom, web-cap or cort (red or cinnabar) (*Cortinarius cinnabarinus*)
NAM cp332; NMT cp3

Mushroom, web-cap (other kinds) (*Cortinarius* sp.)
FGM cp36; GSM 211-214; NAM cp298-300+

Mushroom, wood or forest (*Agaricus* or *Psalliota silvicola*)
FGM cp31; SGM cp205

Musk ox (*Ovibos moschatus*)
ASM cp277; BAA 24(bw), 57(bw), 73; GEM I:171(head only), V:561-563, 566; MAE 82-83; MEM 5, 585; MIA 157; NAW 58; SWM 167, 168; WOI 52-53

Muskellunge (*Esox masquinongy*)
NAF cp34; NAW 305(bw); WNW cp77

Muskflower (*Mimulus moschatus*)
NWE cp267

Muskrat (*Ondatra zibethicus*)
AAL 44; ASM cp222; CGW 177; CMW 178(bw); GEM III:250-251; GMP 210(bw); MEM 653; MIA 173; MNC 219(bw); MNP 218(bw); MPS 247(bw); NAW 59; PGT 234; RFA 32-41; SWM 55; WMC 159; WNW cp609

Muskrat, round-tailed (*Neofiber alleni*)
ASM cp221; WNW cp608

Mussel, blue or edible (*Mytilus edulis*)
AAL 495(bw); AWR 156; EAL 268; MSC cp293

Mussel, european or freshwater swan (*Anodonta cygnaea*)
AWR 44; PGT 109

Mussel (other kinds)
AGC cp132; MSC cp292, cp294, cp295; SAS cpB20

Mustard, black (*Brassica nigra*)
NWE cp346

Mustard, hedge (*Sisymbrium officinale*)
SWW cp332

Mynah, common (*Acridotheres tristis*)
EOB 430

Myna, crested (*Acridotheres cristatellus*)
FWB 296

Myrtle, sand (*Leiophyllum buxifolium*)
NWE cp456; SEF cp473

Myrtle, sea See Groundsel tree

Myrtle, wax See Bayberry, Southern

N

Nannyberry (*Viburnum lentago*)
NTE cp173, cp414, cp582; SEF cp80, cp154, cp195

Narwhal (*Monodon monocerus*)
BAA 51(bw), 147(bw); CRM 209; EAL 312, 314; GEM IV:378; HWD 31, 71; MEM 200; MIA 117; NAF cp674; WDP 72; WOW 163

Nautilus, chambered (*Nautilus macromphalus*)
BMW 146

Nautilus, paper (*Argonauta* sp.)
HSG 69(bw)

Nautilus, pearly (*Nautilus pompilus*)
EAL 268; HSG cp34

Nayan See Argali

Needlefish (*Strongylura* sp.)
ACF p17; AGC cp412; NAF cp435; PCF p12

Needletail, brown (*Chaetura gigantea*)
MIA 277; WGB 87

Nematode or roundworm (Nematoda)
PGT 99

Nettle, false (*Boehmaria cylindrica*)
NWE cp24

Nettle, stinging (*Urtica dioica*)
NWE cp22; PPM cp1

Nettle, wood (*Laportea canadensis*)
NWE cp23; WNW cp317

Newt, black-spotted (*Notophthalmus meridionalis*)
ARA cp28; ART p[3]; WNW cp124

Newt, California (*Taricha torosa*)
ARA cp32, cp33; WFW cp602; WNW cp115; WRA cp1

Newt, red-bellied (*Taricha rivularis*)
ARA cp34, cp36; ERA 33; WRA cp1

Newt, red-spotted or eastern (*Notophthalmus viridescens*)
ARA cp26, cp27, cp29, cp30; ARC 47; ARN 20(bw), 21(bw); CGW 49; ERA 12, 25, 27, 32; MIA 463; NAW 290; SEF cp547, cp550; WNW cp122, cp123

Newt, roughskin (*Taricha granulosa*)
AAL 270(bw); ARA cp31, cp35; ERA 35; MIA 463; WFW cp601

Newt, smooth (*Triturus vulgaris*)
ERA 27, 33

Newt, warty (*Triturus cristatus*)
AAL 270(bw), 271(bw); MIA 463; PGT 207

Night heron, black-crowned (*Nycticorax nycticorax*)
AGC cp572; BWB 16; EAB 523; FEB 132; FWB 131; HHH 189; MAE 93; MIA 209; NAB cp20, cp22; NAW 153; PBC 69; UAB cp8, cp10; WBW 57; WGB 19; WNW cp529

Night heron, Japanese (*Gorsachius goisagi*)
EOB 66; HHH 205

Night heron, Malayan (*Gorsachius melanolophus*)
HHH 209

Night heron, yellow-crowned (*Nyctanassa* or *Nycticorax violacea*)
AGC cp573; EAB 526; FEB 141; HHH 185; NAB cp19, cp21; PBC 70; WNW cp530

Nighthawk (*Chordeiles virginianus*)
NAW 226

Nighthawk, common (*Chordeiles minor*)
EAB 578, 579; FEB 260; FWB 252; JAD cp561; LBG cp515; NAB cp275; PBC 207; UAB cp250; WON 123, 124

Nighthawk, lesser (*Chordeiles acutipennis*)
EAB 578, 579; FEB 256; FWB 253; JAD cp560; UAB cp251

Nightingale (*Luscinia megarhynchos*)
BAS 13; MIA 333; NNW 37; RBB 130; SPN 17(bw); WGB 143

Nightjar, European (*Caprimulgus europaeus*)
AAL 140(bw); EOB 250; MIA 275; NHU 48; RBB 109; WGB 85

Nightjar, standard-winged (*Semiophorus longipennis*)
EOB 249; MIA 275; WGB 85

Nightshade, bittersweet (*Solanum dulcamara*)
CWN 201; FWF 85; NWE cp448, cp653; PPC 102; PPM cp14; SWW cp661; WTV 175

Nightshade, black (*Solanum nigrum*)
PPC 103

Nightshade, common (*Solanum americanum*)
NWE cp79

Nightshade, deadly (*Solanum belladonna*)
PPC 64

Nightshade, enchanter's (*Circaea quadrisulcata*)
NWE cp170

Nilgai (*Boselaphus tragocamelus*)
MEM 544; MIA 141

Ninebark (*Physocarpus opulifolius*)
FWF 87; TSV 166, 167

Ningaui (*Ningaui* sp.)
CRM 19; MEM 838

Noddy, black or lesser (*Anous tenuirostris*)
BWB 162; EAB 436; EOB 194; HSW 144; NAB cp74

Noddy, blue-gray (*Proceloterna cerulea*)
EOB 194

Noddy, brown (*Anous stolidus*)
AGC cp499; EAB 436; FEB 47; HSW 21; MIA 249; NAB cp73; OBL 95; WGB 59

Noddy, white-capped (*Anous minutus*)
OBL 203

Nonesuch or black medick (*Medicago lupulina*)
CWN 161

Norfolk Island Pine (*Araucaria heterophylla*)
OET 87; TSP 13

Nudibranch or Sea slug (various genera)
AAL 492; AGC cp208-214; BCS 109; BMW 100-101, 115, 148-149; EAL 258, 263; HSG cp6; MPC cp415-432; MSC cp199-208, cp220-cp233, cp234; NOA 105; RUP 50, 53, 58; SAS cpB12, cpB15, cpB18; SCS cp48, cp49; WPR 136; WWW 64-69

Nudibranch (*Notodoris* sp.)
RUP 55; WWF cp49

Nudibranch, harlequin (*Polycera chilluna*)
SAS cpB14

Nudibranch, salmon-gilled (*Coryphella salmonacea*)
BCS 42; MSC cp203, cp204

Nudibranch, Spanish dancer (*Hexabranchus* sp.)
RCK 59, 128; RUP 58

Nudibranch, spanish shawl (*Flabellina iodinea*)
BCS 70, 71

Numbat (*Myrmecobius fasciatus*)
CRM 19; MEM 844; MIA 19; NOA 151; WOI 127

Nunbird, black-fronted (*Monasa nigrifrons*)
MIA 289; WGB 99

Nutcracker (*Nucifraga caryocatactes*)
EOB 445; NHU 105; SAB 17

Nutcracker, Clark's (*Nucifraga columbiana*)
AAL 198(bw); EAB 128; FWB 291; NAW 100; UAB cp470; WFW cp259; WOB 208

Nuthatch (*Sitta europaea*)
RBB 164

Nuthatch, brown-headed (*Sitta pusilla*)
EAB 581; FEB 313; NAB cp356; PBC 258; SEF cp328

Nuthatch, pygmy (*Sitta pygmaea*)
AAL 177; EAB 580; EOB 391; FWB 340; UAB cp389; WFW cp267

Nuthatch, red-breasted (*Sitta canadensis*)
EAB 581; FEB 312; FWB 338; MIA 359; NAB cp353; PBC 256; SEF cp297; SPN 70; UAB cp386; WFW cp265; WGB 169

Nuthatch, white-breasted (*Sitta carolinensis*)
AAL 178(bw); EAB 582; FEB 311; FWB 339; NAB cp354; PBC 255; SEF cp334; SPN 71; UAB cp387; WFW cp266; WOB 167

Nutmeg, California See Torreya, California

Nutria (*Myocastor coypus*)
AAL 49; ASM cp223; GEM III:357; MEM 701; MIA 187; MNC 299(bw); NAW 62(bw); RFA 45-49(bw); WMC 173; WNW cp610

Nyala (*Tragelaphus buxtoni*)
AAL 82(bw); CRM 183; GEM V:346; MIA 141

O

Oak, bear (*Quercus ilicifolia*)
NTE cp289, cp606; TSV 142, 143

Oak, black (*Quercus velutina*)
NTE cp285, cp516, cp587; SEF cp113, cp208

Oak, black (California) (*Quercus kelloggii*)
LBG cp438; NTW cp249; WFW cp103

Oak, blackjack (*Quercus marilandica*)
LBG cp437; NTE cp291, cp590; SEF cp117

Oak, blue (*Quercus douglassii*)
NTW cp242, cp513; WFW cp177

Oak, bur (*Quercus macrocarpa*)
LBG cp439; NTE cp281, cp525; PMT cp[116]; SEF cp110, cp205

Oak, chestnut (*Quercus prinus*)
NTE cp276, cp604; SEF cp107

Oak, chestnut (swamp) (*Quercus michauxii*)
NTE cp277; WNW cp473

Oak, Chinkapin (*Quercus muehlenbergii*)
NTE cp275, cp381, cp522; NTW cp170, cp342, cp510; SEF cp106, cp210

Oak, Dunn (*Quercus dunnii*)
NTW cp178, cp503

Oak, Emory (*Quercus emoryi*)
NTW cp172, cp511; WFW cp83, cp171

Oak, English or pedunculate (te (*Quercus robur*)
NTE cp279, cp375; NTW cp243, cp341; OET 10; PMT cp[117]

Oak, gambel (*Quercus gambelii*)
NTW cp246, cp507; WFW cp174

Oak, laurel (*Quercus laurifolia*)
NTE cp51; SEF cp41; WNW cp447

Oak, live (canyon) (*Quercus chrysolepis*)
LBG cp420, cp482; NTW cp177; WFW cp173

Oak, live (coast) (*Quercus agrifolia*)
LBG cp425; NTW cp175; WFW cp84

Oak, live (interior) (*Quercus wislizeni*)
NTW cp168; WFW cp82

Oak, live (island) (*Quercus tomentella*)
NTW cp169

Oak, live (Virginia) (*Quercus virginiana*)
NTE cp47, cp521; SEF cp49; TSV 146, 147

Oak, Mohr (*Quercus mohriana*)
NTE cp274; NTW cp238

Oak, myrtle (*Quercus myrtifolia*)
NTE cp48, cp520; SEF cp42

Oak, netleaf (*Quercus rugosa*)
NTW cp174, cp502

Oak, overcup (*Quercus lyrata*)
NTE cp278; SEF cp108; WNW cp475

Oak, pin (*Quercus palustris*)
NTE cp288, cp517, cp586; SEF cp116; TSV 144, 145; WNW cp477

Oak, pin (Northern) (*Quercus ellipsoidalis*)
NTE cp286

Oak, poison See Poison oak

Oak, post (*Quercus stellata*)
LBG cp441; NTE cp270, cp514; SEF cp111

Oak, red (Northern) (*Quercus rubra*)
NTE cp292, cp370, cp523; SEF cp118, cp209

Oak, red (Southern) (*Quercus falcata*)
NTE cp283; SEF cp112

Oak, scarlet (*Quercus coccinea*)
NTE cp287, cp518, cp591; OET 126; SEF cp115, cp207

Oak, sessile or Durmast (*Quercus petraea*)
OET 125

Oak, shingle (*Quercus imbricaria*)
NTE cp50, cp519, cp603; PMT cp[115]

Oak, Shumard (*Quercus shumardii*)
NTE cp293; SEF cp114

Oak, silk (*Grevillea robusta*)
NTW cp297, cp328

Oak, swamp (white) (*Quercus bicolor*)
NTE cp280, cp515; PMT cp[114]; WNW cp476

Oak, Turbinella (*Quercus turbinella*)
NTW cp179, cp506; WFW cp172

Oak, turkey (*Quercus laevis*)
NTE cp284, cp588

Oak, valley (*Quercus lobata*)
LBG cp440, cp483; NTW cp245, cp512

Oak, water (*Quercus nigra*)
NTE cp271; WNW cp474

Oak, white (*Quercus alba*)
NTE cp282, cp380, cp524, cp597; SEF cp109, cp206; TSV 140, 141

Oak, white (Arizona) (*Quercus arizonica*)
NTW cp173, cp508; WFW cp175

Oak, white (Oregon) (*Quercus garryana*)
NTW cp248, cp509; WFW cp176

Oak, willow (*Quercus phellos*)
NTE cp101; WNW cp452

Oak (other *Quercus* sp.)
NTE cp46, cp49, cp268-cp273, cp290, cp294; NTW cp90, cp106-cp108, cp166, cp239-cp247; PMT cp[113]

Oarfish (*Regalecus glesne*)
ACF cp22; EAL 118-119; MIA 539; PCF cp46

Oats, sea See Wild oats

Obediant plant (*Physostegia virginiana*)
NWE cp536

Ocean spray See Creambush

Ocelot (*Felis pardalis*)
ASM cp273; CRM 153; GEM III:625; MEM 52; MIA 103; NAW 62

Ocelot, tree See Margay

Oconee bells (*Shortia galacifolia*)
NWE cp52

Ocotillo (*Fouquieria splendens*)
JAD cp335; SWW cp426; WAA 145

Octopus (*Octopus* sp.)
AAL 497; AGC cp182, cp183; EAL 146, 251, 271; KCR cp31; MPC cp378-380; MSC cp478-483; NNW 96; RCK 44, 95, 157, 193; RUP 114-122; SCS p46

Octopus (other genera)
NOA 77; WWW 189

Oilbird (*Steatornis caripensis*)
AAL 140(bw); EOB 249; MIA 273; WGB 83

Oilfish (*Ruvettus pretiosus*)
ACF p61; PCF p47

Okapi (*Okapia johnstoni*)
CRM 183; GEM V:262; MEM 541; MIA 139; SWF 120

Old-man's-beard (*Clematis vitalba*)
PPC 69

Oldsquaw See under Duck

Oleander (*Nerium oleander*)
PPM cp17; PPS 31; TSP 69

Olingo (*Bassaricyon* sp.)
CRM 139; GEM III:453; MIA 85

Olive shell (*Oliva* sp.)
MSC cp440-442; SAS cpB6

Olive, Autumn (*Elaeagnus umbellata*)
FWF 99; TSV 68, 69

Olive, common (*Olea europaea*)
NTW cp83, cp497; OET 245

Olive, russian or oleaster (*Elaeagnus angustifolia*)
LBG cp467; NTE cp100, cp392; NTW cp87, cp322

Olm (*Proteus anguinus*)
ERA 27; MIA 467

Onager (*Equus hemionus*)
GEM IV:584; MEM 484; MIA 125

Oncilla See Cat, tiger or little spotted

Onion, wild (*Allium* sp.)
CWN 9; NWE cp551, cp554; WTV 136

Opah (*Lampris guttatus*)
ACF p61; EAL 119; MIA 539; PCF cp46

Opaleye (*Girella nigricans*)
MPC cp270; NAF cp327; PCF cp34

Openbill (*Anastomus oscitans*)
MIA 211; WGB 21

Opossum, mouse (*Marmosa* sp.)
MIA 15

Opossum, Virginia (*Didelphis virginiana*)
AAL 26(bw); ASM cp203; CGW 125; CMW 18(bw); GEM I:208-209, 232; GMP 34(bw); MAE 120; MEM 831, 837; MIA 15; MNC 73(bw); MNP 166(bw); MPS 39(bw); NAW 63; RFA 2-11 (bw); SEF cp600; SWF 140; WMC 42

Opossum (other genera) See also Yapok
CRM 23; GEM I:212, 223-255; MEM 830-833; MIA 15

Orange peel fungus (*Peziza* or *Aleuria aurantia*)
NAM cp603; NMT cp58; SGM cp409

Orange, sour or Seville (*Citrus aurantium*)
NTW cp163

Orangutan (*Pongo pygmaeus*)
CRM 87; GEM I:49, II:20, 357-423; MEM 413, 428-431; MIA 77; NHP 164-167(bw); SWF 90, 91; WWW 99

Orca See Killer whale

Orchid, adder's mouth (*Malaxis* sp.)
GWO 40, 46

Orchid, butterfly (*Epidendrum* or *Encyclia tampense*)
ONA 107; WAA 216; WNW cp265

Orchid, calypso or fairy slipper (*Calypso bulbosa*)
NWE cp491; ONA 96, 97; SEF cp471; SWW cp484; WAA 220; WFW cp487

Orchid, clam shell (*Epidendrum cochleatum*)
WAA 216

Orchid, coral-root (*Corallorhiza* sp.)
GWO 56, 84, 102, 104; NWE cp392, cp422; ONA 113-115; SWW cp7, cp387, cp390; WAA 216, 217; WFW cp465

Orchid, crane-fly (*Tipularia discolor*)
CWN 33; ONA 97

Orchid, Fischer's (*Dactylorhiza aristata*)
ONA 33

Orchid, fringed (bicolor) (*Habenaria x bicolor*)
GWO 68

Orchid, fringed (crested) (*Habenaria cristata*)
GWO 62; ONA 37

Orchid, fringed (large purple) (*Habenaria fimbriata* or *grandiflora*)
NWE cp525; ONA 45; SEF cp481; WNW cp294

Orchid, fringed (purple) (*Habenaria psycodes*)
CWN 35; GWO 92; ONA 45; WAA 219

Orchid, fringed (ragged or green) (*Habenaria lacera*)
CWN 35; GWO 60; LBG cp120; NWE cp8; ONA 41; WAA 219; WTV 3

Orchid, fringed (yellow) (*Habenaria ciliaris*)
CWN 37; GWO 66; NWE cp378; ONA 35; WAA 219; WTV 107

Orchid, fringed (white) (*Habenaria blephariglottis*)
CWN 37; GWO 22; NWE cp151; ONA 35; WAA 219

Orchid, fringeless (purple) (*Habenaria peramoena*)
GWO 96; ONA 45

Orchid, fringeless (yellow) (*Habenaria integra*)
GWO 64; ONA 39

Orchid, ghost or Palm-polly (*Polyrrhiza lindenii*)
ONA 127; WAA 216

Orchid, grass pink (*Calopogon* sp.)
CWN 43; GWO 6, 70, 80; NWE cp498; ONA 65-67; WAA 215, 219; WNW cp269

Orchid, green-fly (*Epidendrum conopseum*)
GWO 44; NWE cp7

Orchid, helleborine (*Epipactis* sp.)
GWO 42; ONA 52

Orchid, ladies' tresses (*Spiranthes* sp.)
CWN 41; GWO 2, 4, 10, 20, 26, 28; NWE cp131; ONA 70-81; SWW cp100

Orchid, lady's slipper See Lady's slipper

Orchid, lawn (*Zeuxine strateumatica*)
NWE cp111; ONA 87

Orchid, leafy white (*Habenaria dilatata*)
ONA 37

Orchid, leafless beaked (*Stenorhynchus orchiodes*)
ONA 81; WAA 216

Orchid, phantom (*Cephalanthera* or *Eburophyton austinae*)
ONA 54; SWW cp91; WFW cp429

Orchid, pogonia (*Isotria* sp.)
GWO 32, 34; NWE cp319

Orchid, pogonia (rose) (*Pogonia ophioglossoides*)
CWN 31; GWO 78; NWE cp495; ONA 59; WAA 219; WNW cp268; WTV 137

Orchid, rattlesnake plantain (*Goodyera* sp.)
GWO 16, 18; NWE cp128; WFW cp420

Orchid, rein (*Habenaria* sp.)
SWW cp99-cp102

Orchid, rosebud (*Cleistes divaricata*)
GWO 8; NWE cp497

Orchid, round-leaved (*Habenaria orbiculata*)
GWO 14; ONA 43; SWW cp98

Orchid, snowy (*Habenaria nivea*)
GWO 12; ONA 43

Orchid, stream (*Epipactis gigantea*)
SWW cp389

Orchid, swamp-pink or dragon's mouth (*Arethusa bulbosa*)
CWN 31; GWO 76; NWE cp492; ONA 65; WAA 218; WNW cp270

Orchid, three birds (*Triphora trianthophora*)
GWO 24; NWE cp503

Orchid, twayblade (*Listera* or *Liparis* sp.)
CWN 43; GWO 36, 38, 86; NWE cp394; ONA 49-51; NWE cp394; SWW cp12; WFW cp403

Orchid, vanilla (*Vanilla* sp.)
ONA 61-63

Orchid, water-spider (*Habenaria repens*)
GWO 54; ONA 47

Orchid, white (*Bauhinia forficata*)
PMT cp[19]

Orchid, wood (*Habenaria clavellata*)
CWN 41; ONA 37; WTV 4

Orchis, showy (*Orchis spectabilis*)
BCF 129; CWN 35; GWO 72; NWE cp105, cp504; ONA 31; SEF cp448; WAA 216; WTV 131

Oregon grape, creeping (*Berberis repens*)
SWW cp348; WFW cp199

Oribi (*Ourebia ourebia*)
CRM 195; GEM V:343; MEM 577; MIA 151; NHA 110(bw)(head only), 136(bw); WPP 86-87

Oriole, Altamira (*Icterus gularis*)
EAB 907; FEB 348; WOB 90

Oriole, Audubon's or black-headed (*Icterus graduacauda*)
FEB 355; NAB cp387

Oriole, black-headed (*Oriolus larvatus*)
EAB 906; WOM 115

Oriole, golden (*Oriolus oriolus*)
EOB 431; MIA 395; WGB 205

Oriole, hooded (*Icterus cucullatus*)
EAB 907; FEB 351; FWB 275; JAD cp614, cp615; NAB cp398; NAW 179; UAB cp435, cp447

Oriole, northern (including Baltimore and Bullock's) (*Icterus galbula*)
AAL 189; EAB 908; FEB 350; FWB 274; MIA 383; NAB cp393; NAW 178; SEF cp309; SPN 150, 151; UAB cp436, cp446; WFW cp307; WGB 193; WON 175, 176

Oriole, orchard (*Icterus spurius*)
EAB 909; FEB 352; NAB cp388, cp396; PBC 345; SPN 152

Oriole, Scott's (*Icterus parisorum*)
EAB 909; FWB 277; JAD cp616; UAB cp427, cp437

Oriole, spot-breasted (*Icterus pectoralis*)
EAB 909; FEB 349; NAB cp395

Oryx, Arabian (*Oryx leucoryx*)
CRM 191; GEM I:171(head only), V:440-446; MAE 25; MEM 573; MIA 149

Oryx, Scimitar (*Oryx dammah*)
CRM 191

Osage orange (*Maclura pomifera*)
LBG cp411, cp459, cp469; NTE cp71, cp391, cp573; NTW cp112, cp334, cp495; OET 159; TSV 122, 123

Osprey (*Pandion haliaetus*)
AGC cp605; BOP 21-28; BWB 243; EAB 647, 648; EOB 106, 114-115; FEB 224; FGH cp3; FWB 214; MIA 217; MPC cp552; NAB cp306; NAW 180; PBC 115; RBB 46; UAB cp304, cp334; WGB 27; WNW cp563; WOB 139

Ostrich (*Struthio camelus*)
AAL 90(bw); EOB 10, 19; MIA 197; SAB 125; WGB 7; WPP 106, 107; WWD 80-81

Oswego tea See Bee-balm

Otter, Cape clawless (*Aonyx capensis*)
GEM III:438; MIA 91

Otter, giant (*Pteronura brasiliensis*)
CRM 145; GEM III:438; MEM 128; MIA 91

Otter, marine (*Lutra felina*)
CRM 143

Otter, river (Eurasian) (*Lutra lutra*)
AWR 71-73; CRM 143; GEM III:432; MIA 91

Otter, river (North American) (*Lutra canadensis*)
ASM cp215; CGW 145; CMW 241(bw); CRM 143; GEM III:436; GMP 298(bw); MEM 129; MNC 271(bw); MPS 300(bw); NAW 64; RFA 255-263 (bw); RMM cp[9]; SWM 119, 120; WFW cp348; WMC 199; WNW cp612

Otter, river (South American) (*Lutra longicaudus*)
CRM 143

Otter, river (Southern) (*Lutra provocax*)
CRM 143

Otter, sea (*Enhydra lutris*)
ASM cp216; BMW 78; CRM 145; GEM III:434-435, 447-449; MIA 91; MPC cp234; NAW 64; RFA 268-275 (bw); SWM 122, 124

Otter, smooth (*Lutra perspicillata*)
ROW cp2

Otter, spotted neck (*Lutra maculicollis*)
GEM III:437

Otter, swamp (*Aonyx* sp.)
CRM 143; GEM III:439; MEM 124-125

Ouakari See Uakari

Ounce See Leopard, snow

Our Lord's candle (*Yucca whipplei*)
SWW cp184; WFW cp197

Ouzel, ring (*Turdus torquatus*)
RBB 136

Ovenbird (*Seiurus aurocapillus*)
EAB 984; FEB 328; FWB 388; MIA 379; NAB cp503; PBC 325; SEF cp346; SPN 114; WGB 189

Owl, barking (*Ninox connivens*)
EOB 241

Owl, barn (*Tyto alba*)
AAL 137(bw); BOP 139, 140, 142; EAB 649; EOB xii, 243; FEB 254; FWB 233; GEM I:17, 157; JAD cp555; LBG cp512; MIA 269; NAB cp291; NAW 181; NNW 75; ONU 19, 105; PBC 198; RBB 104; SAB 27; UAB cp302; WGB 79

Owl, barred (*Strix varia*)
BOP 134, 135; EAB 650; FEB 251; NAB cp285; PBC 202; SEF cp244; WNW cp566; WOB 147

Owl, bay (*Phodilus badius*)
BOP 141; EOB 240; MIA 269

Owl, boreal or Tengmalm's (*Aegolius funerea*)
BOP 113, 119; EAB 651; EOB 241; FEB 245; FWB 249; NAB cp290; ONU 92(bw), 139; UAB cp300

Owl, burrowing (*Athene cunicularia*)
BOP 131; EAB 651; FEB 252; FWB 234; JAD cp559; LBG cp513; MAE 77; MIA 271; NAB cp283; NAW 185; ONU 19; PBC 201; UAB cp301; WGB 81; WON 114, 115; WWD 60-61

Owl, eagle (*Bubo bubo*)
AAL 139; BOP 118; NHU 100; NNW 43; ONU 137(bw), 174; SWF 41; WWD 107

Owl, eagle (Oriental) (*Bubo sumatrana*)
EOB 240; ONU 169(bw)

Owl, elf (*Micrathene whitneyi*)
EAB 651; EOB 240; FEB 243; FWB 245; JAD cp558; MIA 269; NAF cp287; ONU 18; UAB cp294; WGB 79

Owl, ferruginous pygmy (*Glaucidium brasilianum*)
BOP 136; EAB 653; FEB 242; FWB 237; JAD cp557; UAB cp296

Owl, fish (*Ketupa* sp.)
BAS 129(head only); EOB 245(head only); MIA 269; ONU 106; WGB 79

Owl, flammulated (*Otus flammeolus*)
EAB 653; FWB 241; UAB cp297; WFW cp229; WOB 153(head only)

Owl, great gray (*Strix nebulosa*)
BOP 116, 117, 120; EAB 653; FEB 250; FWB 247; NAB cp286; ONU 18, 34(bw), 208; SEF cp245; UAB cp290; WFW cp233; WON 117

Owl, great horned (*Bubo virginianus*)
BOP 114, 115, 124, 125, 138; EAB 652; EOB 239; FEB 249; FWB 243; JAD cp556; MIA 269; NAB cp282; NAW 182, 183; ONU 46(bw), 99(bw), 140; PBC 200; SEF cp242; UAB cp288; WFW cp230; WGB 79; WOB 156-157; WON 107; WPP 26-27

Owl, hawk (*Surnia ulula*)
BOP 126; EAB 654; FEB 247; FWB 244; MIA 271; NAB cp288; UAB cp292; WGB 81

Owl, little (*Athene noctua*)
BAS 109; BOP 109; EOB 11; MIA 271; NNW 57; ONU 17, 89(bw), 171(bw), 206; RBB 105; WGB 81

Owl, long-eared (*Asio otus*)
AAB 88; BOP 120, 121; EAB 654, 655; FEB 248; FWB 242; MIA 271; NAB cp281; ONU 64(bw), 71; RBB 107; SEF cp243; SWF 42; UAB cp289; WFW cp231; WGB 81

Owl, morepork or boobook (*Ninox novaeseelandiae*)
MIA 271; WGB 81

Owl, Pel's fishing (*Scotopelia peli*)
BWB front.; EOB 241

Owl, pygmy (*Glaucidium gnoma*)
EAB 655; FWB 236; MIA 271; UAB cp295; WFW cp232

Owl, pygmy (Eurasian) (*Glaudidium passerinum*)
MIA 271; WGB 81

Owl, saw-whet (*Aegolius acadicus*)
BOP 122, 127; EAB 655; EOB 238; FEB 244; FWB 248; NAB cp289; PBC 204; SEF cp246; UAB 298, cp299; WFW cp234; WON 119, 120

Owl, scops (*Otus* sp.)
AAL 137(bw); EOB 241; NHU 124; ONU 37, 151(bw)

Owl, screech (*Otus asio*)
BAS 44; BOP 137; EAB 656; FEB 246; FWB 238; MIA 269; NAB cp279, cp280; NAW 184; ONU 18, 173; PBC 199; SEF cp241; UAB cp286, 287; WGB 79; WOB 36

Owl, short-eared (*Asio flammeus*)
BOP 123, 133; EAB 657; EOB 247; FEB 253; FWB 235; LBG cp514; NAB cp284; ONU 26(bw), 38, 75(bw); RBB 108; UAB cp291; WNW cp565; WOB 191

Owl, snowy (*Nyctea scandiaca*)
BAA 119; BAS 132; BOP 128, 129, 130; EAB 658, 659; FEB 255; FWB 232; MIA 269; NAB cp292; NAW 185; NHU 84; NNW 63; ONU 19, 72; UAB cp303; WGB 79; WON 111; WPR 86, 87

Owl, spectacled (*Pulsatrix perspicillata*)
EOB 240

Owl, spotted (*Strix occidentalis*)
AAL 137(bw); BOP 132; EAB 658, 659; FWB 246; UAB cp293; WOB 243

Owl, spotted wood (*Strix seloputo*)
EOB 241

Owl, tawny (*Strix aluco*)
BAS 92; MIA 271; ONU 20, 130(bw); RBB 106; WGB 81

Owl, whiskered screech (*Otus trichopsis*)
EAB 659; FWB 240

Owlet, nightjar (*Aegotheles* sp.)
EOB 249, 253

Owlet, pearl-spotted (*Glaucidium perlatum*)
SAB 62

Ox, musk See Musk ox

Oxpecker (*Buphagus* sp.)
SAB 14

Oxslip (*Primula elatior*)
WOM 50

Oyster (*Ostrea* and other genera)
MSC cp289-cp291, cp349, cp357, cp359; SAS cpB21

Oystercatcher, American (*Haematopus palliatus*)
AAL 88; AGC cp576; EAB 660, 661; FEB 156; NAB cp242; PBC 136

Oystercatcher, black (*Haematopus bachmani*)
BAS 96; BWB 184-187; EAB 661, 662; EOB 178; FWB 122; MPC cp556; UAB cp218; WGB 51

Oystercatcher, European (*Haematopus ostralegus*)
EOB 180, 181; MAE 28; MIA 241; RBB 61; WBW 145

Oystercatcher, Magellanic (*Haematopus leucopodus*)
WBW 143

P

Paca (*Agouti* or *Cuniculus paca*)
CRM 127; GEM III:339; MEM 703; MIA 185

Paca, mountain (*Agouti taczanowskii*)
GEM II:339

Pacarana (*Dinomys branickii*)
CRM 127; GEM III:324; MEM 703; MIA 185

Pack rat See Woodrat, bushy-tailed

Pacu (*Colossoma nigripinnis*)
MIA 511

Paddlefish (*Polyodon spathula*)
AAL 322(bw); AWR 218(bw); EAL 20; MIA 499; NAF cp29; WNW cp79

Pademelon (*Thylogale* sp.)
CRM 33; GEM I:362, 363; MEM 865; MIA 23

Pagoda flower (*Clerodendrum paniculatum*)
TSP 115

Pagoda tree, Japanese (*Sophora japonica*)
NTE cp312; PMT cp[123]

Pale face (*Hibiscus denudatus*)
JAD cp45; SWW cp463

Pale trumpets (*Ipomopsis longiflora*)
JAD cp66; SWW cp580

Palm, coconut (*Cocos nucifera*)
OET 257; PFH 154; TSP 23

Palm, date (*Phoenix dactylifera*)
OET 259

Palm, desert See Washingtonia

Palm, royal (*Roystonea regia*)
OET 259; PMT cp[119]; TSP 39

Palm, traveller's (*Ravenala madagascariensis*)
OET 260; TSP 37

Palm, windmill (*Trachycarpus fortunei*)
OET 258

Palmchat (*Dulus dominicus*)
EOB 353; WGB 137

Palmetto, cabbage (*Sabal palmetto*)
NTE cp363; SEF cp144

Palm-polly See Orchid, ghost

Palometa (*Trachinotus goodei*)
ACF p29; NAF cp549

Paloverde, blue (*Cercidium floridum*)
JAD cp301, cp316; NTW cp254, cp326; SWW cp353

Paloverde, yellow (*Cercidium microphyllum*)
NTW cp253, cp325, cp521

Panchax, blue (*Aplocheilus panchax*)
AWR 215(bw)

Panda, giant (*Ailuropoda melanoleuca*)
AAL 60; CRM 135; DBW 122-141; GEM III:371, 473-476; MEM 104, 105; MIA 85; WOM 132

Panda, red or lesser (*Ailurus fulgens*)
AAL 61; CRM 135; GEM III:469, 470; MEM 106; MIA 85; WOM 133

Pangolin (*Manis* sp.)
AAL 41; CRM 91; GEM II:627, 630-641; MEM 785; MIA 29; ROW 61(bw); SWF 124

Panther See Mountain lion

Panther (mushroom) See Blusher, false

Papaw See Pawpaw

Papaya or pawpaw (*Carica papaya*)
OET 269; TSP 17

Paperflower (*Psilostrophe cooperi*)
JAD cp137; SWW cp245

Papoose-root See Cohosh, blue

Papyrus (*Papyrus antiquorum*)
TSP 101

Parakeet, monk (*Myiopsitta monachus*)
EAB 697; MIA 263; NAB cp484; PBC 192; WGB 73

Paramecium (*Paramecium* sp.)
EAL 163; PGT 90, 91; WWW 46

Parasol tree, Chinese (*Firmiana simplex*)
NTE cp265; NTW cp232; PMT cp[50]

Pardalote, spotted (*Pardalotus punctatus*)
EOB 395

Parrot, crimson rosella (*Platycercus elegans*)
EOB 222; MIA 261; NOA 145; WGB 71

Parrot, eclectus (*Eclectus roratus*)
EOB 223

Parrot, gray (*Psittacus erithacus*)
MIA 261; WGB 71

Parrot, king (Australian) (*Alisterus scapularis*)
EOB 221

Parrot, red-capped (*Purpureicephalus spurius*)
EOB 222

Parrot, yellow-headed (*Amazona ochrocephala*)
MIA 263; WGB 73

Parrot (other kinds) AAL 3, 135; BAS 113; EAB 696; EOB 222; WPP 121

Parrotbill, gray-headed (*Paradoxornis gularis*)
EOB 378

Parrotfish, blue (*Scarus coeruleus*)
ACF cp40; AGC cp345; MIA 565; NAF cp365

Parrotfish, bucktooth (*Sparisoma radians*)
ACF cp40; NAF cp356, cp376

Parrotfish, princess (*Scarus taeniopterus*)
ACF cp40; AGC cp343; NAF cp360, cp375

Parrotfish, queen (*Scarus vetula*)
ACF cp40; AGC cp344; NAF cp362

Parrotfish, rainbow (*Scarus guacamaia*)
ACF cp40; KCR cp24; MIA 565

Parrotfish, redband (*Sparisoma aurofrenatum*)
ACF cp40; NAF cp355, cp358, cp359

Parrotfish, redfin (*Sparisoma rubripinne*)
ACF cp40; NAF cp364

Parrotfish, stoplight (*Sparisoma viridae*)
ACF cp40; KCR cp24; NAF cp361, cp379; WWF cp25

Parrotfish (other *Scarus* sp.)
AAL 364; ACF cp40; RCK 37, 68; RUP 70, 170

Parrotlets, green-rumped (*Forpus passerinus*)
EOB 220

Parsnip (*Pastinaca sativa*)
CWN 133

Parsnip, cow (*Heracleum lantanum*)
BCF 95; CWN 137; NWE cp193; SWW cp168

Partridge See also Chukar

Partridge, gray (*Perdix perdix*)
EAB 706; FEB 203; FWB 184; LBG cp497; NAB cp267; RBB 56; UAB cp285

Partridge, red-legged (*Alectoris rufa*)
MAE 121; MIA 229; RBB 55; WGB 39

Partridge, roulroul or crested wood (*Rollulus rouloul*)
MIA 229; WGB 39; WOI 114

Partridgeberry (*Mitchella repens*)
CWN 209; FWF 69; NWE cp74, cp445; SEF cp444, cp508; WTV 21

Partridge-foot (*Luetkea pectinata*)
SWW cp111

Partridge-pea (*Cassia fasciculata*)
WTV 93

Parula, Northern (*Parula americana*)
EAB 984; FEB 460; FWB 363; MIA 379; NAB cp448; PBC 308; SEF cp296; SPN 104; WGB 189

Pasqueflower (*Anemone patens*)
LBG cp281; NWE cp602; SWW cp591; WAA 191

Pasqueflower, alpine (*Pulsatilla* sp.)
PPC 39; WOM 50

Pasqueflower, mountain or Western (*Anemone occidentalis*)
SWW cp58; WAA 190, 244; WFW cp382

Passionflower or Maypop (*Passiflora* sp.)
FWF 29; NWE cp652; SEF cp494; SWW cp2; TSP 89; WAA 165; WTV 82, 185

Paulownia, royal (*Paulownia tomentosa*)
NTE cp98, cp189, cp461; NTW cp124, cp233, cp395; TSV 114, 115

Pauraque, common (*Nyctidromus albicollis*)
EAB 579; FEB 259; NAB cp276; WGB 85

Pawpaw, common (*Asimina triloba*)
FWF 43; NTE cp87, cp364, cp574; PMT cp[18]; SEF cp59, cp184; TSV 116, 117; WNW cp442

Pea, beach (*Lathyrus* sp.)
AGC cp452; CWN 155; MPC cp617; NWE cp510

Pea, butterfly (*Clitoria* sp.)
NWE cp509; WAA 166; WTV 181

Pea, Chaparral (*Pickeringia montana*)
SWW cp569; WFW cp210

Pea, golden (*Thermopsis montana*)
SWW cp300; WFW cp430

Pea, rosary or crab's-eye (*Abrus precatorius*)
NWE cp449; PPM cp12

Peach (*Prunus persica*)
NTE cp141, cp459, cp569; NTW cp137, cp388, cp492

Peacock, Congo (*Afropavo congensis*)
MIA 231; WGB 41

Peacock, Indian (*Pavo cristatus*)
AAL 119; BAS 29; EOB 140-141; MIA 231; WOI 164-165; WGB 41

Peanut, hog (*Amphicarpa bracteata*)
NWE cp575

Pear (*Pyrus communis*)
NTE cp131, cp455, cp575; NTW cp158, cp369, cp494

Pecan (*Carya illinoensis*)
NTE cp327, cp535; SEF cp139

Peccary, Chacoan (*Catagonus wagneri*)
CRM 171; GEM V:48; MIA 131

Peccary, collared (*Tayassu tajacu*)
ASM cp276; GEM V:50-54; JAD cp525; LBG cp36; MAE 60; MEM 504, 505; MIA 131; MNP 279(bw); NMM cp[29]; WWD 146-147

Peccary, white-lipped (*Tayassu pecari*)
GEM V:49; MIA 131

Peeper See Treefrog, spring peeper

Peewit See Lapwing

Pelican, Australian (*Pelecanus conspicillatus*)
EOB 54-55

Pelican, brown (*Pelecanus occidentalis*)
AAL 96; AGC cp519; BWB 86-87; EAB 698, 699; EOB 52; FEB 66; FWB 68; MIA 205; MPC cp531; NAF cp176; NAW 186; OBL 70, 163, 230; PBC 52; UAB cp155; WGB 15; WOB 17, 32

Pelican, Chilean (*Pelecanus thagus*)
OBL 234

Pelican, white (American) (*Pelecanus erythrorhynchus*)
AGC cp518; BWB 88; EAB 700, 701; FEB 67; FWB 69; NAB cp175; NAW 187; OBL 46, 174-175; UAB cp154; WNW cp517; WOB 117, 118; WON 26-28

Pelican, white (great or European) (*Pelecanus onocrotalus*)
BWB 89; EOB 57; MIA 205; OBL 70; WGB 15

Pencil flower (*Stylosanthes biflora*)
CWN 161; WTV 91

Penguin, adelie (*Pygoscelis adeliae*)
AAL 92; BAS 73; BMW 228-231; BWB 122-131; EOB 31; HSW 48-49; OBL 186-187; WPR 123-127; WWW 193

Penguin, chinstrap (*Pygoscelis antarctica*)
HSW 60-61; SAB 64; WPR 123, 128-129

Penguin, emperor (*Aptenodytes forsteri*)
BAS 73, 89; BWB 117-121; MIA 199; NOA 66; WGB 9; WOI 54-57; WPR 131

Penguin, Galapagos (*Spheniscus mendiculus*)
HSW 52; MIA 199; OBL 39; WGB 9

Penguin, gentoo (*Pygoscelis papua*)
BWB 132-133; HSW 46; WPR 120-121

Penguin, jackass or blackfoot or cape (*Spheniscus demersus*)
BWB 135; EOB 31; HSW 57

Penguin, king (*Aptenodytes patagonica*)
BWB 112; EAB 64, 140; EOB 31, 34-35, 39(chick); HSW 62-65; MAE 17; WPR 64-65, 132; WWW 33

Penguin, little blue or fairy (*Eudyptes minor*)
EOB 30; HSW 51; MIA 199; NOA 82; WGB 9

Penguin, macaroni (*Eudyptes chrysolophus*)
BWB 137; HSW 58

Penguin, Magellan (*Spheniscus magellanicus*)
BMW 44, 45; BWB 134; EOB 36; HSW 54-55

Penguin, Peruvian (*Spheniscus humboldti*)
HSW 53; OBL 222

Penguin, rock-hopper (*Eudyptes chrysocome*)
BWB 136; EOB 30, 33; HSW 45

Penguin, royal (*Eudyptes schlegi*)
HSW 58

Penguin, yellow-eyed (*Megadyptes antipodes*)
BWB 139; EOB 30; HSW 59

Pennycress, field (*Thlaspi arvense*)
CWN 111; FWF 7

Pennywort (*Obolaria virginica*)
NWE cp172

Pennywort, water (*Hydrocotyle americana*)
NWE cp13

Penstemon or beardtongue (*Penstemon* sp.)
NWE cp64, cp519, cp555, cp556;WAA 229, 236

Penstemon, Cascade (*Penstemon serrulatus*)
SWW cp626; WFW cp539

Penstemon, cliff or rock (*Penstemon rupicola*)
SWW cp493; WFW cp510

Penstemon, Jones (*Penstemon dolius*)
JAD cp68; SWW cp622

Penstemon, lowbush (*Penstemon fruticosus*)
SWW cp562; WFW cp218

Penstemon, red shrubby (*Penstemon corymbosus*)
SWW cp406; WFW cp207

Peony, Western (*Paeonia brownii*)
SWW cp375; WFW cp466

Pepperbush, sweet (*Clethra alnifolia*)
NWE cp221; WNW cp235

Peppergrass (*Lepidium* sp.)
JAD cp78, cp113; NWE cp174; SWW cp135, cp340

Peppermint (*Mentha piperita*)
CWN 221; NWE cp560

Peppertree (*Schinus molle*)
NTW cp284, cp483

Perch, black (*Embiotoca jacksoni*)
NAF cp519

Perch, climbing (*Anabas testudineus*)
EAL 39; AWR 215(bw)

Perch, kelp (*Brachyistius frenatus*)
MPC cp268; NAF cp530

Perch, pike See Zander

Perch, pirate (*Aphredoderus sayanus*)
NAF cp217

Perch, river (*Perca fluviatilis*)
AWR 216(bw); EAL 119; MAE 97; MIA 551; PGT 203

Perch, Sacramento (*Archoplites interruptus*)
NAF cp84

Perch, sand (*Diplectrum* sp.)
ACF cp26; AGC cp379; NAF cp513

Perch, shiner (*Cymatogaster aggregta*)
MPC cp267; NAF cp543

Perch, silver (*Bairdiella chrysoura*)
ACF p34

Perch, white (*Morone americana*)
ACF cp24; NAF cp97, cp524; WNW cp39

Perch, yellow (*Perca flavescens*)
NAF cp89; WNW cp43

Periwinkle (mollusk) (various kinds)
MSC cp469-473; NNW 91

Periwinkle or myrtle (plant) (*Vinca minor*)
NWE cp599

Periwinkle, greater (plant) (*Vinca major*)
WTV 153

Periwinkle, rosy or Madagascar (*Catharanthus roseus*)
TSP 115

Persimmon (*Diospyros virginiana*)
LBG cp412, cp468, cp470; NTE cp83, cp393, cp571; PMT cp[46]; SEF cp56, cp185; TSV 126, 127

Persimmon, Texas or Mexican (*Diospyros texana*)
NTE cp40

Petrel, Antarctic (*Thalassoica antarctica*)
HSW 86

Petrel, diving (*Pelecanoides urinatrix*)
EOB 50; MIA 203; WGB 13

Petrel, giant (*Macronectes giganteus*)
EOB 46; OBL 58-59; WPR 52-53

Petrel, mottled or scaled (*Pterodroma inexpectata*)
MIA 201; WGB 11

Petrel, painted or Cape pigeon (*Daption capense*)
BMW 207; OBL 59

Petrel, storm (*Hydrobates pelagicus*)
MIA 203; RBB 13; WGB 13

Petrel, storm (ashy) (*Oceanodroma homochroa*)
EAB 851; UAB cp64

Petrel, storm (black) (*Oceanodroma melania*)
MIA 203; WGB 13

Petrel, storm (gray-backed) (*Garrodia nereis*)
EOB 50

Petrel, storm (Leach's) (*Oceanodroma leucorhoa*)
EAB 850; MPC cp543; UAB cp62; WON 19

Petrel, storm (least) (*Halocyptena microsoma*)
EAB 850; UAB cp63

Petrel, storm (ringed) (*Oceanodroma hornbyi*)
MIA 203; WGB 13

Petrel, storm (white-faced or frigate) (*Pelagodroma marina*)
MIA 203; WGB 13

Petrel, storm (Wilson's) (*Oceanites oceanicus*)
AGC cp498; EAB 851; EOB 50; FEB 57; MIA 203; NAB cp82; WGB 13

Petrel, white-chinned (*Procellaria aequinoctialis*)
OBL 59

Petunia, wild (*Ruellia pedunculata*)
NWE cp476

Pewee, wood (Eastern) (*Contopus virens*)
EAB 370; FEB 281; MIA 307; NAB cp465; SEF cp281; SPN 53; WGB 117

Pewee, wood (Western) (*Contopus sordidulus*)
EAB 307; FWB 392; UAB cp513; WFW cp251

Peyote (*Lophophora williamsii*)
JAD cp37; SWW cp469

Phacelia (*Phacelia calthifolia*) See also Scorpionweed
WAA 205

Phainopepla (*Phainopepla nitens*)
AAL 168; EAB 843; FWB 297; JAD cp595; UAB cp491, cp613

Phalanger See Glider

Phalarope, northern or red-necked (*Phalaropus lobatus*)
BWB 181; EAB 702, 703; EOB 167, 175; FEB 176; FWB
   153; HSW 119; MPC cp559; NAB cp203, cp206; OBL 154;
   PBC 151; RBB 83; UAB cp202, cp205, cp237; WBW 311

Phalarope, red or gray (*Phalaropus fulicaria*)
AAL 128(bw); AGC cp579; BAA 107; EAB 702; FEB 177;
   FWB 152; MIA 243; MPC cp560; NAB cp205, cp208; UAB
   cp201, cp204, cp236; WBW 309; WGB 53

Phalarope, Wilson's (*Phalaropus tricolor*)
EAB 702; FEB 175; FWB 151; NAB cp204, cp207; NAW
   101; UAB cp200, cp203, cp238; WBW 313; WNW cp548

Phascogale See Marsupial mouse, brush-tailed

Pheasant See also Tragopan

Pheasant, blood (*Ithaginis cruentus*)
WOM 124-125

Pheasant, blue-eared (*Crossoptilon auritus*)
WOM 123

Pheasant, golden (*Chrysolophus pictus*)
BAS 17; EOB 128; MIA 231; WGB 41

Pheasant, gray peacock (*Polyplectron bicalcaratum*)
MIA 231; WGB 41

Pheasant, great argus (*Argusianus argus*)
EOB 15; SAB 65

Pheasant, Himalayan monal (*Lophophorus impejanus*)
BAS 16; EOB 127; WOM 122

Pheasant, Lady Amherst's (*Chrysolophus amherstiae*)
EOB 135; WOM 126

Pheasant, ring-necked (*Phasianus colchicus*)
AAL 119(head only); BAS 25; EAB 707; FEB 201; FWB
   197; LBG cp498; MIA 231; NAB cp274; RBB 57; UAB
   cp265, cp267; WGB 41; WOB 238-239

Pheasant, silver (*Lophura nycthemera*)
BAS 17; WOM 122

Phlox, blue (*Phlox divaricata*)
BCF 177; NWE cp644

Phlox, long-leaved (*Phlox longifolia*)
JAD cp58; SWW cp538

Phlox, prickly (*Leptodacylon californicum*)
SWW cp438

Phlox, red (*Phlox drummondii*)
NWE cp417; WAA 29

Phlox, Sweet William (*Phlox maculata*)
WAA 183

Phlox, tufted (*Phlox caespitosa*)
SWW cp577; WFW cp515

Phoebe, black (*Sayornis`nigricans*)
EAB 371; FWB 410; JAD cp570; UAB cp610

Phoebe, eastern (*Sayornis phoebe*)
AAL 157; EAB 371; FEB 282; MIA 305; NAB cp466; PBC
   228; SPN 54; WGB 115; WNW cp573

Phoebe, Say's (*Sayornis saya*)
EAB 371; FEB 295; FWB 409; JAD cp571; SPN 55; UAB
   cp521

Pickerel, chain (*Esox niger*)
NAF cp36

Pickerelweed (*Pontederia cordata*)
BCF 143; CWN 27; NWE cp623; WAA 201; WNW cp295;
   WTV 161

Pichi (*Zaedylus pichiy*)
MEM 781

Piddock (various kinds)
EAL 269; MSC cp310, cp313, cp314

Pig, bearded (*Sus barbatus*)
GEM V:32

Pig, bush (*Potamochoerus porcus*)
GEM V:33-37; MIA 129

Pig, Celebes (*Sus celebensis*)
GEM V:32

Pig, wild See Boar, wild

Pigeon, band-tailed (*Columba fasciata*)
EAB 713; FWB 180; MIA 253; UAB cp347; WFW cp228;
   WGB 63

Pigeon, crested (*Ocyphaps lophotes*)
MIA 257; WGB 67

Pigeon, crowned (*Goura cristata*)
SWF 108

Pigeon, common See Dove, rock

Pigeon, nicobar (*Caloenas nicobarica*)
MIA 255; WGB 65

Pigeon, nutmeg (*Ducula spilorrhoa*)
MIA 255; WGB 65

Pigeon, plumed (*Geophaps plumifera*)
WWD 184

Pigeon, red-billed (*Columba flavirostris*)
FEB 198

Pigeon, Victoria crowned (*Goura victoria*)
EOB 217(head only); MIA 257; WGB 67

Pigeon, white-crowned (*Columba leucocephala*)
EAB 713; FEB 197; NAB cp328; SEF cp269

Pigeon, wood (*Columba palumbus*)
RBB 100

Pigeon, yellow-legged green (*Treron phoenicoptera*)
MIA 255; WGB 65

Pigfish (*Orthopristis chrysoptera*)
ACF cp32

Pigfish (*Verreo oxycephalus*)
RCK 172

Pigweed or green amaranth (*Amaranthus retroflexus*)
NWE cp28

Pika (*Ochotona princeps*)
AAL 42(bw); ASM cp81; CMW 73(bw); GEM IV:247, 314-317; MEM 728, 729; MIA 191; MNP 175(bw); NAW 65; NHU 159; RMM cp[6]; SWM 15; WFW cp330; WOM 32

Pika, Altai (*Ochotona alpina*)
GEM IV:320

Pika, collared (*Ochotona collaris*)
ASM cp79, 80; MEM 726, 727; SWM 14

Pika, large-eared (*Ochotona macrotis*)
MEM 726

Pike (*Esox lucius*)
EAL 44; MIA 507; NAF cp35; NAW 306; PGT 204; WNW cp76

Pikeberry (*Chaenopis* sp.)
WWF cp35

Pilchard See Sardine

Pillbug or woodlouse (*Armadillidium vulgare*)
AAL 402(bw); EAL 230; NNW 168

Pilotfish (*Naucrates ductor*)
HSG cp16; PCF cp31

Pimpernel, scarlet (*Anagallis arvensis*)
CWN 183; NWE cp373; SWW cp369

Pinckneya or fever-tree (*Pinckneya pubens*)
NTE cp81, cp464

Pine, Apache (*Pinus engelmannii*)
NTW cp7

Pine, Austrian (*Pinus nigra*)
NTE cp8; NTW cp12

Pine, bishop (*Pinus muricata*)
NTW cp16, cp412; WFW cp128

Pine, bristlecone (*Pinus aristata*)
NTW cp26, cp423; WFW cp42, cp130

Pine, Chihuahua (*Pinus leiophylla*)
NTW cp8, cp422; WFW cp124

Pine, Coulter (*Pinus coulteri*)
NTW cp6, cp413; WFW cp127

Pine, digger (*Pinus sabiniana*)
NTW cp3, cp416; WFW cp125

Pine, foxtail (*Pinus balfouriana*)
NTW cp25, cp415; WFW cp41

Pine, jack (*Pinus banksiana*)
NTE cp15, cp476; SEF cp28

Pine, Jeffrey (*Pinus jeffreyi*)
NTW cp4, cp414; WFW cp37, cp126

Pine, Korean (*Pinus koraiensis*)
PMT cp[101]

Pine, knobcone (*Pinus attenuata*)
NTW cp14, cp411; WFW cp129

Pine, lacebark (*Pinus bungeana*)
PMT cp[98]

Pine, limber (*Pinus flexilis*)
NTW cp18, cp409; PMT cp[100]; WFW cp122

Pine, loblolly (*Pinus taeda*)
LBG cp404; NTE cp3; SEF cp24; TSV 2, 3

Pine, lodgepole (*Pinus contorta*)
NTW cp20, cp420; WFW cp40

Pine, longleaf (*Pinus palustris*)
LBG cp406; NTE cp2; SEF cp20

Pine, Monterey (*Pinus radiata*)
NTW cp15, cp418

Pine, pinyon (*Pinus edulis* or *monophylla*)
LBG cp480; NTW cp23, cp421; WFW cp136

Pine, pitch (*Pinus rigida*)
NTE cp5, cp472; SEF cp25, cp173; TSV 6, 7

Pine, pond or marsh (*Pinus serotina*)
NTE cp4

Pine, Ponderosa or western yellow (*Pinus ponderosa*)
NTW cp11, cp419; OET 65; WFW cp38

Pine, red or Norway (*Pinus resinosa*)
NTE cp7; SEF cp22

Pine, sand or scrub (*Pinus clausa*)
NTE cp13

Pine, scots (*Pinus sylvestris*)
NTE cp12; NTW cp17; OET 12

Pine, screw (*Pandanus tectorius*)
TSP 71

Pine, shortleaf (*Pinus echinata*)
LBG cp403, cp479; NTE cp9, cp474; SEF cp23, cp174

Pine, slash (*Pinus elliottii*)
LBG cp407; NTE cp6; SEF cp21

Pine, spruce or cedar (*Pinus glabra*)
NTE cp11

Pine, sugar (*Pinus lambertiana*)
NTW cp9, cp406; WFW cp120

Pine, Table Mountain (*Pinus pungens*)
NTE cp10; SEF cp26; TSV 4, 5

Pine, Torrey (*Pinus torreyana*)
NTW cp5, cp417

Pine, Virginia (*Pinus virginiana*)
LBG cp402; NTE cp14, cp473; SEF cp27; TSV 8, 9

Pine, white (eastern) (*Pinus strobus*)
LBG cp405, cp478; NTE cp1, cp475; SEF cp19, cp175;
 TSV 1

Pine, white (Japanese) (*Pinus parviflora*)
PMT cp[102]

Pine, white (Southwestern) (*Pinus strobiformis*)
NTW cp13, cp407; WFW cp39

Pine, white (Western) (*Pinus monticola*)
NTW cp10, cp408; WFW cp121

Pine, whitebark (*Pinus albicaulis*)
NTW cp19, cp410; WFW cp123

Pine, wild (*Tillandsia fasciculata*)
NWE cp438; WAA 75; WNW cp283

Pinedrops (*Pterospora andromedea*)
SWW cp392; WFW cp461

Pinesap (*Monotropa hypopithys*)
CWN 151; NWE cp421; SEF cp469; WAA 160; WTV 119

Pineweed (*Hypericum gentianoides*)
NWE cp268

Pinfish (*Lagodon rhomboides*)
AGC cp372; NAF cp533

Pink, Deptford (*Dianthus armeria*)
BCF 85; CWN 117; LBG cp244; NWE cp450; SWW cp450;
 WAA 183

Pink, fire (*Silene virginica*)
NWE cp415; WAA 159

Pink, Indian (*Silene californica*)
SWW cp393; WFW cp504

Pink, Indian (*Spigela marilandica*)
NWE cp425

Pink, marsh (*Sabatia grandiflora*)
WAA 167

Pink, moss (*Phlox subulata*)
CWN 193; NWW cp458; SWW cp431; WTV 126

Pink, swamp (*Helonias bullata*)
CWN 17; NWE cp514

Pink, mountain (*Centaurium beyrichii*)
WAA 70-72

Pink, rose (*Sabatia angularis*)
CWN 187; WTV 143

Pink, saltmarsh (*Sabatia stellaris*)
CWN 187; NWE cp473

Pink, sea (*Sabatia dodecandra*)
CWN 187

Pink, wild (*Silene caroliniana*)
WTV 130

Pinkgill fungus (*Entoloma* sp.)
FGM cp38

Pintail See under Duck

Pinxter (*Rhododendron nudiflorum*)
BCF 133; NWE cp588; SEF cp482

Pipefish (*Syngnathus* sp.)
ACF p23; AGC cp410; MIA 543; MPC cp296; NAF cp439,
 cp440; PCF p13

Pipewort (*Eriocaulon septangulare*)
NWE cp108; WNW cp239

Pipistrelle See under Bat

Pipit, golden (*Tmetothylacus tenellus*)
MIA 317; WGB 127

Pipit, meadow (*Anthus pratensis*)
EAB 715; MIA 317; RBB 121; WGB 127

Pipit, water or rock (*Anthus spinoletta*)
AAL 165; EAB 714, 715; FEB 339; FWB 266; LBG cp526;
 MIA 317; NAB cp546; SPN 97; UAB cp571; WGB 127

Pipit (other kinds)
EAB 714; EOB 337; FEB 338; RBB 120

Pipsissewa (*Chimaphila* sp.)
CWN 149; SWW cp445; WFW cp500

Piranha (*Serrasalmus natterei*)
AWR 183; MIA 511

Pistache, Texas (*Pistacia texana*)
NTE cp303

Pitaya (*Echinocereus* sp.)
SWW cp5; WAA 209

ILLUSTRATIONS OF ANIMALS AND PLANTS

Pitcher plant (*Sarracenia* sp.)
CWN 75; NWE cp423, cp424, cp311; WAA 194; WNW
  cp202, cp280-282; WTV 62

Pitcher plant, California See Cobra plant

Pitohui (*Pitohui* sp.)
EOB 381; MIA 355; WGB 165

Pitta (*Pitta* sp.)
AAL 158; EOB 320-323; MIA 311; NOA 123; WGB 121;
  WOI 113

Pixie (*Pyxidanthera barbulata*)
CWN 177; NWE cp46

Plaice, American (*Hippoglossoides platessoides*)
MIA 579

Plaice, chameleon (*Pleurmectes platessa*)
MIA 577; WWW 55

Plains wanderer (*Pedimomus torquatus*)
EOB 159; MIA 235; WGB 45

Planarian (various kinds)
EAL 197; PGT 96, 97

Plane tree (*Platanus* sp.)
NTE cp257; NTW cp220; OET 121

Planigale (*Planigale* sp.)
CRM 15; MEM 839; MIA 17; NOA 171; WPP 119

Plant hopper (*Metcalfia pruinosa*)
SGI cp73

Plantain (*Plantago* sp.)
LBG cp143; NWE cp33, cp107; SWW cp94; WAA 55;
  WTV 17

Platypus (*Ornithorhynchus anatinus*)
GEM I:190-191, 197-201; MEM 823; WOI 129; WWW
  168

Plover, American or lesser golden (*Pluvialis dominica*)
EAB 717; FEB 169; MIA 241; NAB cp194, cp239; PBC
  141; UAB cp189, cp235; WGB 51

Plover, banded (*Vanellus tricolor*)
EOB 162

Plover, black-bellied (*Pluvialis squatarola*)
AGC cp584; EAB 717; FEB 168; FWB 155; MPC cp579;
  NAB cp195, cp240; PBC 142; UAB cp190, cp234

Plover, black-fronted (*Charadrius melanops*)
EOB 164

Plover, blacksmith (*Hoplopterus* or *Antibyx armatus*)
WBW 149

Plover, Caspian sand (*Charadrius asiaticus*)
WBW 177

Plover, crab (*Dromas ardeola*)
EOB 179; MIA 245; WGB 55

Plover, crowned (*Vanellus coronatus*)
WBW 153

Plover, Egyptian (*Pluvialus aegyptius*)
EOB 183; MIA 247; WGB 57

Plover, golden (*Pluvialus apricaria*)
EOB 166; FWB 154; RBB 67; WBW 173

Plover, gray (*Pluvialus squatarola*)
RBB 68

Plover, great shore (*Esacus magnirostris*)
MIA 245; WGB 55

Plover, green See Lapwing

Plover, greater sand (*Charadrius leschenaultii*)
NHU 184; WBW 185

Plover, little ringed (*Charadrius dubius*)
AAB 81; RBB 64; WBW 181

Plover, mountain (*Charadrius montanus*)
EAB 718; FWB 159; UAB cp233

Plover, Oriental See Plover, Caspian sand

Plover, pied (*Hoploxypterus cayanus*)
WBW 171

Plover, piping (*Charadrius melodus*)
AGC cp581; EAB 719; FEB 173; NAB cp234; WBW 187

Plover, ringed (*Charadrius hiaticula*)
EAB 719; MIA 241; RBB 65; SAB 113; WBW 183; WGB
  51

Plover, semipalmated (*Charadrius semipalmatus*)
AGC cp582; BWB 178; EAB 720; FEB 172; FWB 157;
  MPC cp580; NAB cp236; PBC 139; UAB cp183

Plover, snowy or Kentish (*Charadrius alexandrinus*)
AGC cp580; EAB 721; FEB 174; FWB 158; MPC cp582;
  NAB cp233; UAB cp182; WBW 175

Plover, three-banded (*Charadrius tricollaris*)
WBW 195

Plover, upland See Sandpiper, upland

Plover, wattled (*Vanellus senegallus*)
EOB 166; WBW 163

Plover, Wilson's (*Charadrius wilsonia*)
EAB 721; FEB 171; NAB cp237; PBC 139

Plover, wrybill (*Anarhynchus frontalis*)
MIA 241; WGB 51

Plum, American (*Prunus americana*)
LBG cp418, cp455; NTE cp125, cp420, cp568; NTW
  cp195, cp367, cp490; SEF cp83, cp151, cp186

Plum, Chickasaw (*Prunus angustifolia*)
NTE cp123

Plum, garden or damson (*Prunus domestica*)
NTE cp128, cp421; NTW cp147, cp365, cp491

Plum, Klamath (*Prunus subcordata*)
NTW cp182, cp359

Plum, Mexican (*Prunus mexicana*)
LBG cp423; NTE cp144

Plum, wildgoose or munson (*Prunus munsoniana*)
NTE cp142

Plum (other *Prunus* sp.)
AGC cp454; NTE cp122, cp127, cp146, cp445

Poacher, sturgeon (*Agonus acipenserinus*)
NAF cp427; PCF cp15

Pochard (*Aythya* or *Netta* sp.)
DNA 78, 79, 83, 85, 93, 94; RBB 33

Pocket gopher, Botta's or Valley (*Thomomys bottae*)
ASM cp34; GEM II:130; JAD cp469; MEM 629, 630; NMM
  cp[12]

Pocket gopher, Northern (*Thomomys talpoides*)
ASM cp35; CMW 120(bw); MEM 631; MIA 163; MNC
  177(bw); MPS 176(bw)

Pocket gopher, plains (*Geomys bursarius*)
ASM cp32; CMW 125(bw); LBG cp78; MEM 628; MIA 163;
  MNC 179(bw); MPS 176(bw)

Pocket gopher, southeastern (*Geomys pinetis*)
ASM cp36; CRM 103

Pocket gopher, yellow-faced (*Pappogeomys castanops*)
ASM cp33; MPS 177(bw)

Pocket gopher (other kinds)
ASM cp31; MEM 629

Podocarp tree (*Podocarpus* sp.)
OET 103

Poincettia, Christmas (*Euphorbia pulcherrima*)
TSP 59

Poinciana (*Poinciana* sp.)
PPC 37; TSP 25, 47

Poinciana, wild (*Euphorbia heterophylla*)
NWE cp435

Poison ivy (*Toxicodendron* or *Rhus radicans*)
FWF 45; NWE cp224; PPC 106; SEF cp433

Poison oak (*Toxicodendron diversiloba*)
WFW cp114

Pokeberry or Pokeweed (*Phytolacca* sp.)
BCF 19; CWN 55; FWF 177; NWE cp143, cp656; PPC 92;
  PPM cp23; WTV 43

Polecat, African striped See Zorilla

Polecat, European or Western (*Mustela putorius*)
AWR 68-69; GEM III:399, 401-403; MEM 110, 115; MIA
  87; NNW 42

Polecat, marbled (*Vormela peregusna*)
CRM 139; GEM III:408-409; MEM 112; NHU 186

Polecat, steppe (*Mustela eversmanni*)
GEM III:405

Pollock (*Pollachius virens*)
ACF p15; AGC cp390; MIA 527; NAF cp489

Polypore, sulfur See Chicken-of-the-woods

Pomfret (*Brama* sp.)
ACF p29; MIA 555

Pompano, African (*Alectis ciliaris*)
ACF p29; NAF cp345

Pompano, Florida (*Trachinotus carolinus*)
ACF p29; AGC cp347; MIA 553; NAF cp550

Pompano, gafftopsail (*Trachinotus rhodopus*)
PCF cp31

Poor-will (*Phalaenoptilus nuttallii*)
FWB 251; JAD cp562; MIA 275; NAW 226; UAB cp249;
  WGB 85

Poplar, balsam (*Populus balsamifera*)
LBG cp431; NTE cp118; NTW cp210; SEF cp101

Poplar, lombardy (*Populus nigra*)
NTE cp187; NTW cp212

Poplar, white (*Populus alba*)
NTE cp188, cp238; NTW cp213, cp236

Poplar, yellow, or tulip tree (*Liriodendron tulipifera*)
NTE cp267, cp390, cp490, cp618; NTW cp234, cp336,
  cp534; SEF cp123, cp162; TSV 20, 21

Poppy, alpine (*Papaver kluanense*)
SWW cp223

Poppy, arctic (*Papaver polare* or *radicatum*)
NHU 77; WAA 248

Poppy, California (*Eschscholtzia californica*)
LBG cp223; SWW cp361; WFW cp452

Poppy, corn (*Papaver rhoeas*)
PPC 89

Poppy, desert (*Kallstroemia grandiflora*)
JAD cp155; SWW cp360

Poppy, fire (*Papaver californicum*)
SWW cp363; WFW cp453

Poppy, great desert (*Arctomecon merriami*)
JAD cp95; SWW cp62

Poppy, Mexican or gold (*Eschscholtzia mexicana*)
JAD cp148; SWW cp227; WAA 1, 203

Poppy, opium (*Papaver somniferum*)
PPM cp5

Poppy, prickly (*Argemone* sp.)
NWE cp84; SWW cp64

Poppy, tree (*Dendromecon rigida*)
SWW cp346

Poppy, wood (*Stylophorum diphyllum*)
NWE cp253

Porcelain fungus (*Mucidula mucida*)
NMT cp14; SGM cp39

Porcupine (*Erethizon dorsatum*)
AAL 48; ASM cp219, cp220; CGW 177; CMW 189(bw);
  CRM 127; GEM III:23; GMP 237(bw); JAD cp507; MAE 49;
  MEM 688, 689; MIA 183; MNC 229(bw); MNP 221(bw);
  MPS 249(bw); NAW 66, 67; NMM cp[19]; SWF 152; SWM
  57, 58; WFW cp355; WMC 172

Porcupine, brush-tailed (*Atherurus* sp.)
MEM 705; MIA 183

Porcupine, Cape or South African (*Hystrix
  africaeaustralis*)
AAL 48; GEM III:292-293, 304; MEM 684

Porcupine, crested or North African (*Hystrix cristata*)
CRM 125; GEM III:301; MEM 705; MIA 183

Porcupine, long-tailed (*Trichys fasciculata*)
GEM III:301; ROW 72(bw)

Porcupine, prehensile-tailed or tree (*Coendou
  prehensilis*)
AAL 48; GEM III:315; MEM 687; MIA 183

Porcupine, short-tailed (*Hystrix brachyura*)
GEM III:22, 301; ROW 71(bw)

Porcupinefish (*Diodon* sp.)
AAL 373; ACF cp60; BMW 142; KCR cp26; MIA 583; NAF
  cp302, cp304

Porgy (*Calamus* sp.)
ACF p33; NAF cp547, cp548; PCF cp35; RUP 172, 173

Porkfish (*Anisotremus virginicus*)
ACF cp32; AGC cp371; KCR cp23; NAF cp537; RCK 40

Porpoise, Burmeister's or black (*Phocoena spinipinnis*)
MEM 198; WDP 171; WOW 194

Porpoise, Common or harbor (*Phocoena phocoena*)
AGC cp320; CRM 207; GEM IV:384; HWD 41; MEM 199;
  MIA 113; MPC cp211; NAF cp653; WDP 164, 171; WOW
  192

Porpoise, Dall's (*Phocoenoides dalli*)
CRM 207; GEM IV:384; HWD 40, 90; MEM 199; MIA 113;
  MPC cp202, cp210; NAF cp630, cp652; WDP 164, 172;
  WOW 202

Porpoise, finless (*Neophocaena phocaenoides*)
HWD 41, 73; MEM 198; MIA 113; WDP 164; WOW 200

Porpoise, Gulf of California (*Phocoena sinus*)
MEM 198; WOW 198

Porpoise, pink See River dolphin, Amazon

Porpoise, spectacled (*Phocoena* or *Australophocaena
  dioptrica*)
HWD 40; MEM 199; WDP 171; WOW 196

Portuguese man-of-war (*Physalia physalis*)
AGC cp218; EAL 173; HSG cp31; KCR cp30; MSC cp512,
  cp513

Possum See Opossum, Virginia

Possum, brushtail (*Trichosurus vulpecula*)
GEM I:310, 311; MEM 854; MIA 21; SWF 105; WOI 127

Possum, feathertail (*Distoechurus pennatus*)
MEM 859

Possum, honey (*Tarsipes rostratus* or *spenserae*)
CRM 27; GEM I:328, 329; MEM 879; MIA 19; NOA 158

Possum, leadbeater's (*Gymnobelideus leadbeateri*)
CRM 27; GEM I:316; MEM 860; NOA 137

Possum, mountain pygmy (*Burramys parvus*)
CRM 27; MEM 859; NOA 119

Possum, pygmy (*Cercartetus nanus*)
MEM 861; NOA 35; WOI 125

Possum, ringtail (various genera)
CRM 27; GEM I:313; MEM 858; NOA 131, 132

Possum, scaly-tailed (*Wyulda squamicaudata*)
CRM 27; MEM 851

Possum, striped (*Dactylopsila trivirgata*)
MEM 858; NOA 53

Possumhaw (*Ilex decidua*)
NTE cp207, cp560; WNW cp469

Potato, Indian See Groundnut

Potato vine (*Solanum* sp.)
TSP 95

Potato vine, wild See Manroot

Potoroo (*Potorous* sp.)
CRM 29; GEM I:356; MEM 867; MIA 23

Potto (*Perodicticus potto*)
MEM 335; MIA 59; NHP 96(bw)

Pottoo, common (*Nyctibius griseus*)
EOB 249; GEM II:35, 82; MIA 273; WGB 83

Potto, golden or Angwantibo (*Arctocebus calabarensis*)
MEM 335; MIA 59; NHP 98 (bw)

Pout, ocean (*Macrozoarces americanus*)
AGC cp430; NAF cp454

Powderpuff (*Calliadra* sp.)
TSP 49

Prairie chicken, greater (*Tympanuchus cupido*)
AAL 116; EAB 414, 415; EOB 134; FEB 210; FWB 199;
    LBG cp500; MIA 233; NAB cp261; SPN 19(bw); UAB
    cp254, cp258; WGB 43; WPP 165-167

Prairie chicken, lesser (*Tympanuchus pallidicinctus*)
EAB 415; FEB 211; LBG cp501; NAB cp262; UAB cp255,
    cp259

Prairie dog, black-tailed (*Cynomys ludovicianus*)
AAL 43(bw); ASM cp23; CMW 108(bw); CRM 95; LBG
    cp87; GEM III:54, 55; MIA 161; MNP cp66; MPS 136(bw);
    NAW 68; NMM cp[8]; SAB 115; WPP 152-153; WWW 190

Prairie dog, white-tailed (*Cynomys leucurus*)
ASM cp21; CMW 110(bw); LBG cp86; NAW 69

Prairie dog (other *Cynomys* sp.)
ASM cp20, cp24; MEM 619-621

Prairie smoke (*Geum triflorum*)
LBG cp249; NWE cp408; SWW cp501

Prairie star (*Lithophragma parviflorum*)
LBG cp137; SWW cp28

Pratincole (*Glareola* sp.)
EOB 182, 184; MIA 247; WGB 57

Pretty face (*Triteleia ixioides*)
SWW cp320

Prickleback (several genera)
NAF cp456; PCF cp40, p41

Prickly pear (*Opuntia* sp.)
AGC cp456; CWN 99; JAD cp121; NTE cp242; SWW
    cp231, cp232; TSP 71; WAA 177

Prickly-ash (*Zanthoxylum americanum*)
LBG cp450; NTE cp339

Prickly-ash, lime (*Zanthoxylum fagara*)
NTE cp301, cp400; SEF cp128

Primrose, beach (*Oenothera cheiranthifolia*)
MPC cp607; SWW cp195

Primrose, bird's-eye (*Primula mistassinica*)
WAA 186

Primrose, desert (*Oenothera brevipes*)
JAD cp156; SWW cp329; WAA 210

Primrose, parry (*Primula parryi*)
WAA 224

Primrose, Sierra (*Primula suffrutescens*)
SWW cp543

Prince's or desert plume (*Stanleya pinnata*)
JAD cp109; LBG cp217; NWE cp325; SWW cp307

Princess tree See Paulownia, royal

Prion, broad-billed (*Pachyptila vittata*)
MIA 201; WGB 11

Prion, fairy (*Pachyptila turtur*)
HSW 84

Privet, Chinese (*Ligustrum sinense*)
NTE cp103, cp444

Privet, Florida (*Forestiera segregata*)
NTE cp104

Pronghorn (*Antilocapra americana*)
ASM cp284; CMW 261(bw); CRM 183; GEM V:13(head
    only), 278-279, 282; MAE 74-75; JAD cp527; LBG cp34;
    MEM 543; MIA 139; MNC 291(bw); MNP cp7b, 292(bw);
    MPS 306(bw); NAW 10; NMM cp[32]; SWM 156-157;
    WPP 148, 149

Protea (*Protea* sp.)
TSP 73, 75

Ptarmigan, rock (*Lagopus mutus*)
AAL 88; BAS 85, 105; EAB 416, 417; FEB 207; FWB 187;
    MIA 233; NAB cp264, cp266; NAW 190; NHU 84; RBB 52;
    UAB cp272, cp274; WGB 43; WOM 86, 87; WPR 51

Ptarmigan, white-tailed (*Lagopus leucurus*)
EAB 416, 417; EOB 130; FWB 185; NAW 191; UAB
    cp273, cp275

Ptarmigan, willow (*Lagopus lagopus*)
EAB 416, 417; FEB 206; FWB 186; NAB cp263, cp265;
    NAW 190; RBB 51; UAB cp268-cp271; WOB 200-203;
    WOM 86; WPR 50

Puccoon, hoary See Gromwell

Puddingwife (*Halichoeres radiatus*)
ACF cp39; NAF cp363, cp373

Pudu (*Pudu* sp.)
CRM 181; GEM V:224, 226; MEM 522; MIA 139; WOM
    187

Puffball See also Earth-ball fungus

Puffball, giant (*Calvatia gigantea* or *Lycoperdon maximum*)
FGM p44; NAM cp647; SGM cp368

Puffball, lead (*Bovista plumbea*)
NMT cp22; SGM cp366

Puffball, meadow (*Lycoperdon pratense* or *perlatum*)
NMT cp23; SEF 61

Puffball, thick-skinned See Earth-ball fungus

Puffer (*Arothron* sp.)
BMW 147; RCK 165; RUP 131

Puffer (*Sphoeroides* sp.)
ACF cp60; AGC cp354; MIA 583; NAF cp305, cp310, cp311

Puffer, sharpnose (*Canthigaster rostrata*)
AGC cp349; NAF cp301

Puffer, smooth (*Lagocephalus laevigatus*)
NAF cp307, cp308

Puffer, spiny (various genera)
RCK 63, 157; RUP 137

Puffin, Atlantic or common (*Fratercula arctica*)
AAL 88; AGC cp521; AOG cp12; BMW 203; BWB 110-111; DBN cp30, cp31; EAB 48; FEB 65; HSW 152; MIA 251; NAB cp95, cp97; NAW 192; OBL 27; WGB 61; WOB 186, 187; WON 97

Puffin, horned (*Fratercula arctica*)
AOG cp11; BMW 203; BWB 108; DBN cp32; EAB 49; FWB 66; MPC cp539; RBB 98; UAB cp83; WPR 143

Puffin, tufted (*Lunda cirrhata*)
AAL 132(bw); AOG cp10; BMW 202; BWB 109; DBN cp29; EAB 48, 49; EOB 211; FWB 67; MPC cp540; UAB cp82, cp84; WON 99; WPR 142

Puma See Mountain lion

Punctureweed (*Tribulus terrestris*)
JAD cp143; NWE cp248

Pupfish, desert (*Cyprinodon macularius*)
JAD cp293; NAF cp209; WNW cp98

Purple mat (*Nama demissum*)
JAD cp54; SWW cp540

Purslane (*Portulaca lutea* or *oleracea*)
CWN 139; NWE cp266; SWW cp201; TSP 105

Pussy paws (*Calyptridium umbellatum*)
SWW cp554; WFW cp512

Pussytoes (*Antennaria* sp.)
NWE cp195; SWW cp144; WFW cp424

Pyrrhuloxia (*Cardinalis sinuatus*)
EAB 311; FEB 388; FWB 430; JAD cp602, cp603; NAB cp405; SPN 119; UAB cp468, cp567

Python (*Calabaria* sp.)
ERA 118; LSW 56

Python (*Liasis* sp.)
LSW 54, 68-73; NOA 221

Python, African (*Python sebae*)
ERA 120, 121

Python, carpet (*Morelia* sp.)
LSW 66-67; MIA 447

Python, Indian (*Python* sp.)
ERA 122-123; LSW 57-64; MIA 447; SWF 83

Python, tree (*Chondropython* sp.)
LSW 74-75; MSW cp22

Python, tree (Green) (*Chondropython viridis*)
AAL 240; MAE ii; WOI 138

Q

Quail, California (*Lophortyx californicus*)
EAB 708; FWB 190; JAD cp548, cp549; MIA 233; NAW 193; UAB cp278, cp279; WFW cp226; WGB 43

Quail, European (*Coturnix coturnix*)
MIA 229; WGB 39

Quail, Gambel's (*Lophortyx gambelii*)
EAB 709; FWB 189; JAD cp547; NAW 193; UAB cp276

Quail, little button See Hemipode

Quail, Montezuma or harlequin (*Cyrtonyx montezumae*)
EAB 708; FWB 194; UAB cp283

Quail, mountain (*Oreortyx pictus*)
EAB 709; FWB 191; UAB cp277; WFW cp227

Quail, painted (*Excalfactoria chinensis*)
MIA 229; WGB 39

Quail plant (*Heliotropium curassavicum*)
SWW cp153

Quail, scaled (*Callipepla squamata*)
EAB 708; FEB 202; FWB 188; JAD cp546; LBG cp504; NAB cp258; UAB cp284

Quaker-ladies See Bluets

Queen Anne's lace (*Daucus carota*)
BCF 63; CWN 137; LBG cp145; NWE cp191; SWW cp169; WAA 38

Queen of the night See Cereus, night-blooming

Queen of the prairie (*Filipendula rubra*)
LBG cp258; NWE cp566

Queen's cup (*Clintonia uniflora*)
SWW cp40; WFW cp393

Quelea (*Quelea quelea*)
AAL 192

Quetzal (*Pharomachrus mocino*)
AAL 143; EOB 264; MIA 281; WGB 91; WOM 174

Quillback (*Carpiodes cyprinus*)
NAF cp102; WNW cp35

Quillfish (*Ptilichthys goodei*)
PCF p42

Quince, common (*Cydonia oblonga*)
OET 183

Quokka (*Setonix brachyurus*)
CRM 31; GEM I:365; MEM 864; MIA 25

Quoll or tiger cat (*Dasyurus* sp.) AAL 27(bw); CRM 17;
GEM I:277; MEM 841; MIA 17; NOA 46, 47; WOI 126

R

Rabbit, Amami (*Pentalagus furnessi*)
CRM 93; MEM 714

Rabbit, brush (*Sylvilagus bachmani*)
ASM cp246; MIA 193; WFW cp349

Rabbit, Bunyoro (*Poelagus marjorita*)
MEM 715

Rabbit, cottontail See Cottontail

Rabbit, european (*Oryctolagus cuniculus*)
ASM cp244; GEM IV:250, 257, 292-297; MAE 2, 29, 139;
MEM 715; MIA 193; NNW 113

Rabbit, jack See Jackrabbit

Rabbit, marsh (*Sylvilagus palustris*)
ASM cp248; NAW 72; WMC 105; WNW cp605

Rabbit, snowshoe See Hare, snowshoe

Rabbit, Sumatran (*Nesolagus netscheri*)
CRM 93; MIA 193

Rabbit, swamp (*Sylvilagus aquaticus*)
ASM cp247; GEM IV:298; MIA 193; MNC 139(bw); MPS
115(bw); WMC 110; WNW cp606

Rabbit, volcano (*Romerolagus diazi*)
CRM 93; GEM IV:299; MIA 191

Rabbitbrush (*Chrysanthemum nauseosus*)
JAD cp340; LBG cp212; SWW cp352; WFW cp206

Raccoon (*Procyon lotor*)
AAL 61; ASM cp202; CGW 145; CMW 212(bw); GEM
III:455-457, 460; JAD cp512; MEM 100, 101; MIA 85;
MNC 249(bw); MNP 253(bw); MPS 272(bw); NAW 70, 71;
NMM cp[25]; NNW 40, 41; PGT 231; RFA 153-164(bw);
SEF cp607; SWF 139; SWM 97, 98; WFW cp353; WMC
187; WNW cp614

Raccoon, crab-eating (*Procyon cancrivorus*)
GEM III:458; MEM 101

Racer (*Coluber constrictor*)
AAL 243(bw), 244; ARA cp468, cp478, cp480;cp486; ARC
196, 197; ARN 68(bw); CGW 85; ERA 125; JAD cp232;
LBG cp618; LSW 136; NAW 245; RNA 191; SEF cp521;
WFW cp600; WRA cp36

Racer, forest (*Drymoluber dichrous*)
LSW 150

Racer, horseshoe (*Coluber hippocrepis*)
LSW 141; MSW cp4

Racer, speckled (*Drymobius margaritiferus*)
ARA cp559; ART p[19]; LSW 149; RNA 189

Racer, striped (*Masticophis lateralis*)
ARA cp518; RNA 191; WFW cp589; WRA cp36

Ragfish (*Icosteus aenigmaticus*)
PCF cp46

Ragged robin (*Lychnis flos-cuculi*)
CWN 117; LBG cp245; NWE cp500; WAA 186

Ragweed (*Ambrosia artemisiifolia*)
NWE cp26

Ragwort, tansy (*Senecio jacobaea*)
PPC 101

Ragwort, golden (*Senecio aureus*)
CWN 257; NWE cp289; WNW cp207; WTV 70

Rail, black (*Laterallus jamaicensis*)
EAB 724; FEB 124; FWB 127; PBC 131; UAB cp247

Rail, clapper (*Rallus longirostris*)
AGC cp598; EAB 724; FEB 128; FWB 123; MPC cp562;
NAB cp253; UAB cp243

Rail, flightless (*Gallirallus australis*)
WOI 146

Rail, king (*Rallus elegans*)
EAB 724; FEB 129; NAB cp254; PBC 126; WNW cp541

Rail, Virginia (*Rallus limicola*)
EAB 725; FEB 127; FWB 124; MPC cp561; NAB cp252;
   PBC 128; UAB cp242; WNW cp540

Rail, water (*Rallus aquaticus*)
AWR 64; EOB 145; MIA 237; RBB 58; WGB 47

Rail, wood (*Aramides ypecaha*)
MIA 237; WGB 47

Rail, yellow (*Coturnicops noveboracensis*)
EAB 724; FEB 125; FWB 126; NAB cp251; PBC 130; UAB
   cp244; WNW cp539

Railroad vine See Morning-glory, beach

Rainbow runner (*Elagatis bipinnulata*)
NAF cp558

Ramp See Leek, wild

Raspberry, purple-flowering (*Rubus odoratus*)
BCF 137; NWE cp583; SEF cp485; TSV 112, 113

Rat, bamboo (*Rhizomys* or *Cannomys* sp.)
GEM III:151

Rat, bandicoot (*Bandicota* sp.)
GEM III:163

Rat, black (*Rattus rattus*)
ASM cp117; GEM III:172; MIA 179; NNW 67; WMC 163

Rat, bush (*Rattus fuscipes*)
MEM 595

Rat, cane (Thryonomyidae)
GEM III:309; MAE 121; MEM 702; MIA 189

Rat, chinchilla (*Abrocoma* sp.)
GEM III:347; MEM 702; MIA 187

Rat, climbing (*Tylomys* sp.)
CRM 105; MEM 645

Rat, cotton (*Sigmodon* sp.)
ASM cp141; GEM III:223; MEM 646

Rat, cotton (hispid) (*Sigmodon hispidus*)
ASM cp136; LBG cp60; MNC 203(bw); MPS 216(bw);
   WMC 149

Rat, crested or maned (*Lophiomys imhausii*)
GEM III:225; MEM 667; MIA 171

Rat, fish-eating (*Ichthyomys* sp.)
MEM 647; MIA 167

Rat, giant pouched (*Oricetomys* sp.)
GEM III:268; MEM 670

Rat, kangaroo (banner-tailed) (*Dipodomys spectabilis*)
ASM cp83; JAD cp483; WWD 53

Rat, kangaroo (chisel-toothed) (*Dipodomys microps*)
ASM cp84; CRM 103; JAD cp481

Rat, kangaroo (Merriam's) (*Dipodomys merriami*)
ASM cp88; JAD cp485; MAE 60

Rat, kangaroo (Ord's) (*Dipodomys ordii*)
ASM cp86; CMW 137(bw); CRM 103; JAD cp480; LBG
   cp72;MPS 180(bw); NAW 74; NMM cp[13]

Rat, kangaroo (other *Dipodomys* sp.)
ASM cp82, cp85, cp87, cp89, cp90, cp109; GEM III:140;
   JAD cp482, cp484; LBG cp73; MEM 633; MIA 163; MPS
   179(bw)

Rat, marsh (African) (*Dasymus incontus*)
MEM 661

Rat, marsh rice (*Oryzomys palustris*)
AAL 47(bw); ASM cp121; GEM III:220 (*O.baui*); GMP
   174(bw); MNC 185(bw); MPS 181(bw); WMC 137; WNW
   cp596

Rat, Norway (*Rattus norvegicus*)
ASM cp137, cp140; CGW 185; CMW 180(bw); GEM I:8-9,
   III:152, 171, 173; GMP 217(bw); MIA 179; MNC 298(bw);
   WMC 165

Rat, rice See Rat, marsh rice

Rat, rock (*Petromus* sp.)
GEM III:308; MEM 703

Rat, spiny (various genera)
CRM 115; GEM III:345; MEM 703

Rat, water (Australian) (*Hydromys chrysogaster*)
GEM III:159; MEM 594, 671; MIA 177

Rat, wood See Woodrat

Ratany (*Krameria parvifolia*)
JAD cp56

Ratel See Badger, honey

Ratfish (*Hydrolagus* sp.) See Chimaera

Ratsnake, Baird's (*Elaphe obsoleta bairdi*)
ARA cp509; ART p[20]; LSW 89

Ratsnake, black (*Elaphe obsoleta obsoleta*)
ARN 70(bw); CGW 85; ERA 117; LBG cp613; LSW 86, 87;
   MIA 451; NAW 4; SEF cp524

Ratsnake, gray (*Elaphe obsoleta spiloides*)
ARA cp524, cp581; LSW 92

Ratsnake, green (*Elaphe triaspis intermedia*)
ARA cp479; RNA 185; LSW 95; WRA cp38

Ratsnake, mangrove (*Gonyosoma* sp.)
LSW 105

Ratsnake, radiated See Copperhead

Ratsnake, red or corn snake (*Elaphe guttata guttata*)
AAL 244; ARA cp570, cp608; ARC 200; LBG cp603; LSW 80, 84; MSW cp41; NAW 239; RNA 185; SEF cp531; WRA cp38

Ratsnake, rosy (*Elaphe guttata rosacea*)
LSW 84

Ratsnake, Trans-Pecos (*Elaphe subocularis*)
ARA cp523; JAD cp214; LSW 94; RNA 185; WRA cp38

Ratsnake, yellow (*Elaphe obsoleta quadravittata*)
ARA cp526, cp540; LSW 82, 90

Ratsnake (other *Elaphe* sp.)
ARC 201, 202; LSW 85-106; RNA 185; WRA p48

Rat-tail or Grenadier (*Odontomacrurus murrayi*)
EAL 89; NNW 147

Rattlebox, showy (*Crotolaria spectabilis*)
NWE cp327

Rattlesnake, Baja California (*Crotalus enyo*)
LSW 430; WRA p48

Rattlesnake, black-tailed (*Crotalus molossus*)
AAL 262(bw); ARA cp626; JAD cp248; LSW 416, 417; RNA 207; WRA cp45

Rattlesnake, diamondback (Eastern) (*Crotalus adamanteus*)
ARA cp624; ARC 243; LSW 398; MIA 459; NAW 246, 247; RNA 203; SEF cp528; WWD 150-151

Rattlesnake, diamondback (red) (*Crotalus ruber*)
AAL 262(bw); ARA cp644; JAD cp254; LSW 421; RNA 205; WRA cp44

Rattlesnake, diamondback (Western) (*Crotalus atrox*)
ARA cp639; ERA 131; JAD cp256; LBG cp599; LSW 400; RNA 205; WRA cp44

Rattlesnake, massasauga (*Sistrurus catenatus*)
AAL 260; ARA cp632, cp633, cp638; JAD cp264; LBG cp600; MIA 459; RNA 201; CGW 97; WRA cp45

Rattlesnake, Mojave (*Crotalus scrutulatus*)
ARA cp622; JAD cp259; LSW 422; RNA 205; WRA cp44

Rattlesnake, pigmy (*Crotalus miliarius* or *Sistrurus miliarius*)
ARA cp625, cp641-cp642, cp645; ARC 246, 247; LSW 431, 432; RNA 203; SEF cp532

Rattlesnake pilot See Ratsnake, black

Rattlesnake, ridge-nosed (*Crotalus willardi*)
ARA cp643; LSW 428; RNA 207; WRA cp45

Rattlesnake, rock (*Crotalus lepidus*)
ARA cp636, cp640; JAD cp250, cp257; LSW 412-414; RNA 207; WRA cp45

Rattlesnake, sidewinder (*Crotalus cerastes*)
AAL 262(bw); ARA cp634, cp647; JAD cp252, cp253; LSW 404, 405; MAE 61; MIA 459; RNA 207; WRA cp44

Rattlesnake, speckled (*Crotalus mitchellii*)
ARA cp635, cp646; JAD cp251, cp255; LSW 415; RNA 207; WRA cp44

Rattlesnake, tiger (*Crotalus tigris*)
ARA cp628; JAD cp249; LSW 423; RNA 205; WRA cp45

Rattlesnake, timber (*Crotalus horridus*)
ARA cp619, cp620, cp653; ARC 244, 245; ARN 73(bw); CGW 97; LSW 409; NAW 247; RNA 207; SEF cp529; WNW cp193, cp197, cp198

Rattlesnake, tropical or South American (*Crotalus durissus*)
LSW 406

Rattlesnake, twin-spotted (*Crotalus pricei*)
ARA cp637; LSW 420; RNA 207; WRA cp45

Rattlesnake, Western or Pacific or Prairie (*Crotalus viridis*)
ARA cp621, cp623, cp627, cp629-cp631, cp648; ERA 126; JAD cp258, cp260; LBG cp605; LSW 425-427; RNA 205; WFW cp597; WRA cp44

Rattlesnake-master (*Eryngium yuccifolium*)
LBG cp116; NWE cp42

Rattlesnake-weed (*Euphorbia albomarginata*)
JAD cp102; SWW cp23

Rattlesnake-weed (*Hieracium venosum*)
NWE cp301

Raven, common (*Corvus corax*)
AAL 198(bw); EAB 124; EOB 445; FEB 367; FWB 286; JAD cp582; MPC cp554; NAB cp581; RBB 173; SEF cp265; UAB cp626; WFW cp260

Raven, white-necked or Chihuahuan (*Corvus cryptoleucus* or *albicollis*)
EAB 125; EOB 445; FEB 366; FWB 287; JAD cp581; NAB cp582; UAB cp627

Ray, bat (*Myliobatis californica*)
MPC cp313; NAF cp258; PCF p5

Ray, blue-spotted (*Taeniura limma*)
RCK 55; RUP 88

Ray, butterfly (*Gymnura* sp.)
NAF cp262; PCF p5

Ray, cownose (*Rhinoptera bonasus*)
NAF cp257, cp261

Ray, eagle (*Myliobatis* sp.)
HSG cp14; MIA 495

Ray, eagle (spotted) (*Aetobatus narinari*)
AGC cp424; BMW front.; EAL 139; NAF cp256

Ray, electric (lesser) (*Narcine brasiliensis*)
AAL 382(bw); NAF cp272

Ray, electric (Pacific) (*Torpedo californica*)
BCS 84; MPC cp317; NAF cp270; PCF p4

Ray, electric (other kinds)
AAB 93; WWW 152

Ray, manta See Manta ray

Ray, sting See Stingray

Ray, thorny (*Urogymnus asperrimus*)
EAL 143

Razorbill See under Auk

Razorfish (*Hemipteronotus* sp.)
ACF cp39; KCR cp26

Red maids (*Calandrina ciliata*)
SWW cp457; WFW cp503

Redbay (*Persea borbonia*)
NTE cp69; SEF cp48; WNW cp451

Redbud or Judas tree (*Cercis canadensis*)
FWF 185; NTE cp99, cp466; NTW cp123, cp389; PMT
    cp[29];SEF cp63, cp146; TSV 176, 177

Redbud, California or Western (*Cercis occidentalis*)
NTW cp122, cp390, cp516; SWW cp568; WFW cp67,
    cp166, cp217

Redhorse, river or jumprock (*Moxostoma* sp.)
NAF cp107, cp110, cp111, cp198

Red-hot-poker tree (*Erythrina abyssinica*)
TSP 27

Redpoll, common (*Acanthis* or *Carduelis flammea*)
EAB 311; FEB 403; FWB 469; MIA 385; NAB cp411; RBB
    183; UAB cp463; WGB 195

Redpoll, hoary (*Carduelis hornemanni*)
EAB 311; FWB 468; UAB cp464

Redshank, common (*Tringa totanus*)
EOB 167; MIA 243; RBB 79; WBW 245; WGB 53

Redshank, spotted (*Tringa erythropus*)
NHU 82; WBW 223

Redstart (*Phoenicurus phoenicurus*)
RBB 132; SAB 3

Redstart, American (*Setophaga ruticilla*)
EAB 984; FEB 444; FWB 376; NAB cp382, cp402; PBC
    335; SEF cp306; SPN 112; UAB cp438, cp441

Redstart, black (*Phoenicurus ochruros*)
MIA 335; RBB 131; WGB 145

Redstart, painted (*Myioborus pictus*)
EAB 985; FWB 377; MIA 379; UAB cp456; WGB 189

Redwing (*Turdus iliacus*)
NHU 104; RBB 140

Redwood See also Sequoia

Redwood (*Sequoia sempervirens*)
NTW cp53, cp437; WFW cp53, cp141

Redwood, dawn (*Metasequoia glyptostroboides*)
PMT cp[85]

Reed, giant (*Phragmites australis*)
NWE cp406; WNW cp323

Reedbuck, Common or Southern (*Redunca arundinum*)
CRM 189; GEM V:454; MEM 562; MIA 147; NHA cp10
    (*R.redunca*); SAB 75

Reedhen, king (*Porphyrio madagascariensis*)
AWR 95

Reeve See Ruff

Reindeer See Caribou

Reindeer moss (*Cladonia rangifera*)
WPR 48

Remora (*Remora remora*) See also Sharksucker
ACF p42; HSG cp19; MIA 553; PCF p12; RUP 101

Rhea, common or gray or greater (*Rhea americana*)
EOB 22; MAE 76; MIA 197; WGB 7; WPP 180

Rhea, Darwin's or lesser (*Pterocnemia pennata*)
EOB 19, 22; WOM 188, 189

Rhebok (*Pelea capreolus*)
MEM 562; MIA 147

Rhinoceros, black (*Diceros bicornis*)
CRM 163; GEM IV:611(head only), 621-634; MEM 493;
    MIA 127; RES 8(bw), cp1, cp6; WWW 82-83

Rhinoceros, Indian (*Rhinoceros unicornis*)
AAL 73; AWR 114-115; CRM 165; GEM IV:612-613, 635-
    637; MEM 491, 493; MIA 127; RES 8(bw), cp13, cp15

Rhinoceros, Javan (*Rhinoceros sondaicus*)
CRM 165; MEM 493; RES 8(bw), cp17

Rhinoceros, Sumatran (*Dicerorhinus sumatrensis*)
CRM 165; MEM 493; MIA 127; RES 8(bw), cp18

Rhinoceros, white (*Cerotatherium simus*)
CRM 163; GEM IV:609, 611, 617-620; MEM xvi-1, 492, 497; MIA 127; RES 8(bw), cp8, cp10; WPP 75

Rhododendron, California (*Rhododendron macrophylum*)
WFW cp209

Rhododendron, catawba or purple, or mountain rosebay (*Rhododendron catawbiense*)
NTE cp62, cp468; NWE cp587; SEF cp43, cp147; TSV 110, 111; WAA 4-5

Rhododendron, white, or rosebay (*Rhododendron maximum*)
NTE cp63, cp439, cp465; NTW cp210; SEF cp47, cp150, cp437; TSV 50, 51; WNW cp230

Rhododendron, white (Western) (*Rhododendron albiflorum*)
SWW cp179; WFW cp187

Ribbonfish (*Zu cristatus*)
ACF cp22; BMW 102-103

Rice, wild (*Zizania aquatica*)
WNW cp321

Rifle bird, magnificent (*Craspedophora magnifica*)
AAL 198(bw)

Rifleman bird (*Acanthisitta chloris*)
AAL 159; EOB 329, 330; MIA 311; NOA 126 (*Ptiloris victoriae*); WGB 121; WOI 147

Ringstem, southwestern (*Anulocaulis leiosolensus*)
JAD cp105

Ringtail See also Possum, ringtail

Ringtail or civet cat (*Brassariscus astutus*)
ASM cp201; CMW 210(bw); GEM III:450, 451; JAD cp511; MNP 251(bw); MPS 272(bw); NAW 75; NMM cp[26]; RFA 144-150(bw); WFW cp354

Ringtail, South American or cacomistle (*Brassariscus sumichrasti*)
GEM III:452; MEM 107

River dolphin, Amazon or Bouto (*Inia geoffrensis*)
AAL 53; AWR 189; CRM 205; GEM IV:359, 362-363; HWD 42, 81; MEM 178; MIA 113; WDP 12, 18(bw); WOW 156

River dolphin, Ganges (*Platanista gangetica*)
GEM IV:359; HWD 42; MEM 178; MIA 113; WDP 12, 14; WWW 195

River dolphin, Indus (*Plantanista minor*)
MEM 179; WOW 151

River dolphin, La Plata (*Pontoporia blainvillei*)
CRM 205; HWD 43; MEM 178; WDP 12, 26(bw); WOW 160

River dolphin, Yangtze or Chinese (*Lipotes vexillifer*)
CRM 205; GEM IV:360; HWD 43; MEM 179; WDP 12, 22; WOW 154

Roach, California (*Hesperoleucus symmetricus*)
NAF cp171

Roach, european (*Rutilus rutilus*)
AWR 46-47, 216(bw); MIA 515

Roach, sea See Sea slater

Roadrunner (*Geococcyx californianus*)
AAL 136(bw); EAB 132, 133; EOB 232; FEB 200; FWB 196; JAD cp554; LBG cp511; MAE 61; MIA 267; NAB cp271; UAB cp264; WGB 77; WOB 128-131; WWD 144

Robin, American (*Turdus migratorius*)
EAB 893; FEB 353; FWB 418; MIA 337; NAB cp400; PBC 272; SEF cp307; SPN 89; UAB cp445, cp543; WFW cp279; WGB 147; WON 153-155

Robin, European (*Erithacus rubecula*)
AAB 141; EOB 365; MIA 333; RBB 129; SAB 3; SPN 15; WGB 143

Robin, flame (*Petroica phoenicea*)
EOB 369

Robin, Persian or white-throated (*Irania gutturalis*)
NHU 157

Robin, red-capped (*Petroica goodenovii*)
EOB 374

Robin-chat or Cape robin (*Cossypha* sp.)
EOB 369; MIA 333; SPN 10; WGB 143

Rock beauty (*Holacanthus tricolor*)
ACF cp37; NAF cp333

Rock crawler (*Grylloblatta campodeiformis*)
SGI cp14

Rock fringe (*Epilobium obcordatum*)
SWW cp460

Rock hind (*Epinephelus adscensionis*)
AGC cp383

Rock nettle (*Eucnide* sp.)
JAD cp150; SWW cp203; WAA 204, 208

Rockfish (*Sebastes* sp.)
AAL 349; BCS 85; MPC cp235-263; NAF cp388-cp410; PCF cp23-cp28

Rock-jasmine See Candelabra, fairy

Rockweed (*Fucus* sp.)
AGC cp473; BMW 22-23; MPC cp499

Roller (*Coracias* sp.)
EOB 277-279

Roller, Eastern broad-billed (*Eurystomus orientalis*)
NHU 128

Ronquil (*Bathymaster* sp.)
PCF p42

Rook (*Corvus frugilegus*)
BAS 149; MIA 401; RBB 171; WGB 211; WPP 34

Roosterfish (*Nematistius pectoralis*)
PCF cp31

Rorqual See Whale, sei

Rose, beach (*Rosa rugosa*)
AGC cp450; NWE cp582

Rose moss (*Portulaca pilosa*)
NWE cp453

Rose, multiflora (*Rosa multiflora*)
LBG cp127; NWE cp211; WAA 60

Rose, nootka (*Rosa nutkana*)
SWW cp571; WFW cp211

Rose, pasture (*Rosa carolina*)
TSV 96, 97

Rose, prairie (*Rosa suffulta*)
LBG cp238; NWE cp584

Rose, rock or sun (*Helianthemum scoparium*)
SWW cp221

Rose, swamp (*Rosa* or *Hibiscus palustris*)
BCF 131; CWN 87; WNW cp273

Rose, Virginia (*Rosa virginiana*)
NWE cp581

Rosebay See Rhododendron, white

Rosebay, lapland (*Rhododendron lapponicum*)
NWE cp590; WAA 253

Rosebay, mountain See Rhododendron, catawba

Rosefinch (*Carpodacus rubicilla*)
NHU 48

Rosemallow (*Hibiscus moscheutos*)
WTV 38

Rosemallow, desert (*Hibiscus coulteri*)
JAD cp151; SWW cp214

Rosemallow, swamp (*Hibiscus palustris*)
NWE cp470; WNW cp274

Rosemary, bog (*Andromeda* sp.)
WAA 250; WNW cp261

Roseroot (*Sedum rosea*)
SWW cp385; WFW cp469

Rosin weed (*Calycadenia truncata*)
LBG cp184; SWW cp253

Rotifer (several genera)
EAL 187; PGT 103; WWW 42, 44, 45;

Rowan See Mountain ash, European

Rubber tree, hardy (*Eucommia ulmoides*)
PMT cp[47]

Rue (*Ruta graveolens*)
PPC 98

Rue anemone (*Anemonella thalictroides*)
BCF 165; CWN 73; NWE cp55; WAA 157

Rue anemone, false (*Isopyrum biternatum*)
NWE cp56

Ruellia, pinelands (*Ruellia pinetorum*) See also Petunia, wild
WAA 166

Ruff (*Philomachus pugnax*)
AAL 127; BAS 24, 25; EAB 794; MIA 243; NAB cp214; NHU 82; RBB 73; WBW 299; WGB 53; WPP 49

Rush, conglomerate or common (*Juncus conglomeratus*)
PGT 42

Rush, flower (*Butomus umbellatus*)
PGT 41

Rush, hard (*Juncus inflexus*)
PGT 42

Rush, soft (*Juncus effusus*)
NWE cp395; WNW cp324

Rush, sweet (*Cyperus retrofractus*)
WTV 5

Russula See Sickener

Rust spot fungus or spotted coincap (*Collybia maculata*)
FGM cp17; NAM cp35; NMT cp15; SGM 45

S

Sable (antelope) See Antelope, sable

Sable (*Martes zibellina*)
GEM III:414-415(head only); MIA 87; NHU 51

Sablefish (*Anoplopoma fimbria*)
MIA 547; MPC cp285; NAF cp432; PCF cp34

Sage See also Salvia

Sage, Autumn (*Salvia greggii*)
SWW cp570

Sage, bladder (*Salazaria mexicana*)
JAD cp75; SWW cp617

Sage, Death Valley (*Salvia fumerea*)
JAD cp77; SWW cp663

Sage, lyre-leaved or cancer-weed (*Salvia lyrata*)
CWN 219; NWE cp628

Sage, thistle (*Salvia carduacea*)
SWW cp563; WAA 190

Sage, wood (*Teucrium canadense*)
NWE cp535; SEF cp479

Sagebrush (*Artemisia* sp.)
JAD cp353, cp357; LBG cp113; WFW cp110

Saguaro See under Cactus

Saiga (*Saiga tatarica*)
CRM 197; GEM I:126, V:486-494; MIA 155; NHU 167;
  WWD 101

Sailfish (*Istiophorus platypterus*)
ACF cp49; MIA 571; NAF cp585

St. Andrew's-cross (*Ascyrum hypericoides*)
CWN 95

St. John's-wort, common (*Hypericum perforatum*)
BCF 43; CWN 91; LBG cp202; NWE cp352; PPC 81;
  SWW cp211

St. John's-wort, marsh (*Hypericum elodes* or *virginicum*)
CWN 95; NWE cp472; PGT 38; WNW cp285; WTV 144

St. John's-wort, shrubby (*Hypericum spathulatum*)
TSV 86, 87

St. John's-wort, spotted (*Hypericum punctatum*)
CWN 91; WTV 89

St. Peter's-wort (*Ascyrum stans*)
CWN 95; NWE cp351

Saki, black-bearded (*Chiropotes satanus*)
CRM 69; GEM II:137, 138; MIA 63; NHP 122(bw)

Saki, hairy or monk or red-bearded (*Pithecia monachus*)
GEM II:137; MIA 63; SWF 186

Saki, white-faced (*Pithecia pithecia*)
GEM II:137; MEM 354, 356; NHP 120

Salal (*Gaultheria shallon*)
SWW cp574; WAA 244; WFW cp214

Salamander, arboreal (*Aneides lugubris*)
ARA cp106; WFW cp603; WRA cp6, p7

Salamander, black (*Aneides flavipunctatus*)
ARA cp142; WRA cp7, p7

Salamander, blind (comal) (*Euryea tridentifera*)
ARA cp2, cp11

Salamander, blind (Georgia) (*Haideotriton wallacei*)
AAL 264, 274(bw); ARA cp7

Salamander, blind (Texas) (*Typhlomolge rathbuni*)
ARA cp3, cp9; ART p[2]

Salamander, blue-spotted (*Ambystoma laterale*)
ARA cp58; ARN 18 (bw)

Salamander, cave (*Eurycea lucifuga*)
AAL 264, 277; ARA cp122; ARC 79; WOM 163

Salamander, cheat mountain (*Plethodon nettingi nettingi*)
ARA cp102

Salamander, Cherokee or seepage (*Desmognathus aeneus*)
ARA cp94; ARC 62

Salamander, clouded (*Aneides ferreus*)
ARA cp83; WFW cp607; WRA p7

Salamander, crevice (*Plethodon longicrus*)
ARA cp143

Salamander, Del Norte (*Plethodon elongatus*)
ARA cp67; WRA cp5

Salamander, Dunn (*Plethodon dunni*)
ARA cp110; WRA cp5

Salamander, dusky (Black Mountain) (*Desmognathus welteri*)
ARA cp92; ARC 73

Salamander, dusky (mountain) (*Desmognathus ochrophaeus*)
ARA cp95, cp112, cp120, cp124, cp137; ARC 69, 70; ARN
  23 (bw); CGW 49; SEF cp548

Salamander, dusky (Northern) (*Desmognathus fuscus*)
ARA cp89; ARC 64; ARN 22(bw); CGW 49; MIA 469;
  NAW 293; WNW cp112; WOM 162

Salamander, dusky (Southern) (*Desmognathus auriculatus*)
ARA cp63; ARC 63

Salamander, dwarf (*Eurycea quadridigitata*)
ARA cp109, cp126; ARC 80; WNW cp109, cp117

Salamander, fire (*Salamandra salamandra*)
AAL 269; AWR 50; ERA 32; MIA 463; WOM 48

Salamander, flatwoods (*Ambystoma cingulatum*)
ARA cp42; ARC 54

Salamander, four-toed (*Hemidactylium scutatum*)
ARA cp103; ARN 26(bw); CGW 49; WNW cp114

Salamander, green (*Aneides aeneus*)
AAL 273; ARA cp85; ARC 61; ERA 24; WOM 160

Salamander, grotto (*Typhlotriton spelaeus*)
AAL 273; ARA cp12

Salamander, imitator (*Desmognathus imitator*)
ARA cp134; ARC 65, 66

Salamander, Japanese giant (*Andrias japonicus*)
ERA 26

Salamander, Jefferson (*Ambystoma jeffersonianum*)
ARA cp56; ARN 16 (bw); ARC 55; CGW 45

Salamander, Jemez Mountains (*Plethodon neomexicanus*)
ARA cp69; WRA cp5

Salamander, Jordan's or Appalachian woodland (*Plethodon jordani*)
AAL 272; ARA cp133, cp136, cp138; ARC 89, 90; ERA 28, 33; SEF cp549; WOM 160

Salamander, Larch Mountain (*Plethodon larselli*)
ARA cp84; WRA cp5

Salamander, large-blotched (*Ensatina eschscholtzii klauberi*)
ARA cp108; WRA cp3

Salamander, Limestone See Salamander, web-toed

Salamander, longtail (*Eurycea longicauda*)
AAL 276-277; ARA cp49, cp113, cp123; ARC 78; CGW 49

Salamander, long-toed (*Ambystoma macrodactylum*)
AAL 268; ARA cp49, cp50, cp52, cp53; WFW cp606; WRA cp2

Salamander, Mabee's (*Ambystoma mabeei*)
ARA cp60; ARC 56

Salamander, many-lined (*Stereochilus marginatus*)
ARA cp90; ARC 102; WNW cp113

Salamander, many-ribbed (*Eurycea multiplicata*)
ARA cp93; ERA 3

Salamander, marbled (*Ambystoma opacum*)
AAL 268; ARA cp44, cp45; ARC 58; ARN 15(bw); CGW 45; MIA 465; SEF cp552; WOM 162-163

Salamander, mole (*Ambystoma talpoideum*)
ARA cp61; ARC 59; ART p[1]

Salamander, mud (*Pseudotriton montanus*)
ARA cp98, cp99, cp128, cp131; ARC 100; WNW cp111, 116, 118, 119

Salamander, Northwestern or brown (*Ambystoma gracile*)
ARA cp57; WRA cp2

Salamander, Oklahoma (*Eurycea tynerensis*)
ARA cp10

Salamander, Olympic (*Rhyacotriton olympicus*)
ARA cp100; WFW cp605; WRA cp1

Salamander, Oregon (*Ensatina eschscholtzii oregonensis*)
ARA cp104; WFW cp604; WRA cp4

Salamander, Pacific giant (*Dicamptodon ensatus*)
AAL 268; ARA cp23, cp41; MIA 465; WFW cp608; WNW cp128; WRA cp1

Salamander, Painted (*Ensatina eschscholtzii picta*)
ARA cp105; WRA cp4

Salamander, pigmy (*Desmognathus wrighti*)
ARA cp66; ARC 74; WOM 162-163

Salamander, ravine (*Plethodon richmondi*)
ARA cp72; ARC 93

Salamander, red (*Pseudotriton ruber*)
AAL 272; ARA cp129, cp132; ARC 101; CGW 45; ERA 12, 27; MIA 469; WNW cp120; WOM 162

Salamander, redback (*Plethodon cinerus*)
ARA cp71, cp117; ARC 84; ARN 24(bw); CGW 49; NAW 295

Salamander, redback (Southern) (*Plethodon serratus*)
ARA cp118; ARC 94

Salamander, redback (Western) (*Plethodon vehiculum*)
ARA cp119; WRA cp5

Salamander, ringed (*Ambystoma annulatum*)
ARA cp47

Salamander, seal (*Desmognathus monticola*)
AAL 272; ARA cp65; ARC 68

Salamander, shovelnose (*Leurognathus marmoratus*)
ARA cp135; ARC 83

Salamander, Sierra Nevada (*Ensatina eschscholtzii platensis*)
ARA cp107; WRA cp3

Salamander, silvery (*Ambystoma platineum*)
ARA cp59; ARN 17 (bw)

Salamander, slender (Black bellied) (*Batrachoseps nigriventris*)
ARA cp64; WRA cp9

Salamander, slender (California) (*Batrachoseps attenuatus*)
ARA cp76; WRA cp6, p7, cp9

Salamander, slender (Oregon) (*Batrachoseps wrighti*)
ARA cp78; WRA p7, cp8

Salamander, slender (other *Batrachoseps* sp.)
ARA cp73-cp75, cp77; WRA cp8, cp9

Salamander, slimy (*Plethodon glutinosus*)
ARA cp141; ARC 86; ARN 25(bw); CGW 49; MIA 469

Salamander, small-mouthed (*Ambystoma texanum*)
ARA cp62

Salamander, spotted (*Ambystoma maculatum*)
AAL 267(bw); ARA cp51, cp54; ARC 57; CGW 45; ERA 23; MIA 465; NAW 296; PGT 207; WNW cp126, cp127; WOM 162

Salamander, spring (*Gyrinophilus porphyriticus*)
AAL 272; ARA cp101, cp127, cp130; ARC 81; ARN 27(bw); CGW 49; MIA 469; WOM 162-163

Salamander, Tennessee cave (*Gyrinophilus palleucus*)
ARA cp6, cp8

Salamander, Texas (*Eurycea neotenes*)
ARA cp4

Salamander, three-lined (*Eurycea guttolineata*)
ARA cp114; ARC 76

Salamander, tiger (*Ambystoma tigrinum*)
ARA cp22, cp37, cp38, cp39, cp40, cp43, cp46, cp48, cp55; ARC 60; ARN 19(bw); CGW 45; ERA 27; JAD cp289-291; LBG cp583; MIA 465; SEF cp551; WFW cp609; WNW cp125, cp129; WRA cp2

Salamander, two-lined (*Eurycea bislineata*)
AAL 276; ARA cp88, cp121; ARC 75; ARN 28(bw); CGW 49; ERA 33; SAB 82; WNW cp110, cp121

Salamander, valley or valley and ridge (*Plethodon hoffmani*)
ARA cp70; ARC 87

Salamander, Van Dyke (*Plethodon vandykei*)
ARA cp111; WRA cp5

Salamander, web-toed (Limestone) (*Hydromantes bruus*)
ARA cp97; WRA cp6

Salamander, web-toed (Mount Lyell) (*Hydromantes platycephalus*)
ARA cp86; WRA cp6

Salamander, web-toed (Shasta) (*Hydromantes shastae*)
AAL 274(bw); ARA cp87; WRA cp6

Salamander, Yellow-blotched (*Ensatina eschscholtzii croceater*)
AAL 276-277; MIA 469; WRA cp3

Salamander, Yonahlossee (*Plethodon yonahlossee*)
ARA cp116; WOM 163

Salamander, zigzag (*Plethodon dorsalis*)
ARA cp95

Salano civet (*Salanoia concolor*)
MIA 95

Salema (*Xenistius californiensis*)
PCF cp35

Salmon, Atlantic (*Salmo salar*)
EAL 50-51; MIA 507; NAF cp58, cp494; WNW cp94

Salmon, Chinook or king (*Oncorhynchus tshawytscha*)
MPC cp286; NAF cp63, cp497; PCF p8, cp9

Salmon, coho or silver (*Oncorhynchus kisutch*)
NAW 307; PCF p8, cp9

Salmon, pink (*Oncorhynchus gorbuscha*)
PCF p8, cp9

Salmon, sockeye or red (*Oncorhynchus nerka*)
MIA 509; MPC cp283; NAF cp499; PCF p8, cp9; WNW cp95

Salmonberry (*Rubus spectabilis*)
WFW cp151, cp212

Salsify (*Tragopogon porrofolius*)
CWN 297; NWE cp236, cp482; SWW cp597

Saltator, buff-throated (*Saltator maximus*)
EOB 407

Saltbush See Shadscale

Saltwort (*Salsola kali*)
WTV 146

Salvia or scarlet sage (*Salvia coccinea*)
NWE cp429

Salvia, blue (*Salvia azurea*)
LBG cp271; NWE cp635

Sambar (*Rusa* sp.)
GEM V:165-170

Sand burr (*Cenchrus tribuloides*)
AGC cp459

Sand diver (*Synodus intermedius*)
AAL 337(bw); NAF cp429

Sand dollar (various genera)
AGC cp204; BCS 38; BMW 43; EAL 276; MPC cp376; MSC cp530-533

Sand hopper (various genera)
EAL 230; MAE 28, 102

Sand tiger See Shark, grey nurse

Sanddab (*Citharichthys* sp.)
MPC cp306; NAF cp284; PCF p43

Sanderling (*Calidris alba*)
AGC cp586; BWB 183; EAB 795; EOB 167; FEB 179;
   FWB 171; MPC cp570; NAB cp193, cp220; RBB 71; UAB
   cp193, cp229

Sandfish (*Scincus* sp.)
ERA 102; MIA 431

Sandgrouse (*Pterocles* sp.)
AAL 133(bw); EOB 212, 213; WWD 76-77

Sandgrouse, Pallas' (*Syrrhaptes paradoxus*)
MIA 253; WGB 63

Sandpiper, Baird's (*Calidris bairdii*)
BAA 10, 105, 107; EAB 797; FEB 187; FWB 165; NAB
   cp225; PBC 160; UAB cp197; WBW 273

Sandpiper, buff-breasted (*Tryngites subruficollis*)
EAB 796; FEB 186; NAB cp224; PBC 163; WBW 297;
   WOB 74, 75

Sandpiper, common (*Actitus* or *Tringa hypoleucos*)
FWB 166; RBB 81; WBW 231; WNW cp542

Sandpiper, curlew (*Calidris ferruginea*)
EAB 797; NAB cp210; WBW 275

Sandpiper, least (*Calidris minutilla*)
AGC cp588; EAB 797; FEB 183; FWB 169; MPC cp568;
   NAB cp199, cp223; PBC 159; UAB cp195; WBW 285;
   WNW cp545

Sandpiper, marsh (*Tringa stagnatilus*)
WBW 243

Sandpiper, pectoral (*Calidris melanotos*)
EAB 797; FEB 181; FWB 164; NAB cp219; PBC 160; UAB
   cp198; WBW 283; WOB 75

Sandpiper, purple (*Calidris maritima*)
AGC cp587; EAB 797; FEB 180; NAB cp198, cp217; PBC
   161; WBW 279

Sandpiper, red-backed See Dunlin

Sandpiper, rock (*Calidris ptilocnemis*)
EAB 799; FWB 163; MPC cp569; UAB cp211

Sandpiper, semipalmated (*Calidris pusilla*)
EAB 799; FEB 182; FWB 168; NAB cp221; UAB cp196,
   cp228; WNW cp543

Sandpiper, sharp-tailed (*Calidris acuminata*)
EAB 798; EOB 168-169; UAB cp199

Sandpiper, solitary (*Tringa solitaria*)
EAB 799; FEB 188; FWB 175; NAB cp216; PBC 147; UAB
   cp191; WBW 241; WNW cp544

Sandpiper, spotted (*Actitis macularia*)
AAL 126; AGC cp591; EAB 798, 799; FEB 189; FWB 174;
   MPC cp574; NAB cp197, cp215; NAB cp197, cp215; PBC
   149; UAB cp192, cp232; WNW cp546; WON 89

Sandpiper, stilt (*Micropalama himantopus*)
AGC cp589; EAB 799; FEB 166; NAB cp200, cp230; PBC
   162; WBW 295

Sandpiper, Terek (*Tringa cinereus*)
WBW 221

Sandpiper, upland (*Bartramia longicauda*)
EAB 801; FEB 165; FWB 167; LBG cp507; NAB cp218;
   UAB cp208; WBW 203

Sandpiper, Western (*Calidris mauri*)
EAB 800; FEB 185; FWB 170; MPC cp567; NAB cp222;
   PBC 158; UAB cp194

Sandpiper, white-rumped (*Calidris fuscicolis*)
EAB 801; FEB 184; NAB cp226; WBW 277

Sandpiper, wood (*Tringa glareola*)
WBW 227

Sandplover See Plover, Caspian sand or Plover, greater
   sand

Sandwort (*Arenaria* sp.)
CWN 121; NWE cp47; SWW cp38; WFW cp395

Sapsucker, Williamson's (*Sphyrapicus thyroideus*)
EAB 1037; FWB 331; UAB cp384; WFW cp243

Sapsucker, yellow-bellied (includes "red-naped" and
   "red-breasted") (*Sphyrapicus varius*)
AAL 153; EAB 1036, 1037; EOB 297; FEB 300; FWB 323,
   332 (*S.nuchalis*); MIA 293; NAB cp346; NAW 198; PBC
   218; SEF cp274; UAB cp378, cp379; WFW cp242; WGB
   103

Sardine (*Harengula* sp.)
ACF p12

Sardine or pilchard (*Sardina pilchardus*)
MIA 503

Sargassum (*Sargassum* sp.)
HSG cp61

Sargassum fish (*Histrio histrio*)
AAL 340-341; ACF p14; AGC cp351; BMW 76-77; EAL
   90; HSG cp21; MIA 531; NAF cp295; WWW 161

Sarsaparilla, wild (*Aralia nudicaulis*)
CWN 131; FWF 133; NWE cp2, cp158; SEF cp427

Saskatoon See Serviceberry, Western

Sassaby (*Damaliscus lunatus*)
MIA 149

Sassafras (*Sassafras albidum*)
FWF 171; LBG cp442, cp466; NTE cp239, cp394, cp544, cp589; SEF cp126; TSV 92, 93

Saucer bug (*Ilyocoris cimicioides*)
EOI 54; PGT 152

Sauger (*Stizostedion canadense*)
NAF cp99; WNW cp42

Sausage tree See Cucumber tree

Sawfish (*Pristis* sp.)
AAL 379(bw); AGC cp422; BMW 164-165; MIA 495; NAF cp608

Saxifrage, alpine (*Saxifraga tolmiei*)
SWW cp138

Saxifrage, early (*Saxifraga virginiensis*)
BCF 121; CWN 123; NWE cp159

Saxifrage, Merten's (*Saxifraga mertensiana*)
SWW cp104; WFW cp406

Saxifrage, purple (*Saxifraga oppositifolia*)
NWE cp455; SWW cp432; WAA 255

Saxifrage, spotted (*Saxifraga bronchialis*)
SWW cp17

Saxifrage, swamp (*Saxifraga pensylvanica*)
NWE cp14

Saxifrage, Western (*Saxifraga occidentalis*)
SWW cp139

Scabious, field See Bluebuttons

Scad (*Decapterus* or *Selar* sp.)
ACF p29; NAF cp559

Scale, cushiony cotton (*Icerya purchesi*)
MIS cp50; SGI cp76

Scale, oyster shell (*Lepidosaphes ulmi*)
MIS cp49

Scallop (various kinds)
EAL 261, 266; MSC cp351-356; SAB 15; SAS cpB23

Scarab See under Beetle

Scarlet creeper (*Ipomoea cristulata*)
SWW cp397

Scaup, greater (*Aythya marila*)
AGC cp543; DNA 102, 103; EAB 212; FEB 89; FWB 100; MIA 215; MPC cp523; NAB cp124, cp158; UAB cp93, cp133; WGB 25; WNW cp502

Scaup, lesser (*Aythya affinis*)
DNA 106, 107; EAB 213; FEB 88; FWB 101; NAB cp123, cp157; NAW 199; UAB cp92, cp132; WNW cp503

Scholartree See Pagoda tree, Japanese

Schoolmaster (*Lutjanus apodus*)
KCR cp23; NAF cp538

Scorpion (*Buthus* and other genera)
DIE cp34; MIS cp635, cp636

Scorpion, water See Water scorpion

Scorpionfish (*Rhinopius aphanes*)
RCK 129; RUP 150

Scorpionfish (*Scorpaena* sp.)
ACF p51; NAF cp422; RCK 138; RUP 151

Scorpionfish, California (*Scorpaena guttata*)
AAL 349; MPC cp253; PCF cp23; NAF cp417

Scorpionfish, spotted (*Scorpaena plumieri*)
NAF cp421; WWW 164-165

Scorpionweed (*Phacelia dubia*)
WTV 150

Scoter, black (*Melanitta nigra*)
AGC cp550; DNA 129, 130; EAB 215; FEB 78; FWB 110; MPC cp520; NAB cp131; RBB 40; UAB cp110, cp138

Scoter, surf (*Melanitta perspicillata*)
AGC cp549; DNA 131-133; EAB 215; FEB 79; FWB 109; MPC cp521; NAB cp129, cp148; UAB cp113, cp141

Scoter, white-winged (*Melanitta fusca*)
AGC cp551; DNA 136, 137; EAB 215; FEB 112; FWB 108; MPC cp519; NAB cp130, cp147; UAB cp112, cp140

Screamer, black-necked or northern (*Chauna chavaria*)
MIA 213; WGB 23

Screamer, crested (*Chauna torquata*)
AAL 108; EOB 93

Scrubfowl (*Megapodius freycinet*)
MIA 227; WGB 37

Scud (*Gammarus oceanicus*)
MSC cp591, cp598

Scullcap (*Scutellaria* sp.)
CWN 231; NWE cp638; WNW cp292; WTV 159

Sculpin (*Cottus* sp.)
NAF cp244-246; WNW cp59

Sculpin, buffalo (*Enophrys bison*)
MPC cp251; NAF cp419

Sculpin, grunt (*Rhamphocottus richardsoni*)
MPC cp236; NAF cp418; PCF cp15

Sculpin, lavendar (*Leiocottus hirundo*)
NAF cp416; PCF cp15

Sculpin, shorthorn (*Myxocephalus scorpius*)
ACF p54; MIA 547; WWW 62-63

Sculpin, snubnose (*Orthonopias triacis*)
MPC cp255; NAF cp414

Sculpin (other kinds)
ACF p54; PCF p16-p18, cp15

Scythebill (*Campylorhamphus* sp.)
AAL 154; EOB 313

Sea anemone (various genera)
AGC cp245; BMW 24-25; KCR p14, p15; MPC cp437-
   441, cp444, cp445; MSC cp12, cp166-172, cp178-197;
   SAS cpA15-cpA17; SCS cp27

Sea anemone (*Tealia* sp.)
BCS 96, 97; MPC cp436; MSC cp183

Sea anemone, beadlet (*Actinia equina*)
EAB 117; EAL 173

Sea anemone, Christmas or dahlia (*Tealia crassicornia*)
AAL 524; BCS 107

Sea anemone, collared sand (*Actinostella* or *Phyllactis
   flosculifera*)
KCR p14; SCS cp27

Sea anemone, dahlia or "Flower of the sea" (*Tealia
   felina* or *coriacea*)
BCS 45; BMW 28-29; EAL 178-179

Sea anemone, giant caribbean or pink-tipped
   (*Condylactis gigantea*)
AGC cp268; KCR p15; MSC cp187, cp188; SCS cp28;
   WWF cp40, cp41

Sea anemone, hermit crab (*Calliactis tricolor*)
SAS cpA18

Sea anemone, jewel (*Corynactis australis*)
NOA 76

Sea anemone, pale (*Aiptasia pallida*)
MSC cp167; SCS cp27

Sea anemone, plumose or feather (*Metridium* sp.)
BCS 43; EAL 173; HSG cp62

Sea anemone, rock (*Anthopleura krebsi*)
KCR p14; SCS cp27

Sea anemone, stinging (*Actinia bermudensis*)
KCR p15; SAS cpA14

Sea anemone, sun (*Stoichactis helianthus*)
KCR p15; MAE 109(close-up)

Sea anemone, warty (*Bunodosoma* sp.)
KCR p14; MSC cp193; SCS cp28

Sea bass (*Centropristis* sp.)
ACF cp24; AGC cp381; NAF cp515

Sea bass, giant (*Stereolepis gigas*)
MIA 549; MPC cp237; NAF cp507; PCF cp29

Sea bass, white (*Atractoscion nobilis*)
MIA 559; PCF p33

Sea butterfly (*Clione limacina*)
BCS 48

Sea cat (*Dolabrifera* sp.)
AGC cp215; MSC cp211

Sea cow See Dugong; Manatee

Sea cucumber (several genera)
MSC cp150-152, cp154-157, cp235-238; NOA 101; SCS
   cp61

Sea dragon (*Phycodurus* or *Phyllopteryx* sp.)
NOA 74, 75; WWW title pg.

Sea egg (*Tripneustes ventricosus*)
KCR p34; MSC cp525; SCS cp62

Sea fan (*Gorgonia* sp.)
AGC cp271; EAL 181; MSC cp64

Sea fig (*Mesembryanthemum chilense*)
MPC cp615; SWW cp482

Sea gooseberry (*Pleurobrachia pileus*)
AGC cp225; BMW 98; MPC cp387; MSC cp496; NNW 100

Sea hare (*Aplysia* or *Dactylomela* sp.)
AGC cp207; EAL 251; MSC cp209, cp210; RCK 174; SAS
   cpB11

Sea hollyhock See Rose, swamp

Sea horse (*Hippocampus* sp.)
AAL 348; ACF p23; AGC cp432; BMW 144; EAL 108; MIA
   541; NAF cp441; NOA 76; PCF p13; RCK 37; RUP 54

Sea horse, Australian See Sea dragon

Sea lemon (Doridae family)
EAL 264; MPC cp422; MSC cp228

Sea lettuce (*Ulva lactuca*)
AGC cp474; MPC cp495

Sea lily (*Nemaster* sp.)
KCR cp18, cp32

Sea lion, Australian (*Neophoca cinerea*)
CRM 159; GEM IV:168, 178, 182-183; KSW 32; MEM
   257; MIA 107

Sea lion, California (*Zalophus californianus*)
AAL 64; ASM cp301, 302; CRM 159; GEM IV:164, 166-
   167, 171; HSG cp59; KSW 25; MEM 255; MIA 107; MPC
   cp227; NAW 76; RCK 136; RUP 85; WWW 110-111

Sea lion, Hooker's or New Zealand (*Phocarctos hookeri*)
CRM 159; KSW 35; MEM 253, 255, 257

Sea lion, northern or Steller (*Eumetopias jubatus*)
ASM cp298, 299; BAA 19(bw), 37, 75; CRM 159; GEM
   IV:169, 172-173; KSW 22; MEM 255; MIA 107; MNP
   249(bw); MPC cp228

Sea lion, Southern (*Otaria byronia*)
KSW 28, 29

Sea lion, South American (*Otaria flavescens*)
AAL 24; CRM 159; GEM IV:177; MEM 255

Sea mouse (*Aphrodite aculeata*)
EAL 212

Sea nettle (*Chrysaora quinquecirrha*)
AAL 518(bw) (*Dactylometra* sp.); AGC cp220; KCR p16;
   MSC cp506, cp510; SCS 208(bw)

Sea oats (*Uniola paniculata*)
AGC cp462; NWE cp405; WTV 97

Sea otter See under Otter

Sea palm (*Postelsia palmaeformis*)
MPC cp496

Sea pansy (*Renilla reniformis*)
SAS cpA13

Sea peach (*Holocynthia pyriformis*)
AGC cp232

Sea pen (*Ptilosarcus* or *Leioptilus* sp.)
AAL 522(bw); BCS 105; MPC cp448; MSC cp44

Sea plume (*Pseudopterogorgia* sp.)
AGC cp264; MSC cp60, cp66

Sea pork (*Aplidium* sp.)
MSC cp98

Sea raven (*Hemitripterus americanus*)
ACF p54; BCS 30, 31

Sea roach (*Ligia* sp.)
AGC cp180; MSC cp580-582

Sea rocket (*Cakile edentula*)
AGC cp449; CWN 111; MPC cp612; NWE cp471

Sea rod (*Plexaura* or *Eunicea* sp.)
MSC cp56, cp57

Sea slater (*Ligia oceanica*)
EAL 231, 232; MPC cp352, cp353

Sea slug See Nudibranch

Sea squirt or tunicate (various genera)
AGC cp230-cp237; BMW 115; EAL 285; HSG cp43, cp45;
   MSC cp26, cp41, cp91-cp111, cp119-cp122; SAS cpA20-
   cpA27; SCS cp23; WWW 210, 211

Sea staghorn (*Codium fragile*)
MPC cp497

Sea star See Starfish

Sea urchin (various genera)
AAL 389; AGC cp196, cp198; EAL 282; KCR cp32, p33,
   p34; MSC cp517-528; SAS cpA31; SCS cp62; WWW 151

Sea urchin (*Strongylocentrotus* sp.)
AGC cp197; BCS 38, 102, 103; BMW 74; MPC cp433-
   cp435

Sea urchin, heart (*Moira* or *Lovenia* sp.)
MPC cp377; SAS cpA32

Sea whip (several kinds)
AGC cp267; MSC cp59, cp62, cp63; SAS cpA12

Seal, Baikal (*Phoca sibirica*)
CRM 161; GEM IV:221; KSW 92

Seal, bearded (*Erignathus barbatus*)
ASM cp312; GEM IV:229; KSW 102; MEM 273; MIA 109;
   MPC cp233

Seal, Caspian (*Phoca caspica*)
KSW 90, 91 (bw); NHU 52

Seal, common See Seal, harbor

Seal, crabeater (*Lobodon carcinophagus*)
KSW 115; MEM 272; MIA 109; WPR 166-167

Seal, elephant (Northern) (*Mirounga angustirostris*)
AAB 66-67; ASM cp304, 305; BMW 50-51; GEM IV:231;
   KSW 125(bw), 127; MEM 279; MIA 111; MPC cp224,
   cp232; NAW 77, 78; SAB 91

Seal, elephant (Southern) (*Mirounga leonina*)
AAL 67(bw); GEM IV:231-233, 236-237; KSW 122, 123;
   MEM 272; SAB 78; WWW 81, 127

Seal, fur (Antarctic) (*Arctocephalus gazella*)
GEM IV:181(head only), 189; KSW 49

Seal, fur (Australian) (*Arctocephalus pusillus doriferus*)
KSW 56; NOA 81

Seal, fur (Cape or South African) (*Arctocephalus pusillus*)
GEM IV:180, 183-184; KSW 53; MEM 256

Seal, fur (Galapagos) (*Arctocephalus galapagoensis*)
KSW 40

Seal, fur (Guadaloupe) (*Arctocephalus townsendi*)
ASM cp303; CRM 161; KSW 38; MPC cp226

Seal, fur (New Zealand) (*Arctocephalus forsteri*)
KSW 58; MEM 256

Seal, fur (Northern) (*Callorhinus ursinus*)
ASM cp300; BAA 29(bw); KSW 62; GEM IV:207-211;
   MEM 255; MIA 107; MNP 247(bw); MPC cp225; RFA 313-
   326(bw); WWW 197

Seal, fur (South American) (*Arctocephalus australis*)
GEM IV:191; KSW 43; MEM 255; MIA 107

Seal, fur (Subantarctic) (*Arctocephalus tropicalis*)
GEM IV:186-187; KSW 45

Seal, gray (*Halichoerus grypus*)
AGC cp336; ASM cp313; GEM IV:227; KSW 78; MEM
249, 272; MIA 109; SAB 23

Seal, harbor or common (*Phoca vitulina*)
AGC cp335; ASM cp308; BAA 44, 76; KSW 81; MIA 109;
MNP cp8b; MPC cp231; NAW 77; SWM 137; WMC 201;
WOI 60-61(head only)

Seal, harp (*Phoca groenlandica*)
AAL 66(bw); AGC cp333; ASM cp311; BAA 27(bw),
31(bw)(pup), 58(bw), 69(pup), 77; BMW 225(pup); GEM
IV:222-223; KSW 95; MEM 273, 287(pup); MIA 109; WPR
158

Seal, hooded (*Crytophora cristata*)
AGC cp332, cp334; ASM cp306, 307; BAA 33(head only),
76, 141(bw); CRM 161; KSW 99, 100; MEM 271, 272; MIA
111; WPR 158, 162-163

Seal, larga (*Phoca largha*)
KSW 86

Seal, leopard (*Hydrurga leptonyx*)
KSW 116; MEM 272, 291; MIA 109; WPR 168-169

Seal, monk (Hawaiian) (*Monachus schauinslandi*)
GEM IV:230; KSW 107; MEM 272, 289; WOI 79

Seal, monk (Mediterranean) (*Monachus monachus*)
CRM 161; KSW 105; MIA 111

Seal, ribbon (*Phoca fasciata*)
ASM cp310; KSW 97; MEM 273; MPC cp229; WPR 164

Seal, ringed (*Phoca hispida*)
AAL 66(bw); ASM cp309; BAA 21(bw), 154(bw); GEM
IV:218-219; KSW 89; MEM 272, 283; MPC cp230

Seal, Ross (*Ommatophoca rossi*)
BMW 226-227; KSW 112; MEM 273

Seal, Weddell (*Leptonychotes weddelli*)
BMW 224; KSW 110; GEM IV:159(head only); MEM 270,
273; MIA 111; WPR 154, 158, 166; WWW 12-13

Seaperch (*Halocynthia pyriformis*)
BCS 54

Seaperch, rainbow (*Hypsurus caryi*)
MPC cp261; NAF cp520

Seaperch, rubberlip (*Rhacochilus toxotes*)
MPC cp271; NAF cp529

Seaperch, sharpnose (*Phanerodon atripes*)
NAF cp527

Seaperch, striped (*Embiotoca lateralis*)
MPC cp269; NAF cp531

Searobin (*Prionotus* sp.)
ACF p53; AGC cp378; MIA 545; NAF cp423-cp426; PCF
cp34

Seasnake, banded (*Laticauda semifasciata*)
AAL 255(bw); MIA 455

Seasnake, Dubois's (*Aepysurus duboisi*)
LSW 287; RUP 53

Seasnake, Stoke's (*Astrotia stokesii*)
LSW 290

Seasnake, yellow-bellied (*Pelamis platurus*)
AAB 65; ERA 124; LSW 296; WRA p48

Seasnake, yellow-lipped (*Laticauda colubrina*)
LSW 282, 285; RCK 161; WWW 153

Seaweed, sponge See Sea staghorn

Secretary bird (*Sagittarius serpentarius*)
AAL 110; BOP 16-20; EOB 105, 107; MIA 217; WGB 27;
WWD 78

Sedge, pond (great) (*Carex riparia*)
PGT 46

Sedge, white-topped (*Dichtromena colorata*)
NWE cp103; WNW cp226

Sedum, Sierra (*Sedum obtusatum*)
SWW cp335; WFW cp437

Seedbox or primrose-willow (*Ludwigia alternifolia*)
CWN 147; NWE cp257, cp413

Seedsnipe (*Thinocorus rumicivorus*)
EOB 182; MIA 247; WGB 57

Self-head See Heal-all

Senna (*Cassia* sp.)
NWE cp330; TSP 19

Senorita (*Oxyjulis californica*)
MPC cp265; NAF cp371; PCF cp30

Sensitive briar (*Schrankia nuttalii*)
NWE cp513

Sensitive plant (*Mimosa pudica*)
TSP 125

Sequoia, giant (*Sequoiadendron giganteum*)
NTW cp61, cp436; WFW cp54

Sergeant major (*Abudefduf saxatilis*)
ACF cp38; AGC cp370; KCR cp22; MIA 563; NAF cp342

Sergeant, night (*Abudefduf taurus*)
ACF cp38

Seriema, crested or redleg (*Cariama cristata*)
EOB 159; MIA 239; WGB 49; WPP 182

Serow (*Capricornis* sp.)
CRM 199; GEM V:504, 505; MEM 585; MIA 155; WOM
142

Serval (*Felis serval*)
AWC 116-123; CRM 155; GEM III:620, 622; MEM 50; MIA
103

Serviceberry, downy (*Amelanchier arborea*)
FWF 67; NTE cp175, cp418, cp595; OET 187; SEF cp79,
cp158; TSV 152, 153

Serviceberry, Western (*Amelanchier alnifolia*)
NTE cp200; NTW cp183, cp364; SWW cp171; WFW cp91,
cp184

Serviceberry (other *Amelanchier* sp.)
NTE cp199; PMT cp[16]

Shad (*Dorosoma* sp.)
ACF p12; NAF cp122; PCF p7

Shad, Alabama (*Alosa alabamae*)
AWR 219(bw)

Shad, American (*Alosa sapidissima*)
ACF p12; AGC cp398; MPC cp276; NAF cp123, cp577;
PCF p7; WNW cp32

Shadbush See Serviceberry

Shadscale (*Atriplex confertifolia*)
JAD cp346

Shag (*Phalacrocorax aristotelis*)
OBL 78; RBB 16

Shapu See Urial

Shark, angel (*Squatina* sp.)
MIA 493; SJS 21; SSW 152(bw)

Shark, angel (Pacific) (*Squatina californica*)
MPC cp318; NAF cp271; PCF p3; SJS 40

Shark, basking (*Cetorhinus maximus*)
HSG cp8; MIA 489; NAF cp602; SJS 26, 60

Shark, blue (*Prionace glauca*)
AGC cp415; BMW 162; MIA 491; MPC cp322; NAF cp593;
SJS 32, 46, 153; SSW 110

Shark, bonnethead (*Sphyrna tiburo*)
AGC cp416; NAF cp592

Shark, bull or Zambezi (*Carcharhinus leucas*)
AGC cp419; MIA 491; NAF cp596; SJS 32

Shark, bullhead or horn (*Heterodontus* sp.)
SJS 22; SSW 78(bw), 79

Shark, carpet See Carpetshark

Shark, cat See Catshark

Shark, frilled (*Chlamydoselachus anguineus*)
EAL 137; SJS 19

Shark, goblin (*Mitsukurina* or *Scapanorhynchus owstoni*)
EAL 137; SJS 27

Shark, great white (*Carcharodon carcharias*)
AGC cp420; EAL 135; HSG cp10; MIA 491; MPC cp320;
NAF cp597; RCK 185-187; RUP 180-183; SJS 28, 57,
108; SSW 87; WWF cp43, cp44

Shark, grey nurse or sand tiger (*Eugomphodus taurus*)
AGC cp421; SJS 26, 42, 118

Shark, hammerhead (*Sphyrna* sp.)
AGC cp423; HSG cp12; MIA 491; NAF cp607; SJS 31

Shark, horn (*Heterodontus francisci*)
NAF cp606

Shark, lemon (*Negaprion* sp.)
SJS 62; SSW 114

Shark, leopard (*Triakis felis* or *semifasciata*)
MPC cp319; NAF cp599; SJS 31; SSW 122

Shark, mako See Mako

Shark, megamouth (*Megachasma* sp.)
SJS 27; SSW 138(bw)

Shark, nurse (*Ginglymostoma cirratum*)
NAF cp604; RCK 189; RUP 181; SJS 38, 130; SSW 118

Shark, Port Jackson (*Heterodontus portusjacksoni*)
EAL 131, 136; MIA 493; RUP 187; SJS 22, 45

Shark, reef (*Carcharhinus* sp.)
ACF p63; RUP 184-185

Shark, reef (blacktip) (*Carcharhinus melanopterus* or
limbatus)
AGC cp417; NAF cp594; SJS 58, 70, 145; SSW 106

Shark, reef (grey) (*Carcharhinus amblyrhynchos*)
EAL 2; RCK 191; SJS 33, 101; SSW 107

Shark, reef (oceanic whitetip) (*Carcharhinus longimanus*)
SJS 93

Shark, reef (whitetip) (*Triaenodon obesus*)
HSG cp11; RCK 167; SJS 33; WWF cp31

Shark, requiem (*Carcharinus* sp.)
EAL 129

Shark, sandbar (*Carchiarius plumbeus*)
AGC cp418; NAF cp595, cp598

Shark, sand tiger (*Odontaspis* sp.)
EAL 133; MIA 489; NAF cp601

Shark, saw (*Pristiophorus* sp.)
EAL 136; MIA 493; SJS 21; SSW 155(bw)

Shark, sevengill (*Notorynchus maculatus*)
NAF cp591; SJS 19, 47; SSW 46

Shark, sharp-nosed (*Rhizoprionodon terraenovae*)
SSW 112(bw)

Shark, six-gilled (*Hexanchus* sp.)
ACF p62; EAL 137; MIA 493

Shark, snaggletooth (*Hemipristis elongatus*)
SJS 30

Shark, soupfin (*Galeorhinus zyopterus*)
NAF cp590

Shark, swell (*Cephaloscyllium ventriosum*)
NAF cp605; SJS 29; SSW 127

Shark, tape (*Galeorhinus galeus*)
SJS 31, 70

Shark, tawny or giant sleepy (*Nebrius ferrugineus*)
SJS 25, 137

Shark, thresher (*Alopias vulpinus*)
HSG cp9; MIA 489; SJS 28

Shark, tiger (*Galeocerdo cuvier*)
NAF cp603; SJS 33

Shark, whale (*Rhiniodon typus*)
HSG cp7; MIA 489; NAF cp600; SJS endpaper, 25, 133

Shark, white See Shark, great white

Shark, zebra (*Stegostoma fasciatum*) See also Catshark
RCK 114; SJS 24; SSW 119

Sharksucker (*Echeneis naucrates*) See also Remora
ACF p42; AGC cp413; MIA 553; NAF cp587; PCF p12; RCK 167; SJS 54

Sharpbill (*Oxyruncus cristatus*)
EOB 329

Shearwater, Audubon's (*Puffinus lherminieri*)
EAB 840; HSW 89

Shearwater, black-vented (*Puffinus opisthomelas*)
FWB 52

Shearwater, Cory's (*Puffinus diomedea*)
AGC cp496; EAB 841; EOB 47; NAB cp79; OBL 58, 107

Shearwater, flesh-footed (*Puffinus carneipes*)
EAB 840; UAB cp68

Shearwater, greater (*Puffinus gravis*)
AGC cp497; BMW 204; EAB 841; EOB 47; NAB cp80; OBL 62; WOB 26

Shearwater, little (*Puffinus assimilis*)
OBL 211(chick)

Shearwater, Manx (*Puffinus puffinus*)
EAB 841; HSW 92; MIA 201; RBB 12; UAB cp67; WGB 11

Shearwater, New Zealand or Buller's (*Puffinus bulleri*)
EAB 841; MPC cp545; UAB cp70

Shearwater, pink-footed (*Puffinus creatopus*)
EAB 841; FWB 54; MPC cp546; UAB cp65

Shearwater, sharp-tailed (*Puffinus tenuirostris*)
EAB 841; UAB cp69

Shearwater, sooty (*Puffinus griseus*)
EAB 840; FWB 53; HSW 95; MPC cp545; NAB cp81; UAB cp66

Shearwater, wedge-tailed (*Puffinus pacificus*)
EAB 841; HSW 90-91

Sheathbill, snowy or yellow-billed (*Chionis alba*)
EOB 183; MIA 247; WGB 57

Sheep, barbary (*Ammotragus lervia*)
ASM cp279; CRM 203; GEM V:539, 540; MEM 585; MIA 157; WOM 102, 103

Sheep, bighorn mountain (*Ovis canadensis*)
ASM cp283; CRM 203; GEM V:554, 555; JAD cp526; MEM 587; MIA 157; MNC cp[33]; MNP 298(bw); MPS 308(bw); NAW 80-82; RMM cp[15]; WFW cp375; WOM 4-5, 164, 166-167; WWD 113(head only)

Sheep, Dall's (*Ovis dalli*)
ASM cp281, 282; GEM V:292-293(head only), 556-559; MAE 87; MNP 297(bw); NAW 82; SWM 173; WPR 103, 104

Sheep, wild See Urial

Sheephead, California (*Semicossyphus pulcher*)
AGC cp353; MIA 557; MPC cp257, cp264; NAF cp377, cp380; PCF cp30

Sheeps-bit (*Jasione montana*)
CWN 207

Sheepshead (*Archosargus probatocephalus*)
ACF p33; NAF cp395

Shelduck (*Tadorna tadorna*)
DNA 21, 22; EOB 96; MIA 213; RBB 26; WGB 23

Shelduck, red-backed radjah (*Tadorna radjah rufitergum*)
AWR 130-131

Shelduck, ruddy (*Tadorna ferruginea*)
DNA 17; MAE 92

Shelf or bracket fungus (various genera) See also Beefsteak fungus
FGM cp14; NAM cp481-549; SGM cp289-324

Shepherd's needle (*Bidens pilosa*)
NWE cp95

Shepherd's purse (*Capsella bursa-pastoris*)
NWE cp173; SWW cp157

Shield bug (various kinds)
DIE cp16, cp21; EOI 54, 60-61; WIW 87

Shiner (*Notropis* sp.)
MIA 517; NAF cp146-cp191; WNW cp67

Shiner, golden (*Notemigonus crysoleucas*)
JAD cp299; NAF cp131, cp141

Shiner, redside (*Richardsonius balteatus*)
NAF cp181

Shinleaf (*Pyrola* sp.)
CWN 149; NWE cp132; SEF cp451

Shooting star (*Dodecatheon meadia*)
NWE cp508; SEF cp476; WAA 80, 230(*D.pauciflorum*)

Shooting star, alpine (*Dodecatheon alpinum*)
SWW cp503

Shooting star, few-flowered (*Dodecatheon pulchellum*)
SWW cp504; WFW cp485

Shoveler, northern (*Anas clypeata*)
AGC cp545; BWB 74-75; DNA 72-74; EAB 215; FEB 92;
    FWB 99; MIA 215; NAB cp108, cp143; PBC 89; RBB 32;
    UAB cp96, cp151; WGB 25; WNW cp493

Shrew, Arctic (*Sorex arcticus*)
ASM cp45; WNW cp594

Shrew, desert (*Notiosorex crawfordi*)
ASM cp59; JAD cp470; MIA 35; MPS 62(bw)

Shrew, dusky (*Sorex monticolus*)
CMW 27 (bw)

Shrew, dwarf (*Sorex nanus*)
CMW 29 (bw); MPS 42(bw)

Shrew, elephant (*Elephantulus* and other genera)
CRM 43; GEM I:520-521, 525-529; MAE 18; MEM 730-
    735; MIA 37

Shrew, European pygmy (*Sorex minatus*)
MEM 759

Shrew, least (*Cryptotis parva*)
ASM cp56; CGW 129; GMP 64(bw); LBG cp76; MNC
    95(bw); MPS 61(bw); WMC 57

Shrew, long-tailed or rock (*Sorex dispar*)
AAL 32; ASM cp58; CGW 129; GMP 55(bw)

Shrew, masked (*Sorex cinereus*)
ASM cp46; CGW 129; CMW 23(bw); GMP 43(bw); MIA 35;
    MNC 81(bw); MPS 40(bw); NAW 83(bw); SWM 7; WMC
    45; WNW cp589

Shrew, otter See Water shrew, giant African

Shrew, Pacific (*Sorex pacificus*)
ASM cp47; WNW cp590

Shrew, pygmy (*Microsorex hoyi*)
ASM cp43; CGW 129; CMW 34(bw); CRM 41; MNC
    85(bw); MPS 41(bw); SEF cp577

Shrew, pygmy white-toothed (*Suneus etruscus*)
MEM 758

Shrew, short-tailed (*Blarina* sp.)
CGW 129; NAW 83(bw)

Shrew, short-tailed (Northern) (*Blarina brevicauda*)
ASM cp48, cp60; GMP 60(bw); MEM 763; MNC 91(bw);
    MPS 60(bw); NAW 83(bw); SEF cp578; WMC 53

Shrew, short-tailed (Southern) (*Blarina carolinensis*)
ASM cp57; MNC 93(bw); MPS 60(bw); WMC 55

Shrew, smoky (*Sorex fumeus*)
ASM cp44; CGW 129; GMP 52(bw); MNC 83(bw); WMC
    50; WNW cp593

Shrew, Southeastern or Bachman's (*Sorex longirostris*)
ASM cp51; CRM 41; MNC 87(bw): WMC 47

Shrew, tree (*Tupaia* and other genera)
GEM II:1-11; MEM 441

Shrew, vagrant (*Sorex vagrans*)
ASM cp55; WFW cp325

Shrew, water See Water shrew

Shrike, crimson-breasted (*Laniarius atrococcineus*)
EOB 353

Shrike, fiscal (*Lanius collaris*)
EOB 346

Shrike, loggerhead (*Lanius ludovicianus*)
EAB 843; FEB 378; FWB 300; JAD cp596; LBG cp528;
    NAB cp422; PBC 287; SPN 99; UAB cp472; WON 160

Shrike, Northern or great gray (*Lanius excubitor*)
EAB 842; EOB 352; FEB 379; FWB 301; LBG cp527; MAE
    27; MIA 325; NAB cp421; SEF cp331; UAB cp473; WGB
    135

Shrike, red-backed (*Lanius collurio*)
EOB 344; RBB 166; SAB 22

Shrike, rufous-backed (*Lanius schach*)
EOB 345

Shrike, vanga (*Leptopterus* sp.)
EOB 350

Shrike-tyrant, great (*Agriornis livida*)
MIA 309; WGB 119

Shrimp, banded or barber pole (*Stenopus hispidus*)
AAL 7; EAL 234; KCR cp29; MSC cp618; RCK 45; RUP
    37; SCS cp51; SJS 55

Shrimp, cleaner (*Periclimenes yucatanicus*)
AGC cp165; KCR cp29; MSC cp617; SCS cp51; WWF
  cp41

Shrimp, coonstripe (*Pandalus danae*)
BCS 106; MPC cp348; MSC cp610

Shrimp, freshwater (*Gammarus pulex*)
PGT 133

Shrimp, ghost (*Callianassa* sp.)
MPC cp344, cp345

Shrimp, grass (*Palaemonetes vulgaris*)
SAS cpC16

Shrimp, harlequin or clown (*Hymenocerus* sp.)
RUP 38; WWW 136

Shrimp, mantis (various genera)
MSC cp595, cp596; SAS cpC13

Shrimp, peppermint or veined (*Lysmata wurdemanni*)
AGC cp169; KCR cp29; MSC cp613; SAS cpC18; SCS
  cp51

Shrimp, pink (*Penaeus duorarum*)
AGC cp166; SAS cpC15

Shrimp, red-backed cleaning or Scarlet Lady (*Lysmata grabhami*)
KCR cp29; SCS cp51

Shrimp, skeleton (various genera)
MPC cp355; MSC cp599-601

Shrimpfish (*Aeoliscus strigatus*)
AAL 347(bw); EAL 109

Siamang (*Hylobates syndactylus*)
CRM 85; GEM II:325, 344, 345; MEM 418, 419; MIA 75;
  NHP 162(bw)

Siamese fighting fish (*Betta splendens*)
AAB 143

Sickener (*Russula emetica*)
FGM cp40; NAM cp328; NMT cp5; PPC 141; SGM cp140

Sicklebill (*Vanga curvirostris*)
MIA 327; WGB 137

Sidewinder See under Rattlesnake

Sidewinder viper See under Viper

Sifaka See under Lemur

Silk cotton tree (*Ceiba pentandra*)
OET 268

Silktassel, wavyleaf (*Garrya elliptica*)
NTW cp120; WFW cp71

Silktree or mimosa (*Albizia julibrissia*)
NTE cp310, cp463; NTW cp266, cp386; TSV 106, 107

Silkworm, common or Chinese (*Bombyx mori*)
SGB cp9, cp51

Silverbell, Carolina (*Halesia carolina*)
NTE cp192, cp427; SEF cp92, cp156

Silverbell (other *Halesia* sp.)
NTE cp195, cp215, cp426, cp489; PMT cp[59]

Silverfish, house (*Lepisma saccharina*)
AAL 406(bw); EOI 23; MIS cp87; SGI cp3; WIW 19

Silverleaf, Texas (*Leucophyllum frutescens*)
JAD cp47; SWW cp576

Silverrod (*Solidago bicolor*)
NWE cp155

Silverside (*Menidia* sp.)
ACF p20; NAF cp568

Silversword (*Argyroxiphium sandwicensi* or *caliginis*)
PFH 32, 126, 127, 133; WOI 85

Silverweed (*Potentilla anserina*)
NWE cp249; SWW cp197

Siphonophore (*Physophora* and other sp.)
AAB 132; BMW 95; MSC cp487; WWW 218

Siren, dwarf (*Pseudobranchus striatus*)
AAL 276; ARA cp17; ARC 44; MIA 467; WNW cp108

Siren, greater (*Siren lacertina*)
ARA cp14; ARC 46; ERA 27; MIA 467; WNW cp106

Siren, lesser (*Siren intermedia*)
ARA cp13; ARC 45; WNW cp107

Siskin, pine (*Carduelis pinus*)
EAB 313; FEB 402; FWB 449; NAB cp557; PBC 371; RBB
  181; SEF cp339; SPN 143; UAB cp598; WFW cp312

Sitatnga, spiral-horned (*Tragelaphus spekei*)
GEM V:346; SWF 121

Skate (*Raja* sp.)
AAL 380; ACF p6; AGC cp426; MIA 495; MPC cp315;
  NAF cp263-cp266; PCF p4

Skilletfish (*Gobiesox strumosus*)
ACF p14; NAF cp300

Skimmer, black (*Rynchops niger*)
AGC cp516; BWB 164, 165; EAB 844, 845; EOB 202; FEB
  22; MIA 249; NAB cp70; OBL 98; PBC 185; WGB 59;
  WOB 144-145

Skink, blue-tongued (*Tiliqua* sp.)
ERA 87, 101; MIA 433

Skink, broadhead (*Eumeces laticeps*)
ARA cp431, cp424; ARC 185, 186; LOW 159; RNA 77;
SEF cp518

Skink, coal (*Eumeces anthracinus*)
ARA cp425, cp429; CGW 97; RNA 77; SEF cp515

Skink, five-lined (*Eumeces fasciatus*)
ARA cp427, cp437, cp443; ARC 181, 182; ARN 56(bw);
CGW 97; LOW 70; NAW 263; RNA 77; SEF cp514

Skink, four-lined (*Eumeces tetragrammus*)
RNA 77

Skink, Gilbert (*Eumeces gilberti*)
ARA cp430, cp434; WRA cp28; RNA 79

Skink, Great Plains (*Eumeces obsoletus*)
ARA cp432; JAD cp213; LBG cp582; MIA 431; WRA cp28;
RNA 81

Skink, ground (*Scincella lateralis*)
ARA cp433; ARC 187; MIA 433; RNA 81; SEF cp517

Skink, many-lined (*Eumeces multivirgatus*)
ARA cp422; LBG cp577; WRA cp28; RNA 79

Skink, mole (*Eumeces egregius*)
AGC cp441; ARA cp435, cp436, cp438; RNA 81; SEF
cp516

Skink, mountain (*Eumeces callicephalus*)
ARA cp439; RNA 77; WRA cp28

Skink, prairie (*Eumeces septentrionalis*)
ARA cp423, cp428; LBG cp579; RNA 77

Skink, sand (*Neoseps reynoldsi*) See also Sandfish
ARA cp450; MIA 431; RNA 81

Skink, Southeastern five-lined (*Eumeces inexpectatus*)
ARA cp426, cp444; ARC 183; RNA 75

Skink, stump-tailed (*Tiliqua rugosa*)
AAL 200, 229

Skink, Western (*Eumeces skiltonianus*)
AAL 228; ARA cp421, cp441, cp442; ERA 101; LBG
cp578; NAW 264; WRA cp28; RNA 79; WFW cp587

Skipper butterfly (various genera)
EOI 107; JAD cp432-439; LBG cp299, cp300, cp311-314;
MIS cp9(larva), cp574; PAB cp133-315; SEF cp363-365;
SGB cp166; SGI cp200-cp202; WNW cp395, cp411,
cp413, cp414, cp420-422

Skua, arctic See Jaeger, parasitic

Skua, great (*Catharacta skua*)
AGC cp495; EAB 847; EOB 198, 201; FEB 48; MIA 249;
MPC cp549; NAB cp84; OBL 130; WGB 59; WPR 150

Skua, South Pole (*Catharacta maccormicki*)
FWB 45

Skunk cabbage (*Symplocarpus foetidus*)
BCF 145; CWN 45; NWE cp391; WAA 194; WNW cp312;
WTV 163

Skunk cabbage, yellow (*Lysichitum americanum*)
SWW cp276; WAA 195; WFW cp432; WNW cp223

Skunk, Andes (*Conepatus chinga*)
GEM III:431

Skunk, hog-nosed (*Concepatus mesoleucus*)
ASM cp199; CRM 141; JAD cp516; MIA 91

Skunk, hooded (*Mephitis macroura*)
ASM cp200

Skunk, spotted (*Spilogale* sp.)
CMW 234(bw); GEM III:430; RFA 239, 247(bw); SWM 116

Skunk, spotted (Eastern) (*Spilogale putorius*)
AAL 63(bw); ASM cp197; GMP 289(bw); LBG cp50; MNC
267(bw); NAW 84; SEF cp606; WMC 195

Skunk, spotted (pygmy) (*Spilogale pygmaea*)
MEM 122

Skunk, spotted (Western) (*Spilogale gracilis*)
MIA 91; MPS 299(bw)

Skunk, striped (*Mephitis mephitis*)
ASM cp198; CGW 153; CMW 238(bw); GEM III:429; GMP
293(bw); JAD cp515; LBG cp49; MIA 91; MNC 269(bw);
MNP 266(bw); MPS 278(bw); NAW 85, 86; RFA 238, 241,
245(bw); SEF cp605; SWM 115; WFW cp356; WMC 197

Skylark (*Alauda arvensis*)
BAS 13; EAB 573; EOB 337; FWB 271; MIA 313; RBB
116; UAB cp574; WGB 123

Skyrocket (*Ipomopsis aggregata*)
JAD cp161; SWW cp413; WFW cp477

Sleeper, fat (*Dormitator maculatus*)
NAF cp88, cp431

Slider See Turtle, red-eared

Slime mold (various genera)
NAM cp550-562

Sloth, brown-throated (*Bradypus variegatus*)
MEM 776-779

Sloth, three-toed (*Bradypus* sp.)
AAL 41; CRM 89; GEM II:598-611; MIA 27; SWF 185

Sloth, two-toed (*Choloepus* sp.)
MEM 776; MIA 27; SWF 184; WWW 96

Slowworm (*Anguis fragilis*)
ERA 103; LOW 169(bw); MIA 439; MSW p2

Slug, banana (*Ariolimax californicus*)
EAL 152-153

Smartweed (*Polygonum* sp.)
CWN 51, 53; NWE cp527, cp529, cp530; SWW cp527;
  WNW cp291

Smelt, European (*Osmerus eperlanus*)
MIA 509

Smelt, rainbow or American (*Osmerus mordax*)
NAF cp137, cp573; PCF p10

Smew (*Mergus albellus*)
DNA 153-155

Smokethorn or desert smoketree (*Dalea spinosa*)
JAD cp305, cp313; NTW cp80, cp398; SWW cp666

Smoketree, American (*Cotinus oboratus*)
NTE cp64, cp460; PMT cp[40]

Snail, agate-shell (*Achatinella sowerbyona*)
AAL 493; WOI 90

Snail, moon (*Lunatia* sp.)
BCS 49; MSC cp461

Snail, top (several genera)
MSC cp457-459

Snail, tulip (*Fasciolaria* sp.)
AGC cp125; MSC cp421, cp422

Snail (other kinds)
BCS 80; HSG cp35, 63(bw); PGT 107, 108; SAS cpB8;
  SWF 170; WWF cp12; WWW 66

Snail-eater, narrow (*Scaphinotus angusticollis*)
SGI cp84

Snailfish or seasnail (*Liparis* sp.)
ACF p52; PCF cp20, p21

Snake, Aesculapian (*Elaphe longissima longissima*)
AAL 200; LSW 98; WOM 92

Snake, beaked (African) (*Rhamphiophis multimaculatis*)
LSW 235

Snake, black-headed (*Tantilla* sp.)
ARA cp460, cp500; ART p[23]; JAD cp227; LBG cp615;
  RNA 173, 175; WRA cp37

Snake, black-striped (*Coniophanes imperialis*)
ART p[18]; RNA 177

Snake, blind (Australian) (*Typhlops* sp.)
ERA 119; LSW 12

Snake, blind (Texas) (*Leptotyphlops dulcis*)
ARA cp464; ERA 118; JAD cp225; LBG cp617; RNA 137;
  WRA cp36

Snake, blind (Western) (*Leptotyphlops humilis*)
ARA cp457; JAD cp226; MIA 445; RNA 137; WRA cp36

Snake, brown or DeKay's (*Storeria dekayi*)
ARA cp550; ARC 231; ARN 59(bw); CGW 93; LSW 172;
  NAW 236; RNA 159; WNW cp189

Snake, bull (*Pituophis melanoleucus*)
ARA cp573; LSW 111; NAW 237; RNA 187; WWD 149

Snake, cat (*Telescopus fallax*)
MSW cp88

Snake, cat-eyed (*Leptodeira septentrionalis*)
ARA cp606; LSW 225; RNA 177

Snake, coachwhip (*Masticophis flagellum*)
ARA cp469, cp491, cp553, cp554, cp556, cp558; ARC
  215, 216; JAD cp223, cp229, cp230, cp233; LBG cp614;
  RNA 193; SEF cp523; WRA cp36; WWD 149

Snake, copperhead See Copperhead

Snake, coral (*Micrurus* sp.)
LSW 273; MSW cp29; SAB 36

Snake, coral (Eastern) (*Micruroides fulvius*)
AAL 252; ARA cp617, cp618; ARC 239; LSW 271; MIA
  455; NAW 239; RNA 197; SEF cp534

Snake, coral (false or two-headed) (*Anilius scytale
  scytale*)
LSW 13; MIA 445

Snake, coral (false) (*Erythrolamprus bizona*)
ERA 118; LSW 222

Snake, coral (Western or Arizona) (*Micruroides
  euryxanthus*)
AAL 253; ARA cp616; JAD cp241; LSW 270; RNA 197;
  WRA cp37

Snake, corn See Ratsnake, red

Snake, crayfish (*Regina* sp.)
ARA cp510, cp519, 227; RNA 157, 159; WNW cp184,
  cp188

Snake, crowned (Florida) (*Tantilla relicta*)
ARA cp458, cp461; RNA 173

Snake, crowned (Southeastern) (*Tantilla coronata*)
ARA cp466; ARC 234; RNA 171

Snake, earth (rough) (*Virginia striatula*)
ARA cp470, cp473; ARC 237

Snake, earth (smooth) (*Virginia valeriae*)
ARA cp467; ARC 238

Snake, eastern milk See Kingsnake, scarlet

Snake, egg-eating (African) (*Dasypeltis scabra*)
AAB 34-35; LSW 181

Snake, fishing (*Herpeton tentaculatum*)
LSW 220; MSW cp19, cp90

Snake, flat-head (*Tantilla gracilis*)
ARA cp463; LBG cp616; RNA 175

Snake, flying (*Chrysopelea ornata*)
WOI 118-119

Snake, four-lined (*Elaphe quatuorlineata*)
MSW cp33, cp34

Snake, fox (*Elaphe vulpina*)
ARA cp564; LBG cp608; LSW 97; RNA 187

Snake, garter See Garter snake

Snake, glass (*Ophisaurus apodus*)
MIA 439

Snake, glossy (*Arizona elegans*)
ARA cp566, cp577, cp587; JAD cp268; LBG cp601; LSW
   113; RNA 183; WRA cp39

Snake, gopher See Snake, pine

Snake, grass or ringed (*Natrix natrix*)
LSW 162; MIA 451; PGT 220; WWW 93

Snake, grass-green vine (*Dryophis prasinus*)
ERA 113

Snake, green (smooth) (*Opheodrys vernalis*)
ARA cp475, cp476; ARC 225; ARN 69 (bw); ART p[22];
   CGW 85; LSW 197; RNA 189; WNW cp176; WRA cp39

Snake, green (rough) (*Opheodrys aestivus*)
ARA cp477; ARC 224; LSW 196; MSW cp46; RNA 189;
   SEF cp520; WRA cp39

Snake, ground (*Sonora semiannulata*)
ARA cp459, cp499, cp502, cp555, cp611; JAD cp228,
   cp231, cp239; LBG cp610; LSW 204; RNA 169; WRA
   cp38

Snake, hognose See Hognose snake

Snake, hooknosed (*Gyalopion* sp.)
ARA cp588; JAD cp266; RNA 175

Snake, hooknosed (Mexican) (*Ficimia streckeri*)
ARA cp547; RNA 175

Snake, indigo or cribo (*Drymarchon corais*)
ARA cp489; LSW 137-140; MSW cp69; NAW 242; RNA
   189

Snake, king See Kingsnake

Snake, Kirtland's (*Clonophis kirtlandi*)
ARA cp551; RNA 157; WNW cp192

Snake, leafnosed (saddled) (*Phyllorhynchus browni*)
AAL 249; ARA cp589; JAD cp245; RNA 165; WRA p40

Snake, leafnosed (spotted) (*Phylorhynchus decurtatus*)
ARA cp571, cp583; JAD cp265, cp269; RNA 165; WRA
   p40

Snake, leopard (*Elaphe situla*)
MSW cp27

Snake, lined (*Tropidoclonion lineatum*)
ARA cp507; LBG cp596; RNA 153

Snake, long-nosed (*Rhinocheilus lecontei*)
ARA cp593, cp609; JAD cp238, cp247; LBG cp609; LSW
   201, 211; RNA 165

Snake, lyre (*Trimorphodon biscutatus*)
ARA cp568, cp582; ART p[25]; JAD cp262, cp270; LSW
   238, 239; MSW p89; RNA 177; WRA cp39

Snake, mangrove (*Boiga dendrophila*)
LSW 212, 213; MIA 453

Snake, milk (*Lampropeltis triangulum* or *gentulis*) See
   also Kingsnake, scarlet
AAL 249, 250(bw); ARA cp597, cp600, cp613-cp615; ARN
   71(bw); CGW 93; LBG cp611; LSW 126-129; MSW cp30;
   NAW 244; RNA 181; SEF cp533; WRA cp37

Snake, mole (*Pseudaspis cana*)
LSW 198

Snake, Montpellier (*Malpolon monspessulanus*)
LSW 232; MAE 27; MSW cp87

Snake, mud (*Farancia abacura*)
AAL 249; ARA cp492; ARC 204; LSW 186, 187; RNA 163;
   WNW cp182

Snake, neck-banded (*Scaphiodontophis annulatus*)
LSW 203

Snake, night (*Eridiphas slevini*)
ARA cp586; WRA p48

Snake, night (*Hypsiglena torquata*)
JAD cp267; RNA 177; WRA cp39

Snake, oak See Ratsnake, gray

Snake, parrot (*Leptophis* sp.)
ERA 127; LSW 226

Snake, patchnose (Big Bend) (*Salvadora deserticola*)
ARA cp514; JAD cp217; RNA 195

Snake, patchnose (mountain) (*Salvadora grahamiae*)
ARA cp516; RNA 195

Snake, patchnose (Western) (*Salvadora hexalepis*)
ARA cp527; JAD cp218; RNA 195

Snake, pilot See Ratsnake, black

Snake, pine or gopher (*Pituophis melanoleucus*)
ARA cp488, cp537, cp575, cp591; ARC 226; JAD cp263;
   LBG cp598; LSW 107, 109, 110; MIA 453; WFW cp596;
   WRA cp39

Snake, pine woods (*Rhadinaea flavilata*)
ARA cp462, cp465; ARC 229; RNA 177; SEF cp522;
    WNW cp178, cp179

Snake, pipe (*Cylindrophis* sp.)
ERA 119; LSW 14

Snake, queen (*Regina septemvittata*)
ARA cp503; ARC 228; CGW 93; LSW 168; RNA 159;
    WNW cp186

Snake, racer See Racer

Snake, rainbow (*Farancia erytrogramma*)
ARA cp546; ARC 205; LSW 188; RNA 163

Snake, rat See Ratsnake

Snake, redbelly (*Storeria occipitomaculata*)
ARA cp501, cp505, cp506; ARC 232, 233; ARN 60 (bw);
    CGW 93; MIA 451; RNA 161; WNW cp180

Snake, ribbon (Eastern) (*Thamnophis sauritus*)
ARA cp520, cp532; ARC 235; ARN 63(bw), 64(bw); CGW
    93; LSW 171; RNA 145; WNW cp187

Snake, ribbon (Western) (*Thamnophis proximus*)
ARA cp531; LBG cp595; RNA 147; WRA cp42

Snake, ringneck (*Diadophis punctatus*)
ARA cp495-cp498; ARC 198, 199; ARN 66(bw); CGW 85;
    LBG cp612; LSW 175; NAW 248(*D.amabilis*); RNA 161;
    MSW cp53; WFW cp590; WRA cp37

Snake, sand (banded) (*Chilomeniscus cinctus* or
    stramineus)
ARA cp605; JAD cp236; RNA 171; WRA p48, cp38

Snake, scarlet (*Cemophora coccinea*)
ARA cp595, cp596, cp607; ARC 195; ART p[16]; LSW
    176, 177; RNA 179

Snake, sea See Seasnake

Snake, sharp-tailed (*Contia tenuis*)
ARA cp471; RNA 163; WFW cp598; WRA cp39

Snake, shieldtail (*Uropeltis* sp.)
ERA 118; LSW 16

Snake, short-tailed (*Stilosoma extenuatum*)
ARA cp584

Snake, shovel-nosed (Sonoran) (*Chionactis palarostris*)
ARA cp610; JAD cp240; RNA 169; WRA cp38

Snake, shovel-nosed (Western) (*Chionactis occipatalis*)
ARA cp604, cp612; JAD cp235, cp237; LSW 178; RNA
    169; WRA cp38

Snake, smooth (*Coronella austriaca*)
LSW 179, 180; MIA 453; MSW p55, p56

Snake, speckled or fire-bellied (*Leimadophis* sp.)
LSW 193; SAB 19

Snake, sunbeam (*Xenopeltis unicolor*)
LSW 15; MIA 445; MSW cp72

Snake, swamp (black) (*Seminatrix pygaea*)
ARA cp487, cp494; ARC 230; LSW 167; RNA 159; WNW
    cp181, cp183

Snake, swamp (striped) (*Liodytes alleni*)
RNA 159

Snake, thread (*Leptotyphlops* sp.)
LSW 11; MSW p70

Snake, tiger (*Notechnis scutaris*)
LSW 277, 278

Snake, tree (*Dryophis* and other genera)
LSW 206; MSW cp25; SWF 84

Snake, trinket (*Elaphe helena*)
LSW 106

Snake, vine (Mexican) (*Oybelis aeneus*)
LSW 233; RNA 189

Snake, viperine (*Natrix maura*)
AWR 56-57

Snake, wart (*Acrochordus* sp.)
ERA 125; LSW 78, 79

Snake, water See Watersnake

Snake, whip See Whipsnake

Snake, wolf (*Lycodon* sp.)
LSW 230, 231

Snake, worm (*Carphophis amoenus*)
ARA cp493; ARC 194; ARN 67(bw); ART p[14]; LSW 174;
    RNA 163

Snakehead (*Malacothrix coulteri*)
JAD cp126; LBG cp180; SWW cp258

Snakeroot, black (*Sanicula canadensis*)
See also Cohosh, black
NWE cp16

Snakeroot, white (*Eupatorium rugosum*)
LBG cp148; NWE cp189

Snakeweed, broom (*Gutierrezia sarothrae*)
LBG cp203, cp211; NWE cp366; SWW cp349

Snapdragon-vine (*Maurandya* sp.)
JAD cp71, 74; SWW cp659, cp660

Snapper (*Lutjanus* sp.)
AAL 356; ACF cp31; AGC cp375; NAF cp522, cp534,
    cp539, cp540; RCK 155; SAB 129

Snapper, glasseye (*Priacanthus cruentatus*)
ACF p21; RUP 169

Snapper, yellowtail (*Ocyurus chrysurus*)
ACF cp31; KCR cp22; NAF cp536; WWF cp15

Snapping turtle (*Chelydra serpentina*)
AAL 202(bw); AGC cp433; ARA cp322-cp324; ARC 146;
ARN 42(bw); CGW 69; LBG cp576; MIA 409; NAW 269;
RNA 39; TOW 130(bw); TTW 111; WNW cp151, 152;
WRA p17

Snapping turtle, Alligator (*Macroclemys temminckii*)
ARA cp325, cp326, cp327; ERA 81; MIA 409; PGT 219;
RNA 39; TOW 132(bw)

Sneezeweed or bitterweed (*Helenium* sp.)
CWN 267; NWE cp279; SWW cp240; WTV 92, 96

Snipe, common or Wilson's (*Gallinago gallinago*)
EAB 803; EOB 167; FEB 159; FWB 166; MIA 243; NAB
cp255; PBC 153; RBB 74; UAB cp214; WBW 249; WGB
53; WNW cp542

Snipe, Jack (*Lymnocryptes minima*)
EOB 170-171

Snipe, painted (*Rostratula benghalensis*)
EOB 178; MIA 241; WGB 51

Snipefish (*Macrorhamphosus* sp.)
ACF p20

Snook (*Centropomus undecimalis*)
ACF cp24; AGC cp406; NAF cp486

Snow on the mountain (*Euphorbia marginata*)
WTV 41

Snow plant (*Sarcodes sanguinea*)
SWW cp404; WFW cp480

Snow queen (*Synthyris reniformis*)
SWW cp647; WFW cp537

Snowbell (*Styrax* sp.)
NTE cp196; PMT cp[129], cp[130]

Snowberry (*Symphoricarpos* sp.)
NWE cp240; WFW cp165

Snowcock (*Tetraogallus* sp.)
MIA 229; NHU 151; WGB 39

Soapberry, Western or chinaberry (*Sapindus
drummondii*)
NTE cp308, cp563; NTW cp283, cp480; PMT cp[120]

Soapfish (*Rypticus* sp.)
ACF cp24; NAF cp506

Soapwort See Bouncing Bet

Soldierfish (*Holocentrus* sp.)
NAF cp382, cp383, cp385

Soldierfish, blackbar (*Myripristis jacobus*)
AAL 344-345; ACF cp21; KCR cp25; NAF cp384; RCK
164

Sole (*Solea solea*)
MIA 579

Sole, bigmouth (*Hippoglossina stomata*)
PCF p43

Sole, C-O (*Pleuronichthys coenosus*)
MPC cp309; NAF cp274; PCF p44

Sole, English (*Parophrys vetulus*)
MPC cp308; NAF cp276; PCF p44

Sole, naked (*Gymnachirus melas*)
ACF p58; AGC cp360; MIA 579; NAF cp291

Sole, rock (*Lepidopsetta bilineata*)
MPC cp310; NAF cp277; PCF p44

Solenodon (*Solenodon* sp.)
CRM 35; GEM I:424, 425; MEM 748; MIA 31

Solitaire, Townsend's (*Myadestes townsendi*)
EAB 893; FWB 424; SPN 91; UAB cp476; WFW cp277

Solomon's seal (*Polygonatum biflorum*)
CWN 11; FWF 193; NWE cp4, cp115; WAA 158; WTV 1

Solomon's seal, false (*Smilacina racemosa*)
BCF 35; CWN 13; FWF 103; NWE cp122; SEF cp432;
SWW cp110; WAA 158; WFW cp412

Sora (*Porzana carolina*)
AAL 122; AGC cp599; BWB 220; EAB 726; FEB 126; FWB
125; NAB cp249; PBC 129; UAB cp245; WNW cp538

Sorrel, lady's (*Oxalis europaea*)
FWF 11; NWE cp265

Sorrel, redwood (*Oxalis oregana*)
SWW cp443; WFW cp508

Sorrel, sheep (*Rumex acetosella*)
LBG cp221; NWE cp440

Sorrel, wood See Wood sorrel

Sotol (*Dasylirion wheeleri*)
JAD cp338; SWW cp185

Soursop (*Annona muricata*)
TSP 11

Sourwood (*Oxydendrum arboreum*)
NTE cp137, cp405, cp511, cp585; PMT cp[92]; SEF cp93;
TSV 72, 73

Souslik See Ground squirrel, European

Sowbug See Pillbug

Spadefish, Atlantic (*Chaetodipterus faber*)
ACF cp37; NAF cp344

Spadefish, Pacific (*Chaetodipterus zonatus*)
PCF cp35

Spadefoot See under Toad

Spanish bayonet (*Yucca aloifolia*)
SCS cp19; TSP 79

Spanish moss (*Tillandsia usneoides*)
NWE cp39; SEF cp458; TSP 131

Spanish needles (*Bidens bipinnata*)
FWF 147

Sparkleberry (*Vaccinium arboreum*)
NTE cp206, cp424

Sparrow, Bachman's (*Aimophila aestivalis*)
FEB 422; SPN 129

Sparrow, Baird's (*Ammodramus bairdii*)
EAB 312; FWB 455; FEB 410

Sparrow, black-chinned (*Spizella atrogularis*)
EAB 313; FWB 447; UAB cp601

Sparrow, black-throated (*Amphispiza bilineata*)
EAB 312; FEB 426; FWB 452; JAD cp610; NAB cp528;
UAB cp593

Sparrow, Brewer's (*Spizella breweri*)
EAB 312; FWB 459; JAD cp608; UAB cp600

Sparrow, cassin's (*Aimophila cassinii*)
EAB 313; FEB 413; FWB 457

Sparrow, chipping (*Spizella passerina*)
EAB 313; FEB 419; FWB 451; MIA 369; NAB cp530; PBC
385; SPN 131; UAB cp590; WFW cp300; WGB 179

Sparrow, clay-colored (*Spizella pallida*)
EAB 314; FEB 418; FWB 444; LBG cp531; NAB cp535;
PBC 386; UAB cp578

Sparrow, English or house (*Passer domesticus*)
EAB 1035; EOB 425; FEB 425; FWB 465; MIA 391; NAB
cp525, cp526; PBC 337; RBB 175; UAB cp580, cp592;
WGB 201; WOB 241

Sparrow, field (*Spizella pusilla*)
EAB 315; FEB 421; LBG cp532; NAB cp532; PBC 387

Sparrow, fox (*Passerella iliaca*)
EAB 315; FEB 428; FWB 448; NAB cp543; PBC 389; SEF
cp340; SPN 134; UAB cp572; WFW cp301

Sparrow, golden-crowned (*Zonotrichia atricapilla*)
EAB 314; FWB 441; UAB cp588

Sparrow, grasshopper (*Ammodramus savannarum*)
EAB 317; FEB 409; FWB 456; LBG cp537; NAB cp536;
PBC 377; UAB cp568

Sparrow, Harris' (*Zonotrichia querula*)
EAB 316; FEB 427; NAB cp529; UAB cp594

Sparrow, hedge See Dunnock

Sparrow, Henslow's (*Ammodramus henslowii*)
EAB 317; FEB 414; LBG cp538; NAB cp537; SPN 133

Sparrow, lark (*Chondestes grammacus*)
EAB 317; FEB 423; FWB 453; LBG cp534; NAB cp527;
UAB cp577

Sparrow, Le Conte's (*Ammodramus leconteii*)
EAB 318; FEB 408; NAB cp538

Sparrow, Lincoln's (*Melospiza lincolnii*)
EAB 318; FEB 416; FWB 460; NAB cp544; UAB cp599;
WFW cp303; WNW cp577

Sparrow, olive (*Arremonops rufivirgatus*)
EAB 318; FEB 399; NAB cp478

Sparrow, rock (*Petronia petronia*)
MIA 391; WGB 201

Sparrow, rufous-crowned (*Aimophila ruficeps*)
EAB 319; FEB 412; FWB 463; UAB cp591

Sparrow, sage (*Amphispiza belli*)
EAB 319; FWB 445; JAD cp611; UAB cp602

Sparrow, savannah or Ipswich (*Passercullus
sandwichensis*)
EAB 319; FEB 411; FWB 454; LBG cp536; MIA 369; NAB
cp548; PBC 377; SAB half-title, 102; UAB cp569; WGB
179

Sparrow, seaside (*Ammodramus maritimus*)
EAB 319; FEB 406; NAB cp534

Sparrow, sharp-tailed (*Ammodramus caudacutus*)
EAB 321; FEB 407; NAB cp533; WNW cp582

Sparrow, song (*Melospiza melodia*)
EAB 320; FEB 417; FWB 461; MIA 369; NAB cp542; NAW
202; PbC 391; SPN 136; UAB cp573; WFW cp302; WGB
179; WON 166

Sparrow, swamp (*Melospiza georgiana*)
AGC cp614; EAB 321; FEB 429; NAB cp541; WNW cp576

Sparrow, tree (American) (*Spizella arborea*)
EAB 320; FEB 420; FWB 450; LBG cp530; NAB cp531;
PBC 384; SPN 130; UAB cp589

Sparrow, tree (European) (*Passer montanus*)
EAB 1035; FEB 424; NAB cp524; RBB 176

Sparrow, vesper (*Pooecetes gramineus*)
EAB 322; FEB 415; FWB 458; JAD cp609; LBG cp533;
NAB cp550; PBC 381; SPN 135; UAB cp575

Sparrow, white-crowned (*Zonotrichia leucophrys*)
EAB 322; FEB 430; FWB 442; NAB cp540; PBC 388; SPN
138; UAB cp586; WFW cp304

Sparrow, white-throated (*Zonotrichia albicollis*)
EAB 323; EOB 407; FEB 431; FWB 443; NAB cp539; PBC
    389; SEF cp338; SPN 137; UAB cp587

Spatterdock (*Nuphar advena*)
CWN 59; WTV 81

Spearmint (*Mentha spicata*)
CWN 221

Spectacle pod (*Dithyrea wislizenii*)
JAD cp79; SWW cp134

Speedwell or brooklime (*Veronica* sp.)
CWN 237; NWE cp523, cp591; PGT 40; SWW cp627,
    cp628; WFW cp521; WTV 152

Spicebush (*Lindera benzoin*)
FWF 101; NWE cp365; SEF cp467; TSV 90, 91; WNW
    cp201

Spider, barn (*Araneus cavaticus*)
MIS cp644

Spider, bird See Tarantula

Spider, black widow (*Latrodectus mactans*)
AAL 397; MIS cp651, cp688; SOW 178

Spider, crab (*Xysticus* sp.)
MIS cp682, cp683; SOW 83

Spider, fisher (*Dolomedes* sp.)
MAE 99; MIS cp663, cp668; PGT 121

Spider, garden (*Araneus diadematus*)
BOI 232; MIS cp645

Spider, house (American) (*Achaearanea* sp.)
MIS cp689

Spider, house (*Tegenaria* sp.)
NNW 72; SOW 13(bw)

Spider, jumping (various genera)
BOI 250-251; EOI 5, 139; LBG cp393; MIS cp652-cp656;
    SEF cp426; SOW 117(bw)

Spider, lynx (*Oxyopes* sp.)
LBG cp394; MIS cp641

Spider, lynx (*Peucetia* sp.)
EOI 138; MIS cp677, cp681; SOW 90, 163

Spider, nursery web (various genera)
MIS cp642, cp666; SOW 86, 94

Spider, orb-web (*Araneus* sp.)
EOI 143; LBG cp395; MIS cp643; SOW 15, 98; WFW
    cp581

Spider, orb-web (*Argiope* sp.)
AAB 49; EOI 136; SAB 25, 42

Spider, sea (various kinds)
EAL 244; MPC cp356, cp357; MSC cp575-579

Spider, spitting (*Scytodes* sp.)
MIS cp675; SOW 44(bw)

Spider, trapdoor (Californian) (*Bothriocyrtum
    californicum*)
MIS cp650

Spider, water (*Argyroneta aquatica*)
PGT 123, 124

Spider, wolf (burrowing) (*Geolycosa* sp.)
BOI 253; MIS cp646

Spider, wolf (Carolina) (*Lycosa carolinensis*)
JAD cp401; MIS cp648

Spider, wolf (forest) (*Lycosa gulosa*)
SEF cp425; WFW cp580

Spider, wolf (other kinds)
EOI 135; PGT 123; SOW 68(bw)

Spider, zebra (*Salticus scenicus*)
SOW 11

Spiderwort (*Tradescantia ohiensis* or *virginiana*)
CWN 27; NWE cp614; WAA 30, 42; WTV 173

Spikedace (*Meda fulgida*)
JAD cp300; NAF cp161

Spindle tree (*Euonymus europaeus*)
PPC 73(fruit)

Spiraea (*Spiraea* sp.)
TSV 108, 109; WFW cp190

Spirogyra (algae) (*Spirogyra* sp.)
PGT 75

Spittle bug (*Philaenus spumarius*)
AAL 421; SGI cp70

Sponge, finger (*Haliclona oculata*)
AGC cp270; MSC cp52

Sponge, finger (red) (*Haliclona rubens*)
KCR cp17, cp19; MSC cp31

Sponge, fire (*Tedania ignis*)
KCR cp19; MSC cp129

Sponge, pipes-of-pan (*Agelas schmidti*)
KCR cp19

Sponge, red beard (*Microciona prolifera*)
MSC cp29; SAS cpA4

Sponge, tube (*Spinosella* and other genera)
EAL 168; KCR p2, cp17, cp19, cp20; MSC cp23

Sponge, vase (*Ircinia campana*)
AGC cp278; MSC cp19, cp21

Sponge, velvety red (*Ophlitaspongia pennata*)
MPC cp473; MSC cp128

Sponge, yellow boring (*Siphonodictyon coralliphagum*)
KCR cp20

Sponge (various kinds)
AAB 8; KCR cp20; MSC cp24, cp30, cp45; PGT 94; RUP
   12, 13; SAS cpA1-cpA6; SCS cp45; WPR 138; WWW 133

Spoonbill, African (*Platalea alba*)
AWR 98; BWB 40; EOB 79; WBW 77

Spoonbill, roseate (*Ajaja ajaja*)
AAL 88, 105; AGC cp559; BMW 64, 65; BWB 41; EAB
   571; EOB 74; FEB 147; NAB cp11; WOB 31, 141; WWW
   130

Spoonbill, white (*Platalea leucorodia*)
WBW 113

Spring beauty (*Claytonia virginica*)
BCF 83; CWN 139; NWE cp459; SEF cp489; WAA 157;
   WFW cp392; WTV 124

Spring beauty (other *Claytonia* sp.)
SWW cp444; WAA 254

Springbuck or springbok (*Antidorcas marsupialis*)
CRM 197; GEM V:309, 484; MEM 569, 574-75; MIA 153

Springfish, white river (*Crenichthys baileyi*)
JAD cp292; NAF cp205

Springhare (*Pedetes capensis*)
GEM III:114-115, 123; MEM 635; MIA 165

Springtail (various kinds)
BOI 72; PGT 135; SGI cp1; WIW 181(bw)

Spruce, black (*Picea mariana*)
NTE cp24, cp481; NTW cp30, cp400; SEF cp30, cp169;
   WFW cp115; WNW cp432

Spruce, blue (*Picea pungens*)
NTW cp29, cp403; WFW cp118

Spruce, Brewer's (*Picea brewerana*)
NTW cp33, cp404; OET 69; WFW cp44

Spruce, Engelmann (*Picea engelmannii*)
WFW cp117

Spruce, Norway (*Picea abies*)
NTE cp26, cp477; NTW cp32

Spruce, Oriental (*Picea orientalis*)
PMT cp[97]

Spruce, red (*Picea rubens*)
NTE cp25, cp478; SEF cp31, cp171; TSV 10, 11

Spruce, sitka (*Picea sitchensis*)
NTW cp27, cp402; WFW cp116

Spruce, white (*Picea glauca*)
NTE cp23, cp479; NTW cp28, cp405; SEF cp29, cp170;
   WFW cp119

Spurfowl, red (*Galloperdix spadicea*)
MIA 229; WGB 39

Spurge (*Euphorbia* sp.)
CWN 57; LBG cp146; PPC 75, 76; NWE cp198

Spurge, cypress (*Euphorbia cyparissias*)
NWE cp350; WTV 69

Squawbush (*Rhus trilobata*)
WFW cp155

Squawfish (*Ptychocheilus* sp.)
MIA 515; NAF cp133, cp134

Squawroot or Cancer root (*Conopholis americana*)
CWN 245; NWE cp307; WTV 57

Squawroot trillium See Wake robin

Squeteague See Weakfish

Squid (*Loligo* sp.)
AAL 497; EAL 259 ; KCR cp31; MSC cp486

Squid, Atlantic oval or reef (*Sepioteuthis sepioidea*)
KCR cp31; SCS p46

Squid, oceanic (*Onychoteuthis banksi*)
HSG cp35

Squid, opalescent (*Loligo opalescens*)
AAB 13; MPC cp381; MSC cp485

Squid, shortfin (*Illex illecebrosus*)
MSC cp484

Squilla (various genera)
MSC cp592, cp597

Squirrel, Abert's or tassel-eared (*Sciurus aberti*)
ASM cp193, cp194; CRM 97; MEM 617; RMM cp[2]; WFW
   cp342

Squirrel, African palm (*Epixerus ebii*)
CRM 97; MIA 159

Squirrel, African pygmy (*Myosciurus pumilio*)
MEM 617

Squirrel, antelope (*Ammospermophilus* sp.)
ASM cp10-cp12; JAD cp498-500; MAE 60; NMM cp[10]

Squirrel, Douglas or Pine (*Tamiasciurus douglassi*)
ASM cp185; GEM III.70, MAE 49, WFW cp340

Squirrel, flying See Flying squirrel

Squirrel, fox (*Sciurus niger*)
ASM cp186; CGW 169; CMW 112(bw); CRM 99; GEM
III:85; GMP 149(bw), cp6; LBG cp80; MNC 169(bw); MNP
cp2a; MPS 137(bw); NAW 87; SEF cp587; WMC 125

Squirrel, giant (*Ratufa* sp.)
CRM 99; MEM 617, 623; MIA 159

Squirrel, gray (*Sciurus carolinensis*)
ASM cp182, cp188, cp191, cp192, cp195; CGW 169; GEM
III:84; GMP 145(bw), cp5; MEM 613, 626, 627; MIA 159;
MNC 167(bw); MPS 137(bw); NAW 88; SEF cp588; WFW
cp339(*S.griseus*); WMC 122

Squirrel, ground See Ground squirrel

Squirrel, long-nosed (*Rhinosciurus laticaudatus*)
MEM 617

Squirrel, palm (*Funambulus pennanti*)
WPP 50

Squirrel, prevost's (*Callosciurus prevosti*)
GEM III:86; MEM 616; MIA 159; ROW cp4

Squirrel, red or Chickaree (*Tamiasciurus hudsonicus*)
ASM cp187; CGW 169; CMW 116(bw); GEM III:24-25, 77;
GMP 153(bw), cp7; MAE 59; MEM 617, 618; MNC
171(bw); MPS 138(bw); NAW 88; NMM cp[6]; RMM cp[1];
SEF cp586; SWF 58, 59; SWM 36; WFW cp341; WMC
127

Squirrel, red (Eurasian) (*Sciurus vulgaris*)
CRM 99; GEM III:79; MIA 159

Squirrel, rock (*Spermophilis variegatus*)
ASM cp190, cp196; JAD cp505, cp506; MPS 135(bw)

Squirrelfish (*Holocentrus spinfer* or *rufus*)
AAL 344; ACF cp21; AGC cp340; MIA 541; NAF cp383;
RCK 22, 63; RUP 174, 175; WWF cp60

Starfish or sea star (various kinds)
AAB 54; AGC cp188-190; BCS 68; EAL 275; KCR p33;
MPC cp363, cp365-372; MSC cp535-cp541, cp547-563;
RUP 31; SCS cp59

Starfish (*Asterias* sp.)
BCS 36, 37; SAS cpA29

Starfish, blood or blood-star (*Henricia sanguinolenta*)
AGC cp187; BCS 35, 57; MSC cp552

Starfish, brittle See Brittlestar

Starfish, reticulated or cushion (*Oreaster reticulatus*)
AGC cp191; KCR p33; MSC cp541; SCS cp59

Starfish, spiny (*Echinaster* sp.)
KCR cp32, p33; SCS cp59

Starfish, sun (*Solaster* and other genera)
AGC cp192, cp193; EAL 277; MPC cp373, cp374; MSC
cp542-546

Starflower (*Trientalis borealis*)
CWN 179; NWE cp61; SEF cp443; WAA 157

Starflower, Western (*Trientalis latifolia*)
SWW cp449; WFW cp501

Stargazer (various genera)
ACF p42; PCF p13

Stargrass, yellow (*Hypoxis hirsuta*)
CWN 21; NWE cp270; WTV 60

Starling (*Sturnus vulgaris*)
EAB 848; EOB 429; FEB 359; FWB 281; JAD cp597; MIA
393; NAB cp565; NAW 204; PBC 289; RBB 174; UAB
cp555, cp611; WGB 203

Star-of-Bethlehem (*Ornithogalum umbellatum*)
BCF 29; CWN 19; PPC 88; WTV 11

Steenbuck (*Raphicerus campestris*) See also Grysbok
AAL 24; MEM 577; NHA 77(bw); WPP 86

Steeplebush or hardhack (*Spiraea tomentosa*)
CWN 87; LBG cp263; NWE cp539

Stentor (*Stentor* sp.)
EAL 161; PGT 92; WWW 49

Sterlet (*Acipenser ruthenus*)
EAL 21

Stewartia (*Stewartia* sp.)
NTE cp193, cp194, cp436; PMT cp[126]-cp[128]

Stick insect (various kinds) See also Water scorpion
EOI 48; WIW 61(head only), 185(bw)

Stickleback, brook (*Culaea inconstans*)
NAF cp219

Stickleback, four-spine (*Apeltes quadracus*)
MIA 543

Stickleback, ten-spined (*Pungitius pungitius*)
PGT 202

Stickleback, three-spined (*Gasterosteus aculeatus*)
AGC cp409; AWR 27-29; MIA 543; NAF cp218, cp354;
PCF p13; PGT 199-201;SAB 82

Stickweed, many-flowered (*Hackelia floribunda*)
SWW cp643

Stilt, black-winged or black-necked (*Himantopus
himantopus*)
AGC cp578; AWR 65; BWB 182-183; EAB 50; EOB 172;
FEB 154; FWB 141; MIA 245; NAB cp243; PBC 137; UAB
cp219; WBW 301; WGB 55

Stingaree See Stingray, Australian

Stingray (*Dasyatis* sp.)
ACF p7, p62; PCF p5

Stingray, Australian (*Urolophus mucosus*)
SJS 63

Stingray, blue-spotted (*Taeniura lymma*)
EAL 138; MAE half-title

Stingray, round (*Urolophus halleri*)
MPC cp314; NAF cp268; PCF p5

Stingray, Southern (*Dasyatis americana*)
AGC cp425; KCR cp27; MIA 495; NAF cp267

Stink bug (various kinds)
MIS cp111; SGI cp63, cp64

Stinkhorn fungus (*Phallus* and other genera)
FGM cp43; GSM 253, 254; NAM cp692, cp694, cp701,
cp706, cp707; NMT cp91; NNW 33; SGM cp353-358

Stinkpot (*Sternotherus odoratus*)
AAL 204; ARA cp319; ARC 149; ARN 43(bw); MIAA 409;
RNA 29; TOW 76(bw); WNW cp155

Stint, Temminck's (*Calidris temminckii*)
WBW 291

Stitchwort (*Stellaria graminea*)
CWN 121

Stoat See Ermine

Stone cat See Madtom

Stonechat (*Saxicola torquata*)
MIA 335; RBB 134; WGB 145

Stonefish (*Synancea* sp.)
MIA 545; RCK 129; RUP 146; WWW 60

Stonewort (*Chara fragilis*)
PGT 55

Stork, black (*Ciconia nigra*)
EOB 74; WBW 87

Stork, black-necked (*Xenorhynchus asiaticus*)
AAL 104

Stork, African open-billed (*Anastomus lamelligerus*)
WBW 81

Stork, hammerhead (*Scopus umbretta*)
EOB 75; MIA 211; WBW 75; WGB 21

Stork, marabou (*Leptoptilos crumeniferus*)
BWB 26-27(head only); EOB 75; WBW 93; WWW 196

Stork, painted (*Ibis leucocephalus*)
AAL 104; AWR 107; EOB 82; WBW 79

Stork, saddlebill (*Ephippiorhynchus senegalensis*)
BWB 25; EOB 77; WBW 91

Stork, whale-headed or shoebill (*Balaeniceps rex*)
EOB 74, 81(head only); MIA 211; WBW 73(bw); WGB 21;
WWW 133

Stork, white (*Ciconia ciconia*)
EOB 75, 77; MIA 211; WBW 83; WGB 21

Stork, white-necked (*Ciconia episcopus*)
WBW 85

Stork, wood or Wood ibis (*Mycteria americana*)
AGC cp560; EAB 849; FEB 146; FWB 139; NAB cp9; PBC
73; SEF cp261; UAB cp13; WNW cp524; WOB 62

Stork, yellow-billed (*Ibis ibis*)
BAS 105; EOB 72; WBW 77

Storksbill See Filaree

Straw agaric (*Psalliota staminea*)
NMT cp66

Strawberry, barren (*Waldsteinia fragarioides*)
NWE cp275

Strawberry, beach (*Fragaria chiloensis*)
MPC cp608; SWW cp45

Strawberry bush (*Euonymus americanus*)
FWF 119; TSV 24, 25

Strawberry, Indian (*Duchesnea indica*)
CWN 85; FWF 77; NWE cp251; WTV 59

Strawberry tree (*Arbutus unedo*)
PMT cp[17]

Strawberry, wild (*Fragaria virginiana*)
BCF 23; CWN 85; LBG cp126; NWE cp58

Strawberry, wood (*Fragaria vesca*)
CWN 85

Sturgeon, Atlantic (*Acipenser oxyrhynchos*)
AAL 322(bw)

Sturgeon, lake (*Acipenser fulvescens*)
NAF cp31

Sturgeon, shovel-nose (*Scaphirhynchus platorynchus*)
AAL 322(bw); NAF cp30, cp32; WNW cp80, cp81

Sturgeon, Siberian (*Acipenser baeri*)
AWR 221(bw)

Sturgeon, white (*Acipenser transmontanus*)
MPC cp324; NAF cp33, cp588; PCF p6

Stylops (Strepsiptera)
WIW 205(bw)

Sucker, blue (*Cycleptus elongatus*)
NAF cp100

Sucker, razorback (*Xyrauchen texanus*)
NAF cp105; NAW 312

Sucker, shark See Sharksucker

Sucker, spotted (*Minytrema melanops*)
NAF cp103; WNW cp74

Sucker, white (*Catostomus commersoni*)
MIA 517; NAF cp106; WNW cp73

Sugar cane (*Saccharum officinarum*)
TSP 129

Sugar glider See Glider, sugar

Sugarberry (*Celtis laevigata*)
NTE cp139; SEF cp78; TSV 164, 165; WNW cp463

Sulphur flower (*Eriogonum umbellatum*)
JAD cp112; SWW cp336; WFW cp436

Sumac, fragrant (*Rhus aromatica*)
FWF 59; TSV 94, 95

Sumac, laurel (*Rhus laurina*)
NTW cp115, cp349

Sumac, lemonade (*Rhus integrifolia*)
NTW cp117, cp479

Sumac, poison (*Toxicodendron vernix*)
NTE cp349; NWE cp414; WNW cp441

Sumac, prairie (*Rhus lanceolata*)
NTE cp307; NTW cp286

Sumac, shining or winged (*Rhus copallina*)
FWF 195; LBG cp449, cp464; NTE cp342, cp387, cp577;
    NWE cp37, cp411; SEF cp141, cp168, cp502

Sumac, smooth (*Rhus glabra*)
FWF 75; LBG cp448, cp465; NTE cp315, cp386, cp566;
    NTW cp287, cp345, cp486; NWE cp36, cp409; SEF
    cp142, cp197; TSV 30, 31

Sumac, staghorn (*Rhus typhina*)
LBG cp447, cp471; NTE cp314, cp567; NWE cp35, cp410;
    OET 236; SEF cp143, cp198

Sumac, sugar (*Rhus ovata*)
NTW cp116, cp379; WFW cp220

Sun animal See Heliozoan

Sunbird (various genera)
EOB 328, 397; MIA 363; SAB 20; WGB 173; WOM 106

Sundew, common or roundleaved (*Drosera rotundifolia*)
CWN 123; NWE cp63; SAB 31; SWW cp27; WNW cp251;
    WTV 36

Sundew, threadleaved (*Drosera filiformis*)
NWE cp475; WNW cp272

Sunfish, oceanic (*Mola mola*)
ACF p61; EAL 119; HSG cp20; MIA 583; MPC cp321;
    NAF cp309; PCF cp46

Sunfish, pumpkinseed (*Lepomis gibbosus*)
MIA 551; NAF cp73; WNW cp53

Sunfish, warmouth (*Lepomis gulosus*)
NAF cp86; WNW cp47

Sunfish (other kinds)
NAF cp74-cp81; WNW cp54

Sunflower, alpine (*Hymenoxysa grandiflora*)
SWW cp256

Sunflower, common (*Helianthus annuus*)
CWN 261; LBG cp175; NWE cp286; SWW cp247

Sunflower, desert (*Gerea canescens*)
JAD cp135; SWW cp254

Sunflower, tall (*Helianthus giganteus*)
BCF 11; LBG cp173; NWE cp284

Sunflower, woodland (*Helianthus strumosus*)
SEF cp464

Sungazer (*Cordylus giganteus*)
ERA 104

Sungrebe (*Heliornis fulica*)
EOB 158; MIA 239; WGB 49

Suni (*Neotragus moschatus*)
CRM 195; GEM V:330; NHA 69(bw)

Sunray (*Enceliopsis nudicaulis*)
JAD cp132; SWW cp237

Sunshinefish (*Chromis insolatus*)
ACF cp38

Surfbird (*Aphriza virgata*)
EAB 802; FWB 173; MPC cp575; UAB cp222, cp224,
    cp227

Surfgrass (*Phyllospadix* sp.)
MPC cp509

Surfperch (several genera)
MPC cp272, cp273; NAF cp525, cp528; PCF cp36, cp37

Surgeonfish (*Paracanthurus* or *Acanthurus* sp.)
ACF cp36; AGC cp365; BMW 145; NAF cp329; RCK 76

Susa See River dolphin, Ganges

Swallow, bank or sand martin (*Riparia riparia*)
EAB 852; EOB 334; FEB 267; FWB 259; MIA 315; NAB
    cp334; RBB 117; UAB cp355; WGB 125; WNW cp571;
    WOB 82, 83

Swallow, barn (*Hirundo rustica*)
AAL 164; EAB 853; EOB 333, 334; FEB 265; FWB 257; LBG cp522; MIA 315; NAB cp329; PBC 241; PGT 229; RBB 118; UAB cp352; WGB 125; WON 140

Swallow, cliff (*Petrochelidon* or *Hirundo pyrrhonota*)
EAB 852; FEB 264; FWB 256; LBG cp521; NAB cp330; NAW 205; PBC 242; UAB cp353

Swallow, rough-winged (*Stelgidopteryx ruficollis*)
EAB 852; NAB cp333; PBC 239; UAB cp354

Swallow, rough-winged (Northern) (*Stelgidopteryx serripennis*)
FEB 266; FWB 258; WNW cp572

Swallow, tree (*Tachycineta* or *Iridoprocne bicolor*)
AGC cp610; EAB 853; FEB 263; FWB 260; NAB cp331; NAW 206; SPN 62; UAB cp356; WOB 63

Swallow, violet-green (*Tachycineta thalassina*)
AAL 164; EAB 852; FWB 255; JAD cp579; UAB cp357; WFW cp254

Swallow, wood (masked) (*Artamus personatus*)
EOB 432

Swallowtail butterfly, anise (*Papilio zelicaon*)
JAD cp446

Swallowtail butterfly, black (eastern) (*Papilio polyxenes*)
BOI 133; LBG cp331; PAB cp324, cp335, cp338; SGB cp188

Swallowtail butterfly, black (western) (*Papilio bairdii*)
PAB cp326, cp334, cp337, cp358

Swallowtail butterfly, desert (*Papilio rudkini*)
AAL 456; JAD cp445

Swallowtail butterfly, giant (*Papilio cresphontes*)
MIS cp584; PAB cp27(larva), cp328, cp331; SGI cp204

Swallowtail butterfly, green or pipeline (*Battus philenor*)
LBG cp320; MIS cp578

Swallowtail butterfly, Old world (*Papilio machaon*)
BOI 133; SGB cp23(larva), cp105; SGI cp203

Swallowtail butterfly, Orchard (*Papilio aegus*)
EOI 14-15

Swallowtail butterfly, Palamedes (*Pterourus palamedes*)
SEF cp371; WNW cp389

Swallowtail butterfly, pale (*Papilio eurymedon*)
MIS cp594; PAB cp342, cp345; SGI cp207

Swallowtail butterfly, spicebush (*Papilio troilus*)
AAL 455(bw); MIS cp576; PAB cp51(larva), cp316, cp319; SEF cp368; SGB cp189

Swallowtail butterfly, tiger (*Papilio* sp.)
AAL 456; BOI 130-131(larva), cp132-135; PAB cp49(larva), cp322, cp348, cp351

Swallowtail butterfly, tiger (Eastern) (*Papilio glaucus*)
LBG cp332; MIS cp22(larva), cp580; SEF cp370; SGB cp187; SGI cp205

Swallowtail butterfly, tiger (Western) (*Papilio rutulus*)
JAD cp444; SGI cp206; WFW cp556; WNW cp388

Swallowtail butterfly, zebra (*Graphium marcellus*)
MIS cp593; PAB cp48(larva);cp340, cp343; WNW cp390

Swamp candles (*Lysimachia terrestris*)
NWE cp324; WNW cp204

Swamp hen, purple (*Porphyrio porphyrio*)
MAE 93

Swamp privet (*Forestiera acuminata*)
NTE cp112; WNW cp456

Swan, black (*Cygnus atratus*)
AWR 134-135; BWB 65-67; EOB 91

Swan, black-necked (*Cygnus melanchoryphus*)
BWB 64; EOB 93

Swan, mute (*Cygnus olor*)
AAL 106; AGC cp558; BWB 60-63; EAB 227; EOB 99; FEB 118; NAB cp173; RBB 19; WNW cp516; WOB 240

Swan, trumpeter (*Olor buccinator*)
EAB 226, 227; FWB 113; UAB cp157; WOB 266-271; WON 41, 42

Swan, tundra (*Cygnus columbianus*)
AGC cp557; FEB 119; FWB 112; MPC cp530; WON 43

Swan, whistling or Bewick's (*Olor columbianus*)
EAB 228, 229; MIA 213; NAB cp174; PBC 78; UAB cp156; WGB 23

Swan, whooper (*Cygnus cygnus*)
AAB 58; EAB 230; EOB 92; WOI 76-77

Sweet bay (*Magnolia virginiana*)
NTE cp85, cp433; PMT cp[79]; SEF cp45, cp159; TSV 48, 49; WNW cp449

Sweet gum (*Liquidambar styraciflua*)
LBG cp436, cp481; NTE cp266, cp509, cp594, cp624; NTW cp222, cp538; SEF cp122, cp182; TSV 128, 129

Sweetbread or miller mushroom (*Clitopilus prunulus*)
NAM cp242; NMT cp37; SGM cp162

Sweetflag See Flag, sweet

Sweet-grass, reed (*Glyceria maxima*)
PGT 47

Sweetleaf (*Symplocos tinctoria*)
NTE cp55, cp401

Sweetlips (*Lethrinus* or *Plectorhynchus* sp.)
AAL 320; RUP 102, 156; SAB 137; WWF cp37

Swift See also Needletail

Swift, alpine (*Apus melba*)
EOB 257

Swift, black (*Cypseloides niger*)
EAB 853; UAB cp360

Swift, chimney (*Chaetura pelagica*)
EAB 854; FEB 261; NAB cp335; UAB cp361

Swift, common (*Apus apus*)
EOB 255, 256; MIA 277; RBB 110; WGB 87

Swift, crested (*Hemiprocne longipennis*)
EOB 257; MIA 277; WGB 87

Swift, Northern white-rumped (*Apus pacificus*)
EOB 254

Swift, palm (*Cypsiurus parvus*)
EOB 257

Swift, white-throated (*Aeronautes saxatilis*)
FWB 254; WGB 87

Swiftlet, edible nest (*Collocalia fuciphaga*)
MIA 277; WGB 87

Swordfish (*Xiphius gladius*)
ACF cp49; EAL 112; HSG cp18; MIA 571; PCF cp32

Sycamore (*Platanus occidentalis*)
NTE cp256, cp508; SEF cp124, cp183; TSV 130, 131;
  WNW cp480

Sycamore, Arizona (*Platanus wrightii*)
NTW cp223

Sycamore, California or Western (*Platanus racemosa*)
NTW cp224; WFW cp101

Syringa, Lewis' or mockorange (*Philadelphus lewisii*)
SWW cp178; WFW cp188

T

Tahr, Himalayan (*Hemitragus jemlahicus*)
GEM V:542; MEM 584; MIA 157; WOM 139

Tahr, Nilgiri (*Hemitragus hylocrinus*)
GEM V:316

Taipan (*Oxyuranus scutellatus*)
LSW 279, 280

Takahe (*Notornis mantelli*)
MAE 114; MIA 237; WGB 47

Takin (*Budorcas taxicolor*)
CRM 199; GEM V:508; MIA 157

Talapoin (*Miopithecus talapoin*)
MEM 381; MIA 71; NHP 144 (bw)

Tallowtree (*Sapium sebiferum*)
NTE cp95, cp580

Tamandua (anteater) (*Tamandua* sp.)
GEM II:593, 596; MAE 72; MEM 774, 775; MIA 27

Tamarack (*Larix laricina*)
NTE cp18, cp482; NTW cp43, cp435; SEF cp35; WNW
  cp430

Tamaraw (*Bubalus mindorensis*)
CRM 185

Tamarillo (*Cyphomandra betacea*)
TSP 55

Tamarin, black and red (*Saguinus nigricollis*)
MIA 61; NHP 111(bw)

Tamarin, cotton-top (*Saguinus oedipus*)
CRM 65; GEM I:176, II:197; MEM 347; NHP 111 (bw);
  SAB 125

Tamarin, emperor (*Saguinus imperator*)
CRM 65; GEM II:183(bw), 196; MEM 348; MIA 61

Tamarin, Geoffrey's (*Saguinus geoffroyi*)
MEM 343

Tamarin, lion (golden) (*Leontopithecus rosalia rosalia*)
CRM 67; GEM II:184, 203-204; MEM 350; MIA 61; NHP
  112

Tamarin, lion (golden-headed) (*Leontopithecus
  chrysomelas*)
CRM 67

Tamarin, lion (golden-rumped or black) (*Leontopithecus
  rosalia chrysomelas*)
CRM 67; MEM 343

Tamarin, mustached (*Saguinus mystax* or *labiatus*)
MEM 343; SWF 186

Tamarin, pied (*Saguinus bicolor*)
CRM 65; GEM II:192

Tamarin, red-handed (*Sanguinus midas niger*)
GEM II:194

Tamarin, saddle-backed or brown-headed (*Saguinus
  fuscicollis*)
MEM 343, 345; WWW 56

Tamarind (*Tamarindus indica*)
PMT cp[134]

Tamarisk (*Tamarix chinensis*)
JAD cp304, cp324; NTW cp62, cp383

Tanager, blue-gray (*Thraupis episcopus*)
EAB 887; NAB cp442

Tanager, hepatic (*Piranga flava*)
EAB 888; FWB 432; UAB cp432, cp450

Tanager, rose-breasted thrush (*Rhodinocichla rosea*)
EOB 406

Tanager, scarlet (*Piranga olivacea*)
EAB 888, 889; FEB 386; MIA 375; NAB cp416, cp474;
NAW 208; PBC 354; SEF cp317; SPN 155; WGB 185;
WOB 107, 168, 169

Tanager, Summer (*Piranga rubra*)
EAB 889; FEB 387; FWB 433; NAB cp417, cp473; PBC
355; SEF cp318; SPN 153; UAB cp433, cp449, cp455;
WOB 41, 43

Tanager, swallow (*Tersina viridis*)
EOB 406

Tanager, Western (*Piranga ludoviciana*)
EAB 888; FWB 435; JAD cp599; NAB cp389, cp475; NAW
209; SPN 154; UAB cp434, cp454; WFW cp294

Tandan (*Tandanus tandanus*)
AWR 213(bw)

Tang, blue See Blue tang

Tanoak (*Lithocarpus densiflorus*)
NTW cp171, cp505; WFW cp170

Tansy (*Tanacetum vulgare*)
BCF 111; CWN 291; LBG cp183; MPC cp604
(*T.douglasii*); NWE cp355

Tapeworm (various genera)
AAL 510(bw); PGT 98(subadult)

Tapir, Baird's (*Tapirus bairdii*)
CRM 167; MEM 488

Tapir, Brazilian or lowland (*Tapirus terrestris*)
AAL 72; CRM 167; GEM IV:597(head only); MEM 489;
MIA 127; SWF 182

Tapir, Malayan (*Tapirus indicus*)
CRM 167; GEM IV:607; MEM 489; MIA 127; ROW 48(bw);
WWW 191

Tapir, mountain (*Tapirus pinchaque*)
CRM 167; GEM IV:607; MEM 489

Tarantula, desert (*Aphonopelma chalcodes*)
BOI 254-255; JAD cp399; MIS cp647

Tarantula hawk See under Wasp

Taro (*Colocasia esculenta*)
TSP 99

Tarpon (*Tarpon* sp.)
ACF p0, AQC cp402; EAL 25; MIA 501; NAF cp581

Tarsier (*Tarsius* sp.)
CRM 59; GEM II:97-103; MEM 319, 339; MIA 59; NHP
105; ROW 26(bw); WOI 111

Tasmanian devil (*Sarcophilus harrisi*)
GWM I:278, 279, 284, 285; MIA 17

Tattler (fish) (*Serranus phoebe*)
NAF cp521

Tattler, wandering (bird) (*Heteroscelus incanus*)
EAB 803; FWB 150; MPC cp576; UAB cp226

Tautog (*Tautoga onitis*)
ACF cp39; AGC cp338; MIA 565; NAF cp381

Tayra (*Eira barbara*)
GEM III:419; MIA 87

Tea, New Jersey (*Ceanothus americanus*)
NWE cp225; TSV 78, 79

Teaberry See Checkerberry

Teal, Baikal (*Anas formosa*)
BWB 75; DNA 42, 43

Teal, blue-winged (*Anas discors*)
AGC cp537; BWB 73, 78-79; DNA 66, 67; EAB 217; FEB
104; FWB 89; NAB cp136, cp142; NAW 210; UAB cp120,
cp153; WNW cp497

Teal, chestnut (*Anas castanea*)
BWB 74

Teal, cinnamon (*Anas cyanoptera*)
DNA 70, 71; EAB 217; FEB 108; FWB 93; NAB cp112,
cp141; UAB cp117, cp152; WNW cp495

Teal, falcated (*Anas falcata*)
DNA 36, 37

Teal, green-winged (*Anas crecca*)
DNA 45, 46; EAB 216; FEB 94; FWB 91; NAB cp105,
cp144; NAW 210; RBB 29; UAB cp121, cp150; WNW
cp490

Teal, Hottentot (*Anas punctata*)
BWB 74-75

Teal, marbled (*Marmaronetta angustirostris*)
DNA 76, 77; MAE 92

Teal, ringed (*Calonetta leucophrys*)
BWB 75

Teasel (*Dipsacus sylvestris*)
BCF 69; CWN 251; FWF 163; LBG cp252; NWE cp490;
SWW cp615; WAA 189; WTV 190

Tegu (*Tupinambis tequixin*)
AAL 229; ERA 100; MIA 429

Teledu (*Mydalus javanensis*)
MEM 130; ROW cp2

Tench, golden (*Tinca tinca*)
AWR 47; MIA 513

Tenrec, aquatic (*Limnogale mergulus*)
MEM 744

Tenrec, hedgehog (greater) (*Setifer setosus*)
MEM 745; MIA 31

Tenrec, hedgehog (lesser) (*Echinops telfairi*)
GEM I:439, 440; MEM 744, 746

Tenrec, long-tailed (*Microgale melanorrachis*)
CRM 37; MEM 745

Tenrec, rice (*Oryzoryctes tetradactylus*)
MEM 745; MIA 31

Tenrec, shrew (*Microgale longicaudata*)
MIA 31

Tenrec, streaked (*Hemicentetes nigriceps*)
MEM 745; MIA 31

Tenrec, tailless (*Tenrec ecaudatus*)
MEM 744, 747; MIA 31

Termite, dampwood or Pacific coast (*Zootermopsis angusticollis*)
MIS cp308; SGI cp39

Termite, subterranean (*Reticulitermes hesperus*)
MIS cp324; SGI cp40

Termite (other kinds)
AAL 415(bw); EOI 34-35; JAD cp392; SAB 121, 122; SWF 127

Tern, Aleutian (*Sterna aleutica*)
EAB 429; UAB cp50

Tern, arctic (*Sterna paradisaea*)
AGC cp513; BAA 13, 91(bw); BAS 133; BWB 162-163; EAB 428, 429; EOB 195; FEB 25; FWB 37; HSW 18, 145; MAE 82; NAB cp63; NAW 211; OBL 115; RBB 93; SAB 47; UAB cp46; WPR 151, 152

Tern, black (*Chlidonias niger*)
BWB 163; EAB 430; EOB 195; FEB 20; FWB 44; HSW 28(bw); NAB cp75; NAW 214; OBL 90; PBC 184; UAB cp43; WNW cp549

Tern, bridled or brown-winged (*Sterna anaethetus*)
EAB 431; HSW 19; NAB 72

Tern, Caspian (*Sterna caspia*)
AGC cp509; BWB 162; EAB 430, 431; EOB 195; FEB 23; FWB 39; HSW 138-139; MPC cp586; NAB cp65; NAW 212; UAB cp47; WNW cp552

Tern, common (*Sterna hirundo*)
AAL 130; AGC cp511; EAB 433; FEB 24; FWB 38; HSW 144; MAE 28; MIA 249; MPC cp585; NAB cp58, cp61; NAW 213; OBL 91; PBC 177; RBB 92; SAB 90-91; UAB cp44; WGB 59; WNW cp550

Tern, crested (*Sterna bergii*)
OBL 91

Tern, elegant (*Sterna elegans*)
BWB 163; EAB 433; FWB 41; MPC cp584; OBL 119; UAB cp49

Tern, Forster's (*Sterna forsteri*)
AGC cp512; BWB 160, 161; EAB 433; FEB 26; FWB 42; MPC cp583, cp588; NAB cp60, cp66; PBC 176; UAB cp45, cp52; WNW cp551

Tern, gull-billed (*Gelochelidon* or *Sterna nilotica*)
EAB 432; FEB 30; FWB 36; NAB cp68; PBC 176; UAB cp41

Tern, Inca (*Larosterna inca*)
BWB 159; EOB 194(head only); HSW 37; OBL 94, 95

Tern, large-billed (*Phaetusa simplex*)
EOB 195; HSW 22

Tern, little or least (*Sterna albifrons*)
AGC cp514; BWB 162; EAB 434, 435; FEB 28; FWB 43; HSW 27(bw); MPC cp587; NAB cp69; OBL 94; PBC 179; RBB 94; UAB cp42; WOB 97

Tern, noddy See Noddy

Tern, roseate (*Sterna dougallii*)
AGC cp508; EAB 436; FEB 31; NAB cp62

Tern, royal (*Sterna maxima*)
AGC cp510; BWB 158; EAB 437; FEB 27; FWB 40; NAB cp64; PBC 181; UAB cp48, cp51

Tern, Sandwich or Cabot's (*Sterna sandvicensis*)
EAB 437; FEB 29; HSW 145(bw); NAB cp59, cp67; PBC 182; RBB 91

Tern, sooty (*Sterna fuscata*)
AGC cp515; BWB 162; EAB 438; EOB 195; FEB 21; HSW 32-33; NAB cp71; PBC 178

Tern, South American (*Sterna hirundinacea*)
OBL 31

Tern, whiskered (*Chlidonias hybrida*)
HSW 143; MAE 93

Tern, white or fairy (*Gygis alba*)
AAB reverse title; EOB 194, 197; HSW 140-142; OBL 235; WOI 163

Tern, white-fronted (*Sterna striata*)
HSW 139

Terrapin, diamondback (*Malaclemys terrapin*)
AAL 205; AGC cp435; ARA cp298-cp300; ARC 160; CGW 69; MIA 405; OTT 54; RNA 61; TOW 225(bw); TTW 139

Terrapin, painted or Callagur (*Callagur borneoensis*)
OTT 56, 57; TOW cp8

Tetra, bleeding-heart (*Hyphessobrycon rubrostigma*)
AWR 185; EAL 73

Tetra, cardinal (*Cheirodon axelrodi*)
AWR 185

Tetra, congo (*Phenacogrammus interruptus*)
EAL 71

Tetra, glass (*Moenkhausia oligolepis*)
AWR 185

Tetra, red-nosed (*Hemigrammas rhodostomus*)
AWR 184

Tetra, serpa (*Hyphessobrycon serpae*)
AWR 185

Teyu (*Teius teyou*)
MIA 429

Thick-knee, great See Plover, great shore

Thick-knee, spotted (*Burhinus capensis*)
WBW 317

Thick-knee, water (*Burhinus vermiculatus*)
EOB 185; WBW 323

Thimbleberry (*Rubus parviflorus*)
WFW cp152, cp181

Thimbleweed (*Anemone virginiana*)
CWN 71; NWE cp66; WTV 29

Thistle, bull (*Cirsium vulgare*)
CWN 283; FWF 31; LBG cp254; NWE cp489

Thistle, Canada (*Cirsium arvense*)
LBG cp256; NWE cp485; WAA 188

Thistle, coyote (*Eryngium leavenworthii*)
LBG cp250; SWW cp497

Thistle, musk or nodding (*Carduus nutans*)
LBG cp255; NWE cp488; SWW cp500; WTV 188

Thistle, Russian See Saltwort

Thistle, showy (*Cirsium pastoris*)
LBG cp251; SWW cp421; WFW cp479

Thistle, sow (spiny-leaved) (*Sonchus asper*)
CWN 299; NWE cp299; WTV 58

Thistle, swamp (*Cirsium muticum*)
CWN 283

Thistle, yellow (*Cirsium horridulum*)
CWN 283; LBG cp162; NWE cp317; WTV 71

Thistle (other *Cirsium* sp.)
BCF 103; WAA 188, 225

Thorn apple See Jimsonweed

Thornback (*Platyrhinoidis triseriata*)
MPC cp316; NAF cp269; PCF p3

Thornbill, yellow-rumped (*Acanthiza chrysorrhoa*)
EOB 378

Thorny devil (*Moloch horridus*)
MIA 421; WWD front., 189; WWW 176

Thornyhead, shortspine (*Sebastolobus alascanus*)
MPC cp256; NAF cp412; PCF cp23

Thrasher, bendire's (*Toxostoma bendirei*)
EAB 577; FWB 415; JAD cp592; UAB cp534

Thrasher, brown (*Toxostoma rufum*)
AAL 170(bw); EAB 577; FEB 334; NAB cp494; NAW 4;
   PBC 270; SEF cp341; SPN 95

Thrasher, California (*Toxostoma vedivivum*)
EAB 576; FWB 414; MIA 331; UAB cp536; WGB 141

Thrasher, Crissal (*Toxostoma crissale*)
FWB 416

Thrasher, curve-billed (*Toxostoa curvirostre*)
EAB 577; FEB 336; FWB 412; JAD cp593; NAB cp496;
   UAB cp535

Thrasher, Le Conte's (*Toxostoma lecontei*)
EAB 577; FWB 417; JAD cp594; UAB cp537

Thrasher, long-billed (*Toxostoma longirostre*)
EAB 577; FEB 335; NAB cp495

Thrasher, sage (*Oreoscoptes montanus*)
EAB 577; FEB 337; FWB 413; JAD cp591; NAB cp497;
   UAB cp533

Threadfin or blue bobo (*Polydactylus* sp.)
ACF p41; PCF cp35

Thrift, California (*Armeria maritima*)
MPC cp618; SWW cp553

Thrip (various kinds)
MIS cp66; WIW 195(bw)

Thrush, Cape or olive (*Turdus olivaceus*)
WGB 147

Thrush, dusky (*Turdus eunous*)
NHU 104

Thrush, gray-backed (*Turdus hortulorum*)
NHU 130

Thrush, gray-cheeked (*Catharus minimus*)
EAB 892; FEB 331; FWB 423; NAB cp498; PBC 277; UAB
   cp539

Thrush, hermit (*Catharus guttatus*)
EAB 893; FEB 329; FWB 422; NAB cp501; PBC 275; SEF
   cp344; SPN 87; UAB cp541; WFW cp278

Thrush, island (*Turdus poliocephalus*)
WGB 147

Thrush, mistle (*Turdus viscivorus*)
RBB 141; SWF 49

Thrush, rock (*Monticola* sp.)
EOB 364; WOI 173; WOM 88

Thrush, song (*Turdus philomelos*)
BAS 116; EOB 369; MAE 122; RBB 139; SAB 143

Thrush, Swainson's or olive-backed (*Catharus ustulatus*)
EAB 892; FEB 330; FWB 421; NAB cp502; PBC 276; SEF cp345; SPN 86; UAB cp538; WOB 103

Thrush, varied (*Ixoreus naevius*)
EAB 895; FWB 419; SPN 90; UAB cp444; WFW cp280

Thrush, water See Waterthrush

Thrush, white's (*Zoothera dauma*)
NHU 114; WGB 147

Thrush, white-crested laughing (*Garrulax leucolophus*)
EOB 369

Thrush, Wilson's See Veery

Thrush, wood (*Hylocichla mustelina*)
AAL 170(bw); EAB 894; FEB 332; NAB cp500; PBC 274; SEF cp342; SPN 13, 88; WOB 7

Ti plant (*Cordyline terminalis*)
PFH 20

Tick, wood (*Dermacentor* sp.)
LBG cp378; MIS cp632; SEF cp423; WFW cp579

Tickseed (*Coreopsis tinctoria*)
WTV 117

Tickseed, giant (*Coreopsis gigantea*)
MPC cp602; SWW cp347

Tickseed, lance-leaved (*Coreopsis lanceolata*)
NWE cp294; WTV 77

Tick-trefoil (*Desmodium* sp.)
CWN 163; NWE cp148, cp533

Tidy tips (*Layia platyglossa*)
WAA 180

Tiger (*Panthera tigris*)
AAB 3; AWC 212-247; AWR 116; CRM 151; GEM III:576, IV:5-19; MEM 36-39; MIA 105; NHU 133 "Siberian"; SWF 77, 78; WOM 136-137

Tigerfish (*Hydrocynus* sp.)
EAL 81; MIA 551

Tilapia, mozambique (*Tilapia mossambica*)
NAF cp77, cp523

Tilefish (several genera)
ACF p41; NAF cp471

Tillandsia See Pine, wild

Timema (*Timema* sp.)
MIS cp84; SGI cp32

Tinamou, crested (*Eudromia elegans*)
EOB 28; WPP 181

Tinamou, great (*Tinamus major*)
MIA 197; WGB 7

Tinamou, undulated (*Crypturelles undulatus*)
EOB 29

Tit, azure (*Parus cyanus*)
EOB 383

Tit-babbler, chestnut-headed (*Alcippe castaneceps*)
EOB 368

Tit, bearded (*Panurus biarmicus*)
RBB 156; WGB 153

Tit, blue (*Parus caeruleus*)
AAB 111; EOB 383; MAE 31; RBB 162

Tit, coal (*Parus ater*)
EOB 384; RBB 161

Tit, crested (*Parus cristatus*)
RBB 160

Tit, great (*Parus major*)
EOB 389; RBB 163; SAB 79

Tit, long-tailed (*Aegithalos caudatus*)
RBB 157

Tit, marsh (*Parus palustris*)
AAL 177; EOB 385; RBB 158

Tit, penduline (*Remiz pendulinus*)
EOB 387

Tit, rufous-bellied (*Parus rufiventris*)
EOB 382

Tit, shrike (*Falcunculus frontatus*)
EOB 380

Tit, southern black (*Parus niger*)
EOB 384

Tit, willow (*Parus montanus*)
RBB 159

Tit, yellow-cheeked (*Parus spilonotus*)
EOB 383

Titi (plant) (*Cyrilla racemiflora*)
NTE cp67, cp404; NWE cp222; WNW cp243, cp450

Titi (monkey) (*Callicebus molochi* or *torquatus*)
GEM II:130, 132-135; MEM 354, 366; MIA 63; NHP 116(bw)

Titi, masked (monkey) (*Callicebus personatus*)
CRM 69

Titmouse, bridled (*Parus wollweberi*)
EAB 898; FWB 351; UAB cp489

Titmouse, plain (*Parus inornatus*)
EAB 898; FWB 350; SPN 69; UAB cp486; WFW cp263

Titmouse, tufted (*Parus bicolor*)
EAB 898; EOB 386; FEB 279; NAB cp432; PBC 253; SEF cp329; SPN 68; WOM 158

Titoki (*Alectryon excelsus*)
PMT cp[14]

Toad, American (*Bufo americanus*)
AAL 301; ARA cp237; ARC 106, 107; ARN 31(bw); CGW 25; MIA 479; NAW 280, 287; SEF cp544

Toad, black (*Bufo exsul*)
ARA cp245; WNW cp146

Toad, Boreal (*Bufo boreas boreas*)
ARA cp226

Toad, burrowing (*Rhinophyrynus dorsalis*)
ARA cp217; ART p[4]; ERA 43; MIA 473

Toad, Canadian (*Bufo hemiophrys*)
ARA cp239; WNW cp145; WRA p12

Toad, cane See Toad, marine

Toad, Colorado River or Sonoran desert (*Bufo alvarius*)
ARA cp225; JAD cp281; WRA cp11

Toad, common (*Bufo bufo*)
AAL 302(bw); FTW p3; MIA 479; PGT 216, 217

Toad, European green (*Bufo viridis*)
FTW cp83

Toad, forest (*Bufo asper*)
FTW cp5

Toad, Fowler's (*Bufo woodhousei fowleri*)
AGC cp437; ARA cp248; ARC 110; ARN 32(bw); CGW 25; SEF cp539

Toad, giant See Toad, marine

Toad, golden (*Bufo periglenes*)
ERA reverse title pg.; SAB 92

Toad, Great Plains (*Bufo cognatus*)
ARA cp247; JAD cp283; LBG cp570; WRA cp11

Toad, green (*Bufo debilio*)
AAL 301; ARA cp254, cp255; JAD cp287; LBG cp573; MIA 479; WRA cp10

Toad, Gulf Coast (*Bufo valliceps*)
AAL 302(bw); AGC cp436; ARA cp246

Toad, Houston (*Bufo houstonensis*)
ARA cp228; ART p[7]

Toad, marine (*Bufo marinus*)
ARA cp242; ERA 46; FTW p74; MIA 479; WWW 192

Toad, midwife (*Alytes* sp.)
ERA 42; FTW p58, cp77; MIA 475

Toad, narrow-mouthed (*Cophixsalis* sp.)
WWW 102

Toad, natterjack (*Bufo calamita*)
AAL 300; MAE 29; MIA 479

Toad, oak (*Bufo quercicus*)
AAL 304; ARA cp250; ARC 108

Toad, Oriental fire-bellied (*Bombina orientalis*)
ERA 12, 42

Toad, red-spotted (*Bufo punctatus*)
ARA cp234; FTW cp82; JAD cp278; LBG cp568; WRA p12

Toad, Sonoran Green (*Bufo retriformis*)
AAL 305; ARA cp251; JAD cp284; WRA cp10

Toad, Southern (*Bufo terrestris*)
ARA cp236, cp241; ARC 109; ERA 49; SEF cp545

Toad, Southwestern (*Bufo microscaphus*)
AAL 305; ARA cp223, cp235; JAD cp279; WRA p12

Toad, spadefoot (*Megophrys monticolo nasuta*)
WWW 158

Toad, spadefoot (Couch) (*Scaphiopus couchii*)
AAL 282(bw); ARA cp252; ERA 43; FTW cp79; JAD cp274; WRA cp10

Toad, spadefoot (Eastern) (*Scaphiopus* or *Pelobates* sp.)
AAL 282(bw), 284; ARA cp233; ARC 104; ARN 30(bw); ERA 43; CGW 25; MIA 475

Toad, spadefoot (Great Basin) (*Scaphiopus intermontanus*)
AAL 284; ARA cp232; JAD cp277; WRA cp10

Toad, spadefoot (Plains) (*Scaphiopus bombifrons*)
ARA cp231; JAD cp275; LBG cp567; WRA cp10

Toad, spadefoot (Western) (*Scaphiopus hammondii*)
ARA cp229, cp230, cp253; JAD cp276; LBG cp566; MIA 475; WRA cp10

Toad, Surinam (*Pipa pipa*)
AAL 280; AWR 175-177; ERA 42; FTW cp61; MIA 473

Toad, Texas (*Bufo speciosus*)
ARA cp238; JAD cp280; WRA p12

Toad, Western (*Bufo boreas*)
ARA cp240; WFW cp617; WRA cp11 -

Toad, woodhouse (*Bufo woodhousei*)
ARA cp224, cp249; LBG cp569; NAW 279; WRA p12

Toad, Yosemite (*Bufo canorus*)
ARA cp227, cp244; WRA cp11

Toadfish (*Opsanus* sp.)
ACF p14; AGC cp427; EAL 89; NAF cp460

Toadflax, common See Butter-and-eggs

Toadflax, blue (*Linaria canadensis*)
NWE cp627; SWW cp621

Tobacco, desert (*Nicotiana trigonophylla*)
JAD cp103; SWW cp95

Tobacco, Indian (*Lobelia inflata*)
NWE cp633

Tobacco, tree (*Nicotiana glauca*)
JAD cp114; SWW cp341

Tobaccoweed (*Atrichoseris platyphylla*)
JAD cp85; SWW cp75

Tody, Jamaica (*Todus todus*)
EOB 270; WOI 183

Tomcod (*Microgadus* sp.)
ACF p15; MPC cp292; NAF cp490; PCF p11

Tomtate (*Haemulon aurolineatum*)
ACF cp32; BMW front.; NAF cp535

Tonguefish (*Symphurus* sp.)
ACF p58; PCF p43

Tooth fungus (*Hericium* sp.)
NAM cp547-549

Toothwort (*Dentaria* sp.)
CWN 103; NTE cp164

Topi or tsessebe (*Damaliscus* sp.)
GEM I:174, V:302, 420, 422; MEM 563; NHA cp21

Topminnow (*Fundulus* sp.)
NAF cp202, cp203, cp211, cp213; WNW cp65

Topminnow, gila (*Poeciliopsis occidentalis*)
JAD cp295; NAF cp204

Topsmelt (*Atherinops affinis*)
MPC cp294; NAF cp570; PCF p12

Tor mahseer (*Barbus tor*)
AWR 214(bw); MIA 513

Torpedo, Atlantic (*Torpedo nobiliana*)
MIA 495; NAF cp273

Torreya, California (*Torreya californica*)
NTW cp50, cp496; WFW cp145

Torreya, Florida (*Torreya taxifolia*)
NTE cp30

Tortoise, Asian brown (*Manouria emys*)
TOW 241(bw)

Tortoise, berlandier's or Texas (*Gopherus berlandieri*)
AAL 207(bw); ARA cp329; RNA 63

Tortoise, desert (*Gopherus agassizii*)
ARA cp328; JAD cp169; RNA 63; WRA p18; WWD 166-167

Tortoise, giant (Aldabran or Seychelles) (*Geochelone gigantea* or *megalochelys*)
OTT 19(bw); TOW 250(bw); TTW 153(bw)

Tortoise, giant (Galapagos) (*Chelonoidis* or *Geochelone elephantopus*)
ERA title, 75; MAE 115; MIA 407; OTT 16, 17, 20(bw); TTW 149(bw); WWW 171

Tortoise, giant (Galapagos) (*Testudo elephantophus*)
AAL 208; WOI 30

Tortoise, gopher (*Gopherus polyphemus*)
AAL 207(bw); ARA cp330; ARC 162; ERA 79; MIA 407; NAW 270; RNA 63; TOW 272(bw); TTW 38

Tortoise, Greek or marginated (*Testudo marginata*)
TTW 160(bw)

Tortoise, Hermann's (*Testudo hermanni*)
TOW 267(bw); TTW 159; WOI 171

Tortoise, hingeback (*Kinixys* sp.)
TTW 39, 158

Tortoise, leopard (*Geochelone pardalis*)
MIA 407; OTT 144(bw); TOW cp16; TTW 23

Tortoise, Madagascar Radiated (*Asterochelys radiata*)
OTT 140, 141

Tortoise, Mediterranean spur-thighed (*Testudo graeca*)
TTW 157(bw)

Tortoise, pancake (*Malacochersus tornieri*)
OTT 99; TTW 22

Tortoise, spurred (*Geochelone sulcata*)
TTW 70, 155

Toucan, Cuvier's (*Ramphastos tucanus*)
EOB 290

Toucan, toco (*Ramphastos toco*)
AAL 151; AWR 166; BAS 124; EOB 287, 289; MIA 291; WGB 101

Toucanet, emerald (*Aulacorhynchus prasinus*)
EOB 286

Toucanet, saffron (*Andigena bailloni*)
EOB 287

Touch-me-not, orange (*Impatiens capensis*)
BCF 169; CWN 167; FWF 17; NWE cp376; WAA 158;
 WNW cp279; WTV 105

Touch-me-not, pale (*Impatiens pallida*)
NWE cp315

Towhee, Abert's (*Pipilo aberti*)
EAB 324; FWB 474; JAD cp606; UAB cp564

Towhee, brown (*Pipilo fuscus*)
EAB 324; FWB 446; JAD cp607; UAB cp563; WFW cp299

Towhee, green-tailed (*Pipilo chlorurus*)
EAB 325; FWB 462; JAD cp605; SPN 127; UAB cp524;
 WFW cp297

Towhee, rufous-sided (*Pipilo erythrophthalmus*)
EAB 324, 325; FEB 400; FWB 475; NAB cp401, cp520;
 NAW 217; PBC 375; SEF cp308; SPN 128; UAB cp442;
 WFW cp298; WGB 181

Toyon (*Heteromeles arbutifolia*)
NTW cp167, cp350, cp476; WFW cp159

Tragopan (*Tragopan* sp.)
AAL 118; BAS 17; MIA 231; WGB 41; WOM 127

Trailing arbutus See Arbutus, trailing

Tread-softly or spurge-nettle (*Cnidoscolus stimulosus*)
NWE cp165

Tree ear fungus (*Auricularia auricula*)
FGM cp5; GSM 41; NAM cp617; SGM cp321

Tree of Heaven (*Ailanthus altissima*)
FWF 89; NTE cp313, cp385, cp496; NTW cp288, cp344,
 cp529; OET 241; TSV 28, 29

Treeduck See Whistling duck

Treefish (*Sebastes serriceps*)
NAF cp394

Treefoil, bird's-foot See Bird's-foot trefoil

Treefrog (*Hyla* sp.)
ARA cp158; ERA 36; SAB reverse title, 67

Treefrog (*Rhacophorus* sp.)
ERA 50; FTW cp29

Treefrog, barking (*Hyla gratiosa*)
AAL 308; ARA cp145, cp151; ARC 120; FTW cp87

Treefrog, bird-voiced (*Hyla avivoca*)
ARA cp156; ARC 114; WNW cp147

Treefrog, burrowing (*Pternohyla fodiens*)
ARA cp186; JAD cp273; WRA cp16

Treefrog, California (*Hyla cadaverina*)
ARA cp158; JAD cp271

Treefrog, canyon (*Hyla arenicolor*)
ARA cp159; JAD cp272; WRA cp16

Treefrog, Cope's Gray (*Hyla versicolor*)
ARA cp160; ARN 34(bw); CGW 25; SEF cp540

Treefrog, Cuban (*Osteopilus septentrionalis*)
ARA cp155, cp178

Treefrog, European (*Hyla arborea*)
AAL 312; AWR 53; ERA 57; NNW 87; PGT 215; SAB 83

Treefrog, Gray (*Hyla chrysoscelis*)
AAL 311(bw); ARA cp152, cp157; ARC 115; FTW cp20;
 MIA 481; NAW 289

Treefrog, green (*Hyla cinerea*)
AAL 312; ARA cp146; ARC 117; SEF cp535

Treefrog, Mexican (*Smilisca baudini*)
ARA cp181, cp184; ART p[6]

Treefrog, mountain (*Hyla eximia*)
ARA cp150; WRA cp16

Treefrog, Pacific (*Hyla regilla*)
AAL 309; ARA cp148, cp170, cp182; FTW p91; WFW
 cp618; WRA cp16

Treefrog, Pine Barrens (*Hyla andersoni*)
AAL 306(bw), 308-309; ARA cp149; ARC 113; NAW 5,
 289; WNW cp138

Treefrog, pine woods (*Hyla femoralis*)
ARA cp164; ARC 119; SEF cp538

Treefrog, red (*Litoria* sp.)
AAB 84-85; AAL 313; AWR 127

Treefrog, squirrel (*Hyla squirella*)
ARA cp147, cp174; ARC 121

Treefrog, spring peeper (*Hyla crucifer*)
AAL 309; ARA cp173; ARC 118; ARN 33(bw); CGW 25;
 MIA 481; NAW 287; SEF cp546; WNW cp150

Treefrog, White's (*Litoria caerulea*)
FTW cp89

Treehopper (various genera)
EOI 54; MIS cp109, cp110; SGI cp67-cp69

Triggerfish (*Cantherhines* sp.)
ACF cp59; RUP 129

Triggerfish (*Balistoides* sp.)
MIA 581; RUP 72; WWF cp23

Triggerfish, blue (*Pseudobalistes fuscus*)
BMW 140-141

Triggerfish, gray (*Balistes capriscus*)
ACF cp59

Triggerfish, queen (*Balistes* sp.)
ACF cp59; HSG cp51; KCR cp26; MIA 581; NAF cp330

Triller, white-winged (*Lalage suerii*)
EOB 348

Trillium (*Trillium pusillum*)
WNW cp443

Trillium, large-flowered (*Trillium grandiflorum*)
CWN 7; NWE cp77; SEF cp446; WAA 154

Trillium, nodding (*Trillium cernuum*)
NWE cp75

Trillium, painted (*Trillium undulatum*)
CWN 7; NWE cp76; SEF cp447; WAA 154

Trillium, red (*Trillium sessile*)
NWE cp383

Trillium, snow (*Trillium nivale*)
WAA 154

Trillium, yellow (*Trillium luteum*)
WAA 155

Tripletail (*Lobotes surinamensis*)
AGC cp376; MIA 557; NAF cp517

Triton trumpet (*Charonia tritonis*)
RUP 50

Trixis (*Trixis californica*)
JAD cp106; SWW cp343

Trogon (*Trogon* sp.)
EAB 900; FWB 265; MIA 281; UAB cp457; WFW cp240;
    WGB 91

Tropicbird, red-billed (*Phaethon aethereus*)
EAB 901; OBL 30, 71

Tropicbird, red-tailed (*Phaethon rubricauda*)
AAL 97; BAS 144; EOB xi, 62; FWB 49; MIA 205; UAB
    cp56; WGB 15

Tropicbird, white-tailed (*Phaethon lepturus*)
EAB 901; EOB 53; FEB 53; HSW 116; NAB cp90; OBL
    230

Trout, Apache (*Salmo apache*)
NAF cp59

Trout, brook (*Salvelinus fontinalis*)
NAF cp65, cp498; NAW 313; WNW cp89

Trout, brown river (*Salmo trutta fario*)
AWR 46-47, 217(bw); MIA 507; NAF cp62, cp496; WNW
    cp90

Trout, cutthroat (*Salmo clarki*)
MPC cp284, cp288; NAF cp60, cp64, cp495; NAW 314;
    PCF p8; WNW cp92

Trout, Dolly Varden (*Salvelinus malma*)
ACF p8

Trout, Indian (*Barilius bola*)
AWR 214(bw)

Trout, lake (*Cristivomer* or *Salvelinus namaycush*)
AAL 331; MIA 507; NAF cp66; WNW cp88

Trout, rainbow or steelhead (*Salmo gairdneri*)
EAL 52; MIA 507; NAF cp61, cp493; NAW 314; PCF p8;
    WNW cp91

Trout, sea (*Cynoscion* sp.)
ACF p35; NAF cp476, cp482

Trout-lily (*Erythronium americanum*)
BCF 148; CWN 5; NWE cp278; SEF cp463; WAA 158

Trout-perch (*Percopsis omiscomaycus*)
EAL 87

Truffle (*Tuber* sp.)
SGM cp388-391

Trumpet creeper (*Campsis radicans*)
NWE cp381

Trumpet tree (*Tabebuia* sp.)
PMT cp[133]; TSP 41

Trumpet vine, golden (*Allamanda cathartica*)
TSP 81

Trumpeter, common (*Psophia crepitans*)
WGB 45

Trumpeter, white-winged (*Psophia leucoptera*)
AAL 121(bw); EOB 145

Trumpetfish (*Aulostomus* sp.)
BMW 132; NAF cp438; WWF cp17, cp18

Trumpets (*Sarracenia flava*)
NWE cp312; WNW cp224

Trunkfish (*Lactophrys* sp.)
ACF cp60; KCR cp24; NAF cp312-cp314; RUP 134; WWF
    cp22

Tsessebe See Topi

Tuatara (*Splenodon punctatus*)
ERA 134, 135; MIA 417

Tube-snout (*Aulorhynchus flavidus*)
MIA 541; NAF cp437

Tuco-tuco (*Cienomys opimus*)
GEM III:347; MEM 703; MIA 187

Tui (*Prosthemadera novaeseelandica*)
AAL 182(bw)

Tulip, Mariposa See Lily, Mariposa

Tulip tree See Poplar, yellow

Tulip tree, African (*Spathodea* sp.)
TSP 39; WOI 86

Tumblebug See Dung roller

Tuna, albacore (*Thunnus albalunga*)
ACF cp48; NAF cp567; PCF cp32

Tuna, blackfin (*Thunnus atlanticus*)
ACF cp49

Tuna, bluefin (*Thunnus thynnus*)
ACF cp49; PCF cp32

Tuna, skipjack (*Euthynnus pelamis*)
AAL 370(bw); ACF cp48; MIA 571; PCF cp32

Tuna, yellowfin (*Thunnus albacares*)
ACF cp49; MIA 571; MPC cp282; NAF cp561; PCF cp32

Tunicate See Sea squirt

Tunny (*Euthynnus alletteratus*)
ACF cp48; NAF cp566

Tupelo, black (*Nyssa sylvatica*)
NTE cp72, cp540, cp583; OET 219; PMT cp89; SEF cp57; TSV 154, 155; WNW cp443

Tupelo, water (*Nyssa aquatica*)
NTE cp198; SEF cp82; WNW cp462

Tur (*Capra cylindricornis*)
NHU 161

Turaco, red-crested (*Tauraco erythrolophus*)
EOB 235; WGB 77

Turbot (*Pleuronichthys* sp.)
PCF p44

Turbot (*Scophthalmus maximus*)
MIA 575

Turbot, diamond (*Hypsopsetta guttulata*)
MPC cp311; NAF cp281

Turkey (*Meleagris gallopavo*)
EAB 910, 911; EOB 131, 135; FEB 213; FWB 203; MIA 233; NAB cp269, cp273; NAW 219; PBC 122; SEF cp249; UAB cp266; WGB 43; WNW cp567; WOB 256, 257

Turkey, Australian brush (*Alectura lathami*)
AAL 115; BAS 73; EOB 135

Turkey beard (*Xerophyllum asphodeloides*)
CWN 15; NWE cp139

Turkey, water See Anhinga

Turkeyfish See Lionfish

Turnstone, black (*Arenaria melanocephala*)
EAB 803; FWB 160; MPC cp577; UAB cp221, cp225

Turnstone, ruddy (*Arenaria interpres*)
AGC cp583; BWB 176-177; EAB 803; FEB 157; FWB 161; MIA 243; NAB cp196, cp241; PBC 150; RBB 82; UAB cp185, cp186; WBW 259; WGB 53

Turtle, African forest (*Pelusios gabonensis*)
TOW cp1

Turtle, big-headed (*Platysternon megachephalum*)
ERA 79; MIA 409; OTT 91(bw); TOW cp8, 134(bw); TTW 167

Turtle, Blanding's (*Emydoidea blandingi*)
AAL 203(bw); ARA cp291; ARN 53(bw); CGW 73; RNA 45; TOW cp12

Turtle, bog (*Clemmys muhlenbergi*)
AAL 203(bw); ARA cp301; ARC 157; ARN 45(bw); CGW 69; RNA 43; TOW cp13, 191(bw); WNW cp165

Turtle, Bornean River or Malaysian giant (*Orlitia borneansis*)
OTT 90(bw); TOW 177(bw)

Turtle, box (Eastern) (*Terrapene carolina*)
AAL 205; ARA cp304, cp306, cp308, cp309; ARC 161; ARN 47(bw); CGW 69; MIA 405; OTT 54; RNA 47; SEF cp512; TOW 195(bw); TTW 131

Turtle, box (ornate or Western) (*Terrapene ornata*)
ARA cp305, cp307; JAD cp173; LBG cp574; NAW 271; RNA 47; TOW 196(bw); WRA p17

Turtle, Caspian (*Mauremys caspica*)
TTW 141(bw)

Turtle, Chinese pond or Reeve's (*Chinemys reevesi*)
OTT 129; TOW 145(bw); TTW 138

Turtle, chicken (long-necked) (*Deirochelys reticularis*)
ARA cp285; ARC 158; OTT 54; RNA 45; TOW 216(bw); TTW 129(bw)

Turtle, cooter See Turtle, red-eared

Turtle, Fitzroy (*Rheodytes leuceps*)
OTT 62(bw); TOW 48(bw)

Turtle, Florida cooter (*Chrysemys floridana*)
ARA cp288; ARC 151; RNA 59; TOW 212(bw)

Turtle, green (*Chelonia mydas*)
ARA cp267; ARC 166; ERA 70; HOO cp50; MAE 21, MIA 411; NAW 271, 272; OTT 12; RCK 134; RNA 37; RUP 87; TOW 120(bw); WRA p19

Turtle, softshell (Florida) (*Apalone* or *Trionyx ferox*)
ARA cp272, cp273; ARC 171; NAW 275; RNA 33; TOW 103(bw)

Turtle, softshell (smooth) (*Apalone* or *Trionyx muticus*)
ARA cp268, cp269; RNA 33; TOW 106(bw)

Turtle, softshell (spiny or Spring) (*Apalone* or *Trionyx spiniferus*)
AAL 210(bw); ARA cp270, cp271; ARC 172; ARN 54(bw); CGW 73; ERA 80; JAD cp174; MIA 413; OTT 139(bw); RNA 33; TOW cp6, 104(bw); WNW cp170, cp171; WRA p18; WWW 194

Turtle, spined (*Heosemys spinosa*)
OTT 143; TOW cp10; TTW 136(bw)

Turtle, spotted (*Clemmys guttata*)
ARA cp290; ARC 155; ARN 44(bw); CGW 69; OTT 103; RNA 43; TOW cp13; WNW cp156

Turtle, twisted-neck or flatshelled (*Platemys platycephala*)
OTT 93; TOW cp3, 61(bw)

Turtle, wood (*Clemmys insculpta*)
ARA cp302; ARC 156; ARN 46(bw); CGW 69; MIA 405; NAW 277; NNW cp166; OTT 53; RNA 43; SEF cp513; TOW 190(bw)

Turtleback See Desert velvet

Turtledove (*Streptopelia turtur*)
EOB 218; RBB 102

Turtledove, ringed (*Streptopelia risoria*)
EAB 711; FEB 193; FWB 178; NAB cp323; UAB cp350

Turtlehead (*Chelone glabra*)
BCF 117; NWE cp113; WNW cp231

Twinberry (*Lonicera involucrata*)
WFW cp164, cp205

Twinflower (*Linnaea borealis*)
NWE cp506; SEF cp472; SWW cp507; WAA 161; WFW cp484

Twinleaf (*Cassia bauhinioides*)
JAD cp144; LBG cp188; SWW cp194

Twinleaf (*Jeffersonia diphylla*)
FWF 5; NWE cp88

Twisted stalk, rosy (*Streptopus rosens*)
NWE cp505; SWW cp379; WFW cp459

Twisted stalk, white See Mandarin, wild

Tyrant, pied water (*Fluvicola pica*)
EOB 319

Tyrannulet, northern beardless (*Camptostoma imberbe*)
FWB 396

U

Uakari, black (*Cacajao melanocephalus*)
GEM II:143

Uakari, red or white or bald (*Cacajao rubicundus* or *calvus*)
AAL 24; CRM 69; GEM II:139-142; MEM 352, 354; MIA 63; NHP 121; SWF 186

Umbrella bird, ornate or long-wattled (*Cephalopterus ornatus*)
AWR 169; MIA 303; SPN 19(bw); WGB 113

Umbrella plant (*Peltiphyllum peltatum*)
SWW cp552; WFW cp498; WNW cp233

Umbrella tree See Magnolia, bigleaf

Umbrella tree, Australian (*Schefflera actinophylla*)
TSP 17

Umbrella tree, Japanese (*Sciadopitys verticillata*)
PMT cp[122]

Unicornfish (*Eumecichthys fiski*)
ACF cp22

Urial or Arkal (*Ovis orientalis*)
CRM 203; GEM V:548(bw); MEM 585

Urutu or wutu (*Bothrops alternatus*)
AAL 261; LSW 385

V

Valerian (*Polemonium reptans*)
NWE cp641

Vampire bat (*Desmodeus rotundus*)
GEM I:568; LOB 56; MEM 812, 813; MIA 47; WWW 59

Vampire bat, false (*Vampyrum spectrum*)
LOB 52

Vampire bat, false (greater) (*Megaderma lyra*)
GEM I:14-15, 562-563; LOB 50(bw); MEM 801; MIA 43

Vanga See Sicklebill

Vanilla See Orchid, vanilla

Vanilla leaf (*Achyls triphylla*)
SWW cp107; WFW cp419

Vase flower (*Clematis hirsutissima*)
LBG cp232, cp233; SWW cp381, cp508; WFW cp464

Veery (*Catharus fuscescens*)
EAB 895; FEB 333; FWB 420; MIA 337; NAB cp499; PBC
277; SEF cp343; SPN 85; UAB cp540; WGB 147

Velvet ant See under Wasp

Velvet leaf (*Abutilon theophrasti*)
CWN 89; NWE cp256

Venus-fly-trap (*Dionaea muscipula*)
NWE cp176; WNW cp249

Venus'-looking-glass (*Triodanis perfoliata*)
CWN 207; NWE cp601; SWW cp629

Verbena, sand (*Abronia* sp.)
JAD cp53; MPC cp605; SWW cp542

Verdin (*Auriparus flaviceps*)
EAB 899; EOB 383; FEB 276; FWB 356; JAD cp583; MIA
357; NAB cp381; UAB cp422; WGB 167

Vervain, blue (*Verbena hastata*)
CWN 215; LBG cp274; NWE cp629; WTV 182

Vervain, rose (*Verbena tampensis*)
NWE cp550; WAA 163

Vervain, Western pink (*Verbena ambrosifolia*)
LBG cp243

Vervet See under Monkey

Vetch, American (*Vicia americana*)
SWW cp506; WFW cp488

Vetch, common (*Vicia angustifolia*)
CWN 153; WTV 169

Vetch, cow or bird or tufted (*Vicia cracca*)
LBG cp266; NWE cp534; SWW cp519

Vetch, crown (*Coronilla varia*)
CWN 167; NWE cp572

Vetch, hairy (*Vicia villosa*)
NWE cp532

Vetchling (*Lathyrus venosus*)
WTV 174

Viburnum (*Viburnum* sp.)
NTE cp80, cp410; NWE cp214; WFW cp189

Vicuna (*Vicugna vicugna*)
CRM 173; GEM V:98, 99; MEM 515; MIA 133; WOM 182

Vieja (*Serranus dewegeri*)
ACF p63; MIA 523

Vinegar weed (*Trichostema lanceolatum*)
LBG cp277; SWW cp640

Violet, arrow-leaved (*Viola sagittata*)
CWN 171

Violet, bird-foot (*Viola pedata*)
CWN 169; LBG cp286; NWE cp598; SEF cp491; WTV 167

Violet, blue (*Viola adunca*)
SWW cp588; WFW cp518

Violet, blue (common) (*Viola papilionacea*)
BCF 5; CWN 167; NWE cp597; SEF cp492

Violet, Canada (*Viola canadensis*)
CWN 173; NWE cp64; SWW cp37; WAA 153; WFW cp384

Violet, dog (*Viola conspersa*)
CWN 173; NWE cp595

Violet, dogtooth (yellow) See Trout-lily

Violet, downy (northern) (*Viola fimbriatula*)
WAA 152

Violet, downy yellow (*Viola pubescens*)
NWE cp274; WAA 153

Violet, evergreen (*Viola sempervirens*)
SWW cp189; WFW cp449

Violet, lance-leaved (*Violet lanceolata*)
CWN 171

Violet, long-spurred (*Viola rostrata*)
WAA 153

Violet, marsh (*Viola palustris*)
AWR 43

Violet, marsh blue (*Viola cucullata*)
CWN 169; WAA 153

Violet, stream (*Viola glabella*)
SWW cp190; WAA 153; WFW cp450

Violet, sweet white (*Viola blanda*)
CWN 171; NWE cp65; SEF cp442; WAA 153

Violet, water See Featherfoil

Violet, yellow (*Viola rotundifolia* or *pensylvanica*)
CWN 171, 173

Violet (other *Viola* sp.)
CWN 173; SWW cp192, cp193, cp586, cp587; WAA 239

Viper See also Adder

Viper, asp (*Vipera aspis*)
ERA 117; LSW 329, 330

Viper, Avicenna (*Cerastes vipera*)
LSW 323

Viper, bush or leaf (*Atheris squamiger*)
LSW 303-305

Viper, Gaboon (*Bitis gabonica*)
ERA 127; LSW 297, 312; MIA 457; MSW cp26; WWW 159

Viper, horned (*Cerastes cerastes*)
LSW 321, 322; MSW p96

Viper, Palestinian (*Vipera palaestinae*)
AAL 255(bw); LSW 339

Viper, pit (various genera)
AAB 90; LSW 361, 364, 369-372, 389-392; MSW cp97

Viper, rhinoceros (*Bitis nasicornis*)
LSW 314

Viper, Russell's (*Vipera russellii*)
AAL 255(bw); LSW 343, 344

Viper, sand (*Vipera ammodytes*)
LSW 327; MIA 457; MSW p94

Viper, saw-scaled (*Echis carinatus*)
NHU 183

Viper, sharp-nosed (*Deinagkistrodon acutus*)
LSW 359, 360

Viper, sidewinder or dwarf puff adder (*Bitis peringueyi*)
AAB 103; AAL 256-257; ERA 1(bw); LSW 316, 317; MIA
    457; WWD 66, 70(head only)

Viper (other kinds)
LSW 299, 301, 326, 336-342, 367, 375-378-381

Viperfish, deepsea or sloane's (*Chaulodius sloani*)
BMW 186-187; EAL 65; MAE 111; MIA 509; WWW 182-
    185

Vireo, Bell's (*Vireo bellii*)
EAB 913; FEB 476; FWB 383; NAB cp449; SPN 102; UAB
    cp508

Vireo, black-capped (*Vireo atricapillus*)
FEB 466

Vireo, black-whiskered (*Vireo altiloquus*)
AGC cp615; EAB 912; FEB 474; NAB cp452

Vireo, gray (*Vireo vicinior*)
FWB 385

Vireo, Hutton's (*Vireo huttoni*)
FWB 382

Vireo, Philadelphia (*Vireo philadelphicus*)
EAB 912; FEB 471; NAB cp451

Vireo, red-eyed (*Vireo olivaceus*)
AAL 187(bw); EAB 913; EOB 413; FEB 473; FWB 387;
    MIA 381; NAB cp453; NAW 221; SEF cp283; SPN 24(bw),
    101; UAB cp515; WGB 191

Vireo, solitary (*Vireo solitarius*)
EAB 914; FEB 465; FWB 375; NAB cp450; SEF cp289;
    UAB cp516; WFW cp283

Vireo, warbling (*Vireo gilvus*)
EAB 915; FEB 475; FWB 386; NAB cp454; SPN 100; UAB
    cp514; WFW cp284

Vireo, white-eyed (*Vireo griseus*)
AAL 187(bw); EAB 915; FEB 467; NAB cp383; SEF cp285

Vireo, yellow-throated (*Vireo flavifrons*)
EAB 914; FEB 470; NAB cp362; SEF cp290

Virginia creeper (*Parthenocissus quinquefolia*)
WTV 2

Virgin's bower (*Clematis* sp.)
CWN 69; FWF 25; NWE cp38, cp206; SWW cp172, cp662

Viscacha, mountain (*Lagidium pervatum*)
CRM 129; WOM 186

Viscacha, plains (*Lagostomus maximus*)
CRM 129; GEM III:317-320; MIA 187

Vole, bank (*Clethrionomys glareolus*)
GEM II:247; MEM 650; MIA 173; NNW 34, 46; SWF 66

Vole, common (*Microtus arvalis*)
GEM III:241

Vole, field (*Microtus agrestis*)
GEM III:242

Vole, ground (*Arvicola terrestris*)
AAL 44

Vole, long-tailed (*Microtus longicaudus*)
ASM cp71; MPS 244(bw); WFW cp326

Vole, meadow (*Microtus pennsylvanicus*)
ASM cp76; CGW 185; CMW 165(bw); GEM III:244; GMP
    cp11, 196(bw); LBG cp56; MEM 652; MIA 173; MNC
    215(bw); MPS 245(bw); NAW 91; SWM 52; WMC 154;
    WNW cp603

Vole, mountain (*Microtus montanus*)
ASM cp75; CMW 167(bw)

Vole, pine (*Microtus subterraneus*)
GEM III:247

Vole, prairie (*Microtus ochrogaster*)
ASM cp77; CMW 171(bw); LBG cp59; MNC 213(bw); MPS
    245(bw)

Vole, red-backed (Northern) (*Clethrionomys rutilus*)
AAL 44; ASM cp61

Vole, red-backed (Southern) (*Clethrionomys gapperi*)
ASM cp62; CGW 185; CMW 161(bw); GMP cp10,
    192(bw); MNC 207(bw);-MPS 244(bw); WMC 153; WNW
    cp601, cp602

Vole, rock (*Microtus chrotorrhinus*)
ASM cp68; CGW 185; GMP 200(bw); MNC 211(bw); WMC
    156

Vole, sagebrush (*Lagurus* or *Lemmiscus curtatus*)
CMW 176(bw); GEM III:248; JAD cp496; LBG cp58; MIA 173; MPS 246(bw)

Vole, tree (*Phenacomys* sp.)
AAL 44; ASM cp64, cp66; CMW 163(bw); CRM 109; MEM 653; WFW cp327

Vole, water (*Arvicola* sp.)
ASM cp73; CMW 173(bw); GEM III:236-239; MAE 96; MEM 653, 655; MIA 173; PGT 234; WNW cp604

Vole, woodland (*Microtus pinetorum*)
ASM cp67; CGW 185; GMP cp12, 203(bw); MNC 217(bw); MPS 246(bw); SEF cp583; WMC 157

Vole (other *Microtus* sp.)
ASM cp69, cp70, cp72, cp74; CRM 111; MEM 653; WOM 82; WPP 40

Volvox (*Volvox* sp.)
AAL 538(bw); EAB 6-7; EAL 160; PGT 73

Vulture, bearded See Lammergeier

Vulture, black (*Coragyps atratus* or *Aegypius monachus*)
BOP 35; EAB 916, 917; FEB 215; FGH cp1; FWB 206; JAD cp529; NAB cp318; NHU 153; PBC 104; UAB cp306, cp338; WOM 77

Vulture, Egyptian (*Neophron pernopterus*)
BAS 60; BOP 92; MIA 219; WGB 29

Vulture, griffon (*Gyps fulvus* or *rupelli*)
BOP 96; EOB half-title, 124; WOM 78-80

Vulture, hooded (*Necrosyrtes monachus*)
BOP 94-95

Vulture, king (*Sarcoramphus papa*)
AWR 170-171; BOP 34; EOB 106; MIA 217; WGB 27; WWW 132(head only)

Vulture, lappet-faced (*Torgos tracheliotus*)
BOP 93(head only); EOB 102; MIA 219; WGB 29

Vulture, palm-nut (*Gypohierax angolensis*)
MIA 221; WGB 31

Vulture, turkey (*Cathartes aura*)
BOP 31-33, 36, 38; EAB 918; FEB 214; FGH cp1; FWB 204; JAD cp530; LBG cp488; MIA 217; NAB cp317; NAW 222; PBC 103; UAB cp307, cp336; WGB 27; WON 57; WPP 188; WWD 158-159

Vulture, white-backed (*Pseudogyps africanus*)
BOP 90, 91

Vulture, white-backed (Asian) (*Gyps bengalensis*)
MAE 120

W

Wagtail, gray (*Motacilla cinerea*)
EOB 341; RBB 123

Wagtail, pied (*Motacilla alba*)
MIA 317; RBB 124; WGB 127

Wagtail, yellow (*Motacilla flava*)
EAB 715; EOB 337; MIA 317; RBB 122; UAB cp425; WGB 127

Wahoo (fish) (*Acanthocybium solanderi*)
ACF cp48; MIA 571

Wahoo, Eastern (plant) See Burningbush

Wake robin or Squawroot (*Trillium erectum*)
BCF 153; CWN 7; NWE cp384; SEF cp460; WAA 154

Wake robin, Western (*Trillium ovatum*)
SWW cp49; WFW cp387

Walkingstick, northern (*Diapheromera femorata*)
MIS cp295, cp297; SEF cp418; SGI cp31

Walkingstick, Timema See Timema

Wallaby, bridled nailtail (*Onychogalea fraenata*)
CRM 31; MEM 864

Wallaby, forest (*Dorcopsis* sp.)
CRM 33; GEM I:361; MEM 865-866; MIA 25

Wallaby, hare (banded) (*Lagostrophus fasciatus*)
CRM 31; MEM 867

Wallaby, hare (rufous) (*Lagorchestes hirsutus*)
CRM 31; MEM 867; NOA 234

Wallaby, hare (spectacled) (*Lagorchestes conspicillatus*)
CRM 31; MEM 862; MIA 23

Wallaby, pretty-face (*Wallabia parryi*)
GEM I:374, 377; WPP 117

Wallaby, rock (*Petrogale* sp.)
CRM 33; GEM I:367; MIA 23; WOM 144

Wallaby, rock (yellow-footed) (*Petrogale xanthopus*)
CRM 33; MEM 865; WOM 144

Wallaby, swamp (*Wallabia bicolor*)
MIA 25; GEM I:375

Wallaby, Tammar or Dama (*Macropus eugenii*)
CRM 33

Wallaroo (*Macropus robustus*)
CRM 33; GEM I:386; MEM 864; WWD 176

Wallcreeper (*Tichodroma muraria*)
MIA 359; WGB 169

Walleye (*Stizostedion vitreum*)
NAF cp98; WNW cp41

Wallflower, alpine (*Erysimum nivale*)
SWW cp325

Wallflower, Menzie's (*Erysimum menziesii*)
MPC cp606

Wallflower, plains (*Erysimum asperum*)
LBG cp210; SWW cp326

Wallflower, Western (*Erysimum capitatum*)
SWW cp367; WFW cp455

Walnut, Arizona (*Juglans major*)
NTW cp292

Walnut, black (*Juglans nigra*)
NTE cp317, cp532; SEF cp140, cp201; TSV 124, 125

Walnut, little (*Juglans microcarpa*)
NTE cp316, cp376; NTW cp293

Walnut, Northern California (*Juglans hindsii*)
NTW cp291

Walnut, Southern California (*Juglans californica*)
NTW cp294

Walrus (*Odobenus rosmarus*)
AAL 65; ASM cp297; BAA 8, 22(bw), 23(bw), 78, 149(bw);
BMW 217; CRM 161; GEM I:24, IV:192-201, 205, 206;
KSW 67; MEM 264-69; MIA 107; MPC cp223; NAW 92;
WOI 68(head only); WPR 174-179; WWW 35(head only)

Wanderoo See Macaque monkey, lion-tailed

Wapiti See Elk, American

Warbler, arctic (*Phylloscopus borealis*)
EAB 1032

Warbler, bay-breasted (*Dendroica castanea*)
EAB 985; FEB 452; NAB cp403, cp455; SEF cp310

Warbler, black and white (*Mniotilta varia*)
EAB 985; FEB 310; MIA 379; NAB cp564; PBC 298; SEF
cp336; SPN 110; WGB 189

Warbler, Blackburnian (*Dendroica fusca*)
EAB 985; FEB 445; NAB cp380, cp404; NAW 223; SEF
cp304

Warbler, blackpoll (*Dendroica striata*)
EAB 985; FEB 453; FWB 359; NAB cp563; SEF cp335;
SPN 109; UAB cp609

Warbler, black-throated blue (*Dendroica caerulescens*)
AAL 184; EAB 986; FEB 455; NAB cp446, cp514; PBC
313; SEF cp322; SPN 107

Warbler, black-throated gray (*Dendroica nigrescens*)
EAB 986; FWB 360; JAD cp598; WFW cp289; UAB cp608

Warbler, black-throated green (*Dendroica virens*)
EAB 986; FEB 436; NAB cp375; SEF cp291

Warbler, blue-winged (*Vermivora pinus*)
AAL 184; EAB 987; FEB 438; NAB cp361; PBC 304

Warbler, Canada (*Wilsonia canadensis*)
EAB 987; FEB 439; NAB cp367; SEF cp295

Warbler, Cape May (*Dendroica tigrina*)
EAB 987; FEB 443; NAB cp368; NAW 224; PBC 312; SEF
cp303

Warbler, cerulean (*Dendroica cerulean*)
EAB 987; FEB 454; NAB cp444, cp445

Warbler, chestnut-headed (*Seicercus castaneiceps*)
MIA 347; WGB 157

Warbler, chestnut-sided (*Dendroica pensylvanica*)
EAB 987; FEB 451; NAB cp377; PBC 319; SEF cp302;
SPN 106

Warbler, Connecticut (*Oporornis agilis*)
EAB 989; FEB 457; NAB cp460; SPN 116

Warbler, Dartford (*Sylvia undata*)
RBB 145

Warbler, garden (*Sylvia borin*)
RBB 148

Warbler, golden-crowned (*Basileuterus culicivorus*)
MIA 379; WGB 189

Warbler, golden-winged (*Vermivora chrysoptera*)
AAL 185; EAB 989; FEB 461; MIA 379; NAB cp378; WGB
189

Warbler, grasshopper (*Locustella naevia*)
MIA 347; RBB 142; WGB 157

Warbler, greenish (*Phylloscopus trochiloides*)
NHU 104

Warbler, hermit (*Dendroica occidentalis*)
EAB 989; FWB 362; UAB cp416; WFW cp291

Warbler, hooded (*Wilsonia citrina*)
AAL 184; EAB 989; FEB 435; NAB cp369; PBC 333; SEF
cp286

Warbler, Kentucky (*Oporornis formosus*)
EAB 989; FEB 463; NAB cp370; PBC 328; SEF cp287;
SPN 103

Warbler, Kirtland's (*Dendroica kirtlandii*)
EAB 989; FEB 449; NAB cp374; SEF cp300; WOB 242

Warbler, Lucy's (*Vermivora luciae*)
EAB 988; FWB 384

Warbler, MacGillivray's (*Oporornis tolmiei*)
EAB 991; FWB 370; UAB cp411; WFW cp292

Warbler, magnolia (*Dendroica magnolia*)
EAB 990; FEB 448; NAB cp366; PBC 311; SEF cp298

Warbler, mourning (*Oporornis philadelphia*)
EAB 990; FEB 458; MIA 379; NAB cp372; WGB 189

Warbler, Nashville (*Vermivora ruficapilla*)
EAB 990; FEB 459; FWB 371; NAB cp373; UAB cp412;
   WFW cp286

Warbler, olive (*Peucedramus taeniatus*)
EAB 991; FWB 364

Warbler, orange-crowned (*Vermivora celata*)
EAB 991; FEB 469; FWB 373; NAB cp456; UAB cp511;
   WFW cp285

Warbler, paddyfield (*Acrocephalus agricola*)
MAE 121

Warbler, pale-legged leaf (*Phylloscopus tenellipes*)
NHU 125

Warbler, palm (*Dendroica palmarum*)
EAB 992; FEB 446; NAB cp365; PBC 324; WNW cp583

Warbler, pine (*Dendroica pinus*)
EAB 993; FEB 442; NAB cp364, cp380, cp404; SEF
   cp292; SPN 108

Warbler, prairie (*Dendroica discolor*)
EAB 992; FEB 437; NAB cp363; PBC 323; SEF cp293

Warbler, prothonotary (*Protonotaria citrea*)
AAL 184; EAB 993; FEB 441; NAB cp360; PBC 299; SPN
   113; WNW cp585

Warbler, red-faced (*Cardellina rubrifrons*)
EAB 993; FWB 378; UAB cp458

Warbler, reed (*Acrocephalus scirpaceus*)
EOB 368; PGT 228; RBB 144

Warbler, sedge (*Acrocephalus schoenobaenus*)
MAE 96; RBB 143; SPN 12

Warbler, spectacled (*Sylvia conspicillata*)
EOB 371

Warbler, subalpine (*Sylvia cantillans*)
EOB 372

Warbler, Swainson's (*Limnothlypis swainsonii*)
EAB 993; FEB 472; PBC 301; WNW cp581

Warbler, Tennessee (*Vermivora peregrina*)
EAB 993; FEB 468; FWB 372; NAB cp457; PBC 307; SEF
   cp288; UAB cp512

Warbler, Townsend's (*Dendroica townsendi*)
EAB 994; FWB 365; UAB cp415; WFW cp290

Warbler, Virginia's (*Vermivora virginiae*)
EAB 994; FWB 374; UAB cp478; WFW cp287

Warbler, willow (*Phylloscopus trachilus*)
MAE 29; MIA 347; RBB 152; SPN 6; WGB 157

Warbler, Wilson's (*Wilsonia pusilla*)
EAB 994; FEB 434; FWB 367; NAB cp358; PBC 334; SPN
   111; UAB cp409, cp439; WFW cp293

Warbler, wood (*Phylloscopus sibilatrix*)
RBB 150

Warbler, worm-eating (*Helmitheros vermivorus*)
EAB 994; FEB 433; NAB cp509; PBC 302

Warbler, yellow (*Dendroica petechia*)
EAB 994; EOB 411; FEB 440; FWB 366; MIA 379; NAB
   cp357; PBC 310; SPN 105; UAB cp410; WGB 189; WOB
   86, 87; WON 163, 164

Warbler, yellow-rumped or myrtle (*Dendroica coronata*)
AGC cp618; EAB 996; FEB 450; FWB 361; NAB cp379;
   PBC 314; SEF cp301; UAB cp414, cp552; WFW cp288

Warbler, yellow-throated (*Dendroica dominica*)
EAB 996; FEB 447; NAB cp376; SEF cp299

Warmouth See under Sunfish

Warthog (*Phacocherus aethiopicus*)
CRM 171; GEM V:16-17, 20, 40-43; MEM 500, 502; MIA
   129; WPP 62

Wartwort See Spurge

Washingtonia, California (*Washingtonia filfera*)
JAD cp327; NTW cp313

Wasp, cicada killer (*Sphecius speciosus*)
MIS cp479; SEF cp402; SGI cp294

Wasp, cow killer (*Dasymutilla occidentalis*)
AGC cp486; MIS cp325

Wasp, digger (*Scolia dubia*)
LBG cp369; MIS cp457

Wasp, ichneumon (Ichneumonidae)
MIS cp445; SEF cp398; SWF 146; WFW cp571

Wasp, ichneumon (*Rhysella* sp.)
EOI 115; SGI cp268; WIW 78

Wasp, mud (*Pseudomasarus vespoides*)
SGI cp286

Wasp, mud-dauber wasp (African) (*Sceliphron spirifex*)
EOI 118

Wasp, mud-dauber wasp (black and yellow) (*Sceliphron caementarium*)
MIS cp466; SGI cp292

Wasp, paper (*Polistes* sp.)
BOI 221; EOI 121; LBG cp368; MIS cp442; SGI cp285

Wasp, paper (European) (*Paravespula* sp.)
DIE cp18

Wasp, potter (*Eumenes fraternus*)
AAL 469, 479(bw); MIS cp467; SGI cp283

Wasp, sand (*Bembix americana*)
AGC cp482; SGI cp295

Wasp, social (*Polistes instabilis*)
WIW 123

Wasp, tarantula hawk (*Hemipepsis* or *Pepsis* sp.)
AAL 479(bw); JAD cp383; MIS cp458; SGI cp287

Wasp, thread waisted (*Ammophila* sp.)
EOI 119; MIS cp455; SGI cp290, cp291; WWD 155

Wasp, velvet ant (*Dasymutilla* sp.)
JAD cp389, cp390; MIS cp326, cp327; SGI cp274, cp275

Wasp, yellow jacket See Yellow jacket

Wasp (other kinds)
AAB 142; EOI 110, 118; SGI cp269-cp276, cp288, cp289, cp293, cp296

Water bear (Tardigrada)
EAL 247; PGT 112

Water boatman (*Corixa* sp.) See also Backswimmer
AAL 417; EOI 54; MIS cp100; WNW cp343

Water bug, creeping (*Pelocoris femoratus*)
BNA 59(bw); SGI cp47

Water bug, giant (*Lethocerus americanus*)
MIS cp101; WNW cp344

Water hyacinth (*Eichornia crassipes*)
NWE cp624; PGT 65; TSP 101; WNW cp296

Water lettuce (*Pistia stratiotes*)
NWE cp10; PGT 65; WNW cp308

Water lily See also Spatterdock

Water lily, bullhead or yellow pond (*Nuphar variegatum*)
NWE cp244; WAA 196; WNW cp214

Water lily, fragrant (*Nymphaea odorata*)
CWN 59; NWE cp82; SWW cp60; WAA 193; WNW cp255; WTV 40

Water lily, yellow or brandy-bottle (*Nuphar lutea*)
PGT 50

Water lily, yellow (*Nymphaea mexicana*)
NWE cp246; WNW cp216

Water lily (other kinds)
PGT 50, 52; SGI cp222; TSP 103; WAA 201

Water measurer (*Hydrometra stagnorum*)
EOI 55; PGT 150

Water moccasin See Cottonmouth

Water parsnip (*Sium suave*)
NWE cp188; WNW cp241

Water scavenger See under Beetle

Water scorpion (*Nepa cinerea*)
EOI 55; PGT 154; WIW 132(bw)

Water scorpion (*Ranatra* sp.)
EOI 55; MIS cp293, cp294; PGT 155; WNW cp357

Water shrew, American (*Sorex palustris*)
ASM cp54; CGW 129; CRM 41; MNC 89(bw); MPS 42(bw); WMC 48; WNW cp591

Water shrew, Eurasian (*Neomys fodiens*)
GEM I:483; MEM 739; PGT 231

Water shrew, giant African (*Potamogale velox*)
CRM 35; MEM 745

Water shrew, Pacific (*Sorex bendirii*)
ASM cp50; WNW cp592

Water soldier (*Stratiotes aloides*)
PGT 65

Water strider (*Gerris* sp.)
AAL 418(bw); PGT 151; MIS cp 292; SGI cp48; WFW cp568; WNW cp356

Water willow (*Justicia americana*)
NWE cp499; WNW cp263

Waterbuck (*Kobus ellipsiprymnus*) See also Lechwe
GEM V:448, 450; MEM 561, 562; MIA 147; NHA cp18, 125(bw), 127(bw)

Watercress (*Nasturtium officinale*)
CWN 109; NWE cp178; SWW cp132; WNW cp245

Waterdog See Mudpuppy

Waterleaf, Fendler's (*Hydrophyllum fendleri*)
SWW cp165; WFW cp426

Waterleaf, Virginia (*Hydrophyllum virginianum*)
NWE cp166

Waterlocust (*Gleditsia aquatica*)
NTE cp009

Watermeal See Duckweed, rootless

Water-milfoil (*Myriophyllum* sp.)
PGT 57

Water-plantain (*Alisma* sp.)
NWE cp199; PGT 52; SWW cp158; WNW cp248

Water-plantain, fringed (*Machaerocarpus californicus*)
SWW cp25

Watersnake, brown (*Nerodia taxispilota*)
ARA cp567; ARC 223; LSW 160; MSW p84; RNA 155;
   WNW cp190

Watersnake, common or Northern (*Nerodia sipedon*)
AAL 248; ARA cp580; ARC 221, 222; ARN 58(bw); ART
   p[21]; CGW 85; LBG cp602; RNA 157; WNW cp195; WRA
   cp43

Watersnake, diamondback (*Nerodia rhombifera*)
ARA cp574; RNA 155

Watersnake, glossy crayfish (*Regina rigida*)
ARA cp474

Watersnake, green (*Nerodia cyclopion*)
ARA cp482; ARC 217; LSW 155; MSW cp24; RNA 155;
   WNW cp185

Watersnake, Pacific (*Nerodia valida*)
WRA cp48

Watersnake, plain-bellied or red-bellied (*Nerodia
   erythrogaster*)
AAL 248; ARA cp481, cp490; ARC 218, 219; RNA 155;
   WRA cp43

Watersnake, Southern or banded (*Nerodia fasciata*)
AGC cp443; ARA cp513, cp552, cp562, cp578, cp579;
   ARC 220; LSW 156-158; RNA 157; WNW cp191, cp196

Waterthrush, Louisiana (*Seiurus motacilla*)
EAB 995; FEB 327; NAB cp504; PBC 327; SPN 115;
   WNW cp578

Waterthrush, Northern (*Seiurus noveboracensis*)
EAB 995; FEB 326; FWB 389; NAB cp505; UAB cp542

Waterweed, canadian (*Elodea canadensis*)
PGT 59

Wattle, black (*Acacia mearnsii*)
TSP 9

Wattle, green (*Acacia decurrens*)
NTW cp265

Wax cap fungus (*Hygrophorous* sp.)
FGM cp22-24; GSM 139; NAM cp53, cp56, cp65+; NMT
   cp53, cp40, cp74, cp109, cp152

Waxweed, clammy (*Cuphea petiolata*)
WTV 194

Waxwing, bohemian (*Bombycilla garrulus*)
EAB 1035; EOB 353, 354; FEB 376; FWB 299; MIA 327;
   NAB cp507; RBB 125; UAB cp565; WGB 137; WON 158

Waxwing, cedar (*Bombycilla cedrorum*)
AAL 166; EAB 1034; EOB 355; FEB 377; FWB 298; NAB
   cp506; NAW 225; PBC 285; SEF cp311; SPN 98; UAB
   cp566; WFW cp282; WOB 100-101; WON 156

Weakfish (*Cynoscion regalis*)
AGC cp387; NAF cp483

Weasel, large See Ermine

Weasel, least or common (*Mustela nivalis*)
ASM cp207; CGW 153; CMW 221(bw); GEM III:389-391,
   394; GMP 276(bw); LBG cp54; MIA 87; MNC 259(bw);
   MPS 276(bw); WMC 191

Weasel, long-tailed (*Mustela frenata*)
ASM cp209; CGW 153; CMW 223(bw); GEM III:392-393,
   398, 400; GMP 280(bw); LBG cp53; MNC 257(bw); MPS
   275(bw); NAW 93; RFA 188, 189(bw); SEF cp601; SWM
   104-106; WMC 192

Weasel, North African banded or Libyan striped
   (*Poecilictis libyca*)
GEM III:410; MEM 112

Weasel, Patagonian (*Lyncodon patagonicus*)
MEM 112

Weasel, short-tailed See Ermine

Weasel, Siberian or Kolinsky (*Mustela sibirica*)
NHU 109

Weasel, water (*Mustela felipei*)
CRM 139

Weasel, white-naped or African striped (*Poecilogale
   albinucha*)
GEM III:410; MEM 112

Weaver (*Ploceus* sp.)
EOB 424; SAB 43

Weaver, grenadier or Napoleon See Bishopbird

Weaver, red-headed (*Anaplectes rubriceps*)
EOB 426

Weaver, social (*Philetairus socius*)
EOB 425

Weaver, white-headed buffalo (*Dinemellia dinemelli*)
EOB 425

Web-spinner (Embioptera)
SGI cp38; WIW 191(bw)

Weevil (various genera)
AAL 439(bw); BNA cp12, cp310(bw)-324(bw); DIE cp6;
   MIS cp130-cp138; SGI cp190, cp191

Weevil, boll (*Anthonous grandis*)
BNA 319(bw); MIS cp131

Weevil, cowpea (*Callosobruchus maculatus*)
BNA 305(bw); SGI cp188

Wels (*Silurus glanis*)
MIA 519

Whale, beaked (Baird's) (*Berardius bairdii*)
CRM 211; GEM IV:365; HWD 33, 69; MEM 213; MIA 119;
NAF cp662; WDP 44, 50(bw); WOW 114

Whale, beaked (Blainville's or dense) (*Mesoplodon densirostris*)
NAF cp627, cp666; WDP 37; WOW 142

Whale, beaked (Cuvier's) (*Ziphius cavirostris*)
AGC cp326; GEM IV:365; HWD 32, 69; MEM 213; MIA
119; NAF cp670; WDP 30, 42(bw); WOW 118

Whale, beaked (Gervais' or Antillean) (*Mesoplodon europaeus*)
NAF cp667; WDP 37; WOW 133

Whale, beaked (ginkgo-toothed) (*Mesoplodon ginkgodens*)
CRM 211; NAF cp669; WDP 41; WOW 136

Whale, beaked (Gray's or Scamperdown) (*Mesoplodon grayi*)
WDP 39(bw); WOW 137

Whale, beaked (Hector's) (*Mesoplodon hectori*)
NAF cp664; WDP 39(bw)

Whale, beaked (Hubb's or Arch) (*Mesoplodon carlhubbsi*)
NAF cp668; WDP 38(bw), 41; WOW 134

Whale, beaked (Layard's or straptoothed) (*Mesoplodon layardii*)
GEM IV:365; HWD 33; WDP 37; WOW 146

Whale, beaked (Shepherd's) (*Tasmacetus shepherdi*)
MEM 212; MIA 119; WDP 44; WOW 112

Whale, beaked (Sowerby's or North Sea) (*Mesoplodon bidens*)
GEM IV:365; HWD 33; MEM 213; MIA 119; NAF cp663;
WDP 37; WOW 144

Whale, beaked (Stejneger's or Bering sea) (*Mesoplodon stejnegeri*)
NAF cp665; WDP 41; WOW 139

Whale, beaked (True's or wonderful) (*Mesoplodon mirus*)
GEM IV:365; NAF cp661; WDP 30, 32(bw), 41; WOW 131

Whale, blue (*Balaenoptera musculus*)
AGC cp327; CRM 213; EAL 330; HWD 28; MEM 225; MIA
121; MPC cp219; NAF cp680; WDP 184, 195; WOW 85

Whale, bottlenose (northern) (*Hyperoodon ampullatus*)
CRM 211; GEM IV:365; HWD 32, 69; MEM 212; MIA 119;
NAF cp671; WDP 44, 52(bw); WOW 120

Whale, bottlenose (southern) (*Hyperoodon planifrons*)
WOW 123

Whale, bowhead (*Balaena mysticetus*)
CRM 215; EAL 337; HWD 25; MEM 233; MIA 121; NAF
cp676; WDP 204; WOW 71, 72

Whale, Bryde's (*Balaenoptera edeni*)
CRM 213; EAL 331; HWD 27; MEM 225; NAF cp684;
WDP 195; WOW 93

Whale, fin (*Balaenoptera physalus*)
AGC cp329; CRM 213; EAL 331; HWD 27; MEM 225;
MPC cp221; NAF cp682; WDP 184, 195; WOW 82

Whale, fourtooth (northern) See Whale, beaked (Baird's)

Whale, fourtooth (southern) (*Berardius arnuxi*)
WOW 116

Whale, gray (*Eschrichtius robustus*)
CRM 215; EAL 322; HWD 26; MEM 216; MIA 121; MPC
cp217; NAF cp624, cp678; WDP 178; WOW 77

Whale, humpback (*Megaptera novaeangliae*)
AAL 52; AGC cp330; CRM 215; EAL 330, 333; GEM
IV:329, 420, 438; HWD 8-9, 29, 144, 165; MEM 224, 229;
MIA 121; MPC cp218; NAF cp679; SWM 70, 71; WDP
184; WMC 209; WOW 95

Whale, killer See Killer whale

Whale, melon-headed (*Peponocephala electra*)
EAL 307; HWD 35; MEM 183; WDP 116; WOW 220

Whale, minke (*Balaenoptera acutorostrata*)
AGC cp328; CRM 213; EAL 330; GEM IV:433 (head only);
HWD 29; MEM 224; MIA 121; MPC cp222; NAF cp683;
WDP 195; WOW 88

Whale, pilot (long-fin) (*Globicephala melaena*)
AGC cp325; CRM 209; MIA 115; NAF cp657; WDP 150;
WOW 207

Whale, pilot (shortfin) (*Globicephala macrorhynchus*)
MPC cp214; NAF cp658; WOW 209

Whale, right (*Eubalaena glacialis*)
AGC cp324; CRM 215; EAL 336; MEM 232; MPC cp216;
NAF cp619, cp677; WDP 204, 210; WOW 68

Whale, right (pygmy) (*Caperea marginata*)
CRM 215; EAL 337; HWD 26; MEM 233; WDP 204; WOW
74

Whale, right (Southern) (*Eubalaena australis*)
HWD 25; WOW 244

Whale, sei, or roqual (*Balaenoptera borealis*)
CRM 213; HWD 28; MIA 121; MPC cp220; NAF cp681;
WDP 195; WOW 90

Whale, sperm (*Physeter catodon* or *macrocephalus*)
CRM 211; EAL 319; GEM IV:374-375; HWD 30, 121;
    MEM 207; MIA 117; MPC cp215; NAF cp675; WDP 56;
    WOW 171

Whale, sperm (dwarf) (*Kogia simus*)
EAL 319; MEM 207; MIA 117; NAF cp659; WDP 69(bw);
    WOW 178

Whale, sperm (pygmy) (*Kogia breviceps*)
EAL 319; HWD 30; MEM 207; MIA 117; NAF cp660; WDP
    56; WOW 176

Whale, white See Beluga

Whalefish (*Cetomimus indagator*)
MAE 111

Wheatear (*Oenanthe oenanthe*)
AAL 170(bw); EAB 895; MIA 335; NAB cp510; RBB 135;
    UAB cp482; WGB 145

Whelk (various genera)
MSC cp394, cp408, cp412, cp413, cp427

Whelk, european (*Buccinum undatum*)
EAL 250

Whiff, spotted (*Citharichthys macrops*)
NAF cp288

Whimbrel or Hudsonian curlew (*Numenius phaeopus*)
AGC cp592; BWB 183; EAB 792, 804; FEB 151; FWB
    142; MIA 243; MPC cp563; NAB cp245; PBC 144; RBB
    77; UAB cp216; WBW 209; WGB 53

Whinchat (*Saxicola rubetra*)
RBB 133

Whip-poor-will (*Caprimulgus vociferus*)
EAB 579; EOB 249; FEB 258; FWB 250; NAB cp277; SEF
    cp252; UAB cp248

Whipsnake (*Coluber* sp.)
MIA 451; MSW p81; WOM 94-95

Whipsnake, Cape or red (*Masticophis aurigulus*)
LSW 143, 144WRA cp47

Whipsnake, Sonoran (*Masticophis bilineatus*)
ARA cp517; JAD cp222; LSW 146; RNA 193; WRA cp36

Whipsnake, striped or ornate (*Masticophis taeniatus*)
ARA cp521; JAD cp221; LSW 147, 148; RNA 193; WFW
    cp588; WRA cp36

Whiptail lizard, canyon spotted (*Cnemidophorus burti*)
ARA cp416; RNA 97; WRA cp34

Whiptail lizard, checkered (*Cnemidophorus tesselatus*)
ARA cp419; ERA 101; JAD cp207; WRA cp32

Whiptail lizard, Chihuahuan (*Cnemidophorus exsanguis*)
ARA cp418; JAD cp210; RNA 99; WRA cp33

Whiptail lizard, desert grassland (*Cnemidophorus
    uniparens*)
ARA cp412; ART p[17]; JAD cp209; LOW 87; RNA 99;
    WRA cp31

Whiptail lizard, gila spotted (*Cnemidophorus
    flagellicaudus*)
RNA 99; WRA cp33

Whiptail lizard, little striped (*Cnemidophorus inornatus*)
ARA cp414; JAD cp208; RNA 97; WRA cp32

Whiptail lizard, New Mexican (*Cnemidophorus
    neomexicanus*)
ARA cp415; JAD cp211; RNA 101; WRA cp32

Whiptail lizard, Orange-throated (*Cnemidophorus
    hyperythrus*)
ARA cp409; RNA 97; WRA cp34

Whiptail lizard, plateau spotted (*Cnemidophorus
    septembittatus*)
ARA cp417

Whiptail lizard, plateau striped (*Cnemidophorus velox*)
ARA cp413; RNA 99; WFW cp586; WRA cp32

Whiptail lizard, sonoran spotted (*Cnemidophorus
    sonorae*)
ARA cp410; RNA 99; WRA cp33

Whiptail lizard, Texas spotted (*Cnemidophorus gularis*)
RNA 97; WRA cp33

Whiptail lizard, Western or marbled (*Cnemidophorus
    tigris*)
ARA cp420; JAD cp206; LBG cp588; RNA 97; WRA cp31

Whispering bells (*Emmenanthe penduliflora*)
JAD cp115; SWW cp207

Whistler, golden (*Pachycephala pectoralis*)
EOB 379

Whistling duck, black-bellied (*Dendrocygna autumnalis*)
BWB 68-69; DNA 13; EAB 214; FEB 106; NAB cp167,
    cp168

Whistling duck, fulvous (*Dendrocygna bicolor*)
AGC cp536; DNA 11; EAB 214; FEB 107; FWB 94; NAB
    cp165, cp166; PBC 82; UAB cp122; WNW cp487

Whistling duck, white-faced (*Dendrocygna viduata*)
AWR 92-93; MIA 213; WGB 23

White-eye (*Zosterops* sp.)
EOB 395; MIA 361; WGB 171

Whitefish (*Coregonus* sp.)
MIA 507; NAF cp69

Whitefish, ocean (*Caulolatilus princeps*)
MPC cp280; NAF cp553; PCF cp31

Whitefish, round (*Prosopium cylindraceum*)
NAF cp68

Whitethroat (*Sylvia communis* or *curruca*)
EOB 373; RBB 146, 147

Whiting (*Merlangius merlangus*)
MIA 527

Whydah (*Vidua* sp.)
EOB 425; MIA 391; WGB 201

Wigeon, American (*Anas americana*)
AGC cp535; DNA 33, 35; EAB 216; FEB 95; FWB 92;
    MPC cp522; NAB cp106, cp140; NAW 102; PBC 89; UAB
    cp119, cp149; WNW cp494

Wigeon, European (*Anas penelope*)
DNA 29, 31; EAB 216; FEB 102; FWB 87; NAB cp132;
    RBB 27; UAB cp118

Wild oats (*Uvularia sessilifolia* or *latifolia*)
FWF 143; NWE cp318; SEF cp465; WAA 159; WTV 61

Wildebeest, black See Gnu, white-tailed

Wildebeest, blue See Gnu, brindled

Wiliwili (*Erythrina sandwicensis*)
PFH 71; WOI 86

Willet (*Catoptrophorus semipalmatus*)
AGC cp590; BWB 182-183; EAB 805; FEB 164; FWB
    145; MPC cp573; NAB cp202, cp229; NAW 101; PBC 148;
    UAB cp215, cp240; WBW 247

Willie-wag-tail (*Rhipidura leucophrys*)
AAL 175(bw)

Willow, Australian (*Geijera parviflora*)
PMT cp[55]

Willow, balsam (*Salix pyrifolia*)
NTE cp113; NTW cp144; WNW cp459

Willow, basket (*Salix viminalis*)
NTE cp105

Willow, bebb (*Salix bebbiana*)
NTE cp117; NTW cp102, cp537; WFW cp73; WNW cp458

Willow, black (*Salix nigra*)
NTE cp106; NTW cp131; SEF cp98; WNW cp453

Willow, coastal plain (*Salix caroliniana*)
NTE cp109; SEF cp97; WNW cp455

Willow, crack (*Salix fragilis*)
NTE cp111, cp378; OET 161

Willow, desert (*Chilopsis linearis*)
JAD cp306; NTW cp85, cp376, cp387; PMT cp[31]; SWW
    cp575

Willow, Northwest or velvet (*Salix sessilifolia*)
NTW cp141

Willow, Pacific (*Salix lasiandra*)
NTW cp132; WFW cp79

Willow, pussy (*Salix discolor*)
LBG cp413, cp460; NTE cp116, cp399; NTW cp150,
    cp319; NWE cp364; WNW cp199, cp457

Willow, river (*Salix fluviatilis*)
NTW cp129

Willow, sandbar or coyote (*Salix exigua*)
NTE cp107; NTW cp127; WNW cp454

Willow, scouler (*Salix scoulerana*)
NTW cp103; WFW cp75

Willow, water (*Decodon verticillatus*)
CWN 143

Willow, weeping (*Salix babylonica*)
NTE cp108; NTW cp128; TSV 40, 41

Willow, white (*Salix alba*)
NTE cp110; NTW cp130

Willow (other *Salix* sp.)
FWF 21; NTE cp114; NTW cp81, cp89, cp97-cp104,
    cp133-cp136, cp141-cp143, cp338; OET 162; TSV 38, 39

Willow-herb (*Epilobium* sp.)
CWN 147; NWE cp465; PGT 39

Windflower See Anemone, wood

Windowpane (*Scophthalmus aquosus*)
AGC cp356; MIA 575; NAF cp289

Wineberry (*Rubus phoenicolasius*)
FWF 71; WTV 19

Winterberry (*Ilex* sp.)
FWF 121; NTE cp143, cp557; NWE cp226, cp451; SEF
    cp192, cp500, cp505; WNW cp244

Wintergreen, bog or large (*Pyrola asarifolia*)
SWW cp528

Wintergreen, creeping See Checkerberry

Wintergreen, one-sided (*Pyrola secunda*)
SWW cp8, cp105; WFW cp407

Wintergreen, spotted (*Chimaphila maculata*)
NWE cp67; SEF cp453

Winter's bark (*Drimys winteri*)
OET 116

Wisent See Bison, European

Wisteria (*Wisteria sinensis*)
TSV 118, 119

Witch hazel (*Hamamelis virginiana*)
FWF 55; NTE cp230, cp395, cp529, cp629; NWE cp359;
   SEF cp95, cp165, cp203; TSV 150, 151

Witch hazel, Chinese (*Hamamelis mollis*)
OET 122

Witherod (*Viburnum cassinoides*)
NWE cp216

Wobbegong See Carpetshark

Wolf, gray or arctic (*Canis lupus*)
ASM cp261, cp262, cp266; CMW 194(bw); GEM IV:53-59,
   70-78; MEM 59-61; MIA 79; MNC 239(bw); MNP
   231(bw); MPS 250(bw); NHU 110; NNW 44; RFA 65-
   69(bw); SEF cp610; SWM 77, 79; WFW cp369, cp372;
   WMC 180; WOI 51; WPR 82

Wolf, maned (*Chrysocyon brachyurus*)
CRM 133; GEM IV:61, 154-156; MEM 82, 83; MIA 81;
   WPP 177

Wolf, red (*Canis rufus*)
ASM cp256; CRM 133; MNC 241(bw); MPS 250(bw);
   WMC 179

Wolf, Tasmanian (*Thylacinus cynocephalus*)
CRM 17

Wolffish (*Anarhichas* sp.)
ACF p45; AGC cp429; MIA 569; NAF cp457

Wolverine (*Gulo gulo*)
AAL 63(bw); ASM cp206; CMW 230(bw); CRM 141; GEM
   III:420; MEM 120; MIA 89; MNC 263(bw); MNP 261(bw);
   MPS 277(bw); NAW 97; NHU 108; NNW 42; RFA 217-
   224(bw); SEF cp608; SWM 111; WFW cp358; WPR 84-85

Wombat (*Vombatus ursinus*)
CRM 25; GEM I:343; MEM 878; MIA 19; NOA 117

Wombat, hairy-nosed (*Lasiorhinus latifrons*)
GEM I:339; NOA 162

Wood nymph (*Moneses uniflora*)
SWW cp20; WFW cp398

Wood sorrel (*Oxalis montana*)
BCF 179; NWE cp460; SEF cp488; WTV 125(*O.rubra*)
Wood sorrel, mountain (*Oxalis alpina*)
SWW cp447

Wood sorrel, violet (*Oxalis violacea*)
NWE cp463

Wood sorrel, yellow (*Oxalis stricta*)
CWN 131; WTV 74

Woodchuck (*Marmota monax*)
ASM cp227; CGW 125; GMP 141(bw); LBG cp79; MIA
   161; MNC 159(bw); MNP 189(bw); MPS 132(bw); NAW
   99; SEF cp595; SWM 27; WMC 119

Woodcock, American (*Scolopax minor*)
EAB 805; FEB 158; MIA 243; NAB cp256; NAW 227; PBC
   152; SEF cp250; WGB 51; WON 91

Woodcock, Eurasian (*Scolopax rusticola*)
RBB 75; SWF 49; WBW 267

Woodcreeper, buff-throated (*Xiphorhynchus guttatus*)
EOB 313

Woodcreeper, plain brown (*Dendrocincla fuliginosa*)
EOB 313

Woodlark (*Lullula arborea*)
RBB 115

Woodpecker, acorn (*Melanerpes formicivorus*)
EAB 1036; FWB 330; SAB 17; UAB cp376; WFW cp241;
   WOB 166

Woodpecker, Arizona (*Picoides arizonae*)
EAB 1037; UAB cp380

Woodpecker, black-backed (*Picoides arcticus*)
EAB 1037; FEB 304; FWB 337; NAB cp341, cp343; SEF
   cp279; UAB cp383

Woodpecker, downy (*Picoides pubescens*)
EAB 1037; FEB 302; FWB 334; NAB cp337, cp339; NAW
   228; PBC 219; SEF cp275; UAB cp362, cp364; WON 132

Woodpecker, gila (*Melanerpes uropygialis*)
BAS 37; EAB 1036; EOB 303; FWB 325; JAD cp565;
   NAW 228; UAB cp375

Woodpecker, golden-fronted (*Melanerpes aurifrons*)
AAL 152; EAB 1036; FEB 306; FWB 324; JAD cp566;
   NAB cp350; UAB cp374

Woodpecker, golden-tailed (*Campethera abingoni*)
AAL 153

Woodpecker, great spotted (*Picoides major*)
EOB 297, 300; MIA 293; RBB 113; WGB 103

Woodpecker, green (*Picus viridis*)
EOB 297; MIA 295; RBB 112; WGB 105

Woodpecker, hairy (*Picoides villosus*)
EAB 1037; FEB 301; FWB 335; NAB cp338, cp340; PBC
   219; SEF cp276; UAB cp363, cp365; WFW cp245

Woodpecker, ivory-billed (*Campephilus principalis*)
MIA 293; WGB 103

Woodpecker, ladder-backed (*Picoides scalaris*)
EAB 1037; FEB 309; FWB 333; JAD cp567; NAB cp345;
   UAB cp366, cp368

Woodpecker, Lewis (*Melanerpes lewis*)
EAB 1039; FWB 322; UAB cp377

Woodpecker, Nuttall's (*Picoides nuttallii*)
EAB 1039; UAB cp367, cp369; WFW cp244

Woodpecker, pileated (*Dryocopus pileatus*)
AAL 153; EAB 1038; EOB 297; FEB 299; FWB 329; NAB cp352; PBC 215; SEF cp271; UAB cp381; WFW cp249; WNW cp569; WOB 19

Woodpecker, red-bellied (*Melanerpes* or *Centurus carolinus*)
EAB 1039; FEB 307; NAB cp349; PBC 216; SEF cp273; SWF 150; WNW cp570

Woodpecker, red-cockaded (*Picoides borealis*)
EAB 1041; FEB 303; NAB cp347; PBC 221; SEF cp278

Woodpecker, redheaded (*Melanerpes erythrocephalus*)
EAB 1041; EOB 297; FEB 298; NAB cp351; PBC 217; SEF cp272; SWF 150

Woodpecker, three-toed (*Picoides tridactylus*)
EAB 1039; EOB 296; FEB 305; FWB 336; NAB cp342, cp344; UAB cp382; WFW cp247

Woodpecker, three-toed (olive-backed) (*Dinopium rafflesi*)
EOB 297

Woodpecker, white-backed (*Dendrocopos leucotos*)
NHU 103

Woodpecker, white-headed (*Picoides albolarvatus*)
EAB 1040, 1041; FWB 328; UAB cp385; WFW cp246

Woodrat, bushy-tailed (*Neotoma cinerea*)
AAL 46(bw); ASM cp189; CMW 158(bw); GEM III:222; MNP 216(bw); MPS 217(bw); NAW 98; SWM 50; WFW cp337

Woodrat, eastern (*Neotoma floridana*)
ASM cp131; CGW 177; GMP 187(bw); MNC 205(bw); MPS 217(bw); SEF cp582; WMC 151

Woodrat, Mexican (*Neotoma mexicana*)
ASM cp119; MPS 218(bw)

Woodrat, Southern Plains (*Neotoma micropus*)
ASM cp135; MPS 218(bw)

Woodrat, white-throated (*Neotoma albigula*)
ASM cp139; JAD cp494; MIA 167; MPS 216(bw); NMM cp[17]

Woodrat (other *Neotoma* sp.)
ASM cp134, cp138; GEM III:223; JAD cp493; MEM 640, 642, 643

Woolly bear See under Caterpillar

Worm, Christmas tree or serpulid (*Spirobranchus giganteus*)
AAL 485; AGC cp251; KCR cp28; MSC cp130-135; RCK 164; RUP 18; SCS cp26; WWF cp6, cp55

Worm, earth See Earthworm

Worm, fan (elegant) (*Hypsicomus elegans*)
KCR cp28; SCS cp26

Worm, fan (magnificent banded) (*Sabellastarte magnifica*)
KCR cp28; SCS cp26

Worm, fan (red-banded) (*Potamilla fonticula*)
KCR cp28; SCS cp26

Worm, featherduster (*Sabella* sp.)
AAB 11; AGC cp253; KCR cp28; MPC 455; RCK 41; SAS cpC8; SCS cp26; WWF cp56

Worm, featherduster (other kinds)
MPC cp453, cp454, cp456; MSC 138-147

Worm, fire (various genera)
MSC cp241-243; SAS cpC2

Worm, fire (*Hermodice carunculata*)
KCR cp28; RUP 59; SCS cp26

Worm, fire (red-tipped) (*Chloeia viridis*)
KCR cp28; SCS cp26

Worm, flat (various genera)
MSC cp213-219, cp232

Worm, flat (aquatic) (*Pseudoceros* sp.)
AAB 55; SAS cpB33

Worm, fringed (orange) (*Cirriformia grandis*)
SAS cpC3

Worm, maitre d' (*Notomastus lobatus*)
SAS cpC5

Worm, ribbon (banded) (*Tubulamus* sp.)
EAL 199

Worm, ribbon (pink) (*Micrura leidyi*)
SAS cpB32

Worm, sabellid See Worm, featherduster

Worm, terebellid (various genera)
MSC 162-165

Worm, tube (*Serpula* sp.)
EAL 207; MPC cp457, cp458

Worm, tube (lace) (*Filograna implexa*)
MSC cp83, cp149

Worm, tubifex (Tubificidae)
PGT 115

Worm, velvet (*Macroperipatus* sp.)
AAL 484; EOI 2

Wormfish (*Microdesmus* sp.)
ACF cp46

Worm-lizard (*Amphisbaena* sp.)
ERA 133; MIA 443; MSW cp3

Worm-lizard (*Bipes* sp.)
AAL 234(bw); WRA cp47

Worm-lizard, Florida (*Rhineura floridana*)
ARA cp451; MIA 443; RNA 133

Woundwort (*Stachys palustris*)
CWN 227

Wrack, sugar (*Laminaria saccharina*) See also Rockweed
MPC cp494

Wrasse (*Halichoeres* sp.)
ACF cp39; PCF cp30

Wrasse, bluehead (*Thalassoma bifasciatum*)
NAF cp368, cp370

Wrasse, cleaner (*Labroides* sp.)
SAB 137; WWW 129

Wrasse, Creole (*Clepticus parrai*)
ACF cp39; NAF cp366

Wrasse, yellowhead (*Halichoeres garnoti*)
ACF cp39; NAF cp367, cp372

Wren, bewick's (*Thryomanes bewickii*)
EAB 1043; FEB 317; FWB 346; JAD cp587; MIA 329; NAB
cp490; SPN 75; UAB cp526; WFW cp269; WGB 139

Wren, brown-throated (*Troglodytes brunneicollis*)
EAB 1042

Wren, cactus (*Campylorhynchus brunneicapillus*)
AAL 170(bw); EAB 1042; EOB 357; FEB 315; FWB 344;
JAD cp584; MIA 329; NAB cp493; NAW 230; SPN 74;
UAB cp532; WGB 139; WOB 91

Wren, canyon (*Catherpes mexicanus*)
EAB 1043; FEB 322; FWB 343; JAD cp586; NAB cp491;
SPN 73; UAB cp530

Wren, Carolina (*Thryothorus ludovicianus*)
EAB 1043; FEB 316; NAB cp489; PBC 264

Wren, common or winter (*Troglodytes troglodytes*)
AAL 169; EAB 1044; EOB 359; FEB 321; FWB 348; MIA
329; NAB cp487; RBB 127; SEF cp347; SPN 78; UAB
cp527; WFW cp270; WGB 139

Wren, fairy (*Malurus* sp.)
EOB 377, 379

Wren, house (*Troglodytes aedon*)
AAL 169; EAB 1045; FEB 320; FWB 347; MIA 329; NAB
cp486; NAW 231; PBC 261; SPN 76, 77; UAB cp529;
WGB 139; WON 148

Wren, marsh (long-billed) (*Cistothorus palustris*)
AGC cp613; EAB 1044; FEB 318; MIA 329; NAB cp488;
PBC 265; SPN 79; UAB cp528; WGB 139; WNW cp574

Wren, marsh (short-billed) or sedge (*Cistothorus
platensis*)
EAB 1044; FEB 319; NAB cp485; PBC 266; WNW cp575

Wren, rock (*Salpinctes obsoletus*)
EAB 1044; FEB 323; FWB 342; JAD cp585; MIA 329; NAB
cp492; UAB cp531; WGB 139

Wrentit (*Chamaea fasciata*)
EAB 1046; FWB 379; MIA 343; SPN 96; UAB cp484;
WFW cp281; WGB 153

Wrybill See under Plover

Wrymouth (various genera)
MIA 567; PCF p42

Wryneck (*Jynx torquilla*)
EOB 297, 299; MIA 293; WGB 103

Wutu See Urutu

X

Xenosaur, Chinese (*Shinisaurus crocodilurus*)
ERA 105

Y

Yak, wild (*Bos grunniens mutus*)
AAL 80; CRM 187; GEM V:393-396; MIA 143; WOM 140,
141

Yam, wild (*Dioscorea villosa*)
FWF 47

Yapok or water opposum (*Chironectes minimus*)
MEM 832; NOA 45

Yarrow (*Achillea* sp.)
CWN 279; LBG cp129; MWP 315; NWE cp192; PPC 52;
SWW cp170; WFW cp423

Yarrow, golden (*Eriophyllum lanatum*)
SWW cp252; WFW cp441

Yaupon (*Ilex vomitoria*)
NTE cp208, cp561; NWE cp450; SEF cp104, cp191; WNW
cp329

Yellow bell (*Fritillaria pudica*)
LBG cp198; SWW cp282; WFW cp447

Yellow jacket (*Vespula* sp.)
LBG cp364; MIS cp486; SEF cp401; WFW cp577

Yellow rocket See Watercress

Yellow-eyed grass (*Xyris torta* or *iridifolia*)
CWN 23; NWE cp269; WNW cp210

Yellowhammer (*Emberiza citrinella*)
BAS 13; MIA 369; RBB 188; WGB 179

Yellowhead (*Trichoptilium incisum*)
JAD cp124; SWW cp274

Yellowlegs, greater (*Tringa melanoleuca*)
EAB 805; FEB 162; FWB 146; MPC cp572; NAB cp228;
UAB cp207

Yellowlegs, lesser (*Tringa flavipes*)
AGC cp593; EAB 806; FEB 163; FWB 147; MPC cp571;
NAB cp227; UAB cp206, cp209; WBW 225

Yellowtail (*Seriola lalandei*)
MPC cp281; NAF cp557; PCF cp31

Yellowthroat, common (*Geothlypis trichas*)
EAB 998; FEB 462; FWB 368; NAB cp371; PBC 330; SPN
117; UAB cp419, cp440; WNW cp584

Yellowwood (*Cladrastis* sp.)
NTE cp153; PMT cp[34]

Yerba buena (*Satureja douglassi*)
SWW cp143; WFW cp421

Yerba de Selva (*Whipplea modesta*)
SWW cp162; WFW cp425

Yew, common (*Taxus baccata*)
OET 106, 107; PPC 105; PPM cp15(berries)

Yew, Florida (*Taxus floridana*)
NTE cp29

Yew, Pacific (*Taxus brevifolia*)
NTW cp51, cp481; WFW cp52, cp146

Yucca (*Yucca* sp.) See also Joshua tree, Our Lord's
Candle
JAD cp328-330, cp336; LBG cp408; MWP 319; NTE
cp361, cp362, cp428, cp429, cp576; NTW 304-309, 381;
NWE cp140; SWW cp187, cp188; WAA 143, 175; WFW
cp196

Z

Zander (*Stizostedion lucioperca*)
MIA 551

Zebra, common or plains (*Equus burchelli*)
AAL 71(bw); CRM 169; GEM IV:547, 564; MEM 483, 485,
487; MIA 125; WPP 64-65; WWD 36-37

Zebra, Grevy's (*Equus grevyi*)
CRM 169; MEM 485; MIA 125

Zebra, mountain (*Equus zebra*)
CRM 169; GEM IV:579; MEM 485

Zebrafish See Lionfish

Zebraperch (*Hermosilla azurea*)
PCF cp34

Zebu (*Bos indicus*)
GEM V:414-415

Zelkova, Japanese (*Zelkova serrata*)
NTE cp153; NTW cp155

Zenobia (*Zenobia pulverulenta*)
WTV 20

Zokor (*Myospalax aspalax*)
MEM 666; MIA 169

Zorilla (*Ictonyx striatus*)
GEM III:411; MEM 112; MIA 87

Zorro See Fox, crab-eating

Scientific Name Index

Abies amabilis — Fir, Pacific silver or Cascades
Abies balsamea — Fir, balsam
Abies bracteata — Fir, bristlecone
Abies concolor — Fir, white or Colorado
Abies fraseri — Fir, Fraser
Abies grandis — Fir, grand
Abies koreana — Fir, Korean
Abies lasiocarpa — Fir, subalpine or Rocky Mountain
Abies magnifica — Fir, red
Abies nordmanniana — Fir, Nordman or Caucasian
Abies procera — Fir, noble
Abramis brama — Bream
Abrocoma sp. — Rat, chinchilla
Abronia sp. — Verbena, sand
Abrus precatorius — Pea, rosary
Abudefduf saxatilis — Sergeant major
Abudefduf taurus — Sergeant, night
Abutilon theophrasti — Velvet leaf
Acacia decurrens — Wattle, green
Acacia farnesiana — Huisache
Acacia mearnsii — Wattle, black
Acacia sp. — Acacia
Acacia sp. — Catclaw
Acanthaster planci — Crown-of-thorns starfish
Acanthis cannabina — Linnet
Acanthis flammea — Redpoll, common
Acanthisitta chloris — Rifleman bird
Acanthiza chrysorrhoa — Thornbill, yellow-rumped
Acanthocybium solanderi — Wahoo fish
Acanthodactylus sp. — Lizard, fringetoed
Acanthophis antarcticus — Adder, death
Acanthurus coeruleus — Blue tang
Acanthurus sp. — Surgeonfish
Acarina sp. — Mite, water
Accipiter cooperii — Hawk, Cooper's
Accipiter gentilis — Goshawk, northern
Accipiter melanchoryphus — Hawk, sparrow black-mantled
Accipiter nisus — Hawk, sparrow
Accipiter striatus — Hawk, sharp-shinned
Acer barbatum — Maple, Florida
Acer campestre — Maple, hedge or field
Acer circinatum — Maple, vine
Acer glabrum — Maple, Rocky Mountain or dwarf
Acer griseum — Maple, paperbark
Acer macrophyllum — Maple, bigleaf

Acer negundo — Boxelder
Acer nigrum — Maple, black
Acer pensylvanicum — Maple, striped
Acer platanoides — Maple, Norway
Acer pseudoplatanus — Maple, planetree or sycamore
Acer rubrum — Maple, red
Acer saccharinum — Maple, silver
Acer saccharum — Maple, sugar
Acer spicatum — Maple, mountain
Acer sp. — Maple (other)
Achaearanea sp. — Spider, house American
Achatinella sowerbyona — Snail, agate-shell
Acherontia atropos — Hawkmoth, death's head
Acheta domestica — Cricket, house
Achillea sp. — Yarrow
Achorutes nivicola — Flea, snow
Achlys triphylla — Vanilla leaf
Acinonyx jubatus — Cheetah
Acipenser baeri — Sturgeon, Siberian
Acipenser fulvescens — Sturgeon, lake
Acipenser oxyrhynchos — Sturgeon, Atlantic
Acipenser ruthenus — Sterlet
Acipenser transmontanus — Sturgeon, white
Acleisanthes longiflora — Angel trumpets
Acomys sp. — Mouse, spiny
Aconitum sp. — Monkshood
Acorus calamus — Flag, sweet
Acridotheres cristatellus — Myna, crested
Acridotheres tristis — Mynah, common
Acris crepitans — Frog, cricket (Northern)
Acris gryllus — Frog, cricket (Southern)
Acrobates pygmaeus — Glider, pygmy or feathertail
Acrocephalus agricola — Warbler, paddyfield
Acrocephalus schoenobaenus — Warbler, sedge
Acrocephalus scirpaceus — Warbler, reed
Acrochordus sp. — Snake, wart
Acrocinus longimanus — Beetle, Harlequin
Acropora cervicornis — Coral, staghorn
Acropora palmata — Coral, elkhorn
Acryllium vulturinum — Guinea fowl, vulturine
Actaea pachypoda — Doll's eyes
Actaea rubra — Baneberry, red
Actaea spicata — Herb-Christopher
Actias luna — Moth, luna or moon
Actinia bermudensis — Sea anemone, stinging
Actinia equina — Sea anemone, beadlet

Actinophryidae — Heliozoan
Actinostella flosculifera — Sea anemone, collared sand
Actitus hypoleucos — Sandpiper, common
Actitis macularia — Sandpiper, spotted
Actophilornis africanas — Jacana, African
Adansonia digitata — Baobab
Addax nasomaculatus — Addax
Adelpha bredowii — Butterfly, sister (Western)
Adelpha fessonia — Butterfly, sister (Mexican)
Adlumia fungosa — Allegheny-vine
Aechmophorus occidentalis — Grebe, western
Aedes sp. — Mosquito, salt marsh
Aegithalos caudatus — Tit, long-tailed
Aegithina tiphia — Iora, common
Aegolius acadicus — Owl, saw-whet
Aegolius funerea — Owl, boreal or Tengmalm's
Aegotheles sp. — Owlet, nightjar
Aeoliscus strigatus — Shrimpfish
Aepyceros melampus — Impala
Aepyprymnus rufescens — Kangaroo, rat rufous
Aepysurus duboisi — Seasnake, Dubois's
Aeronautes saxatilis — Swift, white-throated
Aesculus californica — Buckeye, California
Aesculus hippocastanum or *carnea* — Horsechestnut
Aesculus octandra — Buckeye, yellow
Aesculus sp. — Buckeye (other)
Aethia cristatella — Auklet, crested
Aethia pusilla — Auklet, least
Aethia pygmaea — Auklet, whiskered
Aetobatus narinari — Ray, eagle spotted
Afropavo congensis — Peacock, Congo
Afrotis atra — Bustard, little black
Agalinis purpurea — Gerardia, purple
Agalychnis spurrelli — Frog, flying
Agamia agami — Heron, Agami
Agapornis fischeri — Lovebird, Fischer's
Agapostemon texanus — Bee, sweat
Agaricia sp. — Coral, lettuce-leaf
Agaricus bisporus — Mushroom, cultivated
Agaricus campestris — Mushroom, meadow
Agaricus silvicola — Mushroom, wood or forest
Agave sp. — Century plant
Agelaius phoeniceus — Blackbird, red-winged
Agelaius tricolor — Blackbird, tricolored
Agelas schmidti — Sponge, pipes-of-pan
Agkistrodon bilineatus — Moccasin, Mexican
Agkistrodon contortrix — Copperhead
Agkistrodon piscivorus — Cottonmouth or Water moccasin
Agonus acipenserinus — Poacher, sturgeon
Agouti paca — Paca
Agouti taczanowskii — Paca, mountain
Agraulis vanillae — Fritillary, gulf
Agrimonia sp. — Agrimony
Agriornis livida — Shrike-tyrant, great
Agrocybe sp. — Earth-scale fungus
Agrostemma githago — Corn cockle
Agrostis alba — Grass, redtip
Ailanthus altissima — Tree of Heaven
Ailuropoda melanoleuca — Panda, giant
Ailurus fulgens — Panda, red or lesser
Aimophila aestivalis — Sparrow, Bachman's
Aimophila cassinii — Sparrow, cassin's
Aimophila ruficeps — Sparrow, rufous-crowned
Aiptasia pallida — Sea anemone, pale
Aix galericulata — Duck, mandarin
Aix sponsa — Duck, wood

Ajaja ajaja — Spoonbill, roseate
Ajuga reptans — Bugle
Alaria sp. — Kelp, winged
Alauda arvensis — Skylark
Albizzia julibrissin — Mimosa
Albula vulpes — Bonefish
Alca torda — Auk, razorbill
Alcedo atthis — Kingfisher, eurasian
Alcedo vintsioides — Kingfisher, Malagasy
Alcedo sp. — Kingfisher (other)
Alces alces — Moose
Alchephalus busephalus — Hartebeest, Coke's or Kongoni
Alcippe castaneceps — Tit-babbler, chestnut-headed
Alcyonium sp. — Dead man's fingers (coral)
Alectis ciliaris — Pompano, African
Alectoris chukar — Chukar
Alectoris graeca — Chukar
Alectoris rufa — Partridge, red-legged
Alectryon excelsus — Titoki
Alectura lathami — Turkey, Australian brush
Aletris farinosa — Colic-root
Aleuria sp. — Orange peel fungus
Alisma sp. — Water-plantain
Alisterus scapularis — Parrot, king (Australian)
Allactaga sp. — Jerboa
Allamanda cathartica — Trumpet vine, golden
Alle alle — Dovekie
Allenopithecus nigroviridis — Monkey, Allen's swamp
Alligator mississippiensis — Alligator, American
Allionia incarnata — Four o'clock, trailing
Allium sp. — Garlic, wild
Allium sp. — Onion, wild
Allium tricoccum — Leek, wild or ramp
Allotropa virgata — Candystick
Alnus glutinosa — Alder, black or European
Alnus maritima — Alder, seaside
Alnus oblongifolia — Alder, Arizona
Alnus rhombifolia — Alder, white
Alnus rubra — Alder, red
Alnus rugosa — Alder, speckled
Alnus serrulata — Alder, common
Alnus sinuata — Alder, sitka
Alnus tenuifolia — Alder, mountain
Aloe sp. — Aloe
Alopex lagopus — Fox, arctic
Alopias vulpinus — Shark, thresher
Alopochen aegyptiacus — Goose, Egyptian
Alosa alabamae — Shad, Alabama
Alosa pseudoharengus — Alewife
Alosa sapidissima — Shad, American
Alouatta caraya — Howler monkey, black
Alouatta palliata — Howler monkey, mantled
Alouatta seniculus — Howler monkey, red
Alouatta sp. — Howler monkey (other)
Alternanthera philoxeroides — Alligator weed
Aluterus sp. — File fish
Alytes sp. — Toad, midwife
Amanita caesarea — Mushroom, amanita Caesar's
Amanita citrina — Death cap, false
Amanita muscaria — Fly agaric
Amanita pantherina — Blusher, false or Panther
Amanita phalloides — Death cap
Amanita rubescens — Blusher
Amanita sp. — Mushroom, amanita
Amanita verna — Mushroom, amanita (Spring or fool's)
Amanita virosa — Destroying angel

Amaranthus retroflexus — Pigweed or green amaranth
Amazilia beryllina — Hummingbird, Berylline
Amazilia violiceps — Hummingbird, violet-crowned
Amazilia yucatanensis — Hummingbird, buff-bellied
Amazona ochrocephala — Parrot, yellow-headed
Ambloplites rupestris — Bass, rock
Amblyrhynchus cristatus — Iguana, marine
Ambrosia artemisiifolia — Ragweed
Ambystoma annulatum — Salamander, ringed
Ambystoma cingulatum — Salamander, flatwoods
Ambystoma gracile — Salamander, Northwestern or brown
Ambystoma jeffersonianum — Salamander, Jefferson
Ambystoma laterale — Salamander, blue-spotted
Ambystoma mabeei — Salamander, Mabee's
Ambystoma macrodactylum — Salamander, long-toed
Ambystoma maculatum — Salamander, spotted
Ambystoma mexicanum — Axolotl
Ambystoma opacum — Salamander, marbled
Ambystoma platineum — Salamander, silvery
Ambystoma talpoideum — Salamander, mole
Ambystoma texanum — Salamander, small-mouthed
Ambystoma tigrinum — Salamander, tiger
Ameiva ameiva — Ameiva, Giant
Amelanchier alnifolia — Serviceberry, Western
Amelanchier arborea — Serviceberry, downy
Amelanchier sp. — Serviceberry (other)
Amia calva — Bowfin
Ammocrypta sp. — Darter, crystal or sand fish
Ammodorcas clarkei — Dibatag
Ammodramnus caudacutus — Swallow, sharp-tailed
Ammodramnus maritimus — Swallow, seaside
Ammodramus bairdii — Sparrow, Baird's
Ammodramus henslowii — Sparrow, Henslow's
Ammodramus leconteii — Sparrow, Le Conte's
Ammodramus savannarum — Sparrow, grasshopper
Ammomanes deserti — Lark, desert
Ammophila sp. — Wasp, thread waisted
Ammospermophilus sp. — Squirrel, antelope
Ammotragus lervia — Sheep, barbary
Amorpha canescens — Leadplant
Amorpha fruticosa — Indigobush
Amphicarpa bracteata — Peanut, hog
Amphiprion sp. — Clownfish
Amphisbaena sp. — Worm-lizard
Amphispiza belli — Sparrow, sage
Amphispiza bilineata — Sparrow, black-throated
Amphiuma means — Amphiuma, two-toed
Amsinckia retrorsa — Fiddleneck
Anabaena oscillarioides — Algae, blue-green
Anabas testudineus — Perch, climbing
Anabrus simplex — Cricket, Mormon
Anacardium occidentale — Cashew
Anagallis arvensis — Pimpernel, scarlet
Anaphalis margaritacea — Everlasting, pearly
Anaplectes rubriceps — Weaver, red-headed
Anarhichas sp. — Wolffish
Anarhynchus frontalis — Plover, wrybill
Anarrhichthys ocellatus — Eel, wolf
Anartia jatrophae — Butterfly, peacock white
Anas acuta — Duck, pintail
Anas americana — Wigeon, American
Anas castanea — Teal, chestnut
Anas clypeata — Shoveler, northern
Anas crecca — Teal, green-winged
Anas cyanoptera — Teal, cinnamon
Anas diazi — Duck, Mexican

Anas discors — Teal, blue-winged
Anas falcata — Teal, falcated
Anas formosa — Teal, Baikal
Anas fulvigula — Duck, mottled
Anas penelope — Wigeon, European
Anas peocilorhyncha — Duck, spotbill
Anas platyrhynchos — Duck, mallard
Anas punctata — Teal, Hottentot
Anas querquedula — Garganey
Anas rubripes — Duck, black
Anas strepera — Duck, gadwall
Anastomus lamelligerus — Stork, African open-billed
Anastomus oscitans — Openbill
Anax junius — Dragonfly, darner green
Anchoa sp. — Anchovy
Ancylopsetta dilecta — Flounder, three-eye
Andigena bailloni — Toucanet, saffron
Andrena vicina — Bee, mining
Andrias japonicus — Salamander, Japanese giant
Andromeda sp. — Rosemary, bog
Andropogon sp. — Grass, bluestem
Andropogon virginicus — Broomsedge
Androsace septentrionalis — Candelabra, fairy
Aneides aeneus — Salamander, green
Aneides ferreus — Salamander, clouded
Aneides flavipunctatus — Salamander, black
Aneides lugubris — Salamander, arboreal
Anemone caroliniana — Anemone, Carolina
Anemone nemorosa — Anemone, wood
Anemone occidentalis — Pasqueflower, mountain or Western
Anemone oregana — Anemone, blue
Anemone patens — Pasqueflower
Anemone quinquefolia or *nemorosa* — Anemone, wood
Anemone tuberosa — Anemone, desert
Anemone virginiana — Thimbleweed
Anemonella thalictroides — Rue anemone
Anguilla anguilla — Eel, European
Anguilla rostrata — Eel, American
Anguis fragilis — Slowworm
Anhinga anhinga — Anhinga or water turkey
Anhinga melanogaster — Anhinga, African
Anilius scytale scytale — Snake, coral (false or two-headed)
Anisotremus surinamensis — Margate, black
Anisotremus virginicus — Porkfish
Anniella sp. — Lizard, legless
Annona muricata — Soursop
Annona reticulata — Custard apple
Annulipes sp. — Crab, fiddler
Anoa sp. — Anoa or dwarf water-buffalo
Anobiidae — Beetle, death-watch
Anodonta cygnaea — Mussel, european or freshwater swan
Anodorhynchus hyacinthinus — Macaw, hyacinth
Anolis carolinensis — Anole, Carolina or green
Anolis sp. — Anole (other)
Anopheles gambiae — Mosquito, malaria
Anoplopoma fimbria — Sablefish
Anous minutus — Noddy, white-capped
Anous stolidus — Noddy, brown
Anous tenuirostris — Noddy, black or lesser
Anser albifrons — Goose, white-fronted
Anser anser — Goose, graylag
Anser caerulescens — Goose, snow or blue
Anser canagicus — Goose, emperor
Anser fabalis brachyrhyncus — Goose, pink-footed

Anser indicus — Goose, bar-headed
Anseranas semipalmata — Goose, magpie
Antechinomys laniger — Kultarr
Antechinus sp. — Marsupial mouse or antechinus
Antennaria sp. — Pussytoes
Antennarius sp. — Frogfish
Antheraea eucalypti — Moth, Emperor gum
Antheraea polyphemus — Moth, polyphemus
Anthericum torreyi — Lily, amber
Anthocaris sp. — Butterfly, orange tip
Anthonous grandis — Weevil, boll
Anthopleura krebsi — Sea anemone, rock
Anthoxanthum odoratum — Grass, sweet vernal
Anthracoceros albirostris — Hornbill, Northern pied
Anthrenus sp. — Beetle, carpet
Anthropoides paradisea — Crane, blue
Anthropoides virgo — Crane, demoiselle
Anthus pratensis — Pipit, meadow
Anthus spinoletta — Pipit, water or rock
Anthus sp. — Pipit (other kinds)
Antidorcas marsupialis — Springbuck or springbok
Antilocapra americana — Pronghorn
Antilope cervicapra — Blackbuck
Antrozous pallidus — Bat, pallid
Anulocaulis leiosolensus — Ringstem, southwestern
Aonyx capensis — Otter, Cape clawless
Aonyx sp. — Otter, swamp
Aotus trivirgatus — Monkey, night
Apalone ferox — Turtle, softshell Florida
Apalone muticus — Turtle, softshell (smooth)
Apalone spiniferus — Turtle, softshell (spiny or Spring)
Apantesis ornata — Moth, tiger (ornate)
Apantesis phyllira — Moth, tiger (southern)
Apeltes quadracus — Stickleback, four-spine
Aphelocoma coerulescens — Jay, scrub
Aphelocoma ultramarina — Jay, Mexican
Aphredoderus sayanus — Perch, pirate
Aphriza virgata — Surfbird
Aphrodite aculeata — Sea mouse
Apiomerus sp. — Bee assassin
Apios americana — Groundnut
Apis mellifera — Bee, honey
Aplidium sp. — Sea pork
Aplocheilus panchax — Panchax, blue
Aplodinotus grunniens — Drum, freshwater
Aplodontia rufa — Beaver, mountain
Aplysia sp. — Sea hare
Apocynum androsaemifolium — Dogbane, spreading
Apocynum cannabinum — Indian hemp
Apodanthera undulata — Loco, melon
Apodemus flavicollis — Mouse, yellow-necked
Apodemus sylvaticus — Mouse, wood
Apodichthys flavidus — Gunnel, penpoint
Apogon maculatus — Flamefish
Apogon sp. — Cardinalfish
Aptenodytes forsteri — Penguin, emperor
Aptenodytes patagonica — Penguin, king
Apteryx australis — Kiwi
Apus apus — Swift, common
Apus melba — Swift, alpine
Apus pacificus — Swift, Northern white-rumped
Aquila chrysaetos — Eagle, golden
Aquila heliaca adalberti — Eagle, Spanish imperial
Aquila rapax — Eagle, steppe or tawny
Aquila verreauxii — Eagle, black or Verreaux's
Aquilegia caerulea — Columbine, blue
Aquilegia canadensis — Columbine, wild or Eastern

Aquilegia chrysantha — Columbine, yellow
Aquilegia formosa — Columbine, red or Western
Aquilegia vulgaris — Columbine
Ara ararauna — Macaw, blue and yellow
Ara chloroptera — Macaw, military or red-blue-and-green
Ara macao — Macaw, scarlet
Aralia nudicaulis — Sarsaparilla, wild
Aralia spinosa — Devil's walking-stick
Aramides ypecaha — Rail, wood
Aramus guarauna — Limpkin
Araneus cavaticus — Spider, barn
Araneus diadematus — Spider, garden
Araneus sp. — Spider, orb-web
Arapaima gigas — Arapaima
Araucaria heterophylla — Norfolk Island Pine
Arbutus menziesii — Madrone, Pacific
Arbutus texana — Madrone, Texas
Arbutus unedo — Strawberry tree
Archilochus alexandri — Hummingbird, black-chinned
Archilochus colubris — Hummingbird, ruby-throated
Archoplites interruptus — Perch, Sacramento
Archosargus probatocephalus — Sheepshead
Arctictis binturong — Binturong
Arctium minus — Burdock, lesser
Arctocebus calabarensis — Potto, golden or Angwantibo
Arctocephalus australis — Seal, fur (South American)
Arctocephalus forsteri — Seal, fur (New Zealand)
Arctocephalus galapagoensis — Seal, fur (Galapagos)
Arctocephalus gazella — Seal, fur (Antarctic)
Arctocephalus pusillus doriferus — Seal, fur (Australian)
Arctocephalus pusillus — Seal, fur (Cape or South African)
Arctocephalus townsendi — Seal, fur (Guadaloupe)
Arctocephalus tropicalis — Seal, fur (Subantarctic)
Arctogalidia sp. — Civet, palm
Arctomecon merriami — Poppy, great desert
Arctostaphylos manzanita — Manzanita
Arctostaphylos nova-ursi or *rubra* — Bearberry or kinnikinnick
Ardea cinerea — Heron, gray
Ardea goliath — Heron, goliath
Ardea herodias — Heron, great blue
Ardea humbolti — Heron, Malagasy
Ardea pacifica — Heron, white-necked
Ardea purpurea — Heron, purple
Ardea sumatrana — Heron, Sumatran
Ardeola bacchus — Heron, pond (Chinese)
Ardeola grayii — Heron, pond (Indian)
Ardeola ibis — Egret, cattle
Ardeola melanocephala — Heron, black-headed
Ardeola ralloides — Heron, Squacco
Ardeola sp. — Heron, pond
Ardeotis kori — Bustard, kori
Arenaria interpres — Turnstone, ruddy
Arenaria melanocephala — Turnstone, black
Arenaria sp. — Sandwort
Arethusa bulbosa — Orchid, swamp-pink or dragon's mouth
Argemone sp. — Poppy, prickly
Argiope sp. — Spider, orb-web
Argonauta sp. — Nautilus, paper
Argulus sp. — Louse, fish
Argusianus argus — Pheasant, great argus
Argyroneta aquatica — Spider, water
Argyropelecus sp. — Hatchet fish
Argyroxiphium sandwicensi — Silversword

Ariolimax californicus — Slug, banana
Arisaema dracontium — Green dragon
Arisaema triphyllum — Jack-in-the-pulpit
Aristolochia durion — Dutchman's pipe
Arius felis — Catfish, hardhead
Arizona elegans — Snake, glossy
Armadillidium vulgare — Pillbug or woodlouse
Armeria maritima — Thrift, California
Armillaria mellea — Mushroom, honey
Arnica sp. — Arnica
Arothron sp. — Puffer
Arremonops rufivirgatus — Sparrow, olive
Artamus personatus — Swallow, wood (masked)
Artemisia sp. — Sagebrush
Artemisia stelleriana — Dusty miller
Artocarpus altilis — Breadfruit
Artocarpus scortechinii — Jackfruit
Arum maculatum — Lords-and-ladies
Aruncus sp. — Goatsbeard
Arundinaria gigantea — Cane
Arvicola sp. — Vole, water
Arvicola terrestris — Vole, ground
Asarum canadense — Ginger, wild
Asarum caudatum — Ginger, wild (long-tailed or Western)
Ascalapha odorata — Butterfly, black witch
Ascaphus truei — Frog, tailed
Asclepias incarnata — Milkweed, swamp
Asclepias lanceolata — Milkweed, red
Asclepias purpurascens — Milkweed, purple
Asclepias speciosa — Milkweed, showy
Asclepias subverticillata — Milkweed, poison
Asclepias syriaca — Milkweed, common
Asclepias tuberosa — Butterflyweed
Asclepias variegator or *albicans* — Milkweed, white
Ascyrum hypericoides — St. Andrew's-cross
Ascyrum stans — St. Peter's-wort
Asellus sp. — Louse, water
Asimina triloba — Papaw, common
Asio flammeus — Owl, short-eared
Asio otus — Owl, long-eared
Assa sp. — Frog, marsupial
Astacus pallipes — Crayfish, freshwater
Aster engelmannii — Aster, Engelmann
Aster laevis — Aster, smooth
Aster latiflorus — Aster, calico
Aster novae-angliae — Aster, New England
Aster novi-belgii — Aster, New York
Aster simplex — Aster, panicled
Aster tenuifolius — Aster, salt-marsh
Aster undulatus — Aster, wavy-leaved
Aster sp. — Aster (other)
Asterias sp. — Starfish
Asterochelys radiata — Tortoise, Madagascar Radiated
Astragalus mollissimus — Locoweed, woolly
Astragalus sp. — Milkvetch
Astrophyton muricatum — Basketstar, Caribbean
Astrotia stokesii — Seasnake, Stoke's
Ateles sp. — Monkey, spider
Athene cunicularia — Owl, burrowing
Athene noctua — Owl, little
Atherinops affinis — Topsmelt
Atherinopsis californiensis — Jacksmelt
Atheris squamiger — Viper, bush or leaf
Atherurus sp. — Porcupine, brush-tailed
Atilax paludinosus — Mongoose, marsh or water
Atractoscion nobilis — Sea bass, white

Atractosteus sp. — Gar
Atrichoseris platyphylla — Tobaccoweed
Atriplex confertifolia — Shadscale
Atta sp. — Ant, leaf-cutter
Attacus atlas — Moth, Asian atlas
Aulacorhynchus prasinus — Toucanet, emerald
Aulorhynchus flavidus — Tube-snout
Aulostomus sp. — Trumpetfish
Aurelia aurita — Jellyfish, moon or common
Auricularia auricula — Tree ear fungus
Auriparus flaviceps — Verdin
Automeris io — Moth, io
Automeris sp. — Moth, silk
Avahi laniger — Lemur, woolly
Avicennia germinans — Mangrove, black
Axis axis — Deer, axis
Axis calamianensis — Deer, hog (Calamian)
Axis porcinus — Deer, hog
Aythya affinis — Scaup, lesser
Aythya americana — Duck, redhead
Aythya collaris — Duck, ring-necked
Aythya fuligula — Duck, tufted
Aythya marila — Scaup, greater
Aythya nyroca — Duck, ferruginous
Aythya sp. — Pochard
Aythya valisineria — Duck, canvasback
Azolla filiculoides — Fern, water
Babyrousa babyrussa — Babirusa
Baccharis halimifolia — Groundsel tree
Bagre marinus — Catfish, gafftopsail
Baileya multiradiata — Marigold, desert
Baiomys sp. — Mouse, pygmy
Bairdiella chrysoura — Perch, silver
Balaena mysticetus — Whale, bowhead
Balaeniceps rex — Stork, whale-headed or shoebill
Balaenoptera acutorostrata — Whale, minke
Balaenoptera borealis — Whale, sei
Balaenoptera edeni — Whale, Bryde's
Balaenoptera musculus — Whale, blue
Balaenoptera physalus — Whale, fin
Balanophyllia elegans — Coral, orange cup
Balanus nubilis — Barnacle, acorn (giant)
Balanus sp. — Barnacle
Balearica pavonina — Crane, black-crowned (including West African)
Balearica regulorum — Crane, crowned (including South African)
Balistes capriscus — Triggerfish, gray
Balistes sp. — Triggerfish, queen
Balistoides sp. — Triggerfish
Balsamorhiza sagittaria — Balsam-root, arrowleaf
Bandicota sp. — Rat, bandicoot
Banksia sp. — Banksia
Baptisia leucantha — Indigo, false prairie
Baptisia tinctoria — Indigo, wild
Barbarea vulgaris — Cress, Winter
Barbastella barbastellus — Bat, barbastelle
Barbus tor — Tor mahseer
Barilius bola — Trout, Indian
Bartramia longicauda — Sandpiper, upland
Basilarchia astyanax — Butterfly, red-spotted purple
Basileuterus culicivorus — Warbler, golden-crowned
Basiliscus sp. — Basilisk
Bassaricyon sp. — Olingo
Bathymaster sp. — Ronquil
Batrachoseps attenuatus — Salamander, slender (California)

Batrachoseps nigriventris — Salamander, slender (black-bellied)
Batrachoseps wrighti — Salamander, slender (Oregon)
Batrachoseps sp. — Salamander, slender (other)
Batrachospermum sp. — Algae, red
Battus philenor — Swallowtail butterfly, green or pipeline
Bauhinia forficata — Orchid, white
Bdeogale crassicauda — Mongoose, bushy-tailed
Belamcanda chinensis — Lily, blackberry
Bellis perennis — Daisy, English
Belone belone — Garfish
Beloperone californica — Chuparosa
Bembix americana — Wasp, sand
Berardius arnuxi — Whale, fourtooth (Southern)
Berardius bairdii — Whale, beaked (Baird's)
Berberis repens — Oregon grape, creeping
Berberis vulgaris — Barberry
Betta splendens — Siamese fighting fish
Bettongia lesueur — Boodie
Bettongia penicillata — Bettong
Betula alleghaniensis — Birch, yellow
Betula glandulosa — Birch, bog
Betula lenta — Birch, black or sweet or cherry
Betula niger — Birch, Virginia roundleaf
Betula nigra — Birch, river
Betula occidentalis — Birch, water
Betula papyrifera — Birch, paper
Betula pendula — Birch, silver or white
Betula populifera — Birch, gray
Bibio sp. — Fly, black
Bidens bipinnata — Spanish needles
Bidens pilosa — Shepherd's needle
Bignonia capreolata — Crossvine
Bipes sp. — Worm-lizard
Bison bison — Bison
Bison bonasus — Bison, European or wisent
Biston strataria — Moth, oak beauty
Bitis arientans — Adder, puff
Bitis caudalis or *cornuta* — Adder, horned
Bitis gabonica — Viper, Gaboon
Bitis nasicornis — Viper, rhinoceros
Bitis peringueyi — Viper, sidewinder or dwarf puff adder
Blarina brevicauda — Shrew, short-tailed (Northern)
Blarina carolinensis — Shrew, short-tailed (Southern)
Blarina sp. — Shrew, short-tailed (other)
Blastocerus dichotomus — Deer, marsh
Blatta orientalis — Cockroah, Oriental or Asiatic
Blattella germanica — Cockroach, german or Croton bug
Bloomeria crocea — Golden stars
Boa caninus — Boa, tree emerald
Boa constrictor — Boa constrictor
Bodianus rufus — Hogfish, Spanish
Boehmaria cylindrica — Nettle, false
Boiga dendrophila — Snake, mangrove
Boletus sp. — Mushroom, bolete
Bombina orientalis — Toad, Oriental fire-bellied
Bombus sp. — Bumblebee
Bombycilla cedrorum — Waxwing, cedar
Bombycilla garrulus — Waxwing, bohemian
Bombylius sp. — Fly, bee
Bombyx mori — Silkworm, common or Chinese
Bonasa umbellus — Grouse, ruffled
Boocerus eurycerus — Bongo
Bos gaurus — Gaur or gayal
Bos grunniens mutus — Yak, wild
Bos indicus — Zebu
Bos javanicus — Banteng

Bos sauveli — Kouprey
Boschniakia strobilacea — Ground cone, California
Boselaphus tragocamelus — Nilgai
Bostrychia olivacea — Ibis, olive or green
Botaurus lentiginosus — Bittern, American
Botaurus poiciloptilus — Bittern, Australian
Botaurus stellaris — Bittern, Eurasian
Bothriocyrtum californicum — Spider, trapdoor (Californian)
Bothrops alternatus — Urutu or wutu
Bothrops atrox — Fer-de-lance
Bothus lunatus — Flounder, peacock
Bothus sp. — Flounder
Botia sp. — Loach
Bouteloua gracilis — Grass, gramma or mesquite
Bovista plumbea — Puffball, lead
Boyeria vinosa — Dragonfly, darner (brown)
Brachinus sp. — Beetle, bombardier
Brachyistius frenatus — Perch, kelp
Brachylophus vitiensis — Iguana, crested
Brachyramphus brevirostre — Murrelet, Kittlitz
Brachyramphus marmoratus — Murrelet, marbled
Brachystola sp. — Grasshopper, lubber
Brachyteles arachnoides — Muriqui
Bradypus sp. — Sloth, three-toed
Bradypus variegatus — Sloth, brown-throated
Brama sp. — Pomfret
Branchiostoma lanceolatum — Lancelet
Branta bernica — Brant
Branta canadensis — Goose, Canada
Branta leucopsis — Goose, barnacle
Branta sandvicensis — Goose, Hawaiian
Brassariscus astutus — Ringtail or civet cat
Brassariscus sumichrasti — Ringtail, South American or cacomistle
Brassica nigra — Mustard, black
Brephidium exilis — Butterfly, blue (Western pigmy)
Brevoortia sp. — Menhaden
Brodiaea sp. — Brodiaea
Brosme brosme — Cusk
Broussonetia papyrifera — Mulberry, paper
Brugmansia sp. — Angel's trumpet
Bryonia dioica — Bryony, white
Bubalus bubalis — Buffalo, water
Bubalus mindorensis — Tamaraw
Bubalus sp. — Anoa
Bubo bubo — Owl, eagle
Bubo sumatrana — Owl, eagle (Oriental)
Bubo virginianus — Owl, great horned
Bubulcus ibis — Egret, cattle
Buccinum undatum — Whelk, european
Bucephala albeola — Duck, bufflehead
Bucephala clangula — Goldeneye, common or American
Bucephala islandica — Goldeneye, Barrow's
Buceros bicornis — Hornbill, great Indian
Bucorvus sp. — Hornbill, ground
Budorcas taxicolor — Takin
Bufo alvarius — Toad, Colorado River
Bufo americanus — Toad, American
Bufo asper — Toad, forest
Bufo boreas — Toad, Western
Bufo boreas boreas — Toad, Boreal
Bufo bufo — Toad, common
Bufo calamita — Toad, natterjack
Bufo canorus — Toad, Yosemite
Bufo cognatus — Toad, Great Plains
Bufo debilis — Toad, green

Bufo exsul — Toad, black
Bufo hemiophrys — Toad, Canadian
Bufo houstonensis — Toad, Houston
Bufo marinus — Toad, marine
Bufo microscaphus — Toad, Southwestern
Bufo periglenes — Toad, golden
Bufo punctatus — Toad, red-spotted
Bufo quercicus — Toad, oak
Bufo retriformis — Toad, Sonoran Green
Bufo speciosus — Toad, Texas
Bufo terrestris — Toad, Southern
Bufo valliceps — Toad, Gulf Coast
Bufo viridis — Toad, European green
Bufo woodhousei — Toad, woodhouse
Bufo woodhousei fowleri — Toad, Fowler's
Bugeranus carunculatus — Crane, wattled
Bumelia sp. — Bumelia
Bungarus sp. — Krait
Bunodosoma sp. — Sea anemone, warty
Bunolagus monticularis — Hare, bushman
Buphagus sp. — Oxpecker
Buprestidae — Beetle, wood-boring metallic
Burhinus capensis — Thick-knee, spotted
Burhinus magnirostris — Curlew, bush stone
Burhinus oedicneus — Curlew, stone
Burhinus vermiculatus — Thick-knee, water
Burramys parvus — Possum, mountain pygmy
Bursera microphylla — Elephant-tree
Busarellus nigricollis — Buzzard, fishing
Butastur indicus — Buzzard, gray-faced eagle
Buteo albicaudatus — Hawk, white-tailed
Buteo albonotatus — Hawk, zone-tailed
Buteo brachyurus — Hawk, short-tailed
Buteo buteo — Buzzard, common or steppe
Buteo galapagoensis — Hawk, Galapagos
Buteo jamaicensis — Hawk, red-tailed
Buteo lagopus — Hawk, rough-legged
Buteo lineatus — Hawk, red-shouldered
Buteo magnirostris — Hawk, roadside
Buteo nitidus — Hawk, gray
Buteo platypterus — Hawk, broad-winged
Buteo regalis — Hawk, ferruginous
Buteo rufinus — Buzzard, long-legged
Buteo rufofuscus — Buzzard, auger or jackal
Buteo swainsonii — Hawk, Swainson
Buteogallus anthracinus — Hawk, black
Buteogallus urubitinga — Hawk, great black
Buthus sp. — Scorpion
Butomus umbellatus — Rush, flower
Butorides striatus — Heron, green or green-backed
Cabomba caroliniana — Fanwort
Cacajao melanocephalus — Uakari, black
Cacajao rubicundus or *calvus* — Uakari, red or white or bald
Cacatua galerita — Cockatoo, sulfur-crested
Cacatua roseicapilla — Galah
Cactospiza pallida — Finch, woodpecker
Caiman crocodilus — Caiman, spectacled
Cairina moschata — Duck, muscovy
Cakile edentula — Sea rocket
Calabaria sp. — Python
Calamospiza melanocorys — Bunting, lark
Calamus sp. — Porgy
Calandrella cinerea — Lark, short-toed or red-capped
Calandrina ciliata — Red maids
Calcarius lapponicus — Longspur, lapland
Calcarius mccownii — Longspur, McCown's

Calcarius ornatus — Longspur, chestnut-collared
Calcarius pictus — Longspur, Smith's
Calephelis sp. — Butterfly, metalmark
Calidris acuminata — Sandpiper, sharp-tailed
Calidris alba — Sanderling
Calidris alpina — Dunlin or Red-backed sandpiper
Calidris bairdii — Sandpiper, Baird's
Calidris canutus — Knot, red
Calidris ferruginea — Sandpiper, curlew
Calidris fuscicolis — Sandpiper, white-rumped
Calidris maritima — Sandpiper, purple
Calidris mauri — Sandpiper, Western
Calidris melanotos — Sandpiper, pectoral
Calidris minutilla — Sandpiper, least
Calidris ptilocnemis — Sandpiper, rock
Calidris pusilla — Sandpiper, semipalmated
Calidris temminckii — Stint, Temminck's
Calla palustris — Arum, water or wild calla
Callaeas cinerea — Kokako
Callagur borneoensis — Terrapin, painted or Callagur
Calliactis tricolor — Sea anemone, hermit crab
Calliadra sp. — Powderpuff
Callianassa sp. — Shrimp, ghost
Calliandra eriophylla — Fairy duster
Callicarpa americana — Beautyberry
Callicebus molochi or *torquatus* — Titi (monkey)
Callicebus personatus — Titi, masked (monkey)
Callimico goeldii — Marmoset, Goeldi's
Callimorpha jacobaeae — Moth, cinnabar
Callinectes sapidus — Crab, blue
Callipepla squamata — Quail, scaled
Calliphora vomitoria — Fly, bluebottle
Callisaurus draconoides — Lizard, zebra-tailed
Callistemon citrinus — Bottlebrush
Callithrix argentata — Marmoset, silvery or black-tailed
Callithrix aurita — Marmoset, white-eared
Callithrix flaviceps — Marmoset, buffy-headed
Callithrix geoffroyi — Marmoset, white-fronted
Callithrix humeralifer — Marmoset, Santorem or tassel-ear
Callithrix jacchus — Marmoset, common or tufted
Callophrys rubi — Hairstreak butterfly, green
Callorhinus ursinus — Seal, fur (Northern)
Callosamia promethea — Moth, prometheus
Callosciurus prevosti — Squirrel, prevost's
Callosobruchus maculatus — Weevil, cowpea
Calocedrus decurrens — Cedar, incense
Calochortus elegans — Cat's ears, elegant
Calochortus nuttalli — Lily, sego
Calochortus sp. — Lily, Mariposa or globe
Caloenas nicobarica — Pigeon, nicobar
Calonetta leucophrys — Teal, ringed
Calopogon sp. — Orchid, grass pink
Caloprymnus capestris — Kangaroo, rat desert
Calothorax lucifer — Hummingbird, Lucifer
Caltha leptosepala — Marsh marigold, Western
Caltha palustris — Marsh marigold
Calvatia gigantea — Puffball, giant
Calycadenia truncata — Rosin weed
Calypso bulbosa — Orchid, calypso or fairy slipper
Calypte anna — Hummingbird, Anna's
Calypte costae — Hummingbird, costa
Calyptomena sp. — Broadbill, green
Calyptridium umbellatum — Pussy paws
Camassia quamash — Camas, common
Camassia scilloides — Hyacinth, wild
Camellia japonica — Camellia

Camelus bactrianus — Camel, bactrian
Camelus dromedarius — Camel, dromedary
Campanula americana — Bellflower, tall
Campanula rotundifolia — Harebell
Campephaga sp. — Cuckoo-shrike
Campephilus principalis — Woodpecker, ivory-billed
Campethera abingoni — Woodpecker, golden-tailed
Camponotus sp. — Ant, carpenter
Campsis radicans — Trumpet creeper
Camptostoma imberbe — Tyrannulet, northern beardless
Campylomormyrus sp. — Elephant-snout
Campylorhamphus sp. — Scythebill
Campylorhynchus brunneicapillus — Wren, cactus
Cancer borealis — Crab, Jonah
Cancer irroratus or *antennarius* — Crab, rock
Cancer oregonensis — Crab, Oregon
Cancer productus — Crab, red
Canis dingo — Dingo
Canis latrans — Coyote
Canis lupus — Wolf, gray or arctic
Canis mesomelas — Jackal, black-backed or
  silverbacked
Canis rufus — Wolf, red
Canis sp. — Jackal (other)
Cannabis sativa — Marijuana
Cannomys sp. — Rat, bamboo
Canotia holacantha — Crucifixion thorn
Cantharellus cibarius — Chanterelle, yellow
Cantherhines pullus — File fish, orange-spotted or
  taillight
Cantherhines sp. — Triggerfish
Canthigaster rostrata — Puffer, sharpnose
Canthon sp. — Dung roller or tumblebug
Caperea marginata — Whale, right (pygmy)
Capparis spinosa — Caper bush
Capra aegagrus — Goat, wild
Capra cylindricornis — Tur
Capra fulconeri — Markhor
Capra ibex — Ibex
Capreolus capreolus — Deer, roe
Capricornis sp. — Serow
Caprimulgus europaeus — Nightjar, European
Caprimulgus vociferus — Whip-poor-will
Caprolagus hispidus — Hare, hispid
Capromys sp. — Hutia
Capsella bursa-pastoris — Shepherd's purse
Caracara cheriway — Caracara, Audubon's or common
Carabidae — Beetle, ground
Caranx sp. — Jack
Carassius auratus — Goldfish
Carcharhinus amblyrhynchos — Shark, reef grey
Carcharhinus leucas — Shark, bull or Zambezi
Carcharhinus longimanus — Shark, reef (oceanic whitetip)
Carcharhinus melanopterus — Shark, reef (blacktip)
Carcharhinus sp. — Shark, reef
Carcharinus sp. — Shark, requiem
Carcharodon carcharias — Shark, great white
Carchiarius plumbeus — Shark, sandbar
Carcinus maenas — Crab, green
Carcinus sp. — Crab, shore
Cardamine cordifolia — Cress, bitter
Cardamine sp. — Cress
Cardaria draba — Cress, hoary
Cardellina rubrifrons — Warbler, red-faced
Cardinalis cardinalis — Cardinal
Cardinalis sinuatus — Pyrrhuloxia
Cardisoma sp. — Crab, land

Carduelis carduelis — Goldfinch, European
Carduelis chloris — Greenfinch
Carduelis flammea — Redpoll, common
Carduelis hornemanni — Redpoll, hoary
Carduelis lawrencei — Goldfinch, Lawrence's
Carduelis pinus — Siskin, pine
Carduelis psaltria — Goldfinch, lesser
Carduelis tristis — Goldfinch, American
Carduus nutans — Thistle, musk or nodding
Caretta caretta — Turtle, loggerhead
Carettocheyls insculpta — Turtle, pig-nosed
Carex riparia — Sedge, pond (great)
Cariama cristata — Seriema, crested or redleg
Carica papaya — Papaya or pawpaw
Carphophis amoenus — Snake, worm
Carpilius corallinus — Crab, coral
Carpinus betulus — Hornbeam, European
Carpinus caroliniana — Hornbeam
Carpiodes cyprinus — Quillback
Carpodacus cassinii — Finch, Cassin's
Carpodacus mexicanus — Finch, house
Carpodacus purpureus — Finch, purple
Carpodacus rubicilla — Rosefinch
Carya aquatica — Hickory, water
Carya cordiformis — Hickory, bitternut
Carya glabra — Hickory, pignut
Carya illinoensis — Pecan
Carya laciniosa — Hickory, shellbark
Carya myristiciformis — Hickory, nutmeg
Carya ovata — Hickory, shagbark
Carya texana — Hickory, black
Carya tomentosa — Hickory, mockernut
Carybdea sp. — Box jelly or sea wasp
Casmerodius albus — Egret, great or common
Cassia bauhinioides — Twinleaf
Cassia fasciculata — Partridge-pea
Cassia fistula — Golden rain or shower
Cassia sp. — Senna
Cassiope sp. — Heather, mountain
Cassiopeia xamachana — Jellyfish, upside-down
Castanea alnifolia — Chinkapin, Florida
Castanea dentata — Chestnut, American
Castanea ozarkensis — Chinkapin, Ozark
Castanea pumila — Chestnut, dwarf
Castanopsis chrysophylla — Chinkapin, giant
Castanopsis sempervirens — Chinkapin, bush
Castilleia coccinea — Indian paintbrush
Castilleja chromosa — Indian paintbrush, desert
Castilleja miniata — Indian paintbrush, giant red
Castilleja sulphurea — Indian paintbrush, sulfur
Castor canadensis — Beaver
Castor fiber — Beaver, Eurasian
Casuarina cunninghamiana — Casuarina, river-oak
Casuarius sp. — Cassowary
Catagonus wagneri — Peccary, Chacoan
Catalpa bignoides — Catalpa, Southern
Catalpa speciosa — Catalpa, Northern
Catamblyrhynchus diadema — Finch, plush-capped
Catharacta maccormicki — Skua, South Pole
Catharacta skua — Skua, great
Catharanthus roseus — Periwinkle, rosy or Madagascar
Cathartes aura — Vulture, turkey
Catharus fuscescens — Veery
Catharus guttatus — Thrush, hermit
Catharus minimus — Thrush, gray-cheeked
Catharus ustulatus — Thrush, Swainson's or olive-
  backed

Catherpes mexicanus — Wren, canyon
Catoptrophorus semipalmatus — Willet
Catostomus commersoni — Sucker, white
Caulanthus inflatus — Desert candle
Caulolatilus princeps — Whitefish, ocean
Caulophyllum thalictroides — Cohosh, blue
Causus sp. — Adder, night
Cavia porcellus — Guinea pig
Ceanothus americanus — Tea, New Jersey
Ceanothus integerrimus — Deer-brush
Ceanothus thyrsiflorus — Blueblossom
Cebuella pygmaea — Marmoset, pygmy
Cebus albifrons — Monkey, capuchin (white-fronted)
Cebus apella — Monkey, capuchin (black-capped or brown)
Cedrus atlantica — Cedar, blue Atlas
Cedrus deodara — Cedar, deodar
Cedrus libani — Cedar of Lebanon
Ceiba pentandra — Silk cotton tree
Celastrus scandens — Bittersweet
Celtis laevigata — Sugarberry
Celtis occidentalis — Hackberry
Celtis reticulata — Hackberry, netleaf
Cemophora coccinea — Snake, scarlet
Cenchrus tribuloides — Sand burr
Centaurea maculosa — Knapweed, spotted
Centaurium beyrichii — Pink, mountain
Centaurium calycosum — Centaury
Centrarchus macropterus — Flier
Centrocercus urophasianus — Grouse, sage
Centropomus undecimalis — Snook
Centropristis sp. — Sea bass
Centropus sp. — Coucal
Centropyge argi — Cherubfish
Centurus carolinus — Woodpecker, red-bellied
Cephalanthera austinae — Orchid, phantom
Cephalanthus occidentalis — Buttonbush
Cephalopholis fulva — Coney, Caribbean
Cephalopholis miniatus — Grouper
Cephalophus sp. — Duiker
Cephalopterus ornatus — Umbrella bird, ornate or long-wattled
Cephalorhynchus commersonii — Dolphin, Commerson's or piebald
Cephalorhynchus eutropia — Dolphin, Chilean or black
Cephalorhynchus heavisidii — Dolphin, Heaviside's or Benguela
Cephalorhynchus hectori — Dolphin, Hector's or New Zealand
Cephaloscyllium ventriosum — Shark, swell
Cepphus columba — Guillemot, pigeon
Cepphus grylle — Guillemot, black or Tystie
Cerastes cerastes — Viper, horned
Cerastes vipera — Viper, Avicenna
Cerastium arvense — Chickweed, field
Cerastium vulgatum — Chickweed, mouse-ear
Ceratonia siliqua — Carob
Ceratophrys sp. — Frog, horned
Cercartetus nanus — Possum, pygmy
Cercidiphyllum japonicum — Katsura tree
Cercidium floridum — Paloverde, blue
Cercidium microphyllum — Paloverde, yellow
Cercis canadensis — Redbud or Judas tree
Cercis occidentalis — Redbud, California or Western
Cercis siliquastrum — Judas tree
Cercocarpus sp. — Cercocarpus
Cercocarpus sp. — Mahoney, mountain

Cercocebus albigena — Mangabey, gray or white-cheeked
Cercocebus aterrimus — Mangabey, black
Cercocebus galeritus — Mangabey, agile or crested
Cercocebus torquatus — Mangabey, white or collared
Cercopithecus aethiops — Monkey, vervet
Cercopithecus aethiops tantalus — Monkey, tantalus
Cercopithecus ascanius schmidti — Monkey, Schmidt's
Cercopithecus cephus — Monkey, moustached
Cercopithecus diana — Monkey, Diana or Rolaway
Cercopithecus erythrogaster — Monkey, red-bellied
Cercopithecus erythrotis — Monkey, red-eared
Cercopithecus hamlyni — Monkey, owl-faced
Cercopithecus lhoesti — Monkey, L'Hoest's
Cercopithecus mitis — Monkey, blue
Cercopithecus mona — Monkey, mona
Cercopithecus neglectus — Monkey, De Brazza's
Cercopithecus petaurista — Monkey, lesser white-nosed
Cercopithecus sabaeus — Monkey, green
Cercyonis pegala — Butterfly, wood nymph
Cereus giganteus — Cactus, saguaro
Cereus undatus or *greggii* — Cereus, night-blooming
Cerorhinca monocerata — Auklet, rhinoceros
Cerotatherium simus — Rhinoceros, white
Certhia familiaris — Creeper, brown
Certhidea olivacea — Finch, warbler
Cervus albirostris — Deer, Thorold's
Cervus alfredi — Deer, rusa (Prince Albert's spotted)
Cervus axis — Deer, axis
Cervus canadensis — Elk, American or Wapiti
Cervus duvauceli — Barasingha, Indian
Cervus elaphus — Deer, red or elk or wapiti
Cervus eldi — Deer, brow-antlered
Cervus nippon — Deer, sika
Cervus porcinus — Deer, hog
Ceryle alcyon — Kingfisher, belted
Ceryle maxima — Kingfisher, giant
Ceryle rudis — Kingfisher, pied
Cetomimus indagator — Whalefish
Cetorhinus maximus — Shark, basking
Ceuthophilus secretus — Cricket, cave
Ceyx pictus — Kingfisher, pygmy
Chaenopis sp. — Pikeberry
Chaetodipterus faber — Spadefish, Atlantic
Chaetodipterus zonatus — Spadefish, Pacific
Chaetodon striatus — Butterflyfish, banded
Chaetodon sp. — Butterflyfish (other)
Chaetophractus villosus — Armadillo, larger hairy or Patagonian
Chaetura gigantea — Needletail, brown
Chaetura pelagica — Swift, chimney
Chalceria heteronea — Butterfly, copper blue
Chamaea fasciata — Wrentit
Chamaecyparis lawsoniana — Cedar, Port Orford
Chamaecyparis nootkatensis — Cedar, Alaska yellow
Chamaecyparis sp. — Cypress, false
Chamaecyparis thyoides — Cedar, white Atlantic
Chamaedaphne calyculata — Leatherleaf
Chamaeleo dilepsis — Chameleon, common or flap-necked
Chamaeleo sp. — Chameleon
Chamaelirium luteum — Devil's bit
Chanos sp. — Milkfish
Chara fragilis — Stonewort
Charadrius alexandrinus — Plover, snowy or Kentish
Charadrius asiaticus — Plover, Caspian sand
Charadrius dubius — Plover, little ringed

Charadrius hiaticula — Plover, ringed
Charadrius leschenaultii — Plover, greater sand
Charadrius melanops — Plover, black-fronted
Charadrius melodus — Plover, piping
Charadrius montanus — Plover, mountain
Charadrius semipalmatus — Plover, semipalmated
Charadrius tricollaris — Plover, three-banded
Charadrius wilsonia — Plover, Wilson's
Charidrius vociferus — Killdeer
Charina bottae — Boa, rubber
Charonia tritonis — Triton trumpet
Chaulodius sloani — Viperfish, deepsea or sloane's
Chauna chavaria — Screamer, black-necked or northern
Chauna torquata — Screamer, crested
Cheirodon axelrodi — Tetra, cardinal
Cheirogaleus sp. — Lemur, dwarf
Chelidonium major — Celandine, greater
Chelodina expansa — Turtle, snake-necked giant
Chelone glabra — Turtlehead
Chelonia mydas — Turtle, green
Chelonoidis elephantopus — Tortoise, giant Galapagos
Chelus fimbriatus — Matamata
Chelydra serpentina — Snapping turtle
Chen caerulescens — Goose, snow or blue
Chen canagicus — Goose, emperor
Chen rossii — Goose, Ross'
Chenopodium album — Lamb's quarters
Chettusia gregaria — Lapwing, sociable or Sociable Plover
Chichorium intybus — Chicory
Chilomeniscus cinctus — Snake, sand banded
Chilomycterus sp. — Burrfish
Chilopsis linearis — Willow, desert
Chimaphila maculata — Wintergreen, spotted
Chimaphila sp. — Pipsissewa
Chinchilla lanigera — Chinchilla, long-tailed
Chinchillula sahamae — Mouse, chinchilla
Chinemys reevesi — Turtle, Chinese pond or Reeve's
Chionactis occipatalis — Snake, shovel-nosed (Western)
Chionactis palarostris — Snake, shovel-nosed (Sonoran)
Chionanthus virginicus or *retusus* — Fringe tree
Chionis alba — Sheathbill, snowy or yellow-billed
Chironectes minimus — Yapok or water opposum
Chironomus sp. — Midge
Chiropotes satanus — Saki, black-bearded
Chlamydera nuchalis — Bowerbird, great
Chlamydomonas sp. — Chlamydomonas (algae)
Chlamydosaurus kingii — Lizard, frilled or frilled dragon
Chlamydoselachus anguineus — Shark, frilled
Chlamyphorus truncatus — Armadillo, pink or lesser fairy
Chlidonias hybrida — Tern, whiskered
Chlidonias niger — Tern, black
Chloeia viridis — Worm, fire (red-tipped)
Chloroceryle amazona — Kingfisher, Amazon
Chloroceryle americana — Kingfisher, green
Chloropsis sp. — Leafbird
Chlosyne sp. — Butterfly, patch
Choeronycteris or *Glossophaga* sp. — Bat, long-tongued
Choeropsis liberiensis — Hippopotamus, pygmy
Choloepus sp. — Sloth, two-toed
Chondestes grammacus — Sparrow, lark
Chondrohierax uncinatus — Kite, hook-billed
Chondropython sp. — Python, tree
Chondropython viridis — Python, tree (green)
Chondrus crispus — Irish moss
Chordeiles acutipennis — Nighthawk, lesser
Chordeiles minor — Nighthawk, common

Chordeiles virginianus — Nighthawk
Choriotis australis — Bustard, Australian
Chorthippus parallelus — Grasshopper, meadow
Chromis insolatus — Sunshinefish
Chromis punctipinnis — Blacksmith
Chrysanthemum leucanthemum — Daisy, oxeye
Chrysanthemum nauseosus — Rabbitbrush
Chrysaora hydrostatica — Jellyfish, compass
Chrysaora quinquecirrha — Sea nettle
Chrysemys concinna — Turtle, river cooter
Chrysemys floridana — Turtle, Florida cooter
Chrysemys nelsoni — Turtle, red-belly (Florida)
Chrysemys picta — Turtle, painted
Chrysemys rubriventris — Turtle, red-belly
Chrysemys scripta — Turtle, red-eared or cooter or slider
Chrysochloridae — Mole, golden
Chrysocyon brachyurus — Wolf, maned
Chrysolophus amherstiae — Pheasant, Lady Amherst's
Chrysolophus pictus — Pheasant, golden
Chrysopa sp. — Lacewing, green
Chrysopelea ornata — Snake, flying
Chrysops sp. — Fly, deer
Chrysopsis camporum — Aster, golden hairy
Chrysopsis mariana or *villosa* — Aster, golden
Cichlasoma sp. — Cichlid
Ciconia ciconia — Stork, white
Ciconia episcopus — Stork, white-necked
Ciconia nigra — Stork, black
Cicuta douglasii or *maculata* — Hemlock, water
Cienomys opimus — Tuco-tuco
Cimex lectularius — Bedbug
Cimicifuga racemosa — Cohosh, black
Cincindela sp. — Beetle, tiger
Cinclus cinclus — Dipper, Eurasian
Cinclus mexicanus — Dipper, American
Cinnamomum camphora — Camphor tree
Circaea quadrisulcata — Nightshade, enchanter's
Circaetus gallicus — Eagle, short-toed
Circaetus gallicus pectoralis — Eagle, black-breasted snake
Circus assimilis — Harrier, spotted
Circus cyaneus — Harrier, northern or hen
Circus melanoleucus — Harrier, pied
Cirriformia grandis — Worm, fringed orange
Cirsium arvense — Thistle, Canada
Cirsium horridulum — Thistle, yellow
Cirsium muticum — Thistle, swamp
Cirsium pastoris — Thistle, showy
Cirsium vulgare — Thistle, bull
Cirsium sp. — Thistle (other)
Cissa chinensis — Magpie, green
Cistothorus palustris — Wren, marsh (long-billed)
Cistothorus platensis — Wren, marsh (short-billed)
Citellus citellus — Ground squirrel, European
Citharichthys macrops — Whiff, spotted
Citharichthys sp. — Sanddab
Citheronia regalis — Moth, regal
Citrus aurantium — Orange, sour or Seville
Civettictis civetta — Civet, African
Cladonia rangifera — Reindeer moss
Cladrastis sp. — Yellowwood
Clamator glandarius — Cuckoo, great spotted
Clangula hyemalis — Duck, oldsquaw
Clarias batrachus — Catfish, walking
Clarkia amoena — Farewell-to-Spring
Clarkia sp. — Clarkia

Claviceps purpurea — Ergot
Claytonia virginica — Spring beauty
Cleistes divaricata — Orchid, rosebud
Clematis crispa — Jasmine, blue
Clematis hirsutissima — Vase flower
Clematis sp. — Virgin's bower
Clematis vitalba — Old-man's-beard
Clemmys guttata — Turtle, spotted
Clemmys insculpta — Turtle, wood
Clemmys muhlenbergi — Turtle, bog
Clemmys sp. — Turtle, pond
Cleome lutea — Bee plant, yellow
Cleome serrulata — Bee plant, Rocky Mountain
Clepticus parrai — Wrasse, Creole
Cleridae — Beetle, checkered
Clerodendrum paniculatum — Pagoda flower
Clethra alnifolia — Pepperbush, sweet
Clethrionomys gapperi — Vole, red-backed (Southern)
Clethrionomys glareolus — Vole, bank
Clethrionomys rutilus — Vole, red-backed (Northern)
Cliftonia monophylla — Buckwheat tree
Climaciella sp. — Fly, mantid
Clintonia andrewiana — Clintonia, red
Clintonia borealis — Lily, bluebead or corn
Clintonia umbellulata — Lily, wood speckled
Clintonia uniflora — Queen's cup
Clione limacina — Sea butterfly
Clitocybe infundibuliformis — Mushroom, funnel-cup
Clitopilus prunulus — Sweetbread or miller mushroom
Clitoria sp. — Pea, butterfly
Clonophis kirtlandi — Snake, Kirtland's
Clupea harengus — Herring, Atlantic or sea
Clupea harengus pallasi — Herring, Pacific
Clytoceyx rex — Kingfisher, shovel-billed
Cnemidophorus burti — Whiptail lizard, canyon spotted
Cnemidophorus exsanguis — Whiptail lizard, Chihuahuan
Cnemidophorus flagellicaudus — Whiptail lizard, gila spotted
Cnemidophorus gularis — Whiptail lizard, Texas spotted
Cnemidophorus hyperythrus — Whiptail lizard, Orange-throated
Cnemidophorus inornatus — Whiptail lizard, little striped
Cnemidophorus neomexicanus — Whiptail lizard, New Mexican
Cnemidophorus septembittatus — Whiptail lizard, plateau spotted
Cnemidophorus sexlineatus — Lizard, six-lined racerunner
Cnemidophorus sonorae — Whiptail lizard, sonoran spotted
Cnemidophorus tesselatus — Whiptail lizard, checkered
Cnemidophorus tigris — Whiptail lizard, Western or marbled
Cnemidophorus uniparens — Whiptail lizard, desert grassland
Cnemidophorus velox — Whiptail lizard, plateau striped
Cnidoscolus stimulosus — Tread-softly or spurge-nettle
Coccinella novanotata — Beetle, ladybird (nine-spotted)
Coccinella septempunctata — Beetle, ladybird (seven-spotted)
Coccinellidae — Beetle, ladybird
Coccothraustes coccothraustes — Hawfinch
Coccyzus americanus — Cuckoo, yellow-billed
Coccyzus erythropthalmus — Cuckoo, black-billed
Coccyzus minor — Cuckoo, mangrove
Cochlearius cochlearius — Heron, boat-billed
Cochliomyia hominivorax — Fly, screw-worm

Cociella sp. — Crocodilefish
Cocos nucifera — Palm, coconut
Codiaeum variegatum — Croton
Codium fragile — Sea staghorn
Coenagrionidae — Damselfly, narrow-winged
Coendou prehensilis — Porcupine, prehensile-tailed or tree
Coffea sp. — Coffee
Colaptes auratus — Flicker, common, gilded, northern, or yellow-shafted
Colchicum autumnale — Crocus, Autumn
Coleonyx brevis — Gecko, banded Texas
Coleonyx reticulatus — Gecko, reticulated
Coleonyx variegatus — Gecko, banded (Western)
Colias eurytheme — Butterfly, alfafa or orange sulfur
Colias sp. — Butterfly, dog face
Colias sp. — Butterfly, sulfur
Colinus virginianus — Bobwhite
Colius sp. — Mousebird
Colletes sp. — Bee, plasterer or mining
Collinsia heterophylla — Chinese houses, purple
Collinsia verna — Blue-eyed Mary
Collinsonia canadensis — Horse balm
Collocalia fuciphaga — Swiftlet, edible nest
Collybia maculata — Rust spot fungus or spotted coincap
Colobus badius — Colobus monkey, red
Colobus guereza — Colobus monkey, black and white or guereza
Colobus kirki — Colobus monkey, Kirk's or Zanzibar red
Colobus polykomos — Colobus monkey, Western or Western black and white
Colobus satanas — Colobus monkey, black
Colobus verus — Colobus monkey, Olive
Colocasia esculenta — Taro
Colossoma nigripinnis — Pacu
Coluber constrictor — Racer
Coluber hippocrepis — Racer, horseshoe
Coluber sp. — Whipsnake
Columba fasciata — Pigeon, band-tailed
Columba flavirostris — Pigeon, red-billed
Columba leucocephala — Pigeon, white-crowned
Columba livia — Dove, rock
Columba oenas — Dove, stock
Columba palumbus — Pigeon, wood
Columbina inca — Dove, Inca
Columbina passerina — Dove, ground
Commelina sp. — Dayflower
Concepatus mesoleucus — Skunk, hog-nosed
Condalia sp. — Condalia or bluewood
Condylactis gigantea — Sea anemone, giant caribbean or pink-tipped
Condylura cristata — Mole, star-nosed
Conepatus chinga — Skunk, Andes
Conger sp. — Eel, conger
Coniophanes imperialis — Snake, black-striped
Conium maculatum — Hemlock, poison
Connochaetes gnou — Gnu, white-tailed
Connochaetes taurinus — Gnu, brindled
Conocarpus erectus — Mangrove, gray (Buttonwood)
Conolophus subscristatus — Iguana, Galapagos
Conopholis americana — Squawroot or Cancer root
Conopophaga lineata — Gnateater, rufous
Contia tenuis — Snake, sharp-tailed
Contopus borealis — Flycatcher, olive-sided
Contopus pertinax — Flycatcher, Coue's
Contopus sordidulus — Pewee, wood (Western)
Contopus virens — Pewee, wood (Eastern)

Convallaria majalis — Lily-of-the-valley
Convolvulus arvensis — Bindweed, field
Convolvulus sepium — Bindweed, hedge
Cooloola propator — Cooloola monster
Cophixsalis sp. — Toad, narrow-mouthed
Cophosaurus texamus — Lizard, earless (greater)
Coprimulgus carolinensis — Chuck-will's-widow
Coprinus atramentarius — Ink cap, common
Coprinus comatus — Ink cap, shaggy
Coprinus micaceus — Ink cap, mica or glistening
Coptis groenlandica — Goldthread
Coracias sp. — Roller
Coracina sp. — Cuckoo-shrike
Coragyps atratus — Vulture, black
Corallina sp. — Algae, coralline
Corallium rubrum — Coral, precious
Corallorhiza sp. — Orchid, coral-root
Corallus caninus — Boa, tree (emerald)
Corallus sp. — Boa, tree
Cordyline australis — Dracaena, giant
Cordyline terminalis — Ti plant
Cordylus cataphractus — Lizard, armadillo
Cordylus giganteus — Sungazer
Coregonus autumnalis — Cisco
Coregonus sp. — Whitefish
Coreopsis gigantea — Tickseed, giant
Coreopsis lanceolata — Tickseed, lance-leaved
Coreopsis tinctoria — Tickseed
Corixa sp. — Water boatman
Cornus alternifolia — Dogwood, alternate-leaf or pagoda
Cornus canadensis — Bunchberry
Cornus florida — Dogwood
Cornus kousa — Dogwood, Kousa
Cornus nuttallii — Dogwood, Pacific
Cornus stolonifera — Dogwood, red-osier or Western
Coronella austriaca — Snake, smooth
Coronilla varia — Vetch, crown
Cortinarius cinnabarinus — Mushroom, web-cap or cort red or cinnabar
Cortinarius sp. — Mushroom, web-cap (other)
Corvus albus — Crow, pied
Corvus brachyrhynchos — Crow, American or common
Corvus caurinus — Crow, northwestern
Corvus corone — Crow, carrion
Corvus cryptoleucus albicollis — Raven, white-necked or Chihuahuan
Corvus frugilegus — Rook
Corvus monedula — Jackdaw
Corvus ossifragus — Crow, fish
Corydalidae — Dobsonfly
Corydalis caseana — Fitweed, Case's
Corylus americana — Hazelnut, American
Corylus colurna — Hazel, Turkish
Corylus cornuta — Hazelnut, beaked
Corynactis australis — Sea anemone, jewel
Coryphaena sp. — Dolphinfish
Coryphaenoides acrolepis — Grenadier, Pacific
Coryphantha vivipara — Cactus, cushion
Coryphella salmonacea — Nudibranch, salmon-gilled
Corythaixoides sp. — Go-away bird
Corythornis cristatas — Kingfisher, malachite
Coscinocera hercules — Moth, Atlas
Cossypha sp. — Robin-chat or Cape robin
Cotalpa lanigera — Beetle, goldsmith
Cotinus oboratus — Smoketree, American
Cotinus sp. — Beetle, June (green)
Cottus sp. — Sculpin

Coturnicops noveboracensis — Rail, yellow
Coturnix coturnix — Quail, European
Couroupita guianensis — Cannonball tree
Cowania mexicana — Cliffrose
Craseonycteris thonglongyai — Bat, hognosed
Craspedophora magnifica — Rifle bird, magnificent
Crateagus chrysocarpa — Hawthorne, fireberry
Crateagus columbiana — Hawthorne, Columbia
Crateagus crus-galli — Hawthorne, cockspur
Crateagus douglassii — Hawthorne, black
Crateagus flabellata — Hawthorne, fanleaf
Crateagus intricata — Hawthorne, Biltmore
Crateagus mollis — Hawthorne, downy
Crateagus monogyna — Hawthorne, oneseed or common
Crateagus phaenopyrum — Hawthorne, Washington
Crateagus succulenta — Hawthorne, fleshy
Crateagus viridis — Hawthorne, green
Crateagus sp. — Hawthorne (other)
Craterellus sp. — Horn of plenty fungus
Crax rubra — Curassow, great
Creagus furcatus — Gull, swallow-tailed
Crenichthys baileyi — Springfish, white river
Crepis acuminata — Hawk's beard
Crescentia cujete — Calabash tree
Crex crex — Crake, corn
Cricetomys sp. — Rat, giant pouched
Cricetus cricetus — Hamster, common
Crinozoa — Crinoid sea lilies and feather stars
Crinum americanum — Lily, swamp
Cristivomer namaycush — Trout, lake
Crocodylus acutus — Crocodile, American
Crocodylus sp. — Crocodile, African
Crocuta crocuta — Hyena, spotted
Crossoptilon auritus — Pheasant, blue-eared
Crotalus adamanteus — Rattlesnake, diamondback (Eastern)
Crotalus atrox — Rattlesnake, diamondback (Western)
Crotalus cerastes — Rattlesnake, sidewinder
Crotalus durissus — Rattlesnake, tropical or South American
Crotalus enyo — Rattlesnake, Baja California
Crotalus horridus — Rattlesnake, timber
Crotalus intermedius — Rattlesnake, Oaxacan small-headed
Crotalus lepidus — Rattlesnake, rock
Crotalus miliarius — Rattlesnake, pigmy
Crotalus mitchellii — Rattlesnake, speckled
Crotalus molossus — Rattlesnake, black-tailed
Crotalus pricei — Rattlesnake, twin-spotted
Crotalus ruber — Rattlesnake, diamondback (red)
Crotalus scrutulatus — Rattlesnake, Mojave
Crotalus tigris — Rattlesnake, tiger
Crotalus viridis — Rattlesnake, Western or Pacific or Prairie
Crotalus willardi — Rattlesnake, ridge-nosed
Crotaphytus collaris — Lizard, collared (common)
Crotaphytus insularis — Lizard, collared (black or desert)
Crotaphytus reticularis — Lizard, collared (reticulate)
Crotolaria spectabilis — Rattlebox, showy
Crotophaga ani — Ani, smooth-billed
Crotophaga sulcirostris — Ani, groove-billed
Cryptobranchus alleganiensus — Hellbender
Cryptoprocta ferox — Fossa
Cryptotis parva — Shrew, least
Crypturelles undulatus — Tinamou, undulated
Crytophora cristata — Seal, hooded
Ctenocephalides felis — Flea, cat

Ctenodactylus sp. — Gundi
Ctenosaura sp. — Iguana, spiny-tailed
Cuculus canorus — Cuckoo, European
Cucurbita foetidissima — Gourd, buffalo or wild
Culaea inconstans — Stickleback, brook
Culex pipiens — Mosquito, house
Cunila origanoides — Dittany
Cuon alpinus — Dog, dhole or red
Cuphea petiolata — Waxweed, clammy
Cupressus arizonica — Cypress, Arizona
Cupressus bakeri — Cypress, Baker
Cupressus goveniana — Cypress, gowen
Cupressus guadalupensis — Cypress, tecate
Cupressus macnabiana — Cypress, MacNab
Cupressus macrocarpa — Cypress, Monterey
Cupressus sargentii — Cypress, Sargent
Cursorius cursor — Courser, cream-colored
Cuscuta gronovii — Dodder
Cyanea sp. — Haha; Jellyfish, lion's mane
Cyanerpes cyaneus — Honey creeper, blue or red-
legged
Cyanocitta cristata — Jay, blue
Cyanocitta stelleri — Jay, Steller's
Cyanocorax morio — Jay, brown
Cyanocorax yncas — Jay, green
Cyanopica cyana — Magpie, azure-winged
Cyathus sp. — Bird's nest fungus
Cycas sp. — Cycad
Cyclemys dentata — Turtle, leaf
Cycleptus elongatus — Sucker, blue
Cyclopes didactylus — Anteater, silky
Cyclops sp. — Copepod
Cyclopsetta fimbriata — Flounder, spotfin
Cyclopterus lumpus — Lumpfish or Atlantic lumpsucker
Cyclorrhynchus psittacula — Auklet, Parakeet
Cyclura cornuta — Iguana, rhinoceros
Cydonia sinensis — Quince, Chinese
Cygnus atratus — Swan, black
Cygnus columbianus — Swan, tundra
Cygnus cygnus — Swan, whooper
Cygnus melanchoryphus — Swan, black-necked
Cygnus olor — Swan, mute
Cylindrophis sp. — Snake, pipe
Cymatogaster aggregta — Perch, shiner
Cynanthus latirostris — Hummingbird, broad-billed
Cynictis penicillata — Meerkat, red
Cynocephalus variegatus — Lemur, gliding
Cynocephalus volans — Colugo, Philippine (flying lemur)
Cynogale bennetti — Civet, otter
Cynoglossum grande — Hound's-tongue, Western
Cynomys leucurus — Prairie dog, white-tailed
Cynomys ludovicianus — Prairie dog, black-tailed
Cynomys sp. — Prairie dog (other)
Cynopterus sp. — Bat, fruit (short-nosed)
Cynoscion regalis — Weakfish
Cynoscion sp. — Trout, sea
Cynthia cardui — Butterfly, painted lady
Cynthia virginiensis — Butterfly, painted lady (American)
Cyperus retrofractus — Rush, sweet
Cypherotylus californica — Beetle, pleasing fungus
Cyphoma gibbosum — Flamingo tongue
Cyphomandra betacea — Tamarillo
Cypraea spadicea — Cowrie, chestnut
Cypraea tigris — Cowrie, tiger
Cyprinodon macularius — Pupfish, desert
Cyprinodon variegatus — Minnow, sheepshead
Cyprinus carpio — Carp

Cypripedium acaule — Lady's slipper, pink
Cypripedium arietinum — Lady's slipper, ram's head
Cypripedium calceolus — Lady's slipper, yellow
Cypripedium californicum — Lady's slipper, California
Cypripedium candidum — Lady's slipper, white small
Cypripedium fasciculatum — Lady's slipper, clustered
Cypripedium montanum — Lady's slipper, mountain
Cypripedium reginae — Lady's slipper, showy
Cypseloides niger — Swift, black
Cypselurus sp. — Flying fish
Cypsiurus parvus — Swift, palm
Cyrilla racemiflora — Titi (plant)
Cyrtonyx montezumae — Quail, Montezuma or
harlequin
Cysteodemus wislizeni — Beetle, blister desert
Cytisus scoparius — Broom, Scotch
Dacelo sp. — Kookaburra, laughing
Dactylis glomerata — Grass, orchard
Dactylomela sp. — Sea hare
Dactylopsila trivirgata — Possum, striped
Dactylopterus volitans — Gurnard, flying
Dactylorhiza aristata — Orchid, Fischer's
Dactylotum sp. — Grasshopper, lubber
Dalea formosa — Feather plume
Dalea spinosa — Smokethorn or desert smoketree
Dallia pectoralis — Blackfish, Alaskan
Dama dama — Deer, fallow
Damaliscus dorcas — Bontebok
Damaliscus lunatus — Sassaby
Damaliscus sp. — Topi or tsessebe
Danaus chrysippus — Butterfly, monarch (African)
Danaus gilippus berenice — Butterfly, queen
Danaus plexippus — Butterfly, monarch
Daphne cneorum — Garland flower
Daphne mezereum — Mezereon
Daphnia sp. — Flea, water
Daption capense — Petrel, painted or Cape pigeon
Darlingtonia californica — Cobra plant
Dasyatis americana — Stingray, Southern
Dasyatis sp. — Stingray
Dasycercus cristicauda — Mulgara
Dasylirion wheeleri — Sotol
Dasymus incontus — Rat, marsh African
Dasymutilla occidentalis — Wasp, cow killer
Dasymutilla sp. — Wasp, velvet ant
Dasypeltis scabra — Snake, egg-eating (African)
Dasyprocta sp. — Agouti
Dasypus novemcinctus — Armadillo, nine-banded
Dasyuroides byrnei — Kowari
Dasyurus sp. — Quoll
Datura discolor — Indian apple
Datura stramonium — Jimsonweed
Datura wrighti — Jimsonweed or thorn apple,
Southwestern
Daubentonia madagascariensis — Aye-aye
Daucus carota — Queen Anne's lace
Davidia involucrata — Dove tree
Decapterus sp. — Scad
Decodon verticillatus — Willow, water
Deilephila sp. — Hawkmoth, elephant
Deinagkistrodon acutus — Viper, sharp-nosed
Deirochelys reticularis — Turtle, chicken (long-necked)
Delichon urbica — Martin, house
Delphinapterus leucas — Beluga or belukha or white
whale
Delphinium ajacis — Larkspur, rocket
Delphinium nuttallianum — Larkspur, Nuttall's

Delphinium tricorne — Larkspur, spring
Delphinium virescens — Larkspur, prairie
Delphinus delphis — Dolphin, common
Dendeagapus canadensis — Grouse, spruce
Dendragapus obscurus — Grouse, blue
Dendroaspis sp. — Mamba, black or green
Dendrobates auratus — Frog, poison arrow (turquoise)
Dendrobates sp. — Frog, poison arrow or poison dart
Dendrocincla fuliginosa — Woodcreeper, plain brown
Dendrocopos leucotos — Woodpecker, white-backed
Dendrocygna autumnalis — Whistling duck, black-bellied
Dendrocygna bicolor — Whistling duck, fulvous
Dendrocygna viduata — Whistling duck, white-faced
Dendrogyra cylindrus — Coral, pillar
Dendrohyrax sp. — Hyrax, tree
Dendroica caerulescens — Warbler, black-throated blue
Dendroica castanea — Warbler, bay-breasted
Dendroica cerulean — Warbler, cerulean
Dendroica coronata — Warbler, yellow-rumped or myrtle
Dendroica discolor — Warbler, prairie
Dendroica dominica — Warbler, yellow-throated
Dendroica fusca — Warbler, Blackburnian
Dendroica kirtlandii — Warbler, Kirtland's
Dendroica magnolia — Warbler, magnolia
Dendroica nigrescens — Warbler, black-throated gray
Dendroica occidentalis — Warbler, hermit
Dendroica palmarum — Warbler, palm
Dendroica pensylvanica — Warbler, chestnut-sided
Dendroica petechia — Warbler, yellow
Dendroica pinus — Warbler, pine
Dendroica striata — Warbler, blackpoll
Dendroica tigrina — Warbler, Cape May
Dendroica townsendi — Warbler, Townsend's
Dendroica virens — Warbler, black-throated green
Dendroides sp. — Beetle, fire-colored
Dendrolagus sp. — Kangaroo, tree
Dendromecon rigida — Poppy, tree
Dendromus sp. — Mouse, African climbing
Dentaria sp. — Toothwort
Derestidae — Beetle, dermestid
Dermacentor sp. — Tick, wood
Dermatemys mawei — Turtle, river (Central American)
Dermochelys coriacea — Turtle, leatherback
Desmana moschata — Desman, Russian
Desmanthus illinoensis — Mimosa, prairie
Desmodeus rotundus — Vampire bat
Desmodium sp. — Tick-trefoil
Desmognathus aeneus — Salamander, Cherokee
Desmognathus auriculatus — Salamander, dusky (Southern)
Desmognathus brimleyorum — Salamander, dusky (Ouachia)
Desmognathus fuscus — Salamander, dusky (Northern)
Desmognathus imitator — Salamander, imitator
Desmognathus monticola — Salamander, seal
Desmognathus ochrophaeus — Salamander, dusky (mountain)
Desmognathus quadramaculatus — Salamander, blackbelly
Desmognathus welteri — Salamander, dusky (Black Mountain)
Desmognathus wrighti — Salamander, pigmy
Diabrotica sp. — Beetle, cucumber
Diacrisia virginica — Caterpillar, woolly bear (yellow)
Diadophis amabilis — Snake, ringneck (Pacific)
Diadophis punctatus — Snake, ringneck
Dianthus armeria — Pink, Deptford

Diapheromera femorata — Walkingstick, northern
Dicamptodon ensatus — Salamander, Pacific giant
Dicentra cucullaria — Dutchman's breeches
Dicentra eximia — Bleeding-heart, wild
Dicentra formosa — Bleeding-heart, Western
Dicerorhinus sumatrensis — Rhinoceros, Sumatran
Diceros bicornis — Rhinoceros, black
Dichelostemma ida-maia — Firecracker flower
Dichocoenia stokesi — Coral, starlet stoke's
Dichroanassa rufescens — Egret, reddish
Dichtromena colorata — Sedge, white-topped
Dicrostonyx groenlandicus — Lemming, Greenland
Dicrostonyx torquatus — Lemming, collared or arctic
Dicrurus sp. — Drongo
Didelphis virginiana — Opossum, Virginia
Diervilla lonicera — Honeysuckle, bush
Digitaria ischaemum — Crabgrass, smooth
Diglossa cyanea — Flower-piercer
Dinemellia dinemelli — Weaver, white-headed buffalo
Dinocardium robustum — Cockle, giant Atlantic
Dinomys branickii — Pacarana
Dinopium rafflesi — Woodpecker, three-toed olive-backed
Diodon sp. — Porcupinefish
Diomedea albatross — Albatross, short-tailed
Diomedea bulleri — Albatross, Buller's
Diomedea cauta — Albatross, shy or white-headed
Diomedea chlorhynchus — Albatross, yellow-nosed
Diomedea chrysostoma — Albatross, grey-headed
Diomedea epomophora — Albatross, royal
Diomedea exulans — Albatross, wandering
Diomedea immutabilis — Albatross, Laysan
Diomedea irrorata — Albatross, waved
Diomedea melanophrys — Albatross, black-browed
Diomedea nigripes — Albatross, black-footed
Diomedea palpebrata — Albatross, light mantled sooty
Dionaea muscipula — Venus-fly-trap
Dione vanillae — Fritillary, gulf
Dioscorea villosa — Yam, wild
Diospyros virginiana — Persimmon
Diospyros texana — Persimmon, Texas or Mexican
Diplectrum sp. — Perch, sand
Diplodactylus ciliaris — Gecko, spiny-tailed
Diploglossus lessorae — Galliwasp
Diploria sp. — Coral, brain
Dipodomys merriami — Rat, kangaroo (Merriam's)
Dipodomys microps — Rat, kangaroo (chisel-toothed)
Dipodomys ordii — Rat, kangaroo (Ord's)
Dipodomys spectabilis — Rat, kangaroo (banner-tailed)
Dipodomys sp. — Rat, kangaroo (other)
Dipsacus sylvestris — Teasel
Dipsosaurus dorsalis — Iguana, desert
Discina perlata — Mushroom, pig's-ears
Dispholidus typus — Boomslang
Disporum trachycarpum — Fairybell, wartberry
Distoechurus pennatus — Possum, feathertail
Dithyrea wislizenii — Spectacle pod
Dodecatheon alpinum — Shooting star, alpine
Dodecatheon meadia — Shooting star
Dodecatheon pauciflorum — Shooting star
Dodecatheon pulchellum — Shooting star, few-flowered
Dolabrifera sp. — Sea cat
Dolichonyx oryzivorus — Bobolink
Dolichotis sp. — Mara or Cavy, Patagonian
Dolomedes sp. — Spider, fisher
Donax variabilis — Coquina
Dorcatragus megalotis — Antelope, beira

Dorcopsis sp. — Wallaby, forest
Doridae — Sea lemon
Dormitator maculatus — Sleeper, fat
Dorosoma sp. — Shad
Dracaena draco — Dragon tree or dragon blood tree
Draco sp. — Dragon, flying
Drimys winteri — Winter's bark
Dromaius novaehollandiae — Emu
Dromas ardeola — Plover, crab
Dromidia sp. — Crab, sponge
Drosera filiformis — Sundew, threadleaved
Drosera rotundifolia — Sundew, common or roundleaved
Drosophila melanogaster — Fly, fruit
Dryas sp. — Avens, mountain
Drymarchon corais — Snake, indigo
Drymobius margaritiferus — Racer, speckled
Drymoluber dichrous — Racer, forest
Dryocampa rubicunda — Moth, rosy maple
Dryocopus pileatus — Woodpecker, pileated
Dryomys nitedula — Dormouse, forest
Dryophis prasinus — Snake, grass-green vine
Dryophis sp. — Snake, tree
Duchesnea indica — Strawberry, Indian
Ducula spilorrhoa — Pigeon, nutmeg
Dugong dugon — Dugong
Dulus dominicus — Palmchat
Dumetella carolinensis — Catbird, gray
Dusicyon gymnocercus — Fox, pampas
Dusicyon thous — Fox, crab-eating or savannah
Dusicyon vetulus — Fox, hoary
Dynastes hercules — Beetle, hercules or Rhinoceros
Dynastes tityus — Beetle, hercules Eastern
Dytiscidae — Beetle, diving
Dytiscus marginalis — Beetle, diving
Eacles imperialis — Moth, imperial
Eburophyton austinae — Orchid, phantom
Echeneis naucrates — Sharksucker
Echinaster sp. — Starfish, spiny
Echinocactus grusonii — Cactus, barrel
Echinocereus sp. — Cactus, calico or lace or rainbow
Echinocereus sp. — Pitaya
Echinocereus triglochidiatus — Cactus, claret-cup
Echinocystis lobata — Cucumber, wild or burr
Echinops telfairi — Tenrec, hedgehog (lesser)
Echinosorex gymnurus — Moonrat, greater or gymnure
Echis carinatus — Viper, saw-scaled
Echium vulgare — Blueweed or blue-devil
Eciton sp. — Ant, army
Eclectus roratus — Parrot, eclectus
Ectoprocta — Bryozoan
Ectypia clio — Moth, clio
Egretta alba — Egret, great or American white
Egretta ardesiaca — Heron, black
Egretta caerula — Heron, little blue
Egretta eulophotes — Egret, Swinhoe's or Chinese
Egretta garzetta — Egret, little
Egretta gularis or *sacra* — Heron, reef
Egretta intermedia — Egret, intermediate
Egretta novaehollandiae — Heron, white-faced
Egretta picata — Heron, pied
Egretta rufescens — Egret, reddish
Egretta thula — Egret, snowy
Egretta tricolor ruficollis — Heron, Louisiana or tricolored
Egretta tricolor — Heron, tricolored
Ehretia anacua — Anacua
Eichornia crassipes — Water hyacinth
Eira barbara — Tayra

Elaeagnus angustifolia — Olive, russian or oleaster
Elaeagnus umbellata — Olive, Autumn
Elagatis bipinnulata — Rainbow runner
Elanoides forficatus — Kite, swallow-tailed
Elanus caerulus — Kite, black-shouldered
Elanus leucurus — Kite, white-tailed
Elaphe guttata rosacea — Ratsnake, rosy
Elaphe guttata guttata — Ratsnake, red or corn snake
Elaphe helena — Snake, trinket
Elaphe longissima longissima — Snake, Aesculapian
Elaphe obsoleta spiloides — Ratsnake, gray
Elaphe obsoleta bairdi — Ratsnake, Baird's
Elaphe obsoleta quadravittata — Ratsnake, yellow
Elaphe obsoleta obsoleta — Ratsnake, black
Elaphe quatuorlineata — Snake, four-lined
Elaphe situla — Snake, leopard
Elaphe subocularis — Ratsnake, Trans-Pecos
Elaphe triaspis intermedia — Ratsnake, green
Elaphe vulpina — Snake, fox
Elaphe sp. — Ratsnake (other)
Elaphodus cephalophus — Deer, tufted
Elaphurus davidiensis — Deer, Pere David's
Elateridae — Beetle, click
Electrophorus electricus — Eel, electric
Eleodes armata — Beetle, skunk (desert)
Elephantulus sp. — Shrew, elephant
Elephas maximus — Elephant, Asian
Eliomys quercinus — Dormouse, garden or orchard
Elliottia racemosa — Elliottia or Southern-plume
Ellobius fuscocapillus — Mole-vole, Southern
Ellychnia californica — Firefly, western
Elodea canadensis — Waterweed, canadian
Elops saurus — Ladyfish
Emberiza citrinella — Yellowhammer
Emberiza melanocephala — Bunting, black-headed
Emberiza schoeniclus — Bunting, reed
Embioptera — Web-spinner
Embiotoca jacksoni — Perch, black
Embiotoca lateralis — Seaperch, striped
Embothrium coccineum — Fire tree or bush
Emerita sp. — Crab, mole
Emmenanthe penduliflora — Whispering bells
Empidonax alnorum — Flycatcher, alder
Empidonax difficilis — Flycatcher, western
Empidonax flaviventris — Flycatcher, yellow-bellied
Empidonax fulvifrons — Flycatcher, buff-breasted
Empidonax hammondii — Flycatcher, Hammond's
Empidonax minimus — Flycatcher, least
Empidonax oberholseri — Flycatcher, dusky
Empidonax trailli — Flycatcher, willow or traill's
Empidonax virescens — Flycatcher, acadian
Empidonax wrightii — Flycatcher, gray
Empis livida — Fly, empid or dance
Emydoidea blandingi — Turtle, Blanding's
Emys orbicularis — Turtle, pond European
Enallagma sp. — Damselfly
Encelia farinosa — Brittlebrush
Enceliopsis nudicaulis — Sunray
Encyclia tampense — Orchid, butterfly
Engraulis encrastiolus — Anchovy, European
Engraulis mordax — Anchovy, Northern or Pacific
Enhydra lutris — Otter, sea
Enodia portlandia — Butterfly, pearly eye
Enophryo bison — Sculpin, buffalo
Ensatina eschscholtzii oregonensis — Salamander, Oregon

Ensatina eschscholtzii klauberi — Salamander, large-blotched

Ensatina eschscholtzii platensis — Salamander, Sierra Nevada

Ensatina eschscholtzii picta — Salamander, Painted

Ensatina eschscholtzii croceater — Salamander, Yellow-blotched

Ensifera ensifera — Hummingbird, sword-billed

Ensis directus — Clam, razor

Entoloma sp. — Pinkgill fungus

Ephemera danica — Fly, may (Eurasian)

Ephippiorhynchus asiaticus — Jabiru

Ephippiorhynchus senegalensis — Stork, saddlebill

Ephthianura aurifrons — Chat, orange

Ephthianura tricolor — Chat, crimson

Ephydra cinerea — Fly, shore

Epicrates cenchria — Boa, rainbow

Epidendrum cochleatum — Orchid, clam shell

Epidendrum conopseum — Orchid, green-fly

Epidendrum tampense — Orchid, butterfly

Epifagus virginiana — Beechdrops

Epigaea repens — Arbutus, trailing

Epilobium angustifolium — Fireweed

Epilobium latifolium — Fireweed, dwarf

Epilobium obcordatum — Rock fringe

Epilobium sp. — Willow-herb

Epinephelus adscensionis — Rock hind

Epinephelus advensionis — Hind, rock

Epinephelus cruentatus — Graysby

Epinephelus drummondhayi — Hind, speckled

Epinephelus fulvus — Coney

Epinephelus guttatus — Hind, red

Epinephelus itajara — Jewfish

Epinephelus striatus — Grouper, Nassau

Epipactis gigantea — Orchid, stream

Epipactis sp. — Orchid, helleborine

Epixerus ebii — Squirrel, African palm

Eptesicus fuscus — Bat, big or common brown

Equetus lanceolatus — Jackknife fish

Equisetum sp. — Horsetail or scouring rush

Equus africanus — Ass, African

Equus burchelli — Zebra, common or plains

Equus grevyi — Zebra, Grevy's

Equus hemionus — Ass, Asiatic; Onager

Equus kiang — Kiang

Equus przewalskii — Horse, Przewalski's

Equus zebra — Zebra, mountain

Eremophila alpestris — Lark, horned or shore

Eremopterix sp. — Lark, finch

Erethizon dorsatum — Porcupine

Eretmochelys imbricata — Turtle, hawksbill

Eridiphas slevini — Snake, night

Erigeron canadensis — Horseweed

Erigeron divergens — Fleabane, spreading

Erigeron glaucosus — Daisy, seaside

Erigeron philadelphicus — Fleabane, common

Erigeron speciosus — Daisy, showy

Erigeron strigosus or *annuus* — Fleabane, daisy

Erignathus barbatus — Seal, bearded

Erimyzon oblongus — Chubsucker, creek

Erinaceus europaeus — Hedgehog, European

Eriobotrya japonica — Loquat

Eriocaulon septangulare — Pipewort

Eriogonum inflatum — Desert trumpet

Eriogonum sp. — Buckwheat, desert

Eriogonum umbellatum — Sulphur flower

Erioneuron pulchellum — Grass, fluff

Eriophorum sp. — Grass, cotton

Eriophyllum lanatum — Yarrow, golden

Eriophyllum wallacei — Daisy, woolly

Eristalis tenax — Fly, drone or hover

Eristalomyia tenax — Fly, drone

Erithacus rubecula — Robin, European

Erodium cicutarium — Filaree or Storksbill

Erora laeta — Hairstreak butterfly, early

Eryngium leavenworthii — Thistle, coyote

Eryngium yuccifolium — Rattlesnake-master

Erysimum asperum — Wallflower, plains

Erysimum capitatum — Wallflower, Western

Erysimum menziesii — Wallflower, Menzie's

Erysimum nivale — Wallflower, alpine

Erythrina abyssinica — Red-hot-poker tree

Erythrina herbacea — Coralbean, Southeastern

Erythrina sandwicensis — Wiliwili

Erythrina sp. — Coralbean

Erythrocebus patas — Monkey, patas

Erythrolamprus bizona — Snake, coral (false)

Erythronium americanum — Trout-lily

Erythronium grandiflorum — Lily, glacier

Erythronium montanum — Lily, avalanche

Eryx sp. — Boa, sand

Esacus magnirostris — Plover, great shore

Eschrichtius robustus — Whale, gray

Eschscholtzia californica — Poppy, California

Eschscholtzia mexicana — Poppy, Mexican or gold

Esox lucius — Pike

Esox masquinongy — Muskellunge

Esox niger — Pickerel, chain

Estigmene acraea — Moth, acraea

Etheostroma sp. — Darter (fish)

Eubalaena australis — Whale, right (Southern)

Eubalaena glacialis — Whale, right

Eucalyptus sp. — Eucalyptus

Euchaetias egle — Moth, tiger milkweed

Euchroea histrionica — Beetle, rose

Eucinostomus sp. — Mojarra

Eucnide sp. — Rock nettle

Eucommia ulmoides — Rubber tree, hardy

Euderma maculatum — Bat, spotted or pinto

Eudiscopus denticulus — Bat, disc-footed

Eudocimus albus — Ibis, white

Eudocimus ruber — Ibis, scarlet

Eudromia elegans — Tinamou, crested

Eudromias morinellus — Dotterel

Eudynamys scolopacea — Koel

Eudyptes chrysocome — Penguin, rock-hopper

Eudyptes chrysolophus — Penguin, macaroni

Eudyptes minor — Penguin, little blue or fairy

Eudyptes schlegi — Penguin, royal

Eugenes fulgens — Hummingbird, magnificent or Rivoli

Euglena sp. — Euglena

Eugomphodus taurus — Shark, grey nurse or sand tiger

Eumeces anthracinus — Skink, coal

Eumeces callicephalus — Skink, mountain

Eumeces egregius — Skink, mole

Eumeces fasciatus — Skink, five-lined

Eumeces gilberti — Skink, Gilbert

Eumeces inexpectatus — Skink, Southeastern five-lined

Eumeces laticeps — Skink, broadhead

Eumeces multivirgatus — Skink, many-lined

Eumeces obsoletus — Skink, Great Plains

Eumeces septentrionalis — Skink, prairie

Eumeces skiltonianus — Skink, Western

Eumeces tetragrammus — Skink, four-lined

Eumecichthys fiski — Unicornfish
Eumenes fraternus — Wasp, potter
Eumetopias jubatus — Sea lion, northern or Steller
Eumops sp. — Bat, mastiff
Eumorpha pandorus — Moth, sphinx (pandora)
Eunectes murinus or *notaeus* — Anaconda
Euonymous atropurpureus — Burningbush, eastern
Euonymus americanus — Strawberry bush
Euonymus europaeus — Spindle tree
Euoticus elegantulus — Bush baby, Western needle-nailed
Eupatorium rugosum — Snakeroot, white
Eupatorium sp. — Boneset
Eupatorium sp. — Joe-pye-weed
Euphagus carolinus — Blackbird, rusty
Euphagus cyanocephalus — Blackbird, Brewer's
Euphausia superba — Krill
Euphorbia albomarginata — Rattlesnake-weed
Euphorbia candelabrum — Candlelabra tree
Euphorbia cyparissias — Spurge, cypress
Euphorbia heterophylla — Poinciana, wild
Euphorbia marginata — Snow on the mountain
Euphorbia milii — Crown-of-thorns
Euphorbia pulcherrima — Poincettia, Christmas
Euphorbia sp. — Spurge
Euphractus sexcinctus — Armadillo, six-banded
Euphrasia americana — Eyebright
Euphydryas phaeton — Butterfly, baltimore
Euplectes sp. — Bishopbird
Eupleres goudoti — Falanouc
Eurema nicippe — Butterfly, sleepy orange
Eurycea bislineata — Salamander, two-lined
Eurycea guttolineata — Salamander, three-lined
Eurycea longicauda — Salamander, longtail
Eurycea lucifuga — Salamander, cave
Eurycea multiplicata — Salamander, many-ribbed
Eurycea neotenes — Salamander, Texas
Eurycea quadridigitata — Salamander, dwarf
Eurycea tynerensis — Salamander, Oklahoma
Euryceros prevostii — Helmet brid
Euryea tridentifera — Salamander, blind comal
Eurypharynx pelecanoides — Eel, gulper
Eurypyga helias — Bittern, sun
Eurystomus orientalis — Roller, Eastern broad-billed
Eustoma exaltatium — Gentian, seaside or catchfly
Eustoma grandiflorum — Gentian, prairie
Eutamias amoenus — Chipmunk, yellow-pine
Eutamias minimus — Chipmunk, least
Eutamias sp. — Chipmunk
Eutamias umbrinus — Chipmunk, Uinta or Colorado
Euthynnus alletteratus — Tunny
Euthynnus pelamis — Tuna, skipjack
Evolvulus arizonicus — Blue-eyes, Arizona
Excalfactoria chinensis — Quail, painted
Fagus grandifolia — Beech, American
Fagus sylvatica — Beech, European
Falco berigora — Falcon, brown
Falco columbarius — Merlin or Pigeon hawk
Falco eleonorae — Falcon, Eleonora's
Falco femoralis — Falcon, aplomado
Falco longipennis — Falcon, little
Falco mexicanus — Falcon, prairie
Falco peregrinus — Falcon, peregrine
Falco rupicoloides — Kestrel, greater
Falco rusticolus — Gyrfalcon
Falco sparverius — Kestrel, American
Falco subbuteo — Hobby, northern

Falco tinnunculus — Kestrel, common or Eurasian
Falco vespertinus — Falcon, red-footed
Falcunculus frontatus — Tit, shrike
Fallugia paradoxa — Apache plume
Farancia abacura — Snake, mud
Farancia erytrogramma — Snake, rainbow
Fasciolaria sp. — Snail, tulip
Feijoa sellowiana — Guava, pineapple
Felis badia — Cat, bay or Bornean red
Felis bengalensis — Cat, leopard
Felis caracal — Caracal
Felis chaus — Cat, jungle
Felis concolor — Mountain lion
Felis geoffroyi — Cat, Geoffroy's
Felis iriomotensis — Cat, Iriomote
Felis lynx — Lynx
Felis lynx pardina — Lynx, Pardel or Spanish
Felis manul — Cat, Pallas'
Felis margarita — Cat, sand
Felis nigripes — Cat, black-footed
Felis pardalis — Ocelot
Felis planiceps — Cat, flat-headed
Felis rufus — Bobcat
Felis serval — Serval
Felis silvestris or *libyca* — Cat, wild (African)
Felis sylvestris — Cat, wild (European or Scottish)
Felis temmincki — Cat, golden
Felis thinobia — Cat, desert Turkestan
Felis tigrinus — Cat, tiger or little spotted
Felis viverrina — Cat, fishing
Felis wiedi — Margay or Tree ocelot
Felis yagouaroundi — Jaguarundi
Feresa attenuata — Killer whale, pygmy
Ferocactus sp. — Cactus, barrel
Ficedula hypoleuca — Flycatcher, pied
Ficimia streckeri — Snake, hooknosed Mexican
Ficus destruens — Fig, strangler
Ficus sp. — Banyan
Filipendula rubra — Queen of the prairie
Filograna implexa — Worm, tube lace
Firmiana simplex — Parasol tree, Chinese
Fistularia petimba — Cornetfish, red
Fistulina hepatica — Beefsteak fungus
Flabellina iodinea — Nudibranch, spanish shawl
Flammulina velutipes — Mushroom, velvet-foot
Florida caerula — Heron, little blue
Fluvicola pica — Tyrant, pied water
Foeniculum vulgare — Fennel
Forestiera acuminata — Swamp privet
Forestiera segregata — Privet, Florida
Forficula auricularia — Earwig
Foricarius colma — Antthrush, rufous-capped
Formica rufa — Ant, wood
Formica sp. — Ant, red or mound
Forpus passerinus — Parrotlets, green-rumped
Fossa fossa — Civet, malagasy
Fouquieria splendens — Ocotillo
Fragaria chiloensis — Strawberry, beach
Fragaria vesca — Strawberry, wood
Fragaria virginiana — Strawberry, wild
Francolinus afer — Francolin, red-necked
Francolinus francolinus — Francolin, black
Frangula alnus — Alder, black
Franklinia alatamaha — Franklin tree
Frasera speciosa — Monument plant
Fratercula arctica — Puffin, Atlantic or common
Fratercula arctica — Puffin, horned

Fraxinus americana — Ash, white
Fraxinus anomala — Ash, singleleaf
Fraxinus caroliniana — Ash, Carolina
Fraxinus latifolia — Ash, Oregon
Fraxinus nigra — Ash, black
Fraxinus pennsylvanica — Ash, green
Fraxinus quadrangulata — Ash, blue
Fraxinus texensis — Ash, Texas
Fraxinus velutina — Ash, velvet
Fraxinus sp. — Ash other
Fregata magnificens — Frigatebird, magnificent
Fregata minor — Frigatebird, great
Fremontodendron californicum — Fremontia, California
Fringilla coelebs — Chaffinch
Fringilla montifringilla — Brambling
Fritillaria lanceolata — Mission bells
Fritillaria pluriflora — Lily, adobe
Fritillaria pudica — Yellow bell
Fritillaria recurva — Fritillary, scarlet
Fucus sp. — Rockweed
Fulica americana — Coot, American
Fulica atra — Coot, European
Fulica cristata — Coot, crested
Fumaria officinalis — Fumitory
Funambulus pennanti — Squirrel, palm
Fundulus sp. — Killifish
Fundulus sp. — Topminnow
Furmarus glacialis — Fulmar, northern
Furnarius rufus — Hornero, rufous
Gabianus scoresbyi — Gull, dolphin
Gadus morhua — Cod, Atlantic
Gaillardia pulchella — Indian blanket
Galago crassicaudatus — Bush baby, thick-tailed
Galago demidovii — Bush baby, dwarf
Galago senegalensis — Bush baby, lesser or Senegal
Galathea strigosa — Lobster, squat
Galbula ruficauda — Jacamar, rufous-tailed
Galemys pyrenaicus — Desman, Pyrenean
Galeocerdo cuvier — Shark, tiger
Galeorhinus galeus — Shark, tape
Galeorhinus zyopterus — Shark, soupfin
Galerida cristata — Lark, crested
Galictis cuja — Grison
Galidia elegans — Mongoose, ring-tailed
Galium aparine — Cleavers
Galium mollugo — Madder, wild
Galium sp. — Bedstraw
Gallinago gallinago — Snipe, common or Wilson's
Gallinula chloropus — Gallinule, common or Florida or moorhen
Gallirallus australis — Rail, flightless
Galloperdix spadicea — Spurfowl, red
Gallus gallus — Junglefowl, red
Gambelia silus — Lizard, leopard blunt-nosed
Gambelia wislizenii — Lizard, leopard long-nosed
Gambusia affinis — Mosquitofish
Gammarus oceanicus — Scud
Gammarus pulex — Shrimp, freshwater
Ganoderma applanatum — Artist's fungus
Garrodia nereis — Petrel, storm gray-backed
Garrulax leucolophus — Thrush, white-crested laughing
Garrulus glandarius — Jay
Garrya elliptica — Silktassel, wavyleaf
Gasterosteus aculeatus — Stickleback, three-spined
Gastropacha quercifolia — Moth, lappet
Gastrophryne sp. — Frog, narrowmouthed
Gastrotheca sp. — Frog, marsupial

Gaultheria procumbens — Checkerberry
Gaultheria shallon — Salal
Gaura coccinea — Gaura, scarlet
Gavia adamsii — Loon, yellow-billed
Gavia arctica — Loon, Arctic or black-throat
Gavia immer — Loon, common
Gavia pacifica — Loon, Pacific
Gavia stellata — Loon, red-throated
Gavialis gangeticus — Gavial
Gazella dama — Gazelle, Dama
Gazella dorcas — Gazelle, Dorcas
Gazella gazella — Gazelle, Indian or Edmi or mountain
Gazella granti — Gazelle, Grant's
Gazella leptoceros — Gazelle, slender-horned
Gazella subgutturosa — Gazelle, goitered
Gazella thomsoni — Gazelle, Thompson's
Geastrum sp. — Earthstar fungus
Gecarcinus sp. — Crab, land
Geijera parviflora — Willow, Australian
Gelochelidon nilotica — Tern, gull-billed
Gelsemium sempervirens — Jessamine, yellow
Genetta genetta — Genet, common or small spotted
Genetta maculata — Genet, panther
Genetta tigrina — Genet, large-spotted
Gentiana amarella — Gentian, Northern or rose
Gentiana andrewsii — Gentian, closed or bottle
Gentiana calycosa — Gentian, explorer's or blue
Gentiana crinita — Gentian, fringed
Gentiana dentosa — Gentian, fringed Western
Gentiana lutea — Gentian, great yellow
Geochelone elaphantopus — Tortoise, giant (Galapagos)
Geochelone gigantea — Tortoise, giant (Aldabran)
Geochelone pardalis — Tortoise, leopard
Geochelone sulcata — Tortoise, spurred
Geococcyx californianus — Roadrunner
Geolycosa sp. — Spider, wolf (burrowing)
Geomys bursarius — Pocket gopher, plains
Geomys pinetis — Pocket gopher, southeastern
Geopelia cuneata — Dove, diamond
Geophaps plumifera — Pigeon, plumed
Geospixa magnirostris — Finch, large ground
Geothlypis trichas — Yellowthroat, common
Geotrupes sp. — Beetle, dung
Geotrygon sp. — Dove, quail
Geranium maculatum — Geranium, wild
Geranium molle — Cranesbill, dovesfoot
Geranium richarsonii — Geranium, Richardson's
Geranium robertianum — Herb-Robert
Geranium viscosissimum — Geranium, sticky
Gerbillurus sp. — Gerbil, South African or Namib
Gerea canescens — Sunflower, desert
Gerres cinereus — Mojarra, yellowfin
Gerrhonotus coeruleus — Lizard, alligator (Northern)
Gerrhonotus kingi — Lizard, alligator (King or Arizona)
Gerrhonotus kingii — Lizard, alligator (Madrean)
Gerrhonotus liocephalus — Lizard, alligator (Texas)
Gerrhonotus multicarinatus — Lizard, alligator (Southern)
Gerrhonotus panamintinus — Lizard, alligator (Panamint)
Gerris sp. — Water strider
Gersemia rubiformis — Coral, soft red
Geum sp. — Avens
Geum triflorum — Prairie smoke
Gila bicolor — Chub, Tui
Gillenia trifoliata — Bowman's-root
Ginglymostoma cirratum — Shark, nurse
Ginkgo biloba — Ginkgo
Giraffa camelopardalis — Giraffe

Girella nigricans — Opaleye
Glareola sp. — Pratincole
Glaucidium brasilianum — Owl, ferruginous pygmy
Glaucidium gnoma — Owl, pygmy
Glaucidium perlatum — Owlet, pearl-spotted
Glaucomys sabrinus — Flying squirrel, Northern
Glaucomys volans — Flying squirrel, Southern
Glaudidium passerinum — Owl, pygmy Eurasian
Glechoma hederacea — Gill-over-the-ground
Gleditsia aquatica — Waterlocust
Gleditsia triacanthos — Locust, honey tree
Glis glis — Dormouse, edible or fat
Globicephala macrorhynchus — Whale, pilot (shortfin)
Globicephala melaena — Whale, pilot (long-fin)
Gloriosa rothschildiana — Lily, climbing or glory
Glossophaga sp. — Bat, long-tongued
Glyceria maxima — Sweet-grass, reed
Gnaphalium obtusifolium — Everlasting, sweet or catfoot
Gobiesox sp. — Clingfish
Gobiesox strumosus — Skilletfish
Gobio gobio — Gudgeon
Goliathus druryi — Beetle, goliath
Gomphus clavatus — Mushroom, pig's-ears
Gonatodes albogularis — Gecko, yellow-headed
Gonepteryx rhamni — Butterfly, brimstone
Gonyosoma sp. — Ratsnake, mangrove
Goodyera sp. — Orchid, rattlesnake plantain
Gopherus agassizii — Tortoise, desert
Gopherus berlandieri — Tortoise, berlandier's or Texas
Gopherus polyphemus — Tortoise, gopher
Gordonia lasianthus — Bay, loblolly
Gorgonia sp. — Sea fan
Gorgonocephalus arcticus — Basketstar, Northern
Gorilla gorilla — Gorilla
Gorilla gorilla beringei — Gorilla, mountain
Gorsachius goisagi — Night heron, Japanese
Gorsachius melanolophus — Night heron, Malayan
Goura cristata — Pigeon, crowned
Goura victoria — Pigeon, Victoria crowned
Grallaria sp. — Antpitta
Grampus griseus — Dolphin, Risso's or grey
Graphium marcellus — Swallowtail butterfly, zebra
Graphocephala coccinea — Leafhopper, redbanded or scarlet and green
Grapsus grapsus — Crab, sally light-foot
Graptemys geographica — Turtle, map
Graptemys pseudogeographica — Turtle, map (false)
Graptemys sp. — Turtle, map or sawback (other)
Grevillea robusta — Oak, silk
Grifolia frondosus — Hen of the woods
Grindelia squarrosa — Gumweed
Grus americana — Crane, whooping
Grus antigone — Crane, sarus
Grus canadensis — Crane, sandhill
Grus grus — Crane, Eurasian
Grus japonensis — Crane, Japanese or red-crowned
Grus leucogeramus — Crane, white Siberian, Asiatic, or great
Grus monachus — Crane, hooded
Grus nigricollis — Crane, blacknecked
Grus rubicundus — Crane, Australian
Grus vipio — Crane, white-naped
Grylloblatta campodeiformis — Rock crawler
Gryllotalpa hexadactyla — Cricket, mole (Northern)
Gryllotalpa vinae — Cricket, mole (European)
Gryllus sp. — Cricket, field
Guaiacum angustifolium — Lignum vitae, Texas

Guaiacum officinale — Lignum vitae
Guiraca caerulea — Grosbeak, blue
Gulo gulo — Wolverine
Gutierrezia sarothrae — Snakeweed, broom
Gyalopion sp. — Snake, hooknosed
Gygis alba — Tern, white or fairy
Gymnachirus melas — Sole, naked
Gymnobelideus leadbeateri — Possum, leadbeater's
Gymnocladus dioica — Coffee tree, Kentucky
Gymnogyps californianus — Condor, California
Gymnopis multiplicata — Caecilian
Gymnopithys leucaspis — Antbird, bicolored
Gymnorhinus cyanocephalus — Jay, pinyon
Gymnothorax sp. — Eel, moray
Gymnotus carapo — Knifefish, banded
Gymnura sp. — Ray, butterfly
Gypaetus barbatus — Lammergeier
Gypohierax angolensis — Vulture, palm-nut
Gyps bengalensis — Vulture, white-backed Asian
Gyps fulvus or *rupelli* — Vulture, griffon
Gyrinophilus palleucus — Salamander, Tennessee cave
Gyrinophilus porphyriticus — Salamander, spring
Gyrinus sp. — Beetle, whirligig
Gyromitra fastigiata — Elephant-ear fungus
Gyromitra sp. — Morel, false
Habenaria blephariglottis — Orchid, fringed (white)
Habenaria ciliaris — Orchid, fringed (yellow)
Habenaria clavellata — Orchid, wood
Habenaria cristata — Orchid, fringed (crested)
Habenaria dilatata — Orchid, leafy white
Habenaria fimbriata — Orchid, fringed (large purple)
Habenaria integra — Orchid, fringeless (yellow)
Habenaria lacera — Orchid, fringed (ragged or green)
Habenaria nivea — Orchid, snowy
Habenaria orbiculata — Orchid, round-leaved
Habenaria peramoena — Orchid, fringeless (purple)
Habenaria psycodes — Orchid, fringed (purple)
Habenaria repens — Orchid, water-spider
Habenaria sp. — Orchid, rein
Habenaria x bicolor — Orchid, fringed (bicolor)
Hackelia floribunda — Stickweed, many-flowered
Haematopus bachmani — Oystercatcher, black
Haematopus leucopodus — Oystercatcher, Magellanic
Haematopus ostralegus — Oystercatcher, European
Haematopus palliatus — Oystercatcher, American
Haemopsis sanguisuga — Leech, horse
Haemulon album — Margate
Haemulon aurolineatum — Tomtate
Haemulon flavolineatum — Grunt, french
Haemulon plumieri — Grunt, white
Haideotriton wallacei — Salamander, blind (Georgia)
Halcyon chloris — Kingfisher, white-collared or mangrove
Halcyon malimbica — Kingfisher, blue-breasted
Halcyon sancta — Kingfisher, sacred
Halesia carolina — Silverbell, Carolina
Halesia sp. — Silverbell (other)
Haliaectus leucocephalus — Eagle, bald
Haliaeetus sp. — Eagle, sea or white-tailed
Haliaeetus vocifer — Eagle, African fishing
Haliastur indus — Kite, brahminy
Halichoeres garnoti — Wrasse, yellowhead
Halichoeres radiatus — Puddingwife
Halichoeres sp. — Wrasse
Halichoerus grypus — Seal, gray
Haliclona oculata — Sponge, finger
Haliclona rubens — Sponge, finger red
Halietor pygmaeus — Cormorant, pygmy

Haliotis rufescens — Abalone, red
Halisidota maculata — Moth, tiger spotted
Halocynthia pyriformis — Seaperch
Halocyptena microsoma — Petrel, storm (least)
Halosaurus sp. — Halosaur
Hamamelis mollis — Witch hazel, Chinese
Hamamelis virginiana — Witch hazel
Hapalemur griseus — Lemur, gray gentle
Haplopappus spinulosus — Daisy, yellow spiny
Harengula sp. — Sardine
Harpia harpyja — Eagle, harpy
Hebeloma crustuliniforme — Mushroom, poison-pie
Helarctos malayanus — Bear, sun
Helenium sp. — Sneezeweed or bitterweed
Helianthemum canadense — Frostweed
Helianthemum scoparium — Rose, rock or sun
Helianthus annuus — Sunflower, common
Helianthus giganteus — Sunflower, tall
Helianthus strumosus — Sunflower, woodland
Helianthus tuberosus — Jerusalem artichoke
Heliconia rostrata — Lobster claw
Heliconius charitonius — Butterfly, zebra
Heliornis fulica — Sungrebe
Heliothis zea — Earworm, corn
Heliotropium convolvulaceum — Heliotrope, sweet-scented
Heliotropium curassavicum — Quail plant
Helmitheros vermivorus — Warbler, worm-eating
Heloderma suspectum — Gila monster
Helogale parvula — Mongoose, dwarf
Helonias bullata — Pink, swamp
Hemerobius sp. — Lacewing, brown
Hemicentetes nigriceps — Tenrec, streaked
Hemidactylium scutatum — Salamander, four-toed
Hemidactylus garnoti — Gecko, Indo-Pacific or fox
Hemidactylus turcicus — Gecko, Mediterranean
Hemigalus derbyanus — Civet, banded palm
Hemigrammas rhodostomus — Tetra, red-nosed
Hemigrapsus sp. — Crab, shore
Hemilepidotus sp. — Irish lord
Hemipepsis sp. — Wasp, tarantula hawk
Hemiphractus proboscideus — Frog, Casque-headed
Hemipristis elongatus — Shark, snaggletooth
Hemiprocne longipennis — Swift, crested
Hemipteronotus sp. — Razorfish
Hemitragus hylocrinus — Tahr, Nilgiri
Hemitragus jemlahicus — Tahr, Himalayan
Hemitripterus americanus — Sea raven
Henricia sanguinolenta — Starfish, blood or blood-star
Heosemys spinosa — Turtle, spined
Hepatica americana — Hepatica, round-lobed
Heracleum lantanum — Parsnip, cow
Hericium erinaceus — Mushroom, hedgehog
Hericium sp. — Tooth fungus
Hermeuptychia sasybius — Moth, satyr (Carolina)
Hermodice carunculata — Worm, fire
Hermosilla azurea — Zebraperch
Herpestes ichneumon — Mongoose, Egyptian
Herpestes sanguineus — Mongoose, slender
Herpestes sp. — Mongoose, Indian gray
Herpestes urva — Mongoose, crab
Herpeton tentaculatum — Snake, fishing
Herpetotheres cachinnans — Falcon, laughing
Hesperiphona vespertina — Grosbeak, evening
Hesperis matronalis — Dame's rocket
Hesperocallis undulata — Lily, desert
Hesperoleucus symmetricus — Roach, California

Heterodon nasicus — Hognose snake, Western
Heterodon platyrhinos — Hognose snake, Eastern
Heterodon simus — Hognose snake, Southern
Heterodontus francisci — Shark, horn
Heterodontus portusjacksoni — Shark, Port Jackson
Heterodontus sp. — Shark, bullhead or horn
Heterohyrax sp. — Hyrax, bush or yellow spotted
Heteromeles arbutifolia — Toyon
Heteroscelus incanus — Tattler, wandering (bird)
Heterotheca subaxillaris — Camphorweed
Heuchera sanguinea — Coral bells
Heuchera sp. — Alumroot
Hexabranchus sp. — Nudibranch, Spanish dancer
Hexagrammos sp. — Greenling
Hexanchus sp. — Shark, six-gilled
Hibiscus coccineus — Hibiscus
Hibiscus coulteri — Rosemallow, desert
Hibiscus denudatus — Pale face
Hibiscus moscheutos — Rosemallow
Hibiscus palustris — Rosemallow, swamp
Hibiscus rosa-sinensis — China rose
Hibiscus trionum — Flower-of-an-hour
Hieraaetus pennatus — Eagle, booted
Hieracium aurantiacum — Hawkweed, orange
Hieracium pratense — King devil
Hieracium venosum — Rattlesnake-weed
Himantopus himantopus — Stilt, black-winged or black-necked
Himatione sanguinea — Honey creeper, Hawaiian
Hippocamelus sp. — Huemal or guemal
Hippocampus sp. — Sea horse
Hippodamia convergens — Beetle, ladybird (convergent)
Hippoglossina stomata — Sole, bigmouth
Hippoglossoides platessoides — Plaice, American
Hippoglossus hippoglossus — Halibut, Atlantic
Hippoglossus stenolepis — Halibut, Pacific
Hippopotamus amphibius — Hippopotamus
Hipposideros sp. — Bat, leaf-nosed
Hippotragus equinus — Antelope, roan
Hippotragus niger — Antelope, sable
Hippuris vulgaris — Mare's-tail
Hirundo pyrrhonota — Swallow, cliff
Hirundo rustica — Swallow, barn
Histeridae — Beetle, hister
Histrio histrio — Sargassum fish
Histrionicus histrionicus — Duck, harlequin
Hoheria populnea — Lacebark
Holacanthus ciliaris — Angelfish, queen
Holacanthus tricolor — Rock beauty
Holbrookia lacerata — Lizard, earless (spot-tailed)
Holbrookia maculata — Lizard, earless (lesser,Northern, or speckled)
Holbrookia propingua — Lizard, earless (keeled)
Holcus lanatus — Grass, velvet
Holocentrus sp. — Soldierfish
Holocentrus spinfer or *rufus* — Squirrelfish
Holocynthia pyriformis — Sea peach
Holodiscus discolor — Creambush
Homarus americanus — Lobster, northern
Hoplopterus or *Antibyx armatus* — Plover, blacksmith
Hoploxypterus cayanus — Plover, pied
Hordeum jubatum — Grass, squirreltail
Hottonia inflata or *palustris* — Featherfoil
Houstonia caerulea or *serpyllifolia* — Bluets
Hudsonia tomentosa — Heath, beach
Huso huso — Beluga, Russian fish
Hyaena brunnea — Hyena, brown

Hyaena hyaena — Hyena, striped
Hyalophora cecropia — Moth, cecropia
Hyalophora euryalus — Moth, ceanothus silk
Hybopsis sp. — Chub
Hydra sp. — Hydra
Hydrachna sp. — Mite, water
Hydranassa tricolor — Heron, Louisiana
Hydrangea arborescens — Hydrangea, wild
Hydrastis canadensis — Goldenseal
Hydrobates pelagicus — Petrel, storm
Hydrobius sp. — Beetle, water scavenger
Hydrocharis morsus-ranae — Frogbit
Hydrochoerus hydrochaeris — Capybara
Hydrocotyle americana — Pennywort, water
Hydrocynus sp. — Tigerfish
Hydrolagus colliei — Chimaera or ratfish
Hydromantes bruus — Salamander, web-toed Limestone
Hydromantes platycephalus — Salamander, web-toed (Mount Lyell)
Hydromantes shastae — Salamander, web-toed (Shasta)
Hydrometra stagnorum — Water measurer
Hydromys chrysogaster — Rat, water (Australian)
Hydrophasianus chirungus — Jacana, pheasant-tailed
Hydrophyllum fendleri — Waterleaf, Fendler's
Hydrophyllum virginianum — Waterleaf, Virginia
Hydropotes inermis — Deer, water
Hydrurga leptonyx — Seal, leopard
Hygrophorous sp. — Wax cap fungus
Hyla andersoni — Treefrog, Pine Barrens
Hyla arborea — Treefrog, European
Hyla arenicolor — Treefrog, canyon
Hyla avivoca — Treefrog, bird-voiced
Hyla cadaverina — Treefrog, California
Hyla chrysoscelis — Treefrog, Gray
Hyla cinerea — Treefrog, green
Hyla crucifer — Treefrog, spring peeper
Hyla eximia — Treefrog, mountain
Hyla femoralis — Treefrog, pine woods
Hyla gratiosa — Treefrog, barking
Hyla regilla — Treefrog, Pacific
Hyla sp. — Treefrog
Hyla squirella — Treefrog, squirrel
Hyla versicolor — Treefrog, Cope's Gray
Hylactophryne augusti — Frog, barking
Hylobates concolor — Gibbon, crested, black, white-cheeked or concolor
Hylobates hoolock — Gibbon, Hoolock or white-browed
Hylobates klossi — Gibbon, Kloss's or Dwarf
Hylobates lar — Gibbon, white-handed or Lar
Hylobates lar moloch — Gibbon, Moloch or silvery
Hylobates lar mulleri — Gibbon, gray or Muller's
Hylobates lar pileatus — Gibbon, pileated or capped
Hylobates lar agilis — Gibbon, dark-handed or agile
Hylobates syndactylus — Siamang
Hylocharis leucotis — Hummingbird, white-eared
Hylochoerus meinertzhageni — Hog, giant forest
Hylocichla mustelina — Thrush, wood
Hylomys suillus — Moonrat, lesser
Hymenocallis sp. — Lily, spider
Hymenocerus sp. — Shrimp, harlequin or clown
Hymenoxysa grandiflora — Sunflower, alpine
Hyoscyamus niger — Henbane
Hypericum elodes or *virginicum* — St. John's-wort, marsh
Hypericum gentianoides — Pineweed
Hypericum perforatum — St. John's-wort, common
Hypericum punctatum — St. John's-wort, spotted

Hypericum spathulatum — St. John's-wort, shrubby
Hyperolius sp. — Frog, reed or rush
Hyperoodon ampullatus — Whale, bottlenose northern
Hyperoodon planifrons — Whale, bottlenose southern
Hyphessobrycon rubrostigma — Tetra, bleeding-heart
Hyphessobrycon serpae — Tetra, serpa
Hypochoeris radicata — Cat's ear
Hypopachus variolosus — Frog, sheep
Hyporhamphus unifasciatus — Halfbeak
Hypoxis hirsuta — Stargrass, yellow
Hypsicomus elegans — Worm, fan elegant
Hypsiglena torquata — Snake, night
Hypsignathus monstrosus — Bat, hammer-head
Hypsipetes madagascariensis — Bulbul, black
Hypsiprymnodon moschatus — Kangaroo, rat musky
Hypsopsetta guttulata — Turbot, diamond
Hypsurus caryi — Seaperch, rainbow
Hypsypops rubicunda — Garibaldi
Hyssopus officinalis — Hyssop
Hystrix africaeaustralis — Porcupine, Cape or South African
Hystrix brachyura — Porcupine, short-tailed
Hystrix cristata — Porcupine, crested or North African
Hystrix patula — Grass, bottlebrush
Ibidorhyncha struthersii — Ibisbill
Ibis ibis — Stork, yellow-billed
Ibis leucocephalus — Stork, painted
Icerya purchesi — Scale, cushiony cotton
Ichneumia albicauda — Mongoose, white-tailed
Ichneumonidae — Wasp, ichneumon
Ichthyomys sp. — Rat, fish-eating
Icichthys lockingtoni — Medusafish
Icosteus aenigmaticus — Ragfish
Ictalurus sp. — Bullhead
Icteria virens — Chat, yellow-breasted
Icterus cucullatus — Oriole, hooded
Icterus galbula — Oriole, northern (including Baltimore and Bullock's)
Icterus graduacauda — Oriole, Audubon's or black-headed
Icterus gularis — Oriole, Altamira
Icterus parisorum — Oriole, Scott's
Icterus pectoralis — Oriole, spot-breasted
Icterus spurius — Oriole, orchard
Ictinia mississippiensis — Kite, Mississippi
Ictiobus bubalus — Buffalo, smallmouth (fish)
Ictonyx striatus — Zorilla
Idahoa scapigera — Flatpod
Idiurus zenkeri — Flying squirrel, Zenker's
Iguana iguana — Iguana, common
Ilex amelanchier — Holly, Sarvis or serviceberry
Ilex aquifolium — Holly, European or English
Ilex cassina or *myrtifolia* — Dahoon
Ilex decidua — Possumhaw
Ilex opaca — Holly, American
Ilex sp. — Winterberry
Ilex vomitoria — Yaupon
Ilex sp. — Holly (other)
Iliamna rivularis — Globemallow, mountain or stream
Illex illecebrosus — Squid, shortfin
Illicium sp. — Anise-tree
Ilyocoris cimicioides — Saucer bug
Impatiens capensis — Touch-me-not, orange
Impatiens pallida — Touch-me-not, pale
Inachis io — Butterfly, peacock
Incisalia sp. — Butterfly, elfin
Indicator indicator — Honeyguide

Indri indri — Indri
Inia geoffrensis — River dolphin, Amazon or Bouto
Inula helenium — Elecampane
Ipomoea coccinea — Morning-glory, red
Ipomoea cristulata — Scarlet creeper
Ipomoea hederacea — Morning-glory, ivy-leaved
Ipomoea pandurata — Manroot or man-of-the-earth
Ipomoea pes-caprae — Morning-glory, beach
Ipomoea sp. — Morning-glory
Ipomopsis aggregata — Skyrocket
Ipomopsis longiflora — Pale trumpets
Ipomopsis sp. — Gilia
Irania gutturalis — Robin, Persian or white-throated
Ircinia campana — Sponge, vase
Irediparra gallinacea — Jacana, Australian
Iridoprocne bicolor — Swallow, tree
Iris cristata — Iris, dwarf crested
Iris douglasiana — Iris, Douglas'
Iris fulva — Iris, red
Iris pseudacorus — Flag, yellow
Iris tenax — Iris, tough-leaved
Iris versicolor — Flag, blue
Iris sp. — Iris (other)
Isia isabella — Caterpillar, woolly bear
Isomeria arborea — Bladderpod
Isoodon sp. — Bandicoot
Isopyrum biternatum — Rue anemone, false
Isotria sp. — Orchid, pogonia
Istiophorus platypterus — Sailfish
Isurus sp. — Mako
Ithaginis cruentus — Pheasant, blood
Ixobrychus exilis — Bittern, least
Ixobrychus sp. — Bittern
Ixora fulgens — Flame of the woods
Ixoreus naevius — Thrush, varied
Jacana spinosa — Jacana, Northern
Jacaranda mimosaefolia — Jacaranda
Jaculus sp. — Jerboa
Jasione montana — Sheeps-bit
Jeffersonia diphylla — Twinleaf
Jordanella floridae — Flagfish
Juglans californica — Walnut, Southern California
Juglans cinerea — Butternut
Juglans hindsii — Walnut, Northern California
Juglans major — Walnut, Arizona
Juglans microcarpa — Walnut, little
Juglans nigra — Walnut, black
Junco caniceps — Junco, gray-headed
Junco hyemalis — Junco, dark-eyed or slate-colored
Junco phaeonotus — Junco, yellow-eyed
Juncus conglomeratus — Rush, conglomerate or common
Juncus effusus — Rush, soft
Juncus inflexus — Rush, hard
Juniperus californica — Juniper, California
Juniperus communis — Juniper, common
Juniperus deppeana — Juniper, alligator
Juniperus erythrocarpa — Juniper, redberry
Juniperus monosperma — Juniper, oneseed
Juniperus occidentalis — Juniper, Western
Juniperus osteosperma — Juniper, Utah
Juniperus pinchotii — Juniper, pinchot
Juniperus scopulorum — Juniper, Rocky Mountain
Juniperus silicicola — Cedar, red (Southern)
Juniperus virginiana — Cedar, red (Eastern)
Junonia coenis — Butterfly, buckeye
Justicia americana — Water willow
Jynx torquilla — Wryneck

Kallstroemia grandiflora — Poppy, desert
Kalmia angustifolia — Laurel, sheep
Kalmia latifolia — Mountain-laurel
Kalmia microphylla — Laurel, alpine
Kalmia polifolia — Laurel, swamp
Kerodon rupestris — Cavy, rock
Ketupa sp. — Owl, fish
Kinixys sp. — Tortoise, Hingeback
Kinosternon flavescens — Turtle, mud (yellow)
Kinosternon hirtipes — Turtle, mud (Mexican or Big Bend)
Kinosternon sonoriense — Turtle, mud (Sonoran)
Kinosternon subrubrum — Turtle, mud (Eastern or common)
Kinosternon sp. — Turtle, mud (other)
Knautia arvensis — Bluebuttons
Kobus ellipsiprymnus — Waterbuck
Kobus kob — Kob
Kobus leche — Lechwe
Koelreuteria paniculata — Golden rain tree
Kogia breviceps — Whale, sperm (pygmy)
Kogia simus — Whale, sperm (dwarf)
Kosteletzkya virginica — Mallow, seashore
Krameria parvifolia — Ratany
Krigia virginica — Dandelion, dwarf
Kurtus sp. — Humphead or nurseryfish
Kyphosus sectatrix — Chub, Bermuda
Labroides sp. — Wrasse, cleaner
Lacerta muralis — Lizard, wall
Lacerta sp. — Lizard, green
Lacerta vivipara — Lizard, European
Lachesis muta — Bushmaster
Lachnolaimus maximus — Hogfish
Lactarius sp. — Milk cap fungus
Lactophrys sp. — Cowfish
Lactophrys sp. — Trunkfish
Lagenodelphis hosei — Dolphin, Fraser's or shortsnout
Lagenorhynchus acutus — Dolphin, white-sided (Atlantic)
Lagenorhynchus albirostris — Dolphin, white-beaked
Lagenorhynchus australis — Dolphin, Peale's or blackchin
Lagenorhynchus cruciger — Dolphin, hourglass
Lagenorhynchus obliquidens — Dolphin, white-sided (Pacific)
Lagenorhynchus obscurus — Dolphin, dusky
Lagerstroemia sp. — Crape myrtle
Lagidium pervatum — Viscacha, mountain
Lagocephalus laevigatus — Puffer, smooth
Lagodon rhomboides — Pinfish
Lagopus lagopus — Ptarmigan, willow
Lagopus leucurus — Ptarmigan, white-tailed
Lagopus mutus — Ptarmigan, rock
Lagorchestes conspicillatus — Wallaby, hare (spectacled)
Lagorchestes hirsutus — Wallaby, hare (rufous)
Lagostomus maximus — Viscacha, plains
Lagostrophus fasciatus — Wallaby, hare (banded)
Lagothrix lagothricha — Monkey, woolly
Laguncularia racemosa — Mangrove, white
Lagurus curtatus — Vole, sagebrush
Lagurus lagurus — Lemming, steppe
Lalage suerii — Triller, white-winged
Lama guanicoe or *huanacos* — Guanaco
Lama guanicoe f.glama — Alpaca
Lama guanicoe f. glama — Llama
Laminaria saccharina — Wrack, sugar
Lamium amplexicaule — Henbit
Lamium purpureum — Dead nettle, purple
Lampetra fluviatilis — Lampern
Lampetra planeri — Lamprey, brook

Lampetra tridentata — Lamprey, Pacific
Lampetra sp. — Lamprey, brook
Lampornis clemenciae — Hummingbird, blue-throated
Lampribis olivacea — Ibis, olive
Lampris guttatus — Opah
Lampropeltis calligaster — Kingsnake, mole or prairie
Lampropeltis getulus varieties — Kingsnake, common
Lampropeltis mexicana — Kingsnake, gray-banded or Mexican
Lampropeltis pyromelana — Kingsnake, Sonoran or Huachuca mountain
Lampropeltis triangulum — Kingsnake, scarlet
Lampropeltis triangulum — Snake, milk
Lampropeltis zonata — Kingsnake, California mountain or Coral
Lampyridae — Firefly
Langloisia matthewsii — Desert calico
Laniarius atrococcineus — Shrike, crimson-breasted
Lanius collaris — Shrike, fiscal
Lanius collurio — Shrike, red-backed
Lanius excubitor — Shrike, Northern or great gray
Lanius ludovicianus — Shrike, loggerhead
Lanius schach — Shrike, rufous-backed
Laportea canadensis — Nettle, wood
Larix decidua — Larch, European
Larix kaempferi — Larch, Japanese
Larix laricina — Tamarack
Larix lyallii — Larch, subalpine
Larix occidentalis — Larch, Western
Larosterna inca — Tern, Inca
Larrea tridentata — Creosote bush
Larus argentatus — Gull, herring
Larus atricilla — Gull, laughing
Larus belcheri — Gull, simeon
Larus brunnicephalus — Gull, brown-headed
Larus californicus — Gull, Californian
Larus canus — Gull, common or mew
Larus cirrocephalus — Gull, grey-headed
Larus delawarensis — Gull, ring-billed
Larus dominicanus — Gull, black-backed (Southern or kelp)
Larus fuliginosus — Gull, lava or dusky
Larus furcatus — Gull, swallow-tailed
Larus fuscus — Gull, black-backed (lesser)
Larus glaucescens — Gull, glaucous-winged
Larus glaudoides — Gull, Iceland
Larus heermanni — Gull, Heermann's
Larus hemprichi — Gull, sooty or Aden
Larus hyperboreus — Gull, glaucous
Larus leucophthalmus — Gull, white-eyed
Larus livens — Gull, yellow-footed
Larus maculipennis — Gull, black-headed (Patagonian)
Larus marinus — Gull, great or great black-backed
Larus minutus — Gull, little
Larus modestus — Gull, grey
Larus novaehollandiae — Gull, silver
Larus occidentalis — Gull, Western
Larus pacificus — Gull, Pacific
Larus philadelphia — Gull, Bonaparte's
Larus pipixcan — Gull, Franklin's
Larus ridibundus — Gull, black-backed or black-headed
Larus scoresbyi — Gull, dolphin or Magellan
Larus serranus — Gull, Andean
Larus thayeri — Gull, Thayer's
Lasionycteris noctivagans — Bat, silver-haired
Lasiorhinus latifrons — Wombat, hairy-nosed
Lasiurus borealis — Bat, red

Lasiurus cinereus — Bat, hoary
Lasiurus ega — Bat, yellow (Southern)
Lasiurus intermedias — Bat, yellow (Northern)
Lasiurus seminolus — Bat, Seminole
Laspeyresia pomonella — Moth, codling
Lasthenia chrysostoma — Goldfields
Laterallus jamaicensis — Rail, black
Lathyrus sp. — Pea, beach
Lathyrus venosus — Vetchling
Laticauda colubrina — Seasnake, yellow-lipped
Laticauda semifasciata — Seasnake, banded
Latimeria chalumnae — Coelacanth
Latrodectus mactans — Spider, black widow
Lavinia exilicauda — Hitch
Layia platyglossa — Tidy tips
Ledum glandulosum — Labrador tea
Leimadophis sp. — Snake, speckled or fire-bellied
Leiobunum sp. — Daddy-long-legs, eastern
Leiocephalus carinatus — Lizard, curltail
Leiocottus hirundo — Sculpin, lavendar
Leiophyllum buxifolium — Myrtle, sand
Leipoa ocellata — Mallee fowl
Lema trilineata — Beetle, potato
Lemmus lemmus — Lemming, Norway
Lemmus sibiricus — Lemming, brown
Lemna sp. — Duckweed
Lemur catta — Lemur, ring-tailed
Lemur fulvus — Lemur, brown
Lemur macaco — Lemur, black
Lemur mongoz — Lemur, mongoose
Leontideus rosalia — Marmoset, lion-headed
Leontopithecus chrysomelas — Tamarin, lion (golden-headed)
Leontopithecus rosalia rosalia — Tamarin, lion (golden)
Leontopithecus rosalia — Tamarin, lion (golden-rumped or black)
Leontopodium alpinum — Edelweiss
Leonurus cardiaca — Motherwort
Lepas sp. — Barnacle, goose
Lepidium sp. — Peppergrass
Lepidochelys kempi — Turtle, Ridley (Atlantic)
Lepidochelys olivacea — Turtle, Ridley (Pacific or Olive)
Lepidochelys sp. — Turtle, Ridley
Lepidocybium flavobrunneum — Escolar
Lepidopsetta bilineata — Sole, rock
Lepidosaphes ulmi — Scale, oyster shell
Lepilemur mustelinus — Lemur, sportive or weasel
Lepiota sp. — Mushroom, parasol
Lepisma saccharina — Silverfish, house
Lepisosteus sp. — Gar
Lepomis gibbosus — Sunfish, pumpkinseed
Lepomis gulosus — Sunfish, warmouth
Lepomis macrochirus — Bluegill
Leptinotarsa decemlineata — Beetle, Colorado potato
Leptodacylon californicum — Phlox, prickly
Leptodeira septentrionalis — Snake, cat-eyed
Leptonychotes weddelli — Seal, Weddell
Leptonycteris sp. — Bat, long-nosed
Leptophis sp. — Snake, parrot
Leptopterus sp. — Shrike, vanga
Leptoptilos crumeniferus — Stork, marabou
Leptosomus discolor — Cuckoo-roller
Leptotila verreauxi — Dove, white-tipped or white-fronted
Leptotyphlops dulcis — Snake, blind (Texas)
Leptotyphlops humilis — Snake, blind (Western)
Leptotyphlops sp. — Snake, thread
Lepus alleni — Jackrabbit, antelope

Lepus americanus — Hare, snowshoe or varying
Lepus arcticus — Hare, Arctic
Lepus brachyurus — Hare, Japanese
Lepus californicus — Jackrabbit, black-tailed
Lepus capensis — Hare, Cape
Lepus europaeus — Hare, European or field
Lepus timidus — Hare, Northern
Lepus townsendii — Jackrabbit, white-tailed
Lespedeza sp. — Bush-clover
Lethe eurydice — Butterfly, brown eyed
Lethe portlandia — Butterfly, pearly eye
Lethocerus americanus — Water bug, giant
Lethrinus sp. — Sweetlips
Leucana retusa — Leucaena, little-leaf
Leucocrinum montanum — Lily, sand
Leucopaxillus sp. — Mushroom, funnel-cap
Leucophyllum frutescens — Silverleaf, Texas
Leucosticte sp. — Finch, rosy
Leuresthes tenuis — Grunion
Leurognathus marmoratus — Salamander, shovelnose
Lewisia rediviva — Bitterroot
Liananthus aureus — Desert gold
Liasis sp. — Python
Liatris punctata — Gayfeather
Liatris sp. — Blazing-star
Libellula sp. — Dragonfly, darter
Libellula sp. — Dragonfly, skimmer
Libinia sp. — Crab, spider
Libocedrus decurrens — Cedar, incense
Lichanura trivirgata — Boa, rosy
Ligia oceanica — Sea slater
Ligia sp. — Sea roach
Ligustrum sinense — Privet, Chinese
Lilium canadense — Lily, Canada or yellow meadow
Lilium catesbaei — Lily, leopard
Lilium columbianum — Lily, tiger
Lilium philadelphicum — Lily, wood or Rocky Mountain
Lilium superbum — Lily, turk's cap
Lilium washingtonianum — Lily, Washington or Cascade
Lima sp. — File shell
Limanda limanda — Dab
Limenitis archippus — Butterfly, viceroy
Limenitis arthemis — Butterfly, admiral (white)
Limenitis populi — Butterfly, admiral (poplar)
Limnanthese douglasii — Meadow foam, Douglas
Limnaoedus ocularis — Frog, grass (little)
Limnocorax flavirostra — Crake, African black
Limnodromus griseus — Dowitcher, shortbilled
Limnodromus scolopaceus — Dowitcher, longbilled
Limnogale mergulus — Tenrec, aquatic
Limnothlypis swainsonii — Warbler, Swainson's
Limonium carolinianum — Sea lavender
Limosa fedoa — Godwit, marbled
Limosa haemastica — Godwit, Hudsonian
Limosa lapponica — Godwit, bar-tailed
Limosa limosa — Godwit, black-tailed
Limulus polyphemus — Crab, horseshoe
Linanthus androsaceus — Baby stars, false
Linanthus montanus — Clover, mustang
Linaria canadensis — Toadflax, blue
Linaria vulgaris — Butter-and-eggs
Lindera benzoin — Spicebush
Linium sp. — Flax
Linnaea borealis — Twinflower
Liocarcinus sp. — Crab, swimming
Liodytes alleni — Snake, swamp (striped)
Liomys irroratus — Mouse, pocket (Mexican spiny)

Liparis sp. — Snailfish
Lipogramma sp. — Basslet
Lipotes vexillifer — River dolphin, Yangtze or Chinese
Liquidambar styraciflua — Sweet gum
Liriodendron tulipifera — Poplar, yellow
Lissodelphis borealis — Dolphin, northern right whale
Listera sp. — Orchid, twayblade
Lithocarpus densiflorus — Tanoak
Lithophragma parviflorum — Prairie star
Lithospermum sp. — Gromwell or Puccoon
Litocranius walleri — Gerenuk
Litoria caerulea — Treefrog, White's
Litoria sp. — Treefrog, red
Lloydia serotina — Lily, alpine
Lobelia cardinalis — Cardinalflower
Lobelia inflata — Tobacco, Indian
Lobelia siphilitica — Lobelia, great
Lobelia spicata — Lobelia, spiked
Lobodon carcinophagus — Seal, crabeater
Lobotes surinamensis — Tripletail
Locustella — Warbler, grasshopper
Loiseleuria procumbens — Azalea, alpine
Loligo opalescens — Squid, opalescent
Loligo sp. — Squid
Lonchura striata — Munia, white-backed
Lonicera involucrata — Twinberry
Lonicera japonica — Honeysuckle, Japanese
Lonicera sempervirens — Honeysuckle, trumpet
Lophiomys imhausii — Rat, crested or maned
Lophius americanus — Goosefish, American
Lophodytes cucullatus — Merganser, hooded
Lopholithodes sp. — Crab, king
Lophophora williamsii — Peyote
Lophophorus impejanus — Pheasant, Himalayan monal
Lophortyx californicus — Quail, California
Lophortyx gambelii — Quail, Gambel's
Lophotus lacepedei — Crestfish
Lophura nycthemera — Pheasant, silver
Lophuromys sikapusi — Mouse, harsh-furred
Loris tardigradus — Loris, slender
Lota lota — Burbot
Lotus corniculatus — Birdsfoot trefoil
Lotus scoparis — Deer-weed
Lovenia sp. — Sea urchin, heart
Loxia curviostra — Crossbill, common or red
Loxia leucoptera — Crossbill, white-winged
Loxia pytopsittacus — Crossbill, parrot
Loxodonta africana — Elephant, African
Lucanus cervus — Beetle, stag (European)
Lucanus elaphus — Beetle, stag (giant)
Lucilia illustris — Fly, blow
Ludwigia alternifolia — Seedbox
Luetkea pectinata — Partridge-foot
Luina hypoleuca — Luina, silvery
Lullula arborea — Woodlark
Lumbricus terrestris — Earthworm
Lunaria annua — Honesty
Lunatia sp. — Snail, moon
Lunda cirrhata — Puffin, tufted
Lupinus arboreus — Lupine, tree
Lupinus bicolor — Lupine, miniature
Lupinus perennis — Lupine, wild
Lupinus polyphyllus — Lupine, blue-pod
Lupinus sp. — Lupine
Lupinus sparsiflorus — Lupine, Coulter's
Lupinus texensis — Bluebonnet, Texas
Luscinia megarhynchos — Nightingale

Luscinia svecica — Bluethroat
Lutjanus apodus — Schoolmaster
Lutjanus sp. — Snapper
Lutra canadensis — Otter, river (North American)
Lutra felina — Otter, marine
Lutra longicaudus — Otter, river (South American)
Lutra lutra — Otter, river (Eurasian)
Lutra maculicollis — Otter, spotted neck
Lutra perspicillata — Otter, smooth
Lutra provocax — Otter, river (Southern)
Lycaeides argyrognomon — Butterfly, blue (Northern)
Lycaena sp. — Butterfly, copper
Lycaon pictus — Dog, hunting or painted
Lychnis alba — Campion, white
Lychnis dioica — Campion, red
Lychnis flos-cuculi — Ragged robin
Lychnis fulgens — Campion, brilliant
Lycidae — Beetle, net-winged
Lycodes sp. — Eelpout
Lycodon sp. — Snake, wolf
Lycoperdon pratense or *perlatum* — Puffball, meadow
Lycosa carolinensis — Spider, wolf (Carolina)
Lycosa gulosa — Spider, wolf (forest)
Lygaeus kalmii — Milkweed bug, small
Lymantria dispar — Moth, gypsy
Lymnocryptes minima — Snipe, Jack
Lynchailurus pajeros — Cat, pampas
Lyncodon patagonicus — Weasel, Patagonian
Lyonia sp. — Lyonia or Maleberry
Lyonothamnus floribundus — Lyontree
Lyophyllum sp. — Mushroom, fried chicken
Lyrurus tetrix — Grouse, black
Lysichitum americanum — Skunk cabbage, yellow
Lysimachia nummularia — Moneywort
Lysimachia sp. — Loosestrife
Lysimachia terrestris — Swamp candles
Lysmata grabhami — Shrimp, red-backed cleaning or Scarlet Lady
Lysmata wurdemanni — Shrimp, peppermint or veined
Lythrum salicaria — Loosestrife, purple
Lytta sp. — Beetle, blister or oil
Macaca fascicularis — Macaque monkey, crab-eating
Macaca fuscata — Macaque monkey, Japanese or red-faced
Macaca mulatta — Macaque monkey, rhesus
Macaca nemestrina — Macaque monkey, pig-tailed
Macaca nigra — Ape, black Sulawesi or Celebes
Macaca radiata — Macaque monkey, bonnet
Macaca silenus — Macaque monkey, lion-tailed
Macaca sylvanus — Ape, Barbary
Machaeranthera bigelovii — Aster, sticky
Machaeranthera tanacetifolia — Daisy, Tahoka
Machaeranthera tortifolia — Aster, Mohave
Machaerocarpus californicus — Water-plantain, fringed
Maclura pomifera — Osage orange
Macroclemys temminckii — Snapping turtle, Alligator
Macrocystis sp. — Kelp, giant or bull
Macroderma gigas — Bat, ghost
Macroglossum stellatarum — Hawkmoth, hummingbird
Macronectes giganteus — Petrel, giant
Macronyx sp. — Longclaw
Macroperipatus sp. — Worm, velvet
Macropus eugenii — Wallaby, Tammar or Dama
Macropus giganteus — Kangaroo, gray (Eastern)
Macropus robustus — Wallaroo
Macropus rufus — Kangaroo, red
Macrorhamphosus sp. — Snipefish

Macrotis lagotis — Bilby or rabbit bandicoot
Macrotus californicus — Bat, leaf-nosed (California)
Macrotus fuliginosus — Kangaroo, gray (Western)
Macrozoarces americanus — Pout, ocean
Madoqua kirkii — Dikdik, Kirk's
Madoqua sp. — Dikdik
Magalops cyprinoides — Herring, ox-eye
Magicicada sp. — Cicada, seventeen year or periodical
Magnolia acuminata — Cucumber tree
Magnolia ashei — Magnolia, Ashe or sandhill
Magnolia fraseri — Magnolia, Fraser
Magnolia grandifolia — Magnolia, Southern
Magnolia macrophylla — Magnolia, bigleaf
Magnolia pyramidata — Magnolia, pyramid or mountain
Magnolia tripetela — Magnolia, umbrella
Magnolia virginiana — Sweet bay
Magnolia x soulangiana — Magnolia, saucer
Maianthemum canadense — Mayflower, Canada
Maianthemum sp. — Lily-of-the-valley, false and wild
Makaira nigricans — Marlin, blue
Malaclemys terrapin — Terrapin, diamondback
Malacochersus tornieri — Tortoise, pancake
Malacosoma sp. — Caterpillar, tent
Malacosteus niger — Loosejaw
Malacothrix coulteri — Snakehead
Malacothrix glabrata — Dandelion, desert
Malapterurus electricus — Catfish, electric
Malaxis sp. — Orchid, adder's mouth
Malayemys subtrijuga — Turtle, snail-eating
Malpolon monspessulanus — Snake, Montpellier
Malurus sp. — Wren, fairy
Malus sp. — Apple, crab
Malus sylvestris — Apple, crab
Malva moschata — Mallow, musk
Malvastrum rotundifolium — Five-spot, desert
Mamestra brassicae — Moth, cabbage
Mamillaria sp. — Cactus, fishhook
Mandevilla splendens — Allamanda
Mandragora officinarum — Mandrake
Manduca quinquemaculata — Hornworm, tomato
Mangifera indica — Mango
Manis sp. — Pangolin
Manorina melanophrys — Miner, bell
Manouria emys — Tortoise, Asian brown
Manta sp. — Manta ray
Mantis religiosa — Mantis, praying
Mantispa sp. — Fly, mantid or false mantid
Marasinius oreades — Mushroom, fairy-ring
Marmaronetta angustirostris — Teal, marbled
Marmosa sp. — Opossum, mouse
Marmota flaviventris — Marmot, yellow-bellied
Marmota monax — Woodchuck
Marmota sp. — Marmot
Martes americana — Marten, American
Martes pennanti — Fisher
Martes sp. — Marten
Martes zibellina — Sable
Mastacembelus congicus — Eel, spiny
Masticophis aurigulus — Whipsnake, Cape or red
Masticophis bilineatus — Whipsnake, Sonoran
Masticophis flagellum — Snake, coachwhip
Masticophis lateralis — Racer, striped
Masticophis taeniatus — Whipsnake, striped or ornate
Maurandya sp. — Snapdragon-vine
Mauremys caspica — Turtle, Caspian
Mazama americana — Deer, brocket
Meda fulgida — Spikedace

Medeola virginiana — Cucumber-root, Indian
Medialuna californiensis — Halfmoon
Medicago lupulina — Nonesuch or black medick
Megaceryle torquata — Kingfisher, ringed
Megachasma sp. — Shark, megamouth
Megachile sp. — Bee, leafcutter
Megaderma lyra — Vampire bat, false (greater)
Megadyptes antipodes — Penguin, yellow-eyed
Megalaima franklinii — Barbet, golden-throated
Megalochelys gigantea — Tortoise, giant (Seychelles)
Megapodius freycinet — Scrubfowl
Megaptera novaeangliae — Whale, humpback
Megarhynchus pitangua — Flycatcher, boat-billed
Megisto cymela — Moth, satyr (little wood)
Megophrys monticolo nasuta — Toad, spadefoot
Megophrys sp. — Frog, horned
Melampodium leucanthum — Daisy, blackfoot or desert
Melanargia galathea — Butterfly, marbled white
Melanerpes aurifrons — Woodpecker, golden-fronted
Melanerpes carolinus — Woodpecker, red-bellied
Melanerpes erythrocephalus — Woodpecker, redheaded
Melanerpes formicivorus — Woodpecker, acorn
Melanerpes lewis — Woodpecker, Lewis
Melanerpes uropygialis — Woodpecker, gila
Melanitta fusca — Scoter, white-winged
Melanitta nigra — Scoter, black
Melanitta perspicillata — Scoter, surf
Melanogrammus aeglefinus — Haddock
Meleagris gallopavo — Turkey
Meles meles — Badger, Eurasian
Melia azedarach — Chinaberry or china tree
Melichthys niger — Durgon, black
Melierax metabates — Goshawk, chanting (dark)
Melierax poliopterus — Goshawk, chanting (pale)
Melilotus sp. — Clover, sweet
Mellisuga helenae — Hummingbird, bee
Mellivora capensis — Badger, honey or ratel
Melolontha melolontha — Beetle, cockchafer
Melopsittacus undulatus — Budgerigar
Melospiza georgiana — Sparrow, swamp
Melospiza lincolnii — Sparrow, Lincoln's
Melospiza melodia — Sparrow, song
Melursus ursinus — Bear, sloth
Menidia sp. — Silverside
Menippe mercenaria — Crab, stone (Florida)
Menispermum canadense — Moonseed
Mentha aquatica — Mint, water
Mentha arvensis — Mint, field
Mentha piperita — Peppermint
Mentha spicata — Spearmint
Menticirrhus sp. — Kingfish
Mentzelia laevicaulis — Blazing-star
Menura novaehollandiae or *superba* — Lyrebird, Superb
Menyanthes trifoliata — Buckbean
Mephitis macroura — Skunk, hooded
Mephitis mephitis — Skunk, striped
Mercurialis perennis — Mercury, dog's or herb
Merganetta armata — Duck, torrent
Mergus albellus — Smew
Mergus merganser — Merganser, common
Mergus serrator — Merganser, red-breasted
Mergus squamatus — Merganser, Chinese
Meriones sp. — Gerbil, Lybian or Mongolian
Merlangius merlangus — Whiting
Merluccius sp. — Hake
Merops apiaster — Bee-eater, European
Merops ornatus — Bee-eater, rainbow

Merops superciliosus — Bee-eater, blue-cheeked
Mertensia alpina — Bluebells, alpine
Mertensia ciliata — Bluebells, mountain
Mertensia virginica — Bluebells, Virginia
Mesembryanthemum chilense — Sea fig
Mesembryanthemum crystallinum — Iceplant, common
Mesitornis sp. — Mesite
Mesocricetus auratus — Hamster, golden
Mesoplodon bidens — Whale, beaked (Sowerby's or North Sea)
Mesoplodon carlhubbsi — Whale, beaked (Hubb's or Arch)
Mesoplodon densirostris — Whale, beaked (Blainville's or dense)
Mesoplodon europaeus — Whale, beaked (Gervais' or Antillean)
Mesoplodon ginkgodens — Whale, beaked)ginkgo-toothed)
Mesoplodon grayi — Whale, beaked (Gray's or Scamperdown)
Mesoplodon hectori — Whale, beaked (Hector's)
Mesoplodon layardii — Whale, beaked (Layard's or straptoothed)
Mesoplodon mirus — Whale, beaked (True's or wonderful)
Mesoplodon stejnegeri — Whale, beaked (Stejneger's or Bering sea)
Mespilus germanica — Medlar
Metasequoia glyptostroboides — Redwood, dawn
Metasyrphus americanus — Fly, hover or flower
Metcalfia pruinosa — Plant hopper
Metridium sp. — Sea anemone, plumose or feather
Micrastur ruficollis — Falcon, barred forest
Micrathene whitneyi — Owl, elf
Microcebus sp. — Lemur, mouse
Microciona prolifera — Sponge, red beard
Microdesmus sp. — Wormfish
Microdipodops sp. — Mouse, kangaroo
Microgadus sp. — Tomcod
Microgale longicaudata — Tenrec, shrew
Microgale melanorrachis — Tenrec, long-tailed
Microglossum — Earthtongue fungus
Microhierax caerulescens — Falconet, collared or red-legged
Micromys minutus — Mouse, harvest
Micropalama himantopus — Sandpiper, stilt
Micropogonias undulatus — Croaker, Atlantic
Micropterus dolomieui — Bass, smallmouth
Micropterus punctulatus — Bass, spotted
Micropterus salmoides — Bass, largemouth
Microsorex hoyi — Shrew, pygmy
Microspathodon chrysurus — Damselfish, yellowtail
Microtus agrestis — Vole, field
Microtus arvalis — Vole, common
Microtus chrotorrhinus — Vole, rock
Microtus longicaudus — Vole, long-tailed
Microtus montanus — Vole, mountain
Microtus ochrogaster — Vole, prairie
Microtus pennsylvanicus — Vole, meadow
Microtus pinetorum — Vole, woodland
Microtus subterraneus — Vole, pine
Microtus sp. — Vole (other)
Micrura leidyi — Worm, ribbon (pink)
Micruroides euryxanthus — Snake, coral (Western or Arizona)
Micruroides fulvius — Snake, coral (Eastern)
Micrurus sp. — Snake, coral

Mictyris sp. — Crab, soldier
Mikania scandens — Boneset, climbing
Miliaria calandra — Bunting, corn
Millepora sp. — Coral, fire
Milvago chimachino — Caracara, yellow-headed
Milvus milvus — Kite, red
Mimosa pudica — Sensitive plant
Mimulus moschatus — Muskflower
Mimulus sp. — Monkeyflower
Mimus polyglottos — Mockingbird
Minytrema melanops — Sucker, spotted
Miopithecus talapoin — Talapoin
Mirabilis multiflora — Four o'clock, desert or Colorado
Mirounga angustirostris — Seal, elephant (Northern)
Mirounga leonina — Seal, elephant (Southern)
Mitchella repens — Partridgeberry
Mitella sp. — Mitrewort or Bishop's cap
Mithrax sp. — Crab, spider
Mitsukurina owstoni — Shark, goblin
Mnemiopsis maccradyi — Jellyfish, comb
Mniotilta varia — Warbler, black and white
Modiola caroliniana — Mallow, Carolina
Moenkhausia oligolepis — Tetra, glass
Mohavea confertiflora — Ghost-flower
Mola mola — Sunfish, oceanic
Moloch horridus — Thorny devil
Molothrus aeneus — Cowbird, bronzed
Molothrus ater — Cowbird, brown-headed
Molva molva — Ling, European
Momotus momota — Motmot, blue-crowned
Monacanthus sp. — File fish
Monachus monachus — Seal, monk (Mediterranean)
Monachus schauinslandi — Seal, monk (Hawaiian)
Monarda didyma — Bee-balm
Monarda fistulosa — Bergamot, wild
Monardella odoratissima — Mint, coyote
Monasa nigrifrons — Nunbird, black-fronted
Moneses uniflora — Wood nymph
Monochamus sp. — Beetle, sawyer
Monodon monocerus — Narwhal
Monomorium minimum — Ant, black (little)
Monoptilon bellioides — Desert star, Mohave
Monotropa hypopithys — Pinesap
Monotropa uniflora — Indian pipe
Montastrea sp. — Coral, star
Montastrea sp. — Coral, boulder
Montia cordifolia — Montia, broad-leaved
Montia perfoliata — Lettuce, miner's
Monticola sp. — Thrush, rock
Montifringilla nivalis — Finch, snow
Morchella sp. — Morel
Mordella sp. — Beetle, flower (tumbling)
Morelia sp. — Python, carpet
Mormoops megalophylla — Bat, ghost-faced
Morone americana — Perch, white
Morone chrysops — Bass, white
Morone mississippiensis — Bass, yellow
Morone saxatilis — Bass, striped
Morpho sp. — Butterfly, morpho
Morus alba — Mulberry, white
Morus bassanus — Gannet, northern
Morus capensis — Gannet, Cape
Morus microphylla — Mulberry, Texas
Morus rubra — Mulberry, red
Moschus sp. — Deer, musk
Motacilla alba — Wagtail, pied
Motacilla cinerea — Wagtail, gray

Motacilla flava — Wagtail, yellow
Moxostoma sp. — Redhorse, river or Jumprock
Mucidula mucida — Porcelain fungus
Mugil sp. — Mullet
Mulloidichthys sp. — Goatfish
Mullus auratus — Goatfish, red
Mungos mungo — Mongoose, banded
Mungotictis decemlineata — Mongoose, narrow-striped
Muntiacus sp. — Muntjac
Murexia sp. — Marsupial mouse
Mus musculus — Mouse, house
Musa x *paradisiaca* — Banana or plantain
Musca domestica — Fly, house
Muscardinus avellanarius — Dormouse, common or hazel
Muscicapa striata — Flycatcher, spotted
Muscivora forficata — Flycatcher, scissor-tailed
Mussa angulosa — Coral, flower
Mustela erminea — Ermine
Mustela eversmanni — Polecat, steppe
Mustela felipei — Weasel, water
Mustela frenata — Weasel, long-tailed
Mustela lutreola — Mink
Mustela nivalis — Weasel, least or common
Mustela putorius — Polecat, European or Western
Mustela sibirica — Weasel, Siberian or Kolinsky
Mustela vison — Mink
Mustella nigripes — Ferret, black-footed
Myadestes townsendi — Solitaire, Townsend's
Mycena sp. — Mushroom, fairy helmet
Mycteria americana — Stork, wood or Wood ibis
Mycteroperca bonaci — Grouper, black
Mycteroperca interstitialis — Grouper, yellow-mouth
Mycteroperca sp. — Grouper
Myctophum sp. — Lanternfish
Mydalus javanensis — Teledu
Myiarchus cinerascens — Flycatcher, ash-throated
Myiarchus crinitus — Flycatcher, great crested
Myiarchus tyrannulus — Flycatcher, brown-crested
Myiarchus tyrannulus — Flycatcher, Wied's crested
Myioborus pictus — Redstart, painted
Myiodynastes luteiventris — Flycatcher, sulphur-bellied
Myiopsitta monachus — Parakeet, monk
Myironis ecaudata — Flycatcher, short-tailed pygmy
Myliobatis californica — Ray, bat
Myliobatis sp. — Ray, eagle
Myocastor coypus — Nutria
Myoprocta acouchy — Acouchi
Myosciurus pumilio — Squirrel, African pygmy
Myosotis scorpoides — Forget-me-not
Myospalax aspalax — Zokor
Myotis auriculus — Bat, Southwestern myotis
Myotis austroriparicus — Bat, Southeastern myotis
Myotis californicus — Bat, California myotis
Myotis daubentoni — Bat, water
Myotis evotis — Bat, long-eared
Myotis grisescens — Bat, gray myotis
Myotis keenii — Bat, Keen's myotis
Myotis leibii — Bat, small-footed myotis
Myotis lucifugus — Bat, little brown myotis
Myotis myotis — Bat, mouse-eared
Myotis natereri — Bat, natterer's
Myotis septentrionalis — Bat, northern myotis
Myotis sodalis — Bat, Indiana or social myotis
Myotis thysanodes — Bat, fringed myotis
Myotis velifer — Bat, cave myotis
Myotis volans — Bat, long-legged myotis
Myotis yumanensis — Bat, Yuma myotis

Myrica californica — Bayberry, Pacific
Myrica cerifera — Bayberry, Southern
Myrica inodora — Bayberry, odorless
Myriophyllum sp. — Water-milfoil
Myripristis jacobus — Soldierfish, blackbar
Myrmecobius fasciatus — Numbat
Myrmecocystus sp. — Ant, honey
Myrmecophaga tetradactyla — Anteater, lesser
Myrmecophaga tridactyla — Anteater, giant
Myrmeleontidae — Antlion
Myrmotherula axillaris — Antwren, white-flanked
Mytilus edulis — Mussel, blue or edible
Myxocephalus scorpius — Sculpin, shorthorn
Naja haje — Cobra, Egyptian
Naja naja — Cobra, Indian
Naja nigrocollis nigrocollis — Cobra, black-necked
   spitting
Nama demissum — Purple mat
Nandinia binotata — Civet, African palm
Nannopterum harrisi — Cormorant, Galapagos or
   flightless
Napaeozapus insignis — Jumping mouse, woodland or
   Northern
Narcine brasiliensis — Ray, electric (lesser)
Nasalis larvatus — Monkey, proboscis
Nasturtium officinale — Watercress
Nasua nasua — Coatimundi, ring-tailed
Natrix maura — Snake, viperine
Natrix natrix — Snake, grass
Naucrates ductor — Pilotfish
Nautilus macromphalus — Nautilus, chambered
Nautilus pompilus — Nautilus, pearly
Nebrius ferrugineus — Shark, tawny or giant sleepy
Necrosyrtes monachus — Vulture, hooded
Necturus sp. — Mudpuppy or water dog
Negaprion sp. — Shark, lemon
Nelumbo lutea — Lotus, American
Nelumbo nucifera — Lotus, sacred
Nemaster sp. — Sea lily
Nemastylis purpurea — Lily, pinewoods
Nematistius pectoralis — Roosterfish
Nematoda — Nematode or roundworm
Nemophila menziesii — Baby blue eyes
Nemorhaedus goral — Goral
Neoceratodus forsteri — Lungfish, Australian
Neoclinus sp. — Fringehead
Neofelis nebulosa — Leopard, clouded
Neofiber alleni — Muskrat, round-tailed
Neomys fodiens — Water shrew, Eurasian
Neophoca cinerea — Sea lion, Australian
Neophocaena phocaenoides — Porpoise, finless
Neophron pernopterus — Vulture, Egyptian
Neoseps reynoldsi — Skink, sand
Neotoma albigula — Woodrat, white-throated
Neotoma cinerea — Woodrat, bushy-tailed
Neotoma floridana — Woodrat, eastern
Neotoma mexicana — Woodrat, Mexican
Neotoma micropus — Woodrat, Southern Plains
Neotoma sp. — Woodrat (other)
Neotragus moschatus — Suni
Neotragus pygmaeus — Antelope, royal
Nepa cinerea — Water scorpion
Nepeta cataria — Catnip
Nerium oleander — Oleander
Nerodia cyclopion — Watersnake, green
Nerodia erythrogaster — Watersnake, plain-bellied or
   red-bellied

Nerodia fasciata — Watersnake, Southern or banded
Nerodia rhombifera — Watersnake, diamondback
Nerodia sipedon — Watersnake, common or Northern
Nerodia taxispilota — Watersnake, brown
Nerodia valida — Watersnake, Pacific
Nesolagus netscheri — Rabbit, Sumatran
Nestor notabilis — Kea
Neurotrichus gibbsii — Mole, shrew
Nicotiana glauca — Tobacco, tree
Nicotiana trigonophylla — Tobacco, desert
Nicrophorus sp. — Beetle, sexton
Ningaui sp. — Ningaui
Ninox connivens — Owl, barking
Ninox novaeseelandiae — Owl, morepork or boobook
Noctilio leporinus — Bat, fisherman
Nomia melanderi — Bee, alkali
Notaden bennetti — Frog, holy cross
Notechnis scutaris — Snake, tiger
Notemigonus crysoleucas — Shiner, golden
Nothocrax urumutum — Curassow, nocturnal
Notiomys sp. — Mouse, mole
Notiosorex crawfordi — Shrew, desert
Notodoris sp. — Nudibranch
Notomastus lobatus — Worm, maitre d'
Notomys sp. — Mouse, hopping
Notonecta sp. — Backswimmer
Notophthalmus meridionalis — Newt, black-spotted
Notophthalmus viridescens — Newt, red-spotted or
   eastern
Notornis mantelli — Takahe
Notoryctes typhlops — Mole, marsupial
Notorynchus maculatus — Shark, sevengill
Notropis sp. — Shiner
Noturus sp. — Madtom or stonecat
Nucifraga caryocatactes — Nutcracker
Nucifraga columbiana — Nutcracker, Clark's
Numenius americanus — Curlew, long-billed
Numenius arquata — Curlew, Eurasian
Numenius phaeopus — Whimbrel or Hudsonian curlew
Numenius tahitiensis — Curlew, bristle-thighed
Numidia meleagris — Guinea fowl, helmeted
Nuphar advena — Spatterdock
Nuphar lutea — Water lily, yellow or brandy-bottle
Nuphar variegatum — Water lily, bullhead or yellow pond
Nyctalus sp. — Bat, noctule
Nyctanassax violacea — Night heron, yellow-crowned
Nyctea scandiaca — Owl, snowy
Nyctereutes procyonoides — Dog, raccoon
Nyctibius griseus — Potoo, common
Nycticebus coucang — Loris, slow
Nycticeius humeralis — Bat, evening
Nycticorax nycticorax — Night heron, black-crowned
Nyctidromus albicollis — Pauraque, common
Nyctimene sp. — Bat, fruit (tube-nosed)
Nymphaea mexicana — Water lily, yellow
Nymphaea odorata — Water lily, fragrant
Nymphalis antiopa — Butterfly, mourning cloak
Nymphicus hollandicus — Cockatiel
Nymphoides aquatica — Floating hearts
Nyssa aquatica — Tupelo, water
Nyssa sylvatica — Tupelo, black
Obolaria virginica — Pennywort
Oceanites oceanicus — Petrel, storm (Wilson's)
Oceanodroma homochroa — Petrel, storm (ashy)
Oceanodroma hornbyi — Petrel, storm (ringed)
Oceanodroma leucorhoa — Petrel, storm (Leach's)
Oceanodroma melania — Petrel, storm (black)

Ochotona alpina — Pika, Altai
Ochotona collaris — Pika, collared
Ochotona macrotis — Pika, large-eared
Ochotona princeps — Pika
Ochrotomys nuttalli — Mouse, golden
Octodon degus — Degu
Octopus sp. — Octopus
Ocyphaps lophotes — Pigeon, crested
Ocypoda sp. — Crab, ghost
Ocyurus chrysurus — Snapper, yellowtail
Odobenus rosmarus — Walrus
Odocoileus hemionus — Deer, mule or black-tailed
Odocoileus virginianus — Deer, white-tailed
Odocoileus virginianus clavium — Deer, Florida key
Odontaspis sp. — Shark, sand tiger
Odontomacrurus murrayi — Rat-tail or Grenadier
Oecanthus fultoni — Cricket, tree
Oecophylla smaragdina — Ant, tree
Oenanthe crocata — Dropwort, water
Oenanthe oenanthe — Wheatear
Oenothera biennis — Evening primrose, common
Oenothera brevipes — Primrose, desert
Oenothera cheiranthifolia — Primrose, beach
Oenothera deltoides — Evening primrose, birdcage
Oenothera hookeri — Evening primrose, Hooker's
Oenothera sp. — Evening primrose
Oenothera speciosa — Evening primrose, showy
Oenothera speciosa — Evening primrose, pink
Ogcocephalus sp. — Batfish
Okapia johnstoni — Okapi
Olea europaea — Olive, common
Oliva sp. — Olive shell
Olla abdominalis — Beetle, ladybird (ash-gray)
Olneya tesota — Ironweed, desert
Olor buccinator — Swan, trumpeter
Olor columbianus — Swan, whistling or Bewick's
Ommatophoca rossi — Seal, Ross
Omphalotus sp. — Jack-o-Lantern fungus
Oncopeltus fasciatus — Milkweed bug, large
Oncorhynchus gorbuscha — Salmon, pink
Oncorhynchus kisutch — Salmon, coho or silver
Oncorhynchus nerka — Salmon, sockeye or red
Oncorhynchus tshawytscha — Salmon, Chinook or king
Ondatra zibethicus — Muskrat
Onychogalea fraenata — Wallaby, bridled nailtail
Onychomys leucogaster — Mouse, grasshopper (Northern)
Onychomys torridus — Mouse, grasshopper (Southern)
Onychorhynchus cornatus — Flycatcher, royal
Onychoteuthis banksi — Squid, oceanic
Opheodrys aestivus — Snake, green (rough)
Opheodrys vernalis — Snake, green (smooth)
Ophichthus sp. — Eel, snake
Ophiodon elongatus — Lingcod
Ophiophagus hannah — Cobra, king
Ophisaurus apodus — Snake, glass
Ophisaurus attenuatus — Lizard, glass (slender)
Ophisaurus compressus — Lizard, glass (Island)
Ophisaurus ventralis — Lizard, glass (Eastern)
Ophlitaspongia pennata — Sponge, velvety red
Opisthocomus hoatzin — Hoatzin
Opistognathus sp. — Jawfish
Oplopanax horridum — Devil's club
Oporornis agilis — Warbler, Connecticut
Oporornis formosus — Warbler, Kentucky
Oporornis philadelphia — Warbler, mourning
Oporornis tolmiei — Warbler, MacGillivray's

Opsanus sp. — Toadfish
Opuntia basilaris — Cactus, beavertail
Opuntia bigelovii — Cholla, teddybear or jumping
Opuntia fulgida — Cholla, jumping
Opuntia imbricata — Cholla, tree
Opuntia leptocaulis — Christmas cactus, desert
Opuntia sp. — Prickly pear
Orcaella brevirostris — Dolphin, Irrawaddy or snubfin
Orchestoidea sp. — Flea, beach
Orchis spectabilis — Orchis, showy
Orcinus orca — Killer whale
Oreamnos americanus — Goat, mountain
Oreaster reticulatus — Starfish, reticulated or cushion
Orectolobus sp. — Carpetshark or Wobbegong
Oreophasis derbianus — Guan, horned
Oreortyx pictus — Quail, mountain
Oreoscoptes montanus — Thrasher, sage
Oreotragus oreotragus — Klipspringer
Orgyia leucostigma — Moth, tussock (white-marked)
Oriolus larvatus — Oriole, black-headed
Oriolus oriolus — Oriole, golden
Orlitia borneansis — Turtle, Bornean River or Malaysian giant
Ornithogalum umbellatum — Star-of-Bethlehem
Ornithoptera sp. — Butterfly, birdwing
Ornithorhynchus anatinus — Platypus
Orobanche sp. — Broomrape or cancer root
Orontium aquaticum — Golden club
Ortalis sp. — Chachlalaca
Orthonopias triacis — Sculpin, snubnose
Orthonyx spaldingi — Logrunner, northern
Orthopristis chrysoptera — Pigfish
Orycteropus afer — Aardvark
Oryctes nasicornis — Beetle, rhinoceros
Oryctolagus cuniculus — Rabbit, european
Oryx dammah — Oryx, Scimitar
Oryx gazella — Gemsbok
Oryx leucoryx — Oryx, Arabian
Oryzomys palustris — Rat, marsh rice
Oryzoryctes tetradactylus — Tenrec, rice
Osmerus eperlanus — Smelt, European
Osmerus mordax — Smelt, rainbow or American
Osmia sp. — Bee, mason
Osmorhiza claytoni — Cicely, sweet
Osmunda cinnamomea — Fern, cinnamon
Osmunda regalis — Fern, royal
Osphronemus goramy — Gourami
Osteoglossum sp. — Arawana
Osteopilus septentrionalis — Treefrog, Cuban
Ostracion tuberculatus — Boxfish
Ostrea sp. — Oyster
Ostrya chisosensis — Hornbeam, hop (chisos)
Ostrya knowltonii — Hornbean, hop (Knowlton)
Ostrya virginiana — Hornbeam, hop
Otaria byronia — Sea lion, Southern
Otaria flavescens — Sea lion, South American
Otis tarda — Bustard, great
Otocyon megalotis — Fox, bat-eared
Otolemur crassicaudatus — Bush baby, greater
Otus asio — Owl, screech
Otus flammeolus — Owl, flammulated
Otus sp. — Owl, scops
Otus trichopsis — Owl, whiskered
Ourebia ourebia — Oribi
Ovalipes sp. — Crab, lady
Ovibos moschatus — Musk ox
Ovis ammon — Argali

Ovis ammon or *orientalis* — Mouflon
Ovis canadensis — Sheep, bighorn mountain
Ovis dalli — Sheep, Dall's
Ovis orientalis — Urial or Arkal
Oxalis alpina — Wood sorrel, mountain
Oxalis europaea — Sorrel, lady's
Oxalis montana — Wood sorrel
Oxalis oregana — Sorrel, redwood
Oxalis stricta — Wood sorrel, yellow
Oxalis violacea — Wood sorrel, violet
Oxydendrum arboreum — Sourwood
Oxyjulis californica — Senorita
Oxyopes sp. — Spider, lynx
Oxyruncus cristatus — Sharpbill
Oxytropis lambertii — Locoweed, purple
Oxytropis sericea — Locoweed, white
Oxytropis splendens — Locoweed, showy
Oxyura dominica — Duck, masked
Oxyura jamaicensis — Duck, ruddy
Oxyura leucocephala — Duck, white-headed
Oxyuranus scutellatus — Taipan
Oybelis aeneus — Snake, vine Mexican
Ozotoceros bezoarticus — Deer, pampas
Pachycephala pectoralis — Whistler, golden
Pachygrapsus sp. — Crab, shore
Pachymedusa sp. — Frog, leaf
Pachyptila turtur — Prion, fairy
Pachyptila vittata — Prion, broad-billed
Pachyramphus sp. — Becard, rose-throated
Paeonia brownii — Peony, Western
Pagetopsis macropterus — Icefish
Pagophila eburnea — Gull, ivory
Paguma larvata — Civet, masked palm
Palaemonetes vulgaris — Shrimp, grass
Paleosuchus palpebrosus — Caiman, smooth-fronted
Palinurus vulgaris or *argus* — Lobster, spiny
Palmatogecko rangei — Gecko, web-footed
Pan paniscus — Chimpanzee, pygmy
Pan troglodytes — Chimpanzee
Panax ginseng — Ginseng
Panax quinquefolium — Ginseng, American
Panax trifolius — Ginseng, dwarf
Pandalus danae — Shrimp, coonstripe
Pandanus tectorius — Pine, screw
Pandion haliaetus — Osprey
Pangasius pangasius — Catfish, Pungas
Panorpa sp. — Fly, scorpion
Panthera leo — Lion
Panthera onca — Jaguar
Panthera pardus — Leopard
Panthera tigris — Tiger
Panthera uncia — Leopard, snow
Pantholops hodgsoni — Antelope, Tibetan or Chiru
Panulirus interruptus — Lobster, rock (California)
Panulirus versicolor — Crayfish, painted
Panurus biarmicus — Tit, bearded
Papaver californicum — Poppy, fire
Papaver kluanense — Poppy, alpine
Papaver polare or *radicatum* — Poppy, arctic
Papaver rhoeas — Poppy, corn
Papaver somniferum — Poppy, opium
Papilio aegus — Swallowtail butterfly, Orchard
Papilio bairdii — Swallowtail butterfly, black (western)
Papilio cresphontes — Swallowtail butterfly, giant
Papilio eurymedon — Swallowtail butterfly, pale
Papilio glaucus — Swallowtail butterfly, tiger (Eastern)
Papilio machaon — Swallowtail butterfly, Old world

Papilio polyxenes — Swallowtail butterfly, black (Eastern)
Papilio rudkini — Swallowtail butterfly, desert
Papilio rutulus — Swallowtail butterfly, tiger (Western)
Papilio sp. — Swallowtail butterfly, tiger
Papilio troilus — Swallowtail butterfly, spicebush
Papilio zelicaon — Swallowtail butterfly, anise
Papio anubis — Baboon, anubis
Papio cynocephalus — Baboon, olive or common
Papio hamadryas — Baboon, hamadryas
Papio leucophacus — Drill
Papio papio — Baboon, guinea
Papio sphinx — Mandrill
Papio ursinus — Baboon, chacma
Pappogeomys castanops — Pocket gopher, yellow-faced
Paprilus sp. — Butterfish
Papyrus antiquorum — Papyrus
Parabuteo unicinctus — Hawk, Harris'
Paracanthurus sp. — Surgeonfish
Paracotalpa sp. — Beetle, bear
Paracynictis selousi — Meerkat
Paradisaea minor — Bird of Paradise, lesser
Paradisaea raggiana — Bird of Paradise
Paradisaea rudolphi — Bird of Paradise, blue
Paradoxornis gularis — Parrotbill, gray-headed
Paralabrax clathratus — Bass, kelp
Paralichthys californicus — Halibut, California
Paralichthys sp. — Flounder
Paralichthys stellatus — Flounder, starry
Paralithodes sp. — Crab, King
Paramecium sp. — Paramecium
Paranthias furcifer — Creole-fish
Parascalops breweri — Mole, hairy tailed
Parascyllium collare — Carpetshark, collared
Paravespula sp. — Wasp, paper (European)
Pardalotus punctatus — Pardalote, spotted
Pardofelis or *Felis marmorata* — Cat, marbled
Paris quadrifolia — Herb-Paris
Parkinsonia aculeata — Jerusalem thorn
Parnassia sp. — Grass of Parnassus
Parnassius apollo — Butterfly, Apollo
Parnassius phoebus — Butterfly, phoebus
Parophrys vetulus — Sole, English
Parthenocissus quinquefolia — Virginia creeper
Parula americana — Parula, Northern
Parus ater — Tit, coal
Parus atricapillus — Chickadee, black-capped
Parus bicolor — Titmouse, tufted
Parus caeruleus — Tit, blue
Parus carolinensis — Chickadee, Carolina
Parus cristatus — Tit, crested
Parus cyanus — Tit, azure
Parus gambeli — Chickadee, mountain
Parus hudsonicus — Chickadee, boreal
Parus inornatus — Titmouse, plain
Parus major — Tit, great
Parus montanus — Tit, willow
Parus niger — Tit, southern black
Parus palustris — Tit, marsh
Parus rufescens — Chickadee, chestnut-backed
Parus rufiventris — Tit, rufous-bellied
Parus spilonotus — Tit, yellow-cheeked
Parus wollweberi — Titmouse, bridled
Passer domesticus — Sparrow, English or house
Passer montanus — Sparrow, tree (European)
Passerculus sandwichensis — Sparrow, savannah or Ipswich
Passerella iliaca — Sparrow, fox

Passerina amoena — Bunting, lazuli
Passerina ciris — Bunting, painted
Passerina cyanea — Bunting, indigo .
Passerina versicolor — Bunting, varied
Passiflora sp. — Passionflower or Maypop
Pastinaca sativa — Parsnip
Patagona gigas — Hummingbird, giant
Patella vulgata — Limpet
Paulownia tomentosa — Paulownia, royal
Pavo cristatus — Peacock, Indian
Pectis papposa — Chinchweed
Pedetes capensis — Springhare
Pedicularis bracteosa — Lousewort, bracted
Pedicularis canadensis — Lousewort, common
Pedicularis groenlandica — Elephant head
Pedicularis lanata — Lousewort, woolly
Pediculus humanus or *capitis* — Louse, head or body
Pedimomus torquatus — Plains wanderer
Pediocactus simpsonii — Cactus, Simpson's hedgehog
Pelagodroma marina — Petrel, storm (white-faced or frigate)
Pelamis platurus — Seasnake, yellow-bellied
Pelargopsis capensis — Kingfisher, stork-billed
Pelea capreolus — Rhebok
Pelecanoides urinatrix — Petrel, diving
Pelecanus conspicillatus — Pelican, Australian
Pelecanus erythrorhynchus — Pelican, white (American)
Pelecanus occidentalis — Pelican, brown
Pelecanus onocrotalus — Pelican, white (great or European)
Pelecanus thagus — Pelican, Chilean
Pelobates sp. — Toad, spadefoot
Pelocoris femoratus — Water bug, creeping
Pelomedusa subrufa — Turtle, helmeted
Peltandra virginica — Arum, arrow
Peltiphyllum peltatum — Umbrella plant
Pelusios gabonensis — Turtle, African forest
Pelusios niger — Turtle, mud (black African)
Penaeus duorarum — Shrimp, pink
Penelope purpurascens — Guan, crested
Penstemon centranthifolius — Bugler, scarlet
Penstemon corymbosus — Penstemon, red shrubby
Penstemon dolius — Penstemon, Jones
Penstemon fruticosus — Penstemon, lowbush
Penstemon newberryi — Mountain pride
Penstemon palmeri — Balloonflower
Penstemon rupicola — Penstemon, cliff or rock
Penstemon serrulatus — Penstemon, Cascade
Penstemon sp. — Penstemon
Pentalagus furnessi — Rabbit, Amami
Peponocephala electra — Whale, melon-headed
Peprilus alepidotus — Harvest fish
Pepsis sp. — Wasp, tarantula hawk .
Perca flavescens — Perch, yellow
Perca fluviatilis — Perch, river
Percina sp. — Darter (fish)
Percina tanasi — Darter, snail (fish)
Percopsis omiscomaycus — Trout-perch
Perdix perdix — Partridge, gray
Periclimenes yucatanicus — Shrimp, cleaner
Pericrotus divaricatus — Minivet, ashy
Periophthalmus sp. — Mudskipper
Periplaneta americana — Cockroach, American or waterbug
Perisoreus canadensis — Jay, gray or Canadian
Pernis apivorus — Buzzard, honey
Perodicticus potto — Potto

Perognathus flavescens — Mouse, pocket (Plains)
Perognathus flavus — Mouse, pocket (silky)
Perognathus sp. — Mouse, pocket (other)
Peromyscus attwateri — Mouse, Texas
Peromyscus boylii — Mouse, brush
Peromyscus californicus — Mouse, California
Peromyscus crinitus — Mouse, canyon
Peromyscus eremicus — Mouse, cactus
Peromyscus floridanus — Mouse, Florida
Peromyscus gossypinus — Mouse, cotton
Peromyscus leucopus — Mouse, white-footed
Peromyscus maniculatus — Mouse, deer
Peromyscus pectoralis — Mouse, white-ankled
Peromyscus polionotus — Mouse, oldfield
Peromyscus truei — Mouse, pinyon
Persea americana — Avocado
Persea borbonia — Redbay
Petalostemon sp. — Clover, prairie
Petaurista petaurista — Flying squirrel, giant
Petauroides volans — Glider, greater
Petaurus australis — Glider, yellow-bellied
Petaurus breviceps — Glider, sugar
Petaurus norfolcensis — Glider, squirrel
Petaurus sp. — Glider (other)
Petrochelidon pyrrhonota — Swallow, cliff
Petrogale sp. — Wallaby, rock
Petrogale xanthopus — Wallaby, rock (yellow-footed)
Petroica goodenovii — Robin, red-capped
Petroica phoenicea — Robin, flame
Petrolisthes sp. — Crab, porcelain
Petromus sp. — Rat, rock
Petromyscus sp. — Mouse, rock
Petromyzon marinus — Lamprey, sea
Petronia petronia — Sparrow, rock
Petrosaurus sp. — Lizard, rock
Peucedramus taeniatus — Warbler, olive
Peucetia sp. — Spider, lynx
Peziza aurantia — Orange peel fungus
Phacelia calthifolia — Phacelia
Phacelia campanularia — Desert bell
Phacelia dubia — Scorpionweed
Phacocherus aethiopicus — Warthog
Phaenicia sericata — Fly, greenbottle
Phaenostictus mcleannani — Antbird, ocellated
Phaethon aethereus — Tropicbird, red-billed
Phaethon lepturus — Tropicbird, white-tailed
Phaethon rubricauda — Tropicbird, red-tailed
Phaetusa simplex — Tern, large-billed
Phainopepla nitens — Phainopepla
Phalacrocorax africanus — Cormorant, reed
Phalacrocorax albiventor — Cormorant, king or rock
Phalacrocorax aristotelis — Shag
Phalacrocorax atriceps — Cormorant, blue-eyed
Phalacrocorax auritus — Cormorant, double-crested
Phalacrocorax carbo — Cormorant, common or great
Phalacrocorax fuscescens — Cormorant, black-faced
Phalacrocorax gaimardi — Cormorant, red-footed
Phalacrocorax olivaceus — Cormorant, olivaceous
Phalacrocorax pelagicus — Cormorant, pelagic
Phalacrocorax penicillatus — Cormorant, brandt's
Phalacrocorax punctatus — Cormorant, spotted
Phalacrocorax sulcirostris — Cormorant, little black
Phalacrocorax urile — Cormorant, red-faced
Phalacrocorax varius — Cormorant, pied
Phalaenoptilus nuttallii — Poor-will
Phalanger maculatus — Cuscus, spotted
Phalanger orientalus — Cuscus, gray

Phalangium opilio — Daddy-long-legs, brown
Phalaropus fulicaria — Phalarope, red or gray
Phalaropus lobatus — Phalarope, northern or red-necked
Phalaropus tricolor — Phalarope, Wilson's
Phallus sp. — Stinkhorn fungus
Phanaeus sp. — Beetle, dung
Phaner furcifer — Lemur, four-marked
Phanerodon atripes — Seaperch, sharpnose
Pharomachrus mocino — Quetzal
Phascogale sp. — Marsupial mouse, brush-tailed
Phascolarctos cinereus — Koala
Phasianus colchicus — Pheasant, ring-necked
Phelsuma laticauda — Gecko, Madagascan day
Phenacogrammus interruptus — Tetra, congo
Phenacomys sp. — Vole, tree
Phengodidae — Glowworm
Pheucticus ludovicianus — Grosbeak, rose-breasted
Pheucticus melanocephalus — Grosbeak, black-headed
Philadelphus lewisii — Syringa, Lewis'
Philaenus spumarius — Spittle bug
Philemon corniculatus — Friarbird, noisy
Philetairus socius — Weaver, social
Philomachus pugnax — Ruff
Phleum pratense — Grass, timothy
Phlox caespitosa — Phlox, tufted
Phlox divaricata — Phlox, blue
Phlox drummondii — Phlox, red
Phlox longifolia — Phlox, long-leaved
Phlox maculata — Phlox, Sweet William
Phlox subulata — Pink, moss
Phoca caspica — Seal, Caspian
Phoca fasciata — Seal, ribbon
Phoca groenlandica — Seal, harp
Phoca hispida — Seal, ringed
Phoca largha — Seal, larga
Phoca sibirica — Seal, Baikal
Phoca vitulina — Seal, harbor or common
Phocarctos hookeri — Sea lion, Hooker's or New Zealand
Phocoena dioptrica — Porpoise, spectacled
Phocoena phocoena — Porpoise, Common or harbor
Phocoena sinus — Porpoise, Gulf of California
Phocoena spinipinnis — Porpoise, Burmeister's or black
Phocoenoides dalli — Porpoise, Dall's
Phodilus badius — Owl, bay
Phoebis sp. — Butterfly, sulfur
Phoenicaulis cheiranthoides — Dagger pod
Phoenicopterus andinus — Flamingo, Andean
Phoenicopterus minor — Flamingo, lesser
Phoenicopterus ruber — Flamingo, greater (including American)
Phoeniculus purpurens — Hoopoe, wood
Phoenicurus ochruros — Redstart, black
Phoenicurus phoenicurus — Redstart
Phoenix dactylifera — Palm, date
Pholiota sp. — Mushroom, scale-cap
Pholis sp. — Gunnel
Phoradendron sp. — Mistletoe, American
Photinus pyralis — Firefly, eastern
Photuris pennsylvanica — Firefly, woods
Phragmites australis — Reed, giant
Phryma leptostachya — Lopseed
Phrynocephalus sp. — Lizard, toad-headed agamid
Phrynosoa mcalli — Horned lizard, flat-tailed
Phrynosoma cornutum — Horned lizard, Texas or California
Phrynosoma coronatum — Horned lizard, coast
Phrynosoma douglassii — Horned lizard, short-horned

Phrynosoma modestum — Horned lizard, round-tailed
Phrynosoma platyrhinos — Horned lizard, desert
Phrynosoma solare — Horned lizard, regal
Phyciodes sp. — Butterfly, crescent-spot
Phycodurus sp. — Sea dragon
Phyllactis flosculifera — Sea anemone, collared sand
Phyllastrephus scandens — Leaflove
Phyllium sp. — Leaf insect
Phyllodactylus xanti — Gecko, leaf-toed
Phyllomedusa sp. — Frog, leaf or leaf-folding
Phyllophaga sp. — Beetle, May or June
Phyllopteryx sp. — Sea dragon
Phyllorhynchus browni — Snake, leafnosed (saddled)
Phylloscopus borealis — Warbler, arctic
Phylloscopus collybita — Chiff-chaff
Phylloscopus sibilatrix — Warbler, wood
Phylloscopus tenellipes — Warbler, pale-legged leaf
Phylloscopus trachilus — Warbler, willow
Phylloscopus trochiloides — Warbler, greenish
Phyllospadix sp. — Surfgrass
Phyllotis sp. — Mouse, leaf-eared
Phylorhynchus decurtatus — Snake, leafnosed spotted
Phymata sp. — Ambush bug
Physalia physalis — Portuguese man-of-war
Physalis heterophylla — Groundcherry
Physalis lobata — Groundcherry, purple
Physeter catodon or *macrocephalus* — Whale, sperm
Physocarpus opulifolius — Ninebark
Physophora sp. — Siphonophore
Physostegia virginiana — Obediant plant
Phytolacca sp. — Pokeberry or Pokeweed
Pica nuttalli — Magpie, yellow-billed
Pica pica — Magpie, black-billed
Picea abies — Spruce, Norway
Picea brewerana — Spruce, Brewer's
Picea engelmannii — Spruce, Engelmann
Picea glauca — Spruce, white
Picea mariana — Spruce, black
Picea orientalis — Spruce, Oriental
Picea pungens — Spruce, blue
Picea rubens — Spruce, red
Picea sitchensis — Spruce, sitka
Pickeringia montana — Pea, Chaparral
Picoides albolarvatus — Woodpecker, white-headed
Picoides arcticus — Woodpecker, black-backed
Picoides arizonae — Woodpecker, Arizona
Picoides borealis — Woodpecker, red-cockaded
Picoides major — Woodpecker, great spotted
Picoides nuttallii — Woodpecker, Nuttall's
Picoides pubescens — Woodpecker, downy
Picoides scalaris — Woodpecker, ladder-backed
Picoides tridactylus — Woodpecker, three-toed
Picoides villosus — Woodpecker, hairy
Picus viridis — Woodpecker, green
Pieris brassicae — Butterfly, large white
Pieris rapae — Butterfly, cabbage (European)
Pinckneya pubens — Pinckneya or fever-tree
Pinguicula lutea — Butterwort, yellow
Pinguicula vulgaris — Butterwort
Pinicola enucleator — Grosbeak, pine
Pinus albicaulis — Pine, whitebark
Pinus aristata — Pine, bristlecone
Pinus attenuata — Pine, knobcone
Pinus balfouriana — Pine, foxtail
Pinus banksiana — Pine, jack
Pinus bungeana — Pine, lacebark
Pinus clausa — Pine, sand or scrub

Pinus contorta — Pine, lodgepole
Pinus coulteri — Pine, Coulter
Pinus echinata — Pine, shortleaf
Pinus edulis or *monophylla* — Pine, pinyon
Pinus elliottii — Pine, slash
Pinus engelmannii — Pine, Apache
Pinus flexilis — Pine, limber
Pinus glabra — Pine, spruce or cedar
Pinus jeffreyi — Pine, Jeffrey
Pinus koraiensis — Pine, Korean
Pinus lambertiana — Pine, sugar
Pinus leiophylla — Pine, Chihuahua
Pinus monticola — Pine, white (Western)
Pinus muricata — Pine, bishop
Pinus nigra — Pine, Austrian
Pinus palustris — Pine, longleaf
Pinus parviflora — Pine, white (Japanese)
Pinus ponderosa — Pine, Ponderosa or western yellow
Pinus pungens — Pine, Table Mountain
Pinus radiata — Pine, Monterey
Pinus resinosa — Pine, red or Norway
Pinus rigida — Pine, pitch
Pinus sabiniana — Pine, digger
Pinus serotina — Pine, pond or marsh
Pinus strobiformis — Pine, white (Southwestern)
Pinus strobus — Pine, white (eastern)
Pinus sylvestris — Pine, scots
Pinus taeda — Pine, loblolly
Pinus torreyana — Pine, Torrey
Pinus virginiana — Pine, Virginia
Pipa pipa — Toad, Surinam
Pipilo aberti — Towhee, Abert's
Pipilo chlorurus — Towhee, green-tailed
Pipilo erythrophthalmus — Towhee, rufous-sided
Pipilo fuscus — Towhee, brown
Pipistrellus hesperus — Bat, pipistrelle (Western)
Pipistrellus subflavus — Bat, pipistrelle (Eastern)
Pipistrellus sp. — Bat, pipistrelle
Pipra mentalis — Manakin, red-capped
Piranga flava — Tanager, hepatic
Piranga ludoviciana — Tanager, Western
Piranga olivacea — Tanager, scarlet
Piranga rubra — Tanager, Summer
Piscicola geometra — Leech, fish
Pistacia texana — Pistache, Texas
Pistia stratiotes — Water lettuce
Pitangus sulphuratus — Kiskadee, great
Pithecia monachus — Saki, hairy or monk or red-bearded
Pithecia pithecia — Saki, white-faced
Pithecophaga sp. — Eagle, monkey-eating
Pithys albifrons — Antbird, white-plumed
Pitohui sp. — Pitohui
Pitta sp. — Pitta
Pituophis melanoleucus — Snake, pine or gopher
Pituophis melanoleucus — Snake, bull
Pizonyx vivesi — Bat, fishing
Planera aquatica — Elm, water
Planigale sp. — Planigale
Plantago sp. — Plantain
Plantanista minor — River dolphin, Indus
Platalea alba — Spoonbill, African
Platalea leucorodia — Spoonbill, white
Platanista gangetica — River dolphin, Ganges
Platanus occidentalis — Sycamore
Platanus racemosa — Sycamore, California or Western
Platanus sp. — Plane tree

Platanus wrightii — Sycamore, Arizona
Platemys platycephala — Turtle, twisted-neck or flatshelled
Platycercus elegans — Parrot, crimson rosella
Platyrhinoidis triseriata — Thornback
Platypsaris sp. — Becard, rose-throated
Platystemon californicus — Cream cup
Platysternon megachephalum — Turtle, big-headed
Plebejus acmon — Butterfly, blue acmon
Plecia nearctica — Lovebug
Plecoglossus altivelis — Ayu
Plecotus rafinesquii — Bat, big-eared (Rafinesque's)
Plecotus townsendii — Bat, big-eared (Townsend's)
Plecotus sp. — Bat, long-eared
Plectorhynchus sp. — Sweetlips
Plectrophenax hyperboreus — Bunting, McKay's
Plectrophenax nivalis — Bunting, snow
Plegadis chihi — Ibis, white-faced
Plegadis falcinellus — Ibis, glossy
Plethodon cinerus — Salamander, redback
Plethodon dorsalis — Salamander, zigzag
Plethodon dunni — Salamander, Dunn
Plethodon elongatus — Salamander, Del Norte
Plethodon glutinosus — Salamander, slimy
Plethodon hoffmani — Salamander, valley or valley and ridge
Plethodon jordani — Salamander, Jordan's or Appalachian woodland
Plethodon larselli — Salamander, Larch Mountain
Plethodon longicrus — Salamander, crevice
Plethodon neomexicanus — Salamander, Jemez Mountains
Plethodon nettingi nettingi — Salamander, cheat mountain
Plethodon richmondi — Salamander, ravine
Plethodon serratus — Salamander, redback (Southern)
Plethodon vandykei — Salamander, Van Dyke
Plethodon vehiculum — Salamander, redback (Western)
Plethodon yonahlossee — Salamander, Yonahlossee
Pleurmectes platessa — Plaice, chameleon
Pleurobrachia pileus — Sea gooseberry
Pleurobrachia sp. — Comb jelly
Pleuronichthys coenosus — Sole, C-O
Pleuronichthys sp. — Turbot
Pleurotus ostreatus — Mushroom, oyster
Plexaura sp. — Sea rod
Ploceus sp. — Weaver
Pluchea purpurascens — Fleabane, marsh
Plumeria sp. — Frangipani
Plusiotis gloriosa — Beetle, glorious
Plusiotis sp. — Beetle, scarab or shining leaf chafer
Pluvialis dominica — Plover, American or lesser golden
Pluvialis squatarola — Plover, black-bellied
Pluvialus aegyptius — Plover, Egyptian
Pluvialus apricaria — Plover, golden
Pluvialus squatarola — Plover, gray
Poa pratensis — Bluegrass, Kentucky
Podarcis sicula — Lizard, ruin
Podarcis sp. — Lizard, wall
Podargus sp. — Frogmouth
Podaxis pistillaris — Ink cap, desert
Podica senegalensis — Finfoot, African
Podiceps auritus — Grebe, horned or Slavonian
Podiceps cristatus — Grebe, great crested
Podiceps grisegena — Grebe, red-necked
Podiceps nigricollis — Grebe, eared or black-necked
Podilymbus podiceps — Grebe, pied-billed

Podocarpus sp. — Podocarp tree
Podocnemis sp. — Turtle, river
Podogymnura truei — Moonrat, Mindano
Podophyllum peltatum — Mayapple, common
Poecilia latipinna — Molly, sailfin
Poecilictis libyca — Weasel, North African banded or Libyan striped
Poeciliopsis occidentalis — Topminnow, gila
Poecilogale albinucha — Weasel, white-naped
Poelagus marjorita — Rabbit, Bunyoro
Poephila guttata — Finch, zebra
Pogonia ophioglossoides — Orchid, pogonia rose
Pogonias cromis — Drum, black
Pogonomyrmex sp. — Ant, harvester
Poiana richardsoni — Linsang, African
Poinciana sp. — Poinciana
Polanisia dodecandra — Clammyweed
Poleamaetus bellicosus — Eagle, martial
Polemonium reptans — Valerian
Polemonium sp. — Jacob's ladder
Polihierax semitorquatus — Falcon, African pygmy
Polioptila caerulea — Gnatcatcher, blue-gray
Polioptila melanura — Gnatcatcher, black-tailed
Polistes instabilis — Wasp, social
Polistes sp. — Wasp, paper
Pollachius virens — Pollock
Polyboroides sp. — Hawk, harrier African
Polyborus plancus — Caracara, crested
Polycera chilluna — Nudibranch, harlequin
Polydactylus sp. — Threadfin
Polygala chamaebuxus — Milkwort, shrubby
Polygala cruciata — Milkwort, cross
Polygala lutea — Candyweed
Polygala paucifolia — Gaywings
Polygala sp. — Milkwort
Polygonatum biflorum — Solomon's seal
Polygonia c-album — Butterfly, comma
Polygonia interrogationis — Butterfly, question mark
Polygonum persicaria — Lady's thumb
Polygonum sp. — Bistort; Smartweed
Polyodon spathula — Paddlefish
Polyommatus icarus — Butterfly, blue common
Polyphylla fullo — Beetle, fuller
Polyplectron bicalcaratum — Pheasant, gray peacock
Polyporus sulphureus — Chicken-of-the-woods fungus
Polypterus ornatipinnis — Birchir
Polyrrhiza lindenii — Orchid, ghost or Palm-polly
Polysticta stelleri — Eider, Steller's
Polystoechotes sp. — Lacewing, giant
Pomacanthus arcuatus — Angelfish, grey
Pomacanthus imperator — Angelfish, emperor or imperial
Pomacanthus paru — Angelfish, French
Pomacanthus sp. — Angelfish (other)
Pomacentrus leucostictus — Beaugregory
Pomacentrus sp. — Damselfish
Pomatomus saltatrix — Bluefish
Pomoxis nigromaculatus — Crappie, black
Pongo pygmaeus — Orangutan
Pontederia cordata — Pickerelweed
Pontoporia blainvillei — River dolphin, La Plata
Pooecetes gramineus — Sparrow, vesper
Popilla japonica — Beetle, Japanese
Populus alba — Poplar, white
Populus angustifolia — Cottonwood, narrowleaf
Populus balsamifera — Poplar, balsam
Populus deltoides — Cottonwood, eastern

Populus fremontii — Cottonwood, Fremont
Populus grandidentata — Aspen, bigtooth
Populus heterophylla — Cottonwood, swamp
Populus nigra — Poplar, lombardy
Populus tremuloides — Aspen, quaking
Populus trichocarpa — Cottonwood, black
Porichthys sp. — Midshipman
Porites astreoides — Coral, porous
Porites porites — Coral, finger
Porphyra umbilicalis — Laver, purple
Porphyrio madagascariensis — Reedhen, king
Porphyrio porphyrio — Swamp hen, purple
Porphyrula martinica — Gallinule, purple
Porpita porpita or *linneana* — Jellyfish, blue button
Portulaca lutea or *oleracea* — Purslane
Portulaca pilosa — Rose moss
Portunus sayi — Crab, sargassum
Portunus sp. — Crab, swimming
Porzana carolina — Sora
Postelsia palmaeformis — Sea palm
Potamilla fonticula — Worm, fan red-banded
Potamochoerus porcus — Pig, bush
Potamogale velox — Water shrew, giant African
Potentilla anserina — Silverweed
Potentilla sp. — Cinquefoil
Potorous sp. — Potoroo
Potorous tridactylus — Kangaroo, rat long-nosed
Potos flavus — Kinkajou
Presbytis entellus — Langur, common or Hanuman
Presbytis sp. — Monkey, leaf
Priacanthus arenatus — Bigeye
Priacanthus cruentatus — Snapper, glasseye
Priacanthus sp. — Catalufa
Primula elatior — Oxslip
Primula mistassinica — Primrose, bird's-eye
Primula parryi — Primrose, parry
Primula suffrutescens — Primrose, Sierra
Priodontes maximus — Armadillo, giant
Prionace glauca — Shark, blue
Prionailurus rubiginosus — Cat, rust
Prionodon linsang — Linsang, banded
Prionodon pardicolor — Linsang, spotted
Prionodura newtoniana — Bowerbird, golden
Prionotus sp. — Searobin
Pristigenys alta — Bigeye, short
Pristiophorus sp. — Shark, saw
Pristis sp. — Sawfish
Proboscidea althaeafolia — Devil's claw
Procavia sp. — Hyrax, rock
Procellaria aequinoctialis — Petrel, white-chinned
Proceloterna cerulea — Noddy, blue-gray
Procnias tricarunculata — Bellbird, three-wattled
Procyon cancrivorus — Raccoon, crab-eating
Procyon lotor — Raccoon
Progne subis — Martin, purple
Promachus fitchii — Bee killer
Pronolagus crassicaudatus — Hare, rock (greater red)
Propithecus sp. — Lemur, sifaka
Propithecus verreauxi — Lemur, sifaka (Verreaux's)
Prosopis glandulosa — Mesquite, honey
Prosopis pubescens — Mesquite, screwbean
Prosopis velutina — Mesquite, velvet
Prosopium cylindraceum — Whitefish, round
Prosthemadera novaeseelandica — Tui
Protea sp. — Protea
Proteles cristatus — Aardwolf
Proteus anguinus — Olm

Protonotaria citrea — Warbler, prothonotary
Protopterus sp. — Lungfish, African
Prunella collaris or *montanella* — Accentor
Prunella modalaris — Dunnock
Prunella vulgaris — Heal-all
Prunus americana — Plum, American
Prunus angustifolia — Plum, Chickasaw
Prunus avium — Cherry, sweet or Mazzard
Prunus caroliniana — Laurelcherry, Carolina
Prunus cerasus — Cherry, sour
Prunus domestica — Plum, garden or damson
Prunus emarginata — Cherry, bitter
Prunus ilicifolia — Cherry, hollyleaf
Prunus lyonii — Cherry, Catalina
Prunus mahaleb — Cherry, Mahaleb
Prunus mexicana — Plum, Mexican
Prunus munsoniana — Plum, wildgoose or munson
Prunus pensylvanica — Cherry, pin
Prunus persica — Peach
Prunus serotina — Cherry, black
Prunus subcordata — Plum, Klamath
Prunus virginiana — Chokecherry
Prunus sp. — Plum
Psalliota arvensis — Mushroom, horse
Psalliota hortensis — Mushroom, cultivated
Psalliota staminea — Straw agaric
Psaltriparus sp. — Bushtit
Psathyrotes ramosissima — Desert velvet
Pseudacris brachyphona — Frog, chorus (Mountain)
Pseudacris brimleyi — Frog, chorus (Brimley's)
Pseudacris nigrita — Frog, chorus (Southern)
Pseudacris ornata — Frog, chorus (ornate)
Pseudacris triseriata — Frog, chorus (striped)
Pseudaspis cana — Snake, mole
Pseudemys rubriventris — Turtle, red-belly
Pseudemys scripta — Turtle, red-eared
Pseudobalistes fuscus — Triggerfish, blue
Pseudobranchus striatus — Siren, dwarf
Pseudoceros sp. — Worm, flat aquatic
Pseudococcus sp. — Mealy bug
Pseudogyps africanus — Vulture, white-backed
Pseudolarix kaempferi — Larch, golden
Pseudolucanus capreolus — Beetle, stag or pinching bug
Pseudomasarus vespoides — Wasp, mud
Pseudomys sp. — Mouse, Australian
Pseudopleuronectes americanus — Flounder, blackback or winter
Pseudopterogorgia sp. — Sea plume
Pseudorca crassidens — Killer whale, false
Pseudopriacanthus sp. — Catalufa
Pseudotriakis microdon — Catshark, false
Pseudotriton montanus — Salamander, mud
Pseudotriton ruber — Salamander, red
Pseudotsuga macrocarpa — Fir, Douglas bigcone
Pseudotsuga menziesii — Fir, Douglas
Psidium guajava — Guava
Psilostrophe cooperi — Paperflower
Psithyrus vestalis — Bumblebee, cuckoo
Psittacus erithacus — Parrot, gray
Psocoptera — Louse, book
Psophia leucoptera — Trumpeter, white-winged
Psychidae — Bagworm
Ptelea trifoliolata — Hop tree
Pteridophora alberti — Bird of Paradise, King of Saxony
Pternohyla fodiens — Treefrog, burrowing
Pterocles sp. — Sandgrouse
Pterocnemia pennata — Rhea, Darwin's or lesser

Pterodroma inexpectata — Petrel, mottled or scaled
Pterois sp. — Lionfish
Pteronura brasiliensis — Otter, giant
Pteropus giganteus — Flying fox, Indian
Pterospora andromedea — Pinedrops
Pterostyrax hispida — Epaulette tree
Pterourus palamedes — Swallowtail butterfly, Palamedes
Ptilichthys goodei — Quillfish
Ptilinopus superbus — Dove, superb fruit or purple-crowned pigeon
Ptilonorhynchus violaceus — Bowerbird, satin
Ptiloris victoriae — Rifleman bird
Ptilosarcus sp. — Sea pen
Ptilothrix bombiformis — Bee, digger
Ptychocheilus sp. — Squawfish
Ptychoraphus aleutica — Auklet, cassin
Pudu sp. — Pudu
Pueraria lobata — Kudzu
Puffinus assimilis — Shearwater, little
Puffinus bulleri — Shearwater, New Zealand or Buller's
Puffinus carneipes — Shearwater, flesh-footed
Puffinus creatopus — Shearwater, pink-footed
Puffinus diomedea — Shearwater, Cory's
Puffinus gravis — Shearwater, greater
Puffinus griseus — Shearwater, sooty
Puffinus lherminieri — Shearwater, Audubon's
Puffinus opisthomelas — Shearwater, black-vented
Puffinus pacificus — Shearwater, wedge-tailed
Puffinus puffinus — Shearwater, Manx
Puffinus tenuirostris — Shearwater, sharp-tailed
Pugettia producta — Crab, kelp
Pulex irritans — Flea, human
Pulsatilla sp. — Pasqueflower, alpine
Pulsatrix perspicillata — Owl, spectacled
Pungitius pungitius — Stickleback, ten-spined
Purpureicephalus spurius — Parrot, red-capped
Purshia tridentata — Antelope-brush
Pycnanthemum sp. — Mint, mountain
Pycnonotus barbatus — Bulbul, black-eyed
Pycnonotus jocosus — Bulbul, red-whiskered
Pygathrix nemaeus — Langur, douc
Pygathrix roxellanae — Monkey, snub-nosed golden
Pygoplites diacanthus — Angelfish, king
Pygoscelis adeliae — Penguin, adelie
Pygoscelis antarctica — Penguin, chinstrap
Pygoscelis papua — Penguin, gentoo
Pyrocephalus rubinus — Flycatcher, vermilion
Pyrola asarifolia — Wintergreen, bog or large
Pyrola secunda — Wintergreen, one-sided
Pyrola sp. — Shinleaf
Pyrophorus sp. — Beetle, fire
Pyrostegia venusta — Flameflower
Pyrota sp. — Beetle, blister (fire)
Pyrrhocorax graculus — Chough
Pyrrhocorax pyrrhocorax — Chough
Pyrrhula pyrrhula — Bullfinch
Pyrus americana — Mountain ash
Pyrus arbutifolia — Chokeberry, red
Pyrus communis — Pear
Python sebae — Python, African
Python sp. — Python, Indian
Pyxicephalus adspersus — Bullfrog, South African
Pyxidanthera barbulata — Pixie
Quamoclit coccinea — Morning-glory, red
Quelea quelea — Quelea
Quercus agrifolia — Oak, live (coast)
Quercus alba — Oak, white

Quercus arizonica — Oak, white (Arizona)
Quercus bicolor — Oak, swamp white
Quercus chrysolepis — Oak, live (canyon)
Quercus coccinea — Oak, scarlet
Quercus douglassii — Oak, blue
Quercus dunnii — Oak, Dunn
Quercus ellipsoidalis — Oak, pin (Northern)
Quercus emoryi — Oak, Emory
Quercus falcata — Oak, red (Southern)
Quercus gambelii — Oak, gambel
Quercus garryana — Oak, white (Oregon)
Quercus ilicifolia — Oak, bear
Quercus imbricaria — Oak, shingle
Quercus kelloggii — Oak, black (California)
Quercus laevis — Oak, turkey
Quercus laurifolia — Oak, laurel
Quercus lobata — Oak, valley
Quercus lyrata — Oak, overcup
Quercus macrocarpa — Oak, bur
Quercus marilandica — Oak, blackjack
Quercus michauxii — Oak, chestnut (swamp)
Quercus mohriana — Oak, Mohr
Quercus muehlenbergii — Oak, Chinkapin
Quercus myrtifolia — Oak, myrtle
Quercus nigra — Oak, water
Quercus palustris — Oak, pin
Quercus petraea — Oak, sessile or Durmast
Quercus phellos — Oak, willow
Quercus prinus — Oak, chestnut
Quercus rubra — Oak, red (Northern)
Quercus rugosa — Oak, netleaf
Quercus shumardii — Oak, Shumard
Quercus stellata — Oak, post
Quercus tomentella — Oak, live (island)
Quercus turbinella — Oak, Turbinella
Quercus velutina — Oak, black
Quercus virginiana — Oak, live (Virginia)
Quercus wislizeni — Oak, live (interior)
Quercus sp. — Oak (other)
Quiscalus major — Grackle, boat-tailed
Quiscalus mexicanus — Grackle, great-tailed
Quiscalus quiscula — Grackle, common or purple
Rachycentron canadum — Cobia
Rafinesquia neomexicana — Chicory, desert
Raja sp. — Skate
Rallus aquaticus — Rail, water
Rallus elegans — Rail, king
Rallus limicola — Rail, Virginia
Rallus longirostris — Rail, clapper
Ramphastos toco — Toucan, toco
Ramphastos tucanus — Toucan, Cuvier's
Ramphocoris clotbey — Lark, thick-billed
Rana areolata — Frog, crawfish
Rana aurora — Frog, red-legged
Rana berlandieri — Leopard frog, Rio Grande
Rana blairi — Leopard frog, Plains
Rana boylii — Frog, foothill yellow-legged
Rana cascadae — Frog, cascades
Rana catesbeiana — Bullfrog
Rana clamitans — Frog, green or bronze
Rana esculenta — Frog, edible
Rana goliath — Frog, goliath
Rana grylio — Frog, pig
Rana heckscheris — Frog, river
Rana muscosa — Frog, mountain yellow-legged
Rana palustris — Frog, pickerel
Rana pipiens — Leopard frog, Northern

Rana pretiosa — Frog, spotted
Rana ridibunda — Frog, marsh
Rana septentrionalis — Frog, mink
Rana sphenocaphala — Leopard frog, Southern
Rana sylvatica — Frog, wood
Rana tarahumarae — Frog, Tarahumara
Rana temporaria — Frog, common or European
Rana virgatipes — Frog, carpenter
Ranatra sp. — Water scorpion
Rangifer tarandus — Caribou
Ranunculus abortivus — Buttercup, kidney-leaf
Ranunculus acris — Buttercup, common
Ranunculus aquatilis — Buttercup, water
Ranunculus bulbosus — Buttercup, bulbous
Ranunculus eschscholtzii — Buttercup, subalpine
Ranunculus fascicularis — Buttercup, early
Ranunculus ficaria — Celandine, lesser
Ranunuclus flabellaris — Buttercup, water
Ranunculus glaberrimus — Buttercup, sagebrush
Ranunculus hispidus — Buttercup, hispid
Ranunculus occidentalis — Buttercup, Western
Ranunculus repens — Buttercup, creeping
Ranunculus sceleratus — Buttercup, celery-leaved
Ranunculus septentrionalis — Buttercup, swamp
Ranunculus sp. — Crowfoot
Raphicerus campestris — Steenbuck
Raphicerus melanotis — Grysbok
Ratibida pinnata — Coneflower, prairie
Rattus fuscipes — Rat, bush
Rattus norvegicus — Rat, Norway
Rattus rattus — Rat, black
Ratufa sp. — Squirrel, giant
Ravenala madagascariensis — Palm, traveller's
Recurvirostra americana — Avocet, America
Recurvirostra avosetta — Avocet, European
Recurvirostra novaehollandiae — Avocet, rednecked or Australian
Redunca arundinum or *redunca* — Reedbuck, Common or Southern
Regalecus glesne — Oarfish
Regina rigida — Watersnake, glossy crayfish
Regina septemvittata — Snake, queen
Regina sp. — Snake, crayfish
Regulus calendula — Kinglet, ruby-crowned
Regulus regulus — Goldcrest
Regulus satraps — Kinglet, golden-crowned
Reithrodontomys sp. — Mouse, harvest
Remiz pendulinus — Tit, penduline
Remora remora — Remora
Renilla reniformis — Sea pansy
Reticulitermes hesperus — Termite, subterranean
Rhacochilus toxotes — Seaperch, rubberlip
Rhacophorus sp. — Treefrog
Rhadinaea flavilata — Snake, pine woods
Rhamnus caroliniana — Buckthorn, Carolina
Rhamnus cathartica — Buckthorn, European
Rhamnus frangula — Buckthorn, glossy
Rhamnus purshiana — Buckthorn, cascara
Rhamphiophis multimaculatis — Snake, beaked (African)
Rhamphocottus richardsoni — Sculpin, grunt
Rhea americana — Rhea, common or gray or greater
Rheodytes leuceps — Turtle, Fitzroy
Rhexia virginica — Meadow beauty
Rhineura floridana — Worm-lizard, Florida
Rhiniodon typus — Shark, whale
Rhinobatus sp. — Guitarfish
Rhinoceros sondaicus — Rhinoceros, Javan

Rhinoceros unicornis — Rhinoceros, Indian
Rhinocheilus lecontei — Snake, long-nosed
Rhinoclemmys pulcherrima — Turtle, red (Mexican)
Rhinocrypta lanceolata — Gallito, crested
Rhinoderma darwini — Frog, Darwin's
Rhinolophus ferrumequinum — Bat, horseshoe (greater)
Rhinolophus luctus — Bat, horseshoe
Rhinophyrynus dorsalis — Toad, burrowing
Rhinopithecus roxellana — Monkey, golden
Rhinopius aphanes — Scorpionfish
Rhinoplex vigil — Hornbill, helmeted
Rhinopoma sp. — Bat, mouse-tailed
Rhinoptera bonasus — Ray, cownose
Rhinosciurus laticaudatus — Squirrel, long-nosed
Rhipidura leucophrys — Willie-wag-tail
Rhipidura sp. — Fantail
Rhizomys sp. — Rat, bamboo
Rhizophora mangle — Mangrove, red
Rhizoprionodon terraenovae — Shark, sharp-nosed
Rhodeus amarus — Bitterling
Rhodinocichla rosea — Tanager, rose-breasted thrush
Rhododendron albiflorum — Rhododendron, white Western
Rhododendron calendulaceum — Azalea, flame
Rhododendron catawbiense — Rhododendron, catawba or purple
Rhododendron ferrugineum — Alpenrose
Rhododendron lapponicum — Rosebay, lapland
Rhododendron macrophylum — Rhododendron, California
Rhododendron maximum — Rhododendron, white
Rhododendron nudiflorum — Pinxter
Rhododendron occidentale — Azalea, Western
Rhododendron roseum — Azalea, early or mountain
Rhododendron viscosum — Honeysuckle, swamp
Rhodostethia rosea — Gull, Ross'
Rhombomys opimus — Gerbil, great
Rhus aromatica — Sumac, fragrant
Rhus copallina — Sumac, shining or winged
Rhus glabra — Sumac, smooth
Rhus integrifolia — Sumac, lemonade
Rhus lanceolata — Sumac, prairie
Rhus laurina — Sumac, laurel
Rhus ovata — Sumac, sugar
Rhus trilobata — Squawbush
Rhus typhina — Sumac, staghorn
Rhyacotriton olympicus — Salamander, Olympic
Rhynchonycteris sp. — Bat, long-nosed
Rhynocheto jubatus — Kagu
Rhysella sp. — Wasp, ichneumon
Ribes rotundifolium or *roezlii* — Gooseberry
Richardsonius balteatus — Shiner, redside
Ricinus communis — Castor bean
Riparia riparia — Swallow, bank
Rissa brevirostris — Kittiwake, red-legged
Rissa tridactyla — Kittiwake, black-legged
Robinia neomexicana — Locust, New Mexico (tree)
Robinia pseudo-acacia — Locust, black (tree)
Robinia sp. — Locust (tree)
Rollulus rouloul — Partridge, roulroul or crested wood
Romerolagus diazi — Rabbit, volcano
Rosa carolina — Rose, pasture
Rosa multiflora — Rose, multiflora
Rosa nutkana — Rose, nootka
Rosa palustris — Rose, swamp
Rosa rugosa — Rose, beach
Rosa suffulta — Rose, prairie

Rosa virginiana — Rose, Virginia
Rosalia sp. — Beetle, longhorn
Rostratula benghalensis — Snipe, painted
Rostrhamus sociabilis — Kite, snail or Everglades
Rousettus aegyptiacus — Bat, fruit (Egyptian)
Roystonea regia — Palm, royal
Rubus hispidus — Dewberry, swamp
Rubus lasiococcus — Bramble, dwarf
Rubus odoratus — Raspberry, purple-flowering
Rubus parviflorus — Thimbleberry
Rubus phoenicolasius — Wineberry
Rubus spectabilis — Salmonberry
Rudbeckia hirta — Black-eyed Susan or Coneflower
Rudbeckia triloba — Coneflower, three-lobed
Ruellia pedunculata — Petunia, wild
Ruellia pinetorum — Ruellia, pinelands
Rufibrenta ruficollis — Goose, red-breasted
Rumex acetosella — Sorrel, sheep
Rumex sp. — Dock
Rupicapra rupicapra — Chamois
Rupicola peruviana — Cock-of-the-rock, Peruvian
Rusa sp. — Sambar
Russula emetica — Sickener
Russula sp. — Mushroom, russula or brittlegill
Ruta graveolens — Rue
Rutilus rutilus — Roach, european
Ruvettus pretiosus — Oilfish
Rynchops niger — Skimmer, black
Rypticus sp. — Soapfish
Sabal palmetto — Palmetto, cabbage
Sabatia angularis — Pink, rose
Sabatia dodecandra — Pink, sea
Sabatia grandiflora — Pink, marsh
Sabatia stellaris — Pink, saltmarsh
Sabella sp. — Worm, featherduster
Sabellastarte magnifica — Worm, fan (magnificent banded)
Saccharum officinarum — Sugar cane
Sagittaria sp. — Arrowhead or Sagittaria
Sagittarius serpentarius — Secretary bird
Saguinus bicolor — Tamarin, pied
Saguinus fuscicollis — Tamarin, saddle-backed or brown-headed
Saguinus geoffroyi — Tamarin, Geoffrey's
Saguinus imperator — Tamarin, emperor
Saguinus mystax or *labiatus* — Tamarin, mustached
Saguinus nigricollis — Tamarin, black and red
Saguinus oedipus — Tamarin, cotton-top
Saiga tatarica — Saiga
Saimiri sp. — Monkey, squirrel
Salamandra salamandra — Salamander, fire
Salanoia concolor — Salano
Salazaria mexicana — Sage, bladder
Salicornia sp. — Glasswort
Salix alba — Willow, white
Salix babylonica — Willow, weeping
Salix bebbiana — Willow, bebb
Salix caroliniana — Willow, coastal plain
Salix discolor — Willow, pussy
Salix exigua — Willow, sandbar or coyote
Salix fluviatilis — Willow, river
Salix fragilis — Willow, crack
Salix lasiandra — Willow, Pacific
Salix nigra — Willow, black
Salix pyrifolia — Willow, balsam
Salix scoulerana — Willow, scouler
Salix sessilifolia — Willow, Northwest or velvet

Salix viminalis — Willow, basket
Salix sp. — Willow (other)
Salmo apache — Trout, Apache
Salmo clarki — Trout, cutthroat
Salmo gairdneri — Trout, rainbow or steelhead
Salmo salar — Salmon, Atlantic
Salmo trutta fario — Trout, brown (river)
Salpinctes obsoletus — Wren, rock
Salsola kali — Saltwort
Saltator maximus — Saltator, buff-throated
Salticus scenicus — Spider, zebra
Salvadora deserticola — Snake, patchnose (Big Bend)
Salvadora grahamiae — Snake, patchnose (mountain)
Salvadora hexalepis — Snake, patchnose (Western)
Salvelinus fontinalis — Trout, brook
Salvelinus malma — Trout, Dolly Varden
Salvelinus sp. — Charr
Salvia azurea — Salvia, blue
Salvia carduacea — Sage, thistle
Salvia coccinea — Salvia or scarlet sage
Salvia columbariae — Chia
Salvia fumerea — Sage, Death Valley
Salvia greggii — Sage, Autumn
Salvia lyrata — Sage, lyre-leaved or cancer-weed
Salvinia natans — Fern, water or water velvet
Sambucus callicarpa — Elder, red
Sambucus canadensis or *nigra* — Elder or elderberry
Sambucus cerulea — Elder, blue
Sambucus mexicana — Elder, Mexican
Samia cynthia — Moth, cynthia
Sanguinaria canadensis — Bloodroot
Sanguinus midas niger — Tamarin, red-handed
Sanguisorba canadensis — Burnet, Canadian
Sanicula canadensis — Snakeroot, black
Sapindus drummondii — Soapberry, Western
Sapium sebiferum — Tallowtree
Saponaria officinalis — Bouncing Bet
Sarcobatus vermiculatus — Greasewood
Sarcodes sanguinea — Snow plant
Sarcophaga sp. — Fly, flesh
Sarcophilus harrisi — Tasmanian devil
Sarcoramphus papa — Vulture, king
Sarcoscypha coccinea — Elf cup fungus, scarlet
Sarcostemma cynanchoides — Milkweed, climbing
Sarda sp. — Bonito
Sardina pilchardus — Sardine or pilchard
Sargassum sp. — Sargassum
Sarracenia flava — Trumpets
Sarracenia sp. — Pitcher plant
Sassafras albidum — Sassafras
Satanellus sp. — Cat, marsupial
Satureia vulgaris — Basil, wild
Satureja douglassi — Yerba buena
Saturnia pyri — Moth, Emperor or giant peacock
Sauromalus obesus — Chuckwalla
Saururus cernuus — Lizard's tail
Saxicola rubetra — Whinchat
Saxicola torquata — Stonechat
Saxifraga bronchialis — Saxifrage, spotted
Saxifraga mertensiana — Saxifrage, Merten's
Saxifraga occidentalis — Saxifrage, Western
Saxifraga oppositifolia — Saxifrage, purple
Saxifraga pensylvanica — Saxifrage, swamp
Saxifraga tolmiei — Saxifrage, alpine
Saxifraga virginiensis — Saxifrage, early
Sayornis nigricans — Phoebe, black
Sayornis phoebe — Phoebe, eastern

Sayornis saya — Phoebe, Say's
Scalopus aquaticus — Mole, eastern
Scalopus latimanus — Mole, broad-footed
Scalopus orarius — Mole, coast
Scaphidiidae — Beetle, shining fungus
Scaphinotus angusticollis — Snail-eater, narrow
Scaphiodontophis annulatus — Snake, neck-banded
Scaphiopus bombifrons — Toad, spadefoot (Plains)
Scaphiopus couchii — Toad, spadefoot (Couch)
Scaphiopus hammondii — Toad, spadefoot (Western)
Scaphiopus intermontanus — Toad, spadefoot (Great Basin)
Scaphiopus sp. — Toad, spadefoot (Eastern)
Scaphirhynchus platorynchus — Sturgeon, shovel-nose
Scarabaeidae — Beetle, scarab
Scartella cristata — Molly Miller
Scarus coeruleus — Parrotfish, blue
Scarus guacamaia — Parrotfish, rainbow
Scarus taeniopterus — Parrotfish, princess
Scarus vetula — Parrotfish, queen
Scarus sp. — Parrotfish (other)
Scatophaga stercoraria — Fly, dung (common yellow)
Sceliphron caementarium — Wasp, mud-dauber (black and yellow)
Sceliphron spirifex — Wasp, mud-dauber (African)
Sceloporus clarkii — Lizard, spiny (Clark)
Sceloporus graciosus — Lizard, sagebrush
Sceloporus grammicus — Lizard, mesquite
Sceloporus jarrovi — Lizard, spiny (Yarrow's)
Sceloporus magister — Lizard, spiny (desert)
Sceloporus merriami — Lizard, canyon
Sceloporus occidentalis — Lizard, fence (Western)
Sceloporus olivaceus — Lizard, spiny (Texas)
Sceloporus orcutti — Lizard, spiny (granite)
Sceloporus poinsetti — Lizard, spiny (crevice)
Sceloporus scalaris — Lizard, bunch-grass
Sceloporus undulatus — Lizard, fence (Eastern)
Sceloporus variabilis — Lizard, rosebelly
Sceloporus virgatus — Lizard, striped plateau
Sceloporus woodi — Lizard, Florida scrub
Schefflera actinophylla — Umbrella tree, Australian
Scheroderma citrinum — Earth-ball fungus
Schinus molle — Peppertree
Schistocerca gregaria — Locust, desert insect
Schoinobates volans — Glider, greater
Schrankia nuttalii — Sensitive briar
Sciadopitys verticillata — Umbrella tree, Japanese
Sciaenops ocellatus — Drum, red
Scincella lateralis — Skink, ground
Scincus sp. — Sandfish
Scirpus cyperinus — Grass, wool
Scirpus lacustris — Club-rush
Sciurus aberti — Squirrel, Abert's or tassel-eared
Sciurus carolinensis or *griseus* — Squirrel, gray
Sciurus niger — Squirrel, fox
Sciurus vulgaris — Squirrel, red (Eurasian)
Sclerurus albigularis — Leafscraper, gray-throated
Scolia dubia — Wasp, digger
Scoliopus bigelovii — Adder's tongue, fetid
Scolopax minor — Woodcock, American
Scolopax rusticola — Woodcock, Eurasian
Scolytus sp. — Beetle, bark
Scomber sp. — Mackerel
Scomberomorus maculatus — Mackerel, Spanish
Scomberomorus regalis — Cero
Scophthalmus aquosus — Windowpane
Scophthalmus maximus — Turbot

Scopus umbretta — Stork, hammerhead
Scorpaena guttata — Scorpionfish, California
Scorpaena plumieri — Scorpionfish, spotted
Scorpaena sp. — Scorpionfish
Scorpaenichthys marmoratus — Cabezon
Scorpaenidae sp. — Lionfish
Scotopelia peli — Owl, Pel's fishing
Scutellaria sp. — Scullcap
Scyliorhinus sp. — Dogfish, sandy or spotted
Scyllarides sp. — Lobster, slipper
Scythrops novaehollandiae — Cuckoo, channel-billed
Scytodes sp. — Spider, spitting
Sebastes serriceps — Treefish
Sebastes sp. — Rockfish
Sebastolobus alascanus — Thornyhead, shortspine
Sedum obtusatum — Sedum, Sierra
Sedum purpureum — Live-forever
Sedum rosea — Roseroot
Seicercus castaneiceps — Warbler, chestnut-headed
Seiurus aurocapillus — Ovenbird
Seiurus motacilla — Waterthrush, Louisiana
Seiurus noveboracensis — Waterthrush, Northern
Selasphorus platycercus — Hummingbird, broad-tailed
Selasphorus rufus — Hummingbird, rufous
Selasphorus sasin — Hummingbird, Allen's
Selenarctos thibetanus — Bear, Asian
Selene sp. — Moonfish
Selene vomer — Lookdown fish
Semibalanus balanoides — Barnacle, acorn
Semicossyphus pulcher — Sheephead, California
Seminatrix pygaea — Snake, swamp black
Semiophorus longipennis — Nightjar, standard-winged
Semnopithecus sp. — Monkey, leaf
Senecio aureus — Ragwort, golden
Senecio confuscus — Flame or fire vine
Senecio jacobaea — Ragwort, tansy
Senecio sp. — Groundsel
Senella attenuata — Dolphin, spotted
Sepia sp. — Cuttlefish
Sepioteuthis sepioidea — Squid, Atlantic oval or reef
Sequoia sempervirens — Redwood
Sequoiadendron giganteum — Sequoia, giant
Sericocarpus asteroides — Aster, white-topped
Sericulus chrysocephalus — Bowerbird, regent
Serinus canaria — Canary, yellow
Seriola dumerili — Amberjack, greater
Seriola lalandei — Yellowtail
Serpula sp. — Worm, tube
Serranus dewegeri — Vieja
Serranus phoebe — Tattler fish
Serranus trigrinus — Bass, harlequin
Serrasalmus natterei — Piranha
Sesarma cinereum — Crab, wharf
Sesarma sp. — Crab, marsh
Setifer setosus — Tenrec, hedgehog (greater)
Setonix brachyurus — Quokka
Setophaga ruticilla — Redstart, American
Shinisaurus crocodilurus — Xenosaur, Chinese
Shortia galacifolia — Oconee bells
Sialia currucoides — Bluebird, mountain
Sialia mexicana — Bluebird, Western
Sialia sialis — Bluebird, eastern
Sialis sp. — Fly, alder
Sicista betulina — Mouse, birch
Sida spinosa — Mallow, prickly
Sidalcea sp. — Checker mallow or checkers
Sigmodon hispidus — Rat, cotton (hispid)

Sigmodon sp. — Rat, cotton
Silene acaulis — Campion, moss
Silene californica — Pink, Indian
Silene caroliniana — Pink, wild
Silene cucubalus — Campion, bladder
Silene stellata — Campion, starry
Silene virginica — Pink, fire
Silphidae — Beetle, carrion
Silphium laciniatum — Compass plant
Silurus glanis — Wels
Simmondsia chinensis — Jojoba
Simulium sp. — Fly, black
Siphonodictyon coralliphagum — Sponge, yellow boring
Siphunculata — Louse, sucking
Siren intermedia — Siren, lesser
Siren lacertina — Siren, greater
Sistrurus catenatus — Rattlesnake, massasauga
Sisymbrium officinale — Mustard, hedge
Sisyrinchium douglasii — Grass widow
Sisyrinchium sp. — Blue-eyed grass
Sitta canadensis — Nuthatch, red-breasted
Sitta carolinensis — Nuthatch, white-breasted
Sitta europaea — Nuthatch
Sitta pusilla — Nuthatch, brown-headed
Sitta pygmaea — Nuthatch, pygmy
Sium suave — Water parsnip
Smerinthus cerisyi — Moth, sphinx willow
Smerinthus ocellata — Hawkmoth, eyed
Smilacina racemosa — Solomon's seal, false
Smilax herbacea — Carrionflower
Smilisca baudini — Treefrog, Mexican
Sminthopsis sp. — Dunnart
Snyodus sp. — Lizardfish
Solanum americanum — Nightshade, common
Solanum belladonna — Nightshade, deadly
Solanum dulcamara — Nightshade, bittersweet
Solanum nigrum — Nightshade, black
Solanum rostratum — Buffalo burr
Solanum sp. — Horsenettle
Solanum sp. — Potato vine
Solaster sp. — Starfish, sun
Solea solea — Sole
Solen sp. — Clam, razor
Solenodon sp. — Solenodon
Solenopsis sp. — Ant, fire
Solidago bicolor — Silverrod
Solidago juncea — Goldenrod, early
Solidago sp. — Goldenrod
Somateria fischeri — Eider, spectacled
Somateria mollissima — Eider, common
Somateria spectabilis — Eider, king
Sonchus asper — Thistle, sow (spiny-leaved)
Sonora semiannulata — Snake, ground
Sooglossus seychellensis — Frog, Seychelles
Sophora japonica — Pagoda tree, Japanese
Sophora secundiflora — Mescalbean
Sorbus americana — Mountain ash, American
Sorbus aucuparia — Mountain ash, European
Sorbus sp. — Mountain ash
Sorex arcticus — Shrew, Arctic
Sorex bendirii — Water shrew, Pacific
Sorex cinereus — Shrew, masked
Sorex dispar — Shrew, long-tailed or rock
Sorex fumeus — Shrew, smoky
Sorex longirostris — Shrew, Southeastern or Bachman's
Sorex minatus — Shrew, European pygmy
Sorex monticolus — Shrew, dusky

Sorex nanus — Shrew, dwarf
Sorex pacificus — Shrew, Pacific
Sorex palustris — Water shrew, American
Sorex vagrans — Shrew, vagrant
Sorghastrum nutans — Grass, Indian
Sotalia fluviatilis — Dolphin, estuarine or little bay
Sousa chinensis —Dolphin, humpbacked (Indo-Pacific or Chinese white)
Sousa teuszii — Dolphin, humpbacked (Atlantic)
Sparganium sp. — Bur-reed
Sparisoma aurofrenatum — Parrotfish, redband
Sparisoma radians — Parrotfish, bucktooth
Sparisoma rubripinne — Parrotfish, redfin
Sparisoma viridae — Parrotfish, stoplight
Spartina alternifolia — Cordgrass, saltmarsh
Spartina patens — Cordgrass, saltmeadow
Spartina pectinata — Cordgrass, prairie
Spathodea sp. — Tulip tree, African
Speothos venaticus — Dog, bush
Spermophilis variegatus — Squirrel, rock
Spermophilus armatus — Ground squirrel, Uinta
Spermophilus beecheyi — Ground squirrel, California
Spermophilus beldingi — Ground squirrel, Belding's
Spermophilus citellus — Ground squirrel, European
Spermophilus columbianus — Ground squirrel, Columbian
Spermophilus elegans — Ground squirrel, Wyoming
Spermophilus franklinii — Ground squirrel, Franklin's
Spermophilus lateralis — Ground squirrel, golden mantled
Spermophilus leucurus — Ground squirrel, antelope
Spermophilus mexicanus — Ground squirrel, Mexican
Spermophilus parryii or *undulatus* — Ground squirrel, Arctic
Spermophilus richardsonii — Ground squirrel, Richardson's
Spermophilus spilosoma — Ground squirrel, spotted
Spermophilus tereticaudus — Ground squirrel, round-tailed
Spermophilus townsendii — Ground squirrel, Townsend's
Spermophilus tridecemlineatus — Ground squirrel, thirteen lined
Speyeria cybele — Fritillary, great spangled
Speyeria diana — Butterfly, diana
Speyeria idalia — Fritillary, regal
Sphaeralcea ambigua — Globemallow, desert
Sphaeralcea coccinea — Globemallow, scarlet
Sphaeralcea coulteri — Globemallow, Coulter's
Sphaerodactylus cinereus — Gecko, ashy
Sphaerodactylus notatus — Gecko, reef
Sphecius speciosus — Wasp, cicada killer
Spheniscus demersus — Penguin, jackass or blackfoot or cape
Spheniscus humboldti — Penguin, Peruvian
Spheniscus magellanicus — Penguin, Magellan
Spheniscus mendiculus — Penguin, Galapagos
Sphoeroides sp. — Puffer
Sphyraena barracuda — Barracuda, great
Sphyraena sp. — Barracuda
Sphyraena sphyraena — Barracuda, European
Sphyrapicus thyroideus — Sapsucker, Williamson's
Sphyrapicus varius — Sapsucker, yellow-bellied
Sphyrna sp. — Shark, hammerhead
Sphyrna tiburo — Shark, bonnethead
Spigela marilandica — Pink, Indian
Spilogale gracilis — Skunk, spotted (Western)
Spilogale putorius — Skunk, spotted (Eastern)
Spilogale pygmaea — Skunk, spotted (pygmy)
Spilogale sp. — Skunk, spotted

Spilornis cheela — Eagle, crested serpent
Spinosella sp. — Sponge, tube
Spiraea latifolia — Meadow-sweet
Spiraea sp. — Spiraea
Spiraea tomentosa — Steeplebush or hardhack
Spiranthes sp. — Orchid, ladies' tresses
Spirobranchus giganteus — Worm, Christmas tree or serpulid
Spirogyra sp. — Spirogyra algae
Spiza americana — Dickcissel
Spizaetus ornatus — Hawk-eagle, ornate
Spizella arborea — Sparrow, tree American
Spizella atrogularis — Sparrow, black-chinned
Spizella breweri — Sparrow, Brewer's
Spizella pallida — Sparrow, clay-colored
Spizella passerina — Sparrow, chipping
Spizella pusilla — Sparrow, field
Splenodon punctatus — Tuatara
Squalus acanthias — Dogfish, piked or spiny
Squatina californica — Shark, angel (Pacific)
Squatina sp. — Shark, angel
Stachys cooleyae — Hedge-nettle, great
Stachys palustris — Woundwort
Stanleya pinnata — Prince's or desert plume
Staphylea trifolia — Bladdernut
Staphylinidae — Beetle, rove
Starnoenas cyanocephala — Dove, quail (blue-headed)
Steatornis caripensis — Oilbird
Stegostoma fasciatum or *tigrinum* — Catshark
Stegostoma fasciatum — Shark, zebra
Stelgidopteryx ruficollis — Swallow, rough-winged
Stelgidopteryx serripennis — Swallow, rough-winged (Northern)
Stellaria graminea — Stitchwort
Stellaria media — Chickweed
Stellula calliope — Hummingbird, calliope
Stenanthium gramineum — Feather bells
Stenella attenuata — Dolphin, brindled
Stenella clymene — Dolphin, clymene or Helmut
Stenella coerulevalba — Dolphin, striped
Stenella longirostris — Dolphin, spinner
Stenella plagiodon — Dolphin, Atlantic spotted
Steno bredanensis — Dolphin, rough-toothed
Stenopelmatus fuscus — Cricket, Jerusalem
Stenopus hispidus — Shrimp, banded or barber pole
Stenorhynchus orchiodes — Orchid, leafless beaked
Stenorhynchus seticornis — Crab, arrow
Stentor sp. — Stentor
Stercorarius longicaudus — Jaeger, long-tailed
Stercorarius parasiticus — Jaeger, parasitic
Stercorarius pomarinus — Jaeger, pomarine
Stereochilus marginatus — Salamander, many-lined
Stereolepis gigas — Sea bass, giant
Sterna albifrons — Tern, little or least
Sterna aleutica — Tern, Aleutian
Sterna anaethetus — Tern, bridled or brown-winged
Sterna bergii — Tern, crested
Sterna caspia — Tern, Caspian
Sterna dougallii — Tern, roseate
Sterna elegans — Tern, elegant
Sterna forsteri — Tern, Forster's
Sterna fuscata — Tern, sooty
Sterna hirundinacea — Tern, South American
Sterna hirundo — Tern, common
Sterna maxima — Tern, royal
Sterna nilotica — Tern, gull-billed
Sterna paradisaea — Tern, arctic

Sterna sandvicensis — Tern, Sandwich or Cabot's
Sterna striata — Tern, white-fronted
Sternocylta cyanopectus — Hummingbird, violet-chested
Sternotherus carinatus — Turtle, musk (razorback)
Sternotherus depressus — Turtle, musk (flattened)
Sternotherus minor — Turtle, musk (loggerhead)
Sternotherus odoratus — Stinkpot
Stewartia sp. — Stewartia
Stilosoma extenuatum — Snake, short-tailed
Stilpnotia salicis — Moth, satin
Stizostedion canadense — Sauger
Stizostedion lucioperca — Zander
Stizostedion vitreum — Walleye
Stoichactis helianthus — Sea anemone, sun
Stonolophus meleagris — Jellyfish, cannonball
Storeria dekayi — Snake, brown or DeKay's
Storeria occipitomaculata — Snake, redbelly
Stratiotes aloides — Water soldier
Strelitzia reginae — Bird of Paradise flower
Strepera sp. — Currawong
Strepsiptera — Stylops
Streptanthus arizonicus — Jewelflower, Arizona
Streptanthus tortuosus — Jewelflower, mountain
Streptopelia chinensis — Dove, spotted
Streptopelia decaocto — Dove, collared
Streptopelia risoria — Turtledove, ringed
Streptopelia turtur — Turtledove
Streptopus amplexifolius — Mandarin, wild
Streptopus rosens — Twisted stalk, rosy
Strigops habroptilus — Kakapo or owl parrot
Strix aluco — Owl, tawny
Strix nebulosa — Owl, great gray
Strix occidentalis — Owl, spotted
Strix seloputo — Owl, spotted wood
Strix varia — Owl, barred
Strombus sp. — Conch
Strongylocentrotus sp. — Sea urchin
Strongylura sp. — Needlefish
Struthidea cinerea — Apostlebird
Struthio camelus — Ostrich
Strymon melinus — Hairstreak butterfly, gray
Strymondia pruni — Hairstreak butterfly, black
Sturnella magna — Meadowlark, eastern
Sturnella neglecta — Meadowlark, Western
Sturnus vulgaris — Starling
Stylophorum diphyllum — Poppy, wood
Stylopidae — Beetle, bee parasite
Stylosanthes biflora — Pencil flower
Styrax sp. — Snowbell
Sula dactylatra — Booby, masked or blue-faced
Sula leucogaster — Booby, brown
Sula nebouxii — Booby, blue-footed
Sula sula — Booby, red-footed
Sula variegata — Booby, Peruvian
Suncus etruscus — Shrew, pygmy white-toothed
Suricata suricatta — Meerkat
Surnia ulula — Owl, hawk
Sus barbatus — Pig, bearded
Sus celebensis — Pig, Celebes
Sus salvanius — Hog, pygmy
Sus scrofa — Boar, wild
Swertia perennis — Felwort
Sylvia atricapilla — Blackcap
Sylvia borin — Warbler, garden
Sylvia cantillans — Warbler, subalpine
Sylvia communis or *curruca* — Whitethroat
Sylvia conspicillata — Warbler, spectacled

Sylvia undata — Warbler, Dartford
Sylvicapra grimmia — Duiker, common or gray
Sylvietta sp. — Crombec
Sylvilagus aquaticus — Rabbit, swamp
Sylvilagus audubonii — Cottontail, desert
Sylvilagus bachmani — Rabbit, brush
Sylvilagus floridanus — Cottontail, Eastern
Sylvilagus nuttalli — Cottontail, Nuttall's or Mountain
Sylvilagus palustris — Rabbit, marsh
Sylvilagus transitionalis — Cottontail, New England
Symphoricarpos orbiculatus — Coralberry
Symphoricarpos sp. — Snowberry
Symphurus sp. — Tonguefish
Symphytum officinale — Comfrey
Symplocarpus foetidus — Skunk cabbage
Symplocos tinctoria — Sweetleaf
Synancea sp. — Stonefish
Synaptomys cooperi — Lemming, Southern bog
Synceros caffer — Buffalo, African or Cape
Syngnathus sp. — Pipefish
Synodus intermedius — Sand diver
Synthliboramphus antiquua — Murrelet, ancient
Synthliboramphus hypoleucus — Murrelet, Xantus
Synthyris reniformis — Snow queen
Syrigma sibilatrix — Heron, whistling
Syrphus sp. — Fly, hover or flower
Syrrhaptes paradoxus — Sandgrouse, Pallas'
Syrrhophus sp. — Frog, chirping
Tabanus sp. — Fly, horse
Tabebuia sp. — Trumpet tree
Tachybaptus dominicus — Grebe, least
Tachybaptus ruficollis — Grebe, little
Tachycineta bicolor — Swallow, tree
Tachycineta thalassina — Swallow, violet-green
Tachyglossus aculeatus — Echidna, short-beaked
Tachyorcyctes sp. — Mole-rat
Tadorna ferruginea — Shelduck, ruddy
Tadorna radjah rufitergum — Shelduck, red-backed radjah
Tadorna tadorna — Shelduck
Taeniopoda eques — Grasshopper, horse-lubber
Taeniura limma — Ray, blue-spotted
Taeniura lymma — Stingray, blue-spotted
Talarida brasiliensis — Bat, free-tailed (Brazilian or Mexican)
Talarida macrotis — Bat, free-tailed (big)
Talorchestia sp. — Flea, beach
Talpa europea — Mole, European
Tamandua sp. — Tamandua anteater
Tamarindus indica — Tamarind
Tamarix chinensis — Tamarisk
Tamias sibiricus — Chipmunk, Siberian or striped squirrel
Tamias striatus — Chipmunk, Eastern
Tamiasciurus douglassi — Squirrel, Douglas or Pine
Tamiasciurus hudsonicus — Squirrel, red or Chickaree
Tamus communis — Bryony, black
Tanacetum vulgare or *douglasii* — Tansy
Tandanus tandanus — Tandan
Tantilla coronata — Snake, crowned (Southeastern)
Tantilla gracilis — Snake, flat-head
Tantilla relicta — Snake, crowned (Florida)
Tantilla sp. — Snake, black-headed
Tanytarsus sp. — Midge, green
Tapirus bairdii — Tapir, Baird's
Tapirus indicus — Tapir, Malayan
Tapirus pinchaque — Tapir, mountain
Tapirus terrestris — Tapir, Brazilian or lowland

Taraba major — Antshrike, great
Taraxacum officinale — Dandelion
Tardigrada — Water bear
Taricha granulosa — Newt, roughskin
Taricha rivularis — Newt, red-bellied
Taricha torosa — Newt, California
Tarpon sp. — Tarpon
Tarsiger cyanurus — Bluetail, red-flanked
Tarsipes rostratus or *spenserae* — Possum, honey
Tarsius sp. — Tarsier
Tasmacetus shepherdi — Whale, beaked (Shepherd's)
Tauraco erythrolophus — Turaco, red-crested
Taurotragus derbiamus — Eland, giant or Lord Derby's
Taurotragus oryx — Eland, Common or cape
Tautoga onitis — Tautog
Tautogolabris adspersus — Cunner
Taxidea taxus — Badger
Taxodium distichum — Cypress, bald
Taxus baccata — Yew, common
Taxus brevifolia — Yew, Pacific
Taxus floridana — Yew, Florida
Tayassu pecari — Peccary, white-lipped
Tayassu tajacu — Peccary, collared
Tealia crassicornia — Sea anemone, Christmas or dahlia
Tealia felina — Sea anemone, dahlia or "Flower of the sea"
Tealia sp. — Sea anemone
Tedania ignis — Sponge, fire
Tegenaria sp. — Spider, house
Teius teyou — Teyu
Telescopus fallax — Snake, cat
Tellina grandiflora — Fringe cups
Tenebrionidae — Beetle, darkling
Tenrec ecaudatus — Tenrec, tailless
Tephrosia virginiana — Goats-rue
Terathopius ecaudatus — Eagle, bateleur
Terebratulina sp. — Lampshell
Terpsiphone sp. — Flycatcher, paradise
Terrapene carolina — Turtle, box Eastern
Terrapene ornata — Turtle, box ornate or Western
Tersina viridis — Tanager, swallow
Testudo elephantophus — Tortoise, giant (Galapagos)
Testudo graeca — Tortoise, Mediterranean spur-thighed
Testudo hermanni — Tortoise, Hermann's
Testudo marginata — Tortoise, Greek or marginated
Tetanocera sp. — Fly, marsh
Tetracerus quadricornis — Antelope, four-horned
Tetrao urogallus — Capercaillie
Tetraogallus sp. — Snowcock
Tetraopes sp. — Beetle, longhorn milkweed
Tetrapterus sp. — Marlin
Tetrastes bonasia — Grouse, hazel or hazelhen
Tetrax tetrax — Bustard, little
Teucrium canadense — Sage, wood
Thalassoica antarctica — Petrel, Antarctic
Thalassoma bifasciatum — Wrasse, bluehead
Thalictrum sp. — Meadow-rue
Thamnophilus doliatus — Antshrike, barred
Thamnophis brachystoma — Garter snake, shorthead
Thamnophis butleri — Garter snake, Butler's
Thamnophis couchii — Garter snake, Western aquatic or Santa Cruz
Thamnophis cyrtopsis — Garter snake, black-nosed or blackneck
Thamnophis elegans — Garter snake, Western terrestrial
Thamnophis eques — Garter snake, Mexican
Thamnophis hammondii — Garter snake, two-striped

Thamnophis marcianus — Garter snake, checkered
Thamnophis ordinoides — Garter snake, Northwestern
Thamnophis proximus — Snake, ribbon (Western)
Thamnophis radix — Garter snake, plains
Thamnophis rufipunctatus — Garter snake, narrow headed
Thamnophis sauritus — Snake, ribbon (Eastern)
Thamnophis sirtalis — Garter snake, eastern or common
Thecla butulae — Hairstreak butterfly, brown
Thelypteris palustris — Fern, marsh
Thermobia domestica — Firebrat
Thermopsis montana — Pea, golden
Theropithecus gelada — Baboon, gelada
Thinocorus rumicivorus — Seedsnipe
Thlaspi arvense — Pennycress, field
Thomomys bottae — Pocket gopher, Botta's or Valley
Thomomys talpoides — Pocket gopher, Northern
Thraupis episcopus — Tanager, blue-gray
Threskiornis aethiopica — Ibis, sacred
Thryomanes bewickii — Wren, bewick's
Thryonomyidae — Rat, cane
Thryothorus ludovicianus — Wren, Carolina
Thuja occidentalis — Cedar, white (Northern)
Thuja orientalis — Arborvitae, Oriental
Thuja plicata — Cedar, red (Western or giant)
Thunbergia alata — Black-eyed-Susan vine
Thunnus albacares — Tuna, yellowfin
Thunnus albalunga — Tuna, albacore
Thunnus atlanticus — Tuna, blackfin
Thunnus thynnus — Tuna, bluefin
Thylacinus cynocephalus — Wolf, Tasmanian
Thylogale sp. — Pademelon
Thymallus arcticus — Grayling, Arctic
Thymallus thymallus — Grayling, European
Tiarella anifoliata — Mitrewort, false
Tiarella cordifolia — Foamflower
Tibicen canicularis — Harvestfly, dogday
Tibicen dorsata — Cicada, grand Western
Tichodroma muraria — Wallcreeper
Tigrisoma sp. — Heron, tiger
Tilapia mossambica — Tilapia, mozambique
Tilia americana — Basswood, American
Tilia heterophylla — Basswood, white
Tilia tomentosa — Linden, silver
Tilia x europaea — Linden, european
Tiliqua rugosa — Skink, stump-tailed
Tiliqua scincoides — Lizard, blue-tongued
Tiliqua sp. — Skink, blue-tongued
Tillandsia fasciculata — Pine, wild
Tillandsia usneoides — Spanish moss
Timema sp. — Timema
Tinamus major — Tinamou, great
Tinca tinca — Tench, golden
Tipula sp. — Fly, crane
Tipularia discolor — Orchid, crane-fly
Tmetothylacus tenellus — Pipit, golden
Tockus erythrorhynchus — Hornbill, red-billed
Tockus flavirostris — Hornbill, yellow-billed
Todus todus — Tody, Jamaica
Tolypeutes tricinctus — Armadillo, three-banded
Tomistoma schlegelii — Gharial, false
Torgos tracheliotus — Vulture, lappet-faced
Torpedo californica — Ray, electric Pacific
Torpedo nobiliana — Torpedo, Atlantic
Torreya californica — Torreya, California
Torreya taxifolia — Torreya, Florida
Townsendia excapa — Daisy, stemless

Toxicodendron diversiloba — Poison oak
Toxicodendron radicans — Poison ivy
Toxicodendron vernix — Sumac, poison
Toxostoa curvirostre — Thrasher, curve-billed
Toxostoma bendirei — Thrasher, bendire's
Toxostoma crissale — Thrasher, Crissal
Toxostoma lecontei — Thrasher, Le Conte's
Toxostoma longirostre — Thrasher, long-billed
Toxostoma rufum — Thrasher, brown
Toxostoma vedivivum — Thrasher, California
Toxotes sp. — Archerfish
Trachemys scripta — Turtle, red-eared
Trachinotus carolinus — Pompano, Florida
Trachinotus goodei — Palometa
Trachinotus rhodopus — Pompano, gafftopsail
Trachurus symmetricus — Mackerel, Jack
Trachycarpus fortunei — Palm, windmill
Trachydosaurus rugosus — Lizard, shingleback
Trachyphonus erythrocephalus — Barbet, red-and-yellow
Tradescantia ohiensis — Spiderwort
Tragelaphus buxtoni — Nyala
Tragelaphus eurycerus — Bongo
Tragelaphus imberbis — Kudu, lesser
Tragelaphus scriptus — Bushbuck
Tragelaphus spekei — Sitatnga, spiral-horned
Tragelaphus strepsiceros — Kudu, greater
Tragopan sp. — Tragopan
Tragopogon porrofolius — Salsify
Tragopogon sp. — Goatsbeard
Tragulus sp. — Deer, mouse or Chevrotain
Tremarctos ornatus — Bear, spectacled
Treron phoenicoptera — Pigeon, yellow-legged green
Triaenodon obesus — Shark, reef (whitetip)
Triakis felis or *semifasciata* — Shark, leopard
Trialeurodes vaporariorum — Fly, white (greenhouse)
Tribulus terrestris — Punctureweed
Trichechus inunguis — Manatee, Amazonian
Trichechus manatus — Manatee
Trichiotinus sp. — Beetle, flower
Trichiurus lepturus — Cutlassfish, Atlantic
Trichoglossus haematodus — Lorikeet, rainbow
Trichoptilium incisum — Yellowhead
Trichostema dichotomum — Blue-curls
Trichostema lanceolatum — Vinegar weed
Trichosurus vulpecula — Possum, brushtail
Trichys fasciculata — Porcupine, long-tailed
Tridacna sp. — Clam, giant
Trientalis borealis — Starflower
Trientalis latifolia — Starflower, Western
Trifolium agrarium — Clover, hop
Trifolium arvense — Clover, rabbit's foot
Trifolium incarnatum — Clover, crimson
Trifolium pratense — Clover, red
Trifolium repens — Clover, white
Trillium cernuum — Trillium, nodding
Trillium erectum — Wake robin or Squawroot
Trillium grandiflorum — Trillium, large-flowered
Trillium luteum — Trillium, yellow
Trillium nivale — Trillium, snow
Trillium ovatum — Wake robin, Western
Trillium sessile — Trillium, red
Trillium undulatum — Trillium, painted
Trimeresurus sp. — Kufah or Habu
Trimorphodon biscutatus — Snake, lyre
Trinectes maculatus — Hogchoker
Tringa cinereus — Sandpiper, Terek

Tringa erythropus — Redshank, spotted
Tringa flavipes — Yellowlegs, lesser
Tringa glareola — Sandpiper, wood
Tringa hypleucos — Sandpiper, common
Tringa melanoleuca — Yellowlegs, greater
Tringa nebularia — Greenshank
Tringa solitaria — Sandpiper, solitary
Tringa stagnatilus — Sandpiper, marsh
Tringa totanus — Redshank, common
Triodanis perfoliata — Venus'-looking-glass
Trionyx ferox — Turtle, softshelled (Florida)
Trionyx muticus — Turtle, softshelled (smooth)
Trionyx spiniferus — Turtle, softshelled (spiny)
Triosteum perfoliatum — Coffee, wild
Triphora trianthophora — Orchid, three birds
Tripneustes ventricosus — Sea egg
Tripsacum dactyloides — Grass, gama
Triteleia ixioides — Pretty face
Triturus cristatus — Newt, warty
Triturus vulgaris — Newt, smooth
Trixis californica — Trixis
Troglodytes aedon — Wren, house
Troglodytes brunneicollis — Wren, brown-throated
Troglodytes troglodytes — Wren, common or winter
Trogon sp. — Trogon
Trollius laxus — Globeflower
Trombidiidae — Mite, velvet
Tropidoclonion lineatum — Snake, lined
Tryngites subruficollis — Sandpiper, buff-breasted
Tsuga canadensis — Hemlock, Eastern
Tsuga caroliniana — Hemlock, Carolina
Tsuga heterophylla — Hemlock, Western
Tsuga mertensiana — Hemlock, mountain
Tubastrea coccinea — Coral, red or orange
Tuber sp. — Truffle
Tubificidae — Worm, tubifex
Tubulamus sp. — Worm, ribbon banded
Tupaia sp. — Shrew, tree
Tupinambis tequixin — Tegu
Turdus eunous — Thrush, dusky
Turdus hortulorum — Thrush, gray-backed
Turdus iliacus — Redwing
Turdus merula — Blackbird, European
Turdus migratorius — Robin, American
Turdus olivaceus — Thrush, Cape or olive
Turdus philomelos — Thrush, song
Turdus pilaris — Fieldfare
Turdus poliocephalus — Thrush, island
Turdus torquatus — Ouzel, ring
Turdus viscivorus — Thrush, mistle
Turnix sylvatica — Hemipode, Andalusian or little buttonquail
Tursiops gilli — Dolphin, bottle-nosed (Pacific)
Tursiops truncatus — Dolphin, bottle-nosed
Tussilago farfara — Coltsfoot
Tylomys sp. — Rat, climbing
Tylosurus crocodilus — Houndfish
Tympanuchus cupido — Prairie chicken, greater
Tympanuchus pallidicinctus — Prairie chicken, lesser
Tympanuchus phasianellus — Grouse, sharp-tailed
Typha latifolia — Cattail
Typhlichthys subterraneus — Cavefish, southern
Typhlomolge rathbuni — Salamander, blind (Texas)
Typhlops sp. — Snake, blind (Australian)
Typhlotriton spelaeus — Salamander, grotto
Tyrannus crassirostris — Kingbird, thick-billed
Tyrannus dominicensis — Kingbird, gray

Tyrannus melancholicus — Kingbird, tropical
Tyrannus tyrannus — Kingbird, eastern
Tyrannus verticalis — Kingbird, Western or Arkansas
Tyrannus vociferans — Kingbird, Cassin's
Tyto alba — Owl, barn
Uca sp. — Crab, fiddler
Ulmus alata — Elm, winged
Ulmus americana — Elm, American
Ulmus parvifolia — Elm, Chinese
Ulmus procera — Elm, English
Ulmus pumila — Elm, Siberian
Ulmus rubra — Elm, slippery
Ulva lactuca — Sea lettuce
Uma inornata — Lizard, fringetoed (Coachella Valley)
Uma notata — Lizard, fringetoed (Colorado desert)
Uma scoparia — Lizard, fringetoed (Mojave)
Umbellularia californica — Laurel, California
Umbra sp. — Mudminnow
Umbrina roncador — Croaker, yellowfin
Ungnadia speciosa — Buckeye, Mexican
Uniola paniculata — Sea oats
Upucerthia dumentaria — Earthcreeper, scale-throated
Upupa epops — Hoopoe
Uraeginthus sp. — Cordon bleu
Uria aalge — Murre, common or guillemot
Uria lomvia — Murre, thick-billed
Urnula craterium — Devil's urn fungus
Urocissa erythrorhyncha — Magpie, red-billed blue
Urocyon cinereoargenteus — Fox, gray
Uroderma bilobatum — Bat, tent-making
Urogymnus asperrimus — Ray, thorny
Urolophus halleri — Stingray, round
Urolophus mucosus — Stingray, Australian
Uropeltis sp. — Snake, shieldtail
Urophycis sp. — Hake
Uroplatus finibriatus — Gecko, leaf-tailed
Urosaurus microscutatus — Lizard, small-scaled
Urosaurus ornatus — Lizard, tree
Urosaurus sp. — Lizard, brush
Ursus americanus — Bear, black
Ursus arctos — Bear, brown or grizzly
Ursus malayanus — Bear, sun
Ursus maritimus — Bear, polar
Urtica dioica — Nettle, stinging
Uta stansburiana — Lizard, side-blotched
Utricularia cornuta — Bladderwort, horned
Utricularia inflata — Bladderwort, swollen
Utricularia vulgaris — Bladderwort, greater
Uvularia perfoliata — Bellwort, perfoliate
Uvularia sessilifolia or *latifolia* — Wild oats
Vaccinium arboreum — Sparkleberry
Vaccinium sp. — Blueberry; Cranberry; Huckleberry
Vaccinium stamineum — Deerberry
Vampyrum spectrum — Vampire bat, false
Vancouveria hexandra — Inside-out flower, Northern
Vanellus coronatus — Plover, crowned
Vanellus senegallus — Plover, wattled
Vanellus tricolor — Plover, banded
Vanellus vanellus — Lapwing
Vanessa atalanta — Butterfly, admiral (red)
Vanessa cardui — Butterfly, painted lady
Vanessa virginiensis — Butterfly, painted lady (American)
Vanga curvirostris — Sicklebill
Vanilla sp. — Orchid, vanilla
Varanus gouldii — Monitor lizard, Gould's or sand goanna

Varanus komodoensis — Komodo dragon
Varanus varius — Monitor lizard, lace
Varecia variegata — Lemur, ruffed or variegated
Velella velella — By-the-wind sailor
Veratrum californicum — Lily, corn (California)
Veratrum viride or *album* — Hellebore, false
Verbascum blattaria — Mullein, moth
Verbascum thapsus — Mullein
Verbena ambrosifolia — Vervain, Western (pink)
Verbena hastata — Vervain, blue
Verbena tampensis — Vervain, rose
Verbesina encelioides — Daisy, cowpen
Vermicella annulata — Bandy-bandy
Vermivora celata — Warbler, orange-crowned
Vermivora chrysoptera — Warbler, golden-winged
Vermivora luciae — Warbler, Lucy's
Vermivora peregrina — Warbler, Tennessee
Vermivora pinus — Warbler, blue-winged
Vermivora ruficapilla — Warbler, Nashville
Vermivora virginiae — Warbler, Virginia's
Vernonia altissima — Ironweed, tall
Vernonia noveboracensis — Ironweed
Veronica sp. — Speedwell or brooklime
Veronicastrum virginicum — Culver's root
Verpa sp. — Mushroom, thimble-cap
Verreo oxycephalus — Pigfish
Vespa crabro germana — Hornet, worker (giant)
Vespa crabro — Hornet, worker or European
Vespula maculata — Hornet, bald-faced
Vespula sp. — Yellow jacket
Viburnum alnifolium — Hobblebush
Viburnum cassinoides — Witherod
Viburnum dentatum — Arrowwood
Viburnum lentago — Nannyberry
Viburnum prunifolium — Haw, black
Viburnum sp. — Viburnum
Vicia americana — Vetch, American
Vicia angustifolia — Vetch, common
Vicia cracca — Vetch, cow or bird or tufted
Vicia villosa — Vetch, hairy
Vicugna vicugna — Vicuna
Vidua sp. — Whydah
Vinca major — Periwinkle, greater (plant)
Vinca minor — Periwinkle or myrtle (plant)
Viola adunca — Violet, blue
Viola blanda — Violet, sweet white
Viola canadensis — Violet, Canada
Viola conspersa — Violet, dog
Viola cucullata — Violet, marsh blue
Viola fimbriatula — Violet, downy (northern)
Viola glabella — Violet, stream
Viola palustris — Violet, marsh
Viola papilionacea — Violet, blue (common)
Viola pedata — Violet, bird-foot
Viola pubescens — Violet, downy yellow
Viola rostrata — Violet, long-spurred
Viola rotundifolia or *pensylvanica* — Violet, yellow
Viola sagittata — Violet, arrow-leaved
Viola sempervirens — Violet, evergreen
Viola sp. — Violet other
Violet lanceolata — Violet, lance-leaved
Vipera ammodytes — Viper, sand
Vipera aspis — Viper, asp
Vipera berus — Adder, European or common viper
Vipera palaestinae — Viper, Palestinian
Vipera russellii — Viper, Russell's
Vireo altiloquus — Vireo, black-whiskered

Vireo atricapillus — Vireo, black-capped
Vireo bellii — Vireo, Bell's
Vireo flavifrons — Vireo, yellow-throated
Vireo gilvus — Vireo, warbling
Vireo griseus — Vireo, white-eyed
Vireo huttoni — Vireo, Hutton's
Vireo olivaceus — Vireo, red-eyed
Vireo philadelphicus — Vireo, Philadelphia
Vireo solitarius — Vireo, solitary
Vireo vicinior — Vireo, gray
Virginia striatula — Snake, earth rough
Virginia valeriae — Snake, earth smooth
Viverra tangalunga or *megaspila* — Civet, Oriental or Malay
Viverricula indica — Civet, Indian
Volvox sp. — Volvox algae
Vombatus ursinus — Wombat
Vormela peregusna — Polecat, marbled
Vulpes cana — Fox, Blandford's
Vulpes chama — Fox, Cape or silver-backed
Vulpes corsac — Fox, Corsac
Vulpes pallida — Fox, pale
Vulpes velox — Fox, kit or swift
Vulpes vulpes — Fox, red
Vulpes zerba — Fox, fennec
Vultur gryphus — Condor, Andean
Waldsteinia fragarioides — Strawberry, barren
Wallabia bicolor — Wallaby, swamp
Wallabia parryi — Wallaby, pretty-face
Washingtonia filfera — Washingtonia, California
Whipplea modesta — Yerba de Selva
Wilsonia canadensis — Warbler, Canada
Wilsonia citrina — Warbler, hooded
Wilsonia pusilla — Warbler, Wilson's
Wislizenia refracta — Clover, jackass
Wisteria sinensis — Wisteria
Wolfia arrhiza — Duckweed, rootless
Wyethia amplexicaulis — Mule's ears
Wyulda squamicaudata — Possum, scaly-tailed
Xanthium strumarium — Cocklebur
Xanthocephalus xanthocephalus — Blackbird, yellow-headed
Xantusia henshawi — Lizard, night (granite)
Xantusia riversiana — Lizard, night (Island)
Xantusia virgilis — Lizard, night (common or desert)
Xema sabini — Gull, Sabine's
Xenistius californiensis — Salema
Xenodon rabdocephalus — Fer-de-lance, false
Xenopeltis unicolor — Snake, sunbeam
Xenopus laevis — Frog, African clawed
Xenorhynchus asiaticus — Stork, black-necked
Xerophyllum asphodeloides — Turkey beard
Xerophyllum tenax — Beargrass

Xerus erythropus — Ground squirrel, African (Western or striped)
Xerus inauris — Ground squirrel, African or Cape
Xiphius gladius — Swordfish
Xiphorhynchus guttatus — Woodcreeper, buff-throated
Xylaria polymorpha — Dead man's fingers (fungus)
Xylocopa sp. — Bee, carpenter African
Xyrauchen texanus — Sucker, razorback
Xyris torta or *iridifolia* — Yellow-eyed grass
Xysticus sp. — Spider, crab
Yucca aloifolia — Spanish bayonet
Yucca brevifolia — Joshua tree
Yucca sp. — Yucca
Yucca whipplei — Our Lord's candle
Zaedylus pichiy — Pichi
Zaglossus bruijni — Echidna, long-beaked
Zalophus californianus — Sea lion, California
Zantedeschia aethiopica — Lily, calla
Zanthoxylum americanum — Prickly-ash
Zanthoxylum clava-herculis — Hercules club
Zanthoxylum fagara — Prickly-ash, lime
Zapus hudsonius — Jumping mouse, meadow
Zapus princeps — Jumping mouse, Western
Zapus trinotatus — Jumping mouse, Pacific
Zarhipis sp. — Beetle, glowworm
Zauschneria californica — Fuchsia, California
Zelkova serrata — Zelkova, Japanese
Zenaida asiatica — Dove, white-winged
Zenaida macroura — Dove, mourning
Zenobia pulverulenta — Zenobia
Zephranthes atamasco — Lily, Atamasco or zephyr
Zephyranthes candida — Lily, zephyr
Zephyranthes longifolia — Lily, rain
Zerene sp. — Butterfly, dog face
Zeus sp. — John Dorie
Zeuxine strateumatica — Orchid, lawn
Zigadenus sp. — Death camus
Ziphius cavirostris — Whale, beaked (Cuvier's)
Zizania aquatica — Rice, wild
Zizia aptera — Alexanders, golden
Zonaria spadicea — Cowrie, chestnut
Zonotrichia albicollis — Sparrow, white-throated
Zonotrichia atricapilla — Sparrow, golden-crowned
Zonotrichia leucophrys — Sparrow, white-crowned
Zonotrichia querula — Sparrow, Harris'
Zootermopsis angusticollis — Termite, dampwood or Pacific coast
Zoothera dauma — Thrush, white's
Zostera marina — Eelgrass
Zosterops sp. — White-eye
Zu cristatus — Ribbonfish
Zygaena sp. — Moth, burnet

Bibliography by Titles

America's Neighborhood Bats. Merlin D. Tuttle. Austin: University of Texas Press, 1988. ISBN 0-292-70403-8, 0-292-70406-2 (paper). ANB

Amphibians and Reptiles of New England: Habitats and Natural History. Richard M. DeGraaf and Deborah D. Rudis. Amherst: University of Massachusetts, 1983. ISBN 0-87023-399-8, 0-87023-400-5 (paper). ARN

Amphibians and Reptiles of Texas. James R. Dixon. College Station: Texas A&M Press, 1987. ISBN 0-89096-293-6, 0-89096-358-4 (paper). ART

Amphibians and Reptiles of the Carolinas and Virginia. Bernard S. Martof, et al. Chapel Hill: University of North Carolina Press, 1980. ISBN 0-8078-1389-3, 0-8078-4252-4 (paper). ARC

Arctic Animals: A Celebration of Survival. Fred Bruemmer. Ashland, WI: NorthWord, Inc., 1986. ISBN 0-7710-1717-0, 0-9428-0253-5. BAA

Atlantic and Gulf Coasts. (Audubon Society Nature Guide.) William H. Amos and Stephen H. Amos. NY: Knopf, 1985. ISBN 0-394-73109-3 (paper). AGC

Audubon Society Book of Insects. Les Line, Lorus Milne, and Margery Milne. NY: Abrams, 1983. ISBN 0-8109-1806-4. BOI

Audubon Society Book of Marine Wildlife. Les Line, with text by George Reiger. NY: Abrams, 1980. ISBN 0-8109-0672-4. BMW

Audubon Society Book of Water Birds. Les Line, Kimball L. Garrett, and Kenn Kaufman. NY: Abrams, 1987. ISBN 0-8109-1863-3. BWB

Audubon Society Book of Wild Cats. Les Line and Edward R. Ricciuti. NY: Abrams, 1985. ISBN 0-8109-1828-5. AWC

Audubon Society Encyclopedia of Animal Life. Ralph Buchsbaum, et al. NY: Clarkson N. Potter, 1982. ISBN 0-5175-4657-4. AAL

Audubon Society Encyclopedia of North American Birds. John K. Terres. NY: Knopf, 1980. ISBN 0-3944-6651-9. EAB

Audubon Society Field Guide to North American Birds (Eastern Region). John Bull and John Farrand. NY: Knopf, 1977. ISBN 0-3944-1405-5. NAB

Audubon Society Field Guide to North American Birds (Western Region). Miklos D.F. Udvardy. NY: Knopf, 1977. ISBN 0-3944-1410-1. UAB

Audubon Society Field Guide to North American Butterflies. Robert Michael Pyle. NY: Knopf, 1981. ISBN 0-3945-1914-0. PAB

Audubon Society Field Guide to North American Fishes, Whales, and Dolphins. Herbert T. Boschung, et al. NY: Knopf, 1983. ISBN 0-3945-3405-0. NAF

Audubon Society Field Guide to North American Insects and Spiders. Lorus Milne and Margery Milne. NY: Knopf, 1980. ISBN 0-3945-0763-0. MIS

Audubon Society Field Guide to North American Mammals. John O. Whitaker. NY: Knopf, 1980. ISBN 0-3945-0762-2. ASM

Audubon Society Field Guide to North American Mushrooms. Gary H. Lincoff. NY: Knopf, 1981. ISBN 0-3945-1992-2. NAM

Audubon Society Field Guide to North American Reptiles and Amphibians. John L. Behler and F.Wayne King. NY: Knopf, 1979. ISBN 0-3945-0824-6. ARA

Audubon Society Field Guide to North American Seashore Creatures. Norman A. Meinkoth. NY: Knopf, 1981. ISBN 0-3945-1993-0. MSC

Audubon Society Field Guide to North American Trees (Eastern Region). Elbert L. Little. NY: Knopf, 1980. ISBN 0-3945-0760-6. NTE

Audubon Society Field Guide to North American Trees (Western Region). Elbert L. Little. NY: Knopf, 1980. ISBN 0-3945-0761-4. NTW

Audubon Society Field Guide to North American Wildflowers (Eastern Region). William A. Niering and Nancy C. Olmstead. NY: Knopf, 1979. ISBN 0-3945-0432-1. NWE

Audubon Society Field Guide to North American Wildflowers (Western Region). Richard Spellenberg. NY: Knopf, 1979. ISBN 0-3945-0431-3. SWW

Auks: An Ornithologist's Guide. Ron Freethy. Poole, Dorset: Blandford Press; NY: Facts on File, 1987. ISBN 0-7137-1597-9 (Blandford), 0-8160-1696-8 (Facts on File). AOG

Bears of the World. Terry Domico. NY: Facts on File, 1988. ISBN 0-8160-1536-8. DBW

Beneath Cold Seas: Exploring Cold-Temperate Waters of North America. Photographs by Jeffrey L. Rotman, text by Barry W. Allen. NY: Van Nostrand Reinhold, 1983. ISBN 0-4422-7058-5. BCS

Birds for All Seasons. Jeffery Boswall. London: BBC, 1986. ISBN 0-5632-0453-2. BAS

Birds of Prey. (Birds of the World.) John P.S. MacKenzie. Ashland, WI: Paper Birch Press, Inc. 1986. ISBN 0-9613-9618-0. BOP

Birds of the Carolinas. Eloise F. Potter, James F. Parnell, and Robert P. Teulings. Chapel Hill: University of North Carolina Press, 1980. ISBN 0-8078-1399-0, 0-8078-4155-2 (paper). PBC

British Birds: A Field Guide. Alan J. Richards. London: David and Charles, 1979. ISBN 0-7153-8016-8. RBB

Collins Guide to the Rare Mammals of the World. John A. Burton and Bruce Pearson. Lexington, MA: Stephen Greene Press, 1988. ISBN 0-8289-0658-0. CRM

Common Wildflowers of the Northeastern United States. (New York Botanical Garden Field Guide). Carol H. Woodward and Harold William Rickett. Woodbury, NY: Barron's, 1979. ISBN 0-8120-0937-1. CWN

Coral Kingdoms. Carl Roessler. NY: Abrams, 1986. ISBN 0-8109-0774-7. RCK

A Country-Lover's Guide to Wildlife: Mammals, Amphibians, and Reptiles of the Northeastern United States. Kenneth A. Chambers. Baltimore: Johns Hopkins, 1979. ISBN 0-8018-2207-6. CGW

A Countryman's Flowers. Hal Borland. Photographs by Les Line. NY: Knopf, 1981. ISBN 0-3945-1893-4. BCF

Cranes of the World. Paul A. Johnsgard. Bloomington, IN: Indiana University Press, 1983. ISBN 0-2531-1255-9. COW

Deserts. (Audubon Society Nature Guide.) James A. MacMahon. NY: Knopf, 1985. ISBN 0-3947-3139-5. JAD

Diving Birds of North America. Paul A. Johnsgard. Lincoln: University of Nebraska, 1987. ISBN 0-8032-2566-0. DBN

Ducks of North America and the Northern Hemisphere. John Gooders and Trevor Boyer. NY: Facts on File, 1986. ISBN 0-8160-1422-1. DNA

Eastern Birds. (Audubon Handbook.) John Farrand. NY: McGraw-Hill, 1988. ISBN 0-07-019976-0. FEB

Eastern Forests. (Audubon Society Nature Guide.) Ann Sutton and Myron Sutton. NY: Knopf, 1985. ISBN 0-394-73126-3 (paper). SEF

Encyclopedia of Animal Behavior. Peter J.B. Slater, ed. NY: Facts on File, 1987. ISBN 0-8160-1816-2. SAB

Encyclopedia of Animal Biology. R. McNeill Alexander, ed. NY: Facts on File, 1987. ISBN 0-8160-1817-0. AAB

Encyclopedia of Animal Ecology. Peter D. Moore, ed. NY: Facts on File, 1987. ISBN 0-8160-1818-9. MAE

Encyclopedia of Aquatic Life. Keith Banister and Andrew Campbell, editors. NY: Facts on File, 1985. ISBN 0-8160-1257-1. EAL

Encyclopedia of Birds. Edited by Christopher M. Perrins and Alex L.A. Middleton. NY: Facts on File, 1985. ISBN 0-8160-1150-8. EOB

Encyclopedia of Insects. Edited by Christopher O'Toole. NY: Facts on File, 1986. ISBN 0-8160-1358-6. EOI

Encyclopedia of Mammals. David MacDonald. NY: Facts on File, 1984. ISBN 0-8719-6871-1. MEM

Encyclopedia of North American Wildlife. Stanley Klein. NY: Facts on File, 1983. ISBN 0-8719-6758-8. NAW

Encyclopedia of Reptiles and Amphibians. Edited by Tim Halliday and Kraig Adler. NY: Facts on File, 1986. ISBN 0-8160-1359-4. ERA

Fall Wildflowers of the Blue Ridge and Great Smoky Mountains. Oscar W. Gupton and Fred C. Swope. Charlottesville: University Press of Virginia, 1987. ISBN 0-8139-1123-0. FWF

A Field Guide to Atlantic Coast Fishes of North America. (Peterson Field Guide.) C. Richard Robins and G. Carleton Ray. Boston: Houghton Mifflin, 1986. ISBN 0-395-31852-1, -395-39198-9 (paper). ACF

A Field Guide to Coral Reefs of the Caribbean and Florida. (Peterson Field Guide.) Eugene H. Kaplan. Boston: Houghton Mifflin, 1982. ISBN 0-395-31661-8, 0-395-46939-2 (paper). KCR

A Field Guide to Hawks: North America. (Peterson Field Guide.) William S. Clark. Boston: Houghton Mifflin, 1987. ISBN 0-395-36001-3, 0-395-44112-9 (paper). FGH

A Field Guide to Mushrooms (North America). (Peterson Field Guide.) Kent H. McKnight and Vera B. McKnight. Boston: Houghton Mifflin, 1987. ISBN 0-395-42101-2, 0-395-42102-0 (paper). FGM

A Field Guide to Pacific Coast Fishes of North America. (Peterson Field Guide.) William N. Eschmeyer and Earl S. Herald. Boston: Houghton Mifflin, 1983. ISBN 0-395-26873-7, 0-395-33188-9 (paper). PCF

Field Guide to Poisonous Plants and Mushrooms of North America. Charles Kingsley Levy and Richard B. Primack. Brattleboro, VT: Stephen Greene Press, 1984. ISBN 0-828-90531-2, 0-828-90530-4 (paper). PPM

A Field Guide to Southeastern and Caribbean Seashores. (Peterson Field Guide). Eugene H. Kaplan. Boston: Houghton Mifflin, 1988. ISBN 0-395-31321-x, 0-395-46811-6 (paper). SCS

A Field Guide to Southern Mushrooms. Nancy Smith Weber and Alexander H. Smith. Ann Arbor: University of Michigan Press, 1985. ISBN 0-472-85615-4. GSM

A Field Guide to the Beetles of North America. (Peterson Field Guide.) Richard E. White. Boston: Houghton Mifflin, 1983. ISBN 0-395-31808-4, 0-395-33953-7 (paper). BNA

Field Guide to the Orchids of North America. John G. Williams and Andrew E. Williams. NY: Universe Books, 1983. ISBN 0-8766-3415-3, 0-8766-3586-9 (paper). ONA

A Field Guide to Tropical and Subtropical Plants. Frances Perry and Roy Hay. NY: Van Nostrand Reinhold, 1982. ISBN 0-4422-6861-0, 0-4422-6859-9 (paper). TSP

A Field Guide to Western Reptiles and Amphibians. (Second edition.) (Peterson Field Guide.) Robert C. Stebbins. Boston: Houghton Mifflin, 1985. ISBN 0-395-38254-8, 0-395-38253-x (paper). WRA

Frogs and Toads of the World. Chris Mattison. NY: Facts on File, 1987. ISBN 0-8160-1602-x. FTW

Furbearing Animals of North America. Leonard Lee Rue III. NY: Crown, 1981. ISBN 0-517-53942-x. RFA

Grasslands. (Audubon Society Nature Guide.) Lauren Brown. NY: Knopf, 1985. ISBN 0-394-73121-2. LBG

Grzimek's Encyclopedia of Mammals. (5 volumes.) Bernhard Grzimek, ed. NY: McGraw-Hill, 1990. ISBN 0-07-909508-9 (set). GEM

Guide to Mammals of the Plains States. J. Knox Jones, David M. Armstrong, and Jerry R. Choate. Lincoln: University of Nebraska Press, 1985. ISBN 0-8032-2562-8, 0-8032-7557-9 (paper). MPS

Guide to the Mammals of Pennsylvania. Joseph F. Merritt. Pittsburgh, PA: University of Pittsburgh for the Carnegie Museum of Natural History, 1987. ISBN 0-8229-3563-5, 0-8229-5393-5 (paper). GMP

Handbook of Mammals of the North-Central States. J. Knox Jones and Elmer C. Birney. Minneapolis: University of Minnesota Press, 1988. ISBN 0-8166-1419-9, 0-8166-1420-2 (paper). MNC

The Herons Handbook. James Hancock and James Kushlan. NY: Harper and Row, 1984. ISBN 0-06-015331-8. HHH

Hummingbirds—Their Life and Behavior: A Photographic Study of the North American Species. Text by Esther Quesada Tyrrell, photographs by Robert A. Tyrrell. NY: Crown, 1985. ISBN 0-517-55336-8. HLB

Insects Etc. An Anthology of Arthropods Featuring a Bounty of Beetles. Painting by Bernard Durin, with a Literary Anthology introduced and selected by Paul Armand Gette. Entomological commentaries by Gerhard Scherer. NY: Hudson Hills Press, 1981; Munich: Schirmer/Mosel Verlag, 1980. ISBN 0-933-92025-3. DIE

Insects of the World. Anthony Wootton. NY: Facts on File, 1984. ISBN 0-87196-991-2. WIW

The Lives of Bats. Wilfred Schober. NY: Arco, 1984. ISBN 0-668-05993-1. LOB

Living Snakes of the World in Color. John M. Mehrtens. NY: Sterling Pub. Co., 1987. ISBN 0-8069-6460-x. LSW

Lizards of the World. Chris Mattison. NY: Facts on File, 1989. ISBN 0-816-01900-2. LOW

Longman World Guide to Birds. Philip Whitfield, ed. London: Longman, 1986. ISBN 0-582-89354-2. WGB

Macmillan Illustrated Animal Encyclopedia. Philip Whitfield, ed. NY: Macmillan, 1984. ISBN 0-02-627680-1. MIA

Mammals in Wyoming. Tim W. Clark and Mark R. Stromberg. Lawrence, KS: University of Kansas Press, 1987. ISBN 0-8933-8026-1. CMW

Mammals of the Carolinas, Virginia and Maryland. Wm. David Webster, James F. Parnell, and Walter C. Biggs. Chapel Hill: University of North Carolina Press, 1985. ISBN 0-8078-1663-9. WMC

Mammals of the National Parks. Richard G. van Gelder. Baltimore: Johns Hopkins Press, 1982. ISBN 0-8018-2688-8, 0-8018-2689-6 (paper). MNP

Mushrooms and Toadstools: A Color Field Guide. U. Nonis. NY: Hippocrene Books, 1982. ISBN 0-882-54755-0. NMT

Natural History of Antelopes. Clive A. Spinage. London: Croom Helm; NY: Facts on File, 1986. ISBN 0-8160-1581-3 (Facts on File). NHA

Natural History of New Mexican Mammals. James S. Findley. Albuquerque: University of New Mexico Press, 1987. ISBN 0-8263-0957-7, 0-8263-0958-5 (paper). NMM

Natural History of the Primates. J.R. Napier and P.H. Napier. Cambridge, MA: MIT Press, 1985. ISBN 0-262-14039-x. NHP

Natural History of the USSR. Algirdas Knystautas. NY: McGraw-Hill, 1987. ISBN 0-07-035409-x. NHU

Nature of Australia: A Portrait of the Island Continent. John Vandenbeld. NY: Facts on File, 1988. ISBN 0-8160-2006-x. NOA

Nightwatch: The Natural World from Dawn to Dusk. John Cloudsley-Thomas et al. NY: Facts on File, 1984. ISBN 0-87196-271-3. NNW

Ocean Birds. Lars Lofgren. NY: Knopf, 1984. ISBN 0-394-53101-9. OBL

Owls: Their Natural and Unnatural History. John Sparks and Tony Soper. NY: Facts on File, 1989. ISBN 0-8160-2154-6. ONU

Oxford Encyclopedia of Trees of the World. Bayard Hora, consulting ed. NY: Oxford, 1981. ISBN 0-192-17712-5. OET

Pacific Coast. (Audubon Society Nature Guide). Bayard H. McConnaughey and Evelyn McConnaughey. NY:Knopf, 1985. ISBN 0-394-73130-1. MPC

Plants and Flowers of Hawai'i. S.H. Sohmer and R. Gustafson. Honolulu: University of Hawaii, 1987. ISBN 0-8248-1096-1. PFH

Plants That Merit Attention: Volume I—Trees. Janet Meakin Poore, ed. Portland, OR: Timber Press, 1984. ISBN 0-917304-75-6. PMT

Poisonous Plants: A Colour Field Guide. Lucia Woodward. Newton Abbot, Devon:

David & Charles; NY: Hippocrene Books, 1985. ISBN 0-715-38628-x (David & Charles), 0-8705-2014-8 (Hippocrene). PPC

The Pond. Gerald Thompson, Jennifer Coldrey, and George Bernard. Cambridge, MA: MIT Press, 1984. ISBN 0-262-20049-x. PGT

Reptiles of North America. (Golden Field Guide.) Hobart M. Smith and Edmund D. Brodie. NY: Golden Press, 1982. ISBN 0-307-13666-3. RNA

Rhinos: Endangered Species. Malcolm Penny. NY: Facts on File, 1988. ISBN 0-8160-1882-0. RES

Riches of the Wild: Land Mammals of South-East Asia. Gathorne Gathorne-Hardy Cranbrook, Earl of Cranbrook. Singapore, NY: Oxford University Press, 1987. ISBN 0-19-582697-3. ROW

Rocky Mountain Mammals: A Handbook of Mammals of Rocky Mountain National Park and Vicinity. (Revised ed.) David M. Armstrong. [Boulder, CO]: Colorado Associated University Press in cooperation with Rocky Mountain Nature Association, 1987. ISBN 0-8708-1168-1. RMM

Sea Guide to Whales of the World. Lyall Watson. NY: Dutton, 1981. ISBN 0-525-93202-x. WOW

Sea Watch: The Seafarer's Guide to Marine Life. Paul V. Horsman. NY: Facts on File, 1985. ISBN 0-8160-1191-5. HSG

Seabirds of the World. Photographs by Eric Hosking, text by Ronald M. Lockley. NY: Facts on File, 1983. ISBN 0-87196-249-7. HSW

Seals of the World. (Second edition.) Judith E. King. London: British Museum (Natural History); NY: Cornell University Press, 1983. ISBN 0-8014-1568-3. KSW

Seashore Animals of the Southeast: A Guide to Common Shallow-Water Invertebrates of the Southeastern Atlantic Coast. Edward Ruppert and Richard Fox. Columbia: University of South Carolina Press, 1988. ISBN 0-87249-534-5, 0-87249-535-3 (paper). SAS

Sharks. John D. Stevens, consulting ed. NY: Facts on File, 1987. ISBN 0-8160-1800-6. SJS

Sharks of the World. Rodney Steel. NY: Facts on File, 1985. ISBN 0-8160-1086-2. SSW

Simon and Schuster's Guide to Butterflies and Moths. Mauro Daccordi, Paolo Triberti, and Adriano Zanetti. NY: Simon and Schuster, 1988. ISBN 0-671-66065-9, 0-671-66066-7 (paper). SGB

Simon and Schuster's Guide to Insects. Ross H. Arnett and Richard L. Jacques. NY: Simon and Schuster, 1981. ISBN 0-671-25013-2, 0-671-25014-0 (paper). SGI

Simon and Schuster's Guide to Mushrooms. Gary H. Lincoff. NY: Simon and Schuster, 1981. ISBN 0-671-42798-9. 0-671-42849-7 (paper). SGM

Snakes of the World. Chris Mattison. NY: Facts on File, 1986. ISBN 0-8160-1082-x. MSW

Songbirds: How to Attract Them and Identify Their Songs. Noble Proctor. Emmaus, PA: Rodale Press, 1988. ISBN 0-878-57773-4. SPN

Spiders of the World. Rod and Ken Preston-Mafham. NY: Facts on File, 1984. ISBN 0-87196-996-3. SOW

Trees and Shrubs of Virginia. Oscar W. Gupton and Fred C. Swope. Charlottesville, Virginia: University Press of Virginia, 1981. ISBN 0-8139-0886-8. TSV

Turtles and Tortoises of the World. David Alderton. NY: Facts on File, 1988. ISBN 0-8160-1733-6. TTW

Turtles of the World. Carl H. Ernst and Roger W. Barbour. Washington, D.C.: Smithsonian, 1989. ISBN 0-87474-414-8. TOW

Turtles, Tortoises and Terrapins. Fritz Jurgen Obst. NY: St. Martin's Press, 1986. ISBN 0-312-82362-2. OTT

Undersea Predators. Carl Roessler. NY: Facts on File, 1984. ISBN 0-871-96893-2. RUP

Wading Birds of the World. Eric Soothill and Richard Soothill. Poole, Dorset: Blandford Press, 1982. ISBN 0-713-70913-8. WBW

Watching Fishes: Life and Behavior on Coral Reefs. Roberta Wilson and James Q. Wilson. NY: Harper and Row, 1985. ISBN 0-06-015371-7. WWF

Weird and Wonderful Wildlife. Michael Marten, John May, and Rosemary Taylor. San Francisco: Chronicle Books, 1983. ISBN 0-87701-295-4. WWW

Western Birds. (An Audubon Handbook.) John Farrand. NY: McGraw-Hill, 1988. ISBN 0-07-019977-9. FWB

Western Forests. (Audubon Society Nature Guide.) Stephen Whitney. NY: Knopf, 1985. ISBN 0-394-73127-1. WFW

Wetlands. (Audubon Society Nature Guide.) William A. Niering. NY: Knopf, 1985. ISBN 0-394-73147-6. WNW

Whales, Dolphins and Porpoises. Richard Harrison and Michael Bryden, consulting eds. Facts on File, 1988. ISBN 0-816-01977-0. HWD

Whales, Dolphins and Porpoises of the World. Mary L. Baker. Garden City, NY: Doubleday, 1987. ISBN 0-385-15366-x. WDP

Wild Mammals of Northwest America. Arthur Savage and Candace Savage. Baltimore: Johns Hopkins University Press, 1981. ISBN 0-8018-2627-6. SWM

Wild Orchids of the Mid-Atlantic States. Oscar W. Gupton and Fred C. Swope. Knoxville, TN: University of Tennessee, 1986. ISBN 0-87049-509-7. GWO

Wildflowers Across America. Lady Bird Johnson and Carlton B. Lees. NY: Abbeville Press, 1988. ISBN 0-89659-770-9. WAA

Wildflowers of Tidewater Virginia. Oscar W. Gupton and Fred C. Swope. Charlottesville: University Press of Virginia, 1982. ISBN 0-8139-0922-8. WTV

Wildlife of the Deserts. Frederic H. Wagner. NY: Abrams, 1980. ISBN 0-8109-1764-5. WWD

Wildlife of the Forests. Ann Sutton and Myron Sutton. NY: Abrams, 1979. ISBN 0-8109-1759-9. SWF

Wildlife of the Islands. William H. Amos. NY: Abrams, 1980. ISBN 0-8109-1763-7. WOI

Wildlife of the Mountains. Edward R. Ricciuti. NY: Abrams, 1979. ISBN 0-8109-1757-2. WOM

Wildlife of the Polar Regions. C. Carleton Ray and M.G. McCormick-Ray. NY: Abrams, 1981. ISBN 0-8109-1768-8. WPR

Wildlife of the Prairies and Plains. Kai Curry-Lindahl. NY: Abrams, 1981. ISBN 0-8109-1766-1. WPP

Wildlife of the Rivers. William H. Amos. NY: Abrams, 1981. ISBN 0-8109-1767-x. AWR

Wings of the North: A Gallery of Favorite Birds. Candace Savage. Minneapolis: University of Minnesota, 1985. ISBN 0-8166-1433-4. WON

The Wonder of Birds. Robert M. Poole, editor. Washington, DC: National Geographic, 1983. ISBN 0-8704-4470-0. WOB